D1824071

ENVIRONMENTAL ASPECTS OF HOUSING FOR ANIMAL PRODUCTION

Proceedings of Previous Easter Schools in Agricultural Science, published by Butterworths, London

*SOIL ZOOLOGY   Edited by D. K. McE. Kevan (1955)
*THE GROWTH OF LEAVES   Edited by F. L. Milthorpe (1956)
*CONTROL OF THE PLANT ENVIRONMENT   Edited by J. P. Hudson (1957)
*NUTRITION OF THE LEGUMES   Edited by E. G. Hallsworth (1958)
*THE MEASUREMENT OF GRASSLAND PRODUCTIVITY   Edited by J. D. Ivins (1959)
*DIGESTIVE PHYSIOLOGY AND NUTRITION OF THE RUMINANT   Edited by D. Lewis (1960)
*NUTRITION OF PIGS AND POULTRY   Edited by J. T. Morgan and D. Lewis (1961)
*ANTIBIOTICS IN AGRICULTURE   Edited by M. Woodbine (1962)
*THE GROWTH OF THE POTATO   Edited by J. D. Ivins and F. L. Milthorpe (1963)
*EXPERIMENTAL PEDOLOGY   Edited by E. G. Hallsworth and D. V. Crawford (1964)
*THE GROWTH OF CEREALS AND GRASSES   Edited by F. L. Milthorpe and J. D. Ivins (1965)
*REPRODUCTION IN THE FEMALE MAMMAL   Edited by G. E. Lamming and E. C. Amoroso (1967)
*GROWTH AND DEVELOPMENT OF MAMMALS   Edited by G. A. Lodge and G. E. Lamming (1968)
*ROOT GROWTH   Edited by W. J. Whittington (1968)
*PROTEINS AS HUMAN FOOD   Edited by R. A. Lawrie (1970)
*LACTATION   Edited by I. R. Falconer (1971)
*PIG PRODUCTION   Edited by D. J. A. Cole (1972)
*SEED ECOLOGY   Edited by W. Heydecker (1973)
 HEAT LOSS FROM ANIMALS AND MAN: ASSESSMENT AND CONTROL   Edited by J. L. Monteith and L. E. Mount (1974)
*MEAT   Edited by D. J. A. Cole and R. A. Lawrie (1975)
*PRINCIPLES OF CATTLE PRODUCTION   Edited by Henry Swan and W. H. Broster (1976)
*LIGHT AND PLANT DEVELOPMENT   Edited by H. Smith (1976)
 PLANT PROTEINS   Edited by G. Norton (1977)
 ANTIBIOTICS AND ANTIBIOSIS IN AGRICULTURE   Edited by M. Woodbine (1977)
 CONTROL OF OVULATION   Edited by D. B. Crighton, N. B. Haynes, G. R. Foxcroft and G. E. Lamming (1978)
 POLYSACCHARIDES IN FOOD   Edited by J. M. V. Blanshard and J. R. Mitchell (1979)
 SEED PRODUCTION   Edited by P. D. Hebblethwaite (1980)
 PROTEIN DEPOSITION IN ANIMALS   Edited by P. J. Buttery and D. B. Lindsay (1981)
 PHYSIOLOGICAL PROCESSES LIMITING PLANT PRODUCTIVITY   Edited by C. Johnson (1981)

These titles are now out of print but are available in microfiche editions

Environmental Aspects of Housing for Animal Production

J. A. CLARK
University of Nottingham

Butterworths
London Boston Sydney Wellington Durban Toronto

First published 1981

© The several contributors named in the list of contents 1981

British Library Cataloguing in Publication Data
*Environmental aspects of housing for animal
  production.*
  1. Animal housing – Environmental aspects
  – Congresses
  I. *Clark, J. A.*
  636.08'31   SF91

  ISBN 0–408–10688–3

Typeset by Scribe Design Limited, Gillingham, Kent
Printed in England by Page Bros (Norwich) Ltd.

PREFACE

'. . . Monsieur de Réamur has in the bringing up of his chickens found the means to raise them both by the warmth of the dung and by that of an ordinary fire: he has even showed the advantages which have resulted from this last expedient; and which are such, that he even thinks there might be a gain, in the taking away from the hens the chickens they should have hatched themselves, in order to bring them up in the new manner which he has discovered. He has reared chickens with great success in his stores heated with dung; wherein they were fully sheltered both from the cold and from all other dangers: and he has also brought up others, in much larger rooms, heated either with dung or with ordinary fires. The heat indeed of these last rooms, cannot be every where so equal as that in a store. It was therefore necessary that there should be in each of them some particular places warmer than the rest, into which the young chickens might occasionally retire as they would under the wings of a hen . . .'

From *The Art of Hatching and Bringing up Fowls by Means of Artificial Heat* by Monsieur de Réamur (Translated by Mr Trembley)
Printed by the Royal Society in 1750

The study and application of environmental aspects of housing farm animals not only has a long history, as shown by the above quotation, but is necessarily multidisciplinary. An understanding of the needs of the animal can be derived from physiological, physical and behavioural studies, and these may need to overlap with agricultural and ecological work. Meteorological data and economic criteria can be used to assess the need for housing, and finally the design of suitable systems draws on information and principles from the applied physical disciplines such as engineering.

Environmental work has been approached from all these points of view, and in the past twenty years an impressive amount of progress has been made. Many people working in this field, both in research and development and at a practical level, have recognized its eclectic nature and have employed multidisciplinary skills; but opportunities for interaction across the whole range of relevant subjects have been rare, and international opportunities even rarer. The 31st Easter School was an attempt to provide such a forum and the chapters of this book constitute a reference text spanning all the relevant disciplines.

The need for the exploitation of a variety of different skills can be illustrated by an example from an important field which received thorough discussion at the Easter School, both inside and outside the conference hall. The heat balance of animals (discussed at a previous School in this series) is central to environmental studies and has been investigated by workers in a number of the disciplines. However, an effective functional knowledge of the thermal requirements of farm animals cannot be gained through any one discipline in isolation. The physiologists have defined the lower and upper critical temperatures, and these are fundamental to an understanding of the temperature requirement of animals; but the conference produced some persuasive evidence that they may not be the whole answer, because there are many complicating factors such as interactions with ventilation rate and because economic optima may not precisely correspond with physiological optima. Also, a simple description of dry-bulb air temperature is a far from adequate assessment of the thermal environment experienced by the animal. Similarly, empirical comparisons of housing systems cannot definitively reveal the best production techniques, because they will not make clear which variables account for observed effects. Thus the proper analysis of thermal requirements needs calorimetric work, followed by population level response experiments; preferably experiments measuring responses to environmental heat demand rather than merely to temperature. The responses can then be appraised economically and relevant housing systems can be designed, developed and compared. This volume provides the reader with the experience of authors who have approached the problem from all these directions, though at the present time the information is not complete for all farm species.

The importance of the subject reflects that of the animal production industry. Throughout the developed and developing world, animal products are used to improve the acceptability, digestibility and nutritional value of plant material to the human population. As a consequence, a financial value is added to the original value of the plants and therefore, in both developed and developing countries, the animal industry is often a substantial contributor to the gross national product. In addition, there is international trade in the products, the means of production, and in production skills.

The relevance of environmental studies to the energy balance of farm animals is almost self-evident, and it is therefore appropriate that this Easter School should be organized by the Department of Physiology and Environmental Studies, which also ran the 1973 Easter School on 'Heat loss in animals and man'. Many of the participants and several speakers were present at both Easter Schools.

D. R. Charles

ACKNOWLEDGEMENTS

The editor wishes to thank those who agreed to present papers at the meeting, both for their individual contributions and for their cooperation with the editing, which has helped in producing a text in which the units and format are consistent throughout. The Dean of the Faculty of Agricultural Science, Professor R. A. Lawrie, opened the meeting and welcomed the delegates, who came from five continents. The smooth progress of the meeting also owed much to the expert chairmanship of J. L. Monteith, G. E. Folk, A. H. Sykes, D. J. A. Cole, A. Berman, D. R. Charles, J. E. Owen and F. G. Clegg. It is also a pleasure to acknowledge the considerable personal contribution of David Charles to the inception of the School and to its philosophy.

Several members of the Department of Physiology and Environmental Studies helped with the arrangements for the meeting, in particular Dr A. J. McArthur and Mr G. D. MacLeod. Miss Edna Lord handled the secretarial side with the easy skill born of long experience of previous meetings in this series.

Financial assistance towards the expenses of the School was provided by the following companies:

Beecham Animal Health, Vitamealo Ltd
J. Bibby Agriculture Ltd
BOCM Silcock Ltd
British Industrial Plastics Ltd
Cherry Valley Farms Ltd
Cyanamid of Great Britain Ltd
The Hydor Company Ltd
Imperial Chemical Industries Ltd, Pharmaceutical Division
Pauls & Whites International (UK) Ltd
RHM Animal Feed Services Ltd
Unifeeds International Ltd

SYMBOLS AND UNITS

The symbols used in this text have been based largely on those employed in the previous related volume in this series, *Heat Loss from Animals and Man* (Monteith and Mount, 1974), though some changes have been necessary to conform with current usage and with a different subject spread. Units of the Système International are used throughout. Temperatures are expressed as degrees Celsius (°C) or kelvin (K), as appropriate. Temperature *differences* are expressed as K, as are the units of quantities such as thermal conductance ($W m^{-2} K^{-1}$), because the thermodynamic unit (kelvin) is that employed in both the absolute and Celsius scales.

In any multidisciplinary area it is difficult to achieve complete consistency, even within the SI, and this is no exception. One apparent contradiction has proved unavoidable in this text. In the literature of environmental physics the resistance to mass transfer is usually defined from the fundamental Fickian diffusion equation, in the form:

Flux = concentration difference/resistance

If the flux is expressed in $kg m^{-2} s^{-1}$ and the concentration is measured in consistent units of $kg m^{-3}$, then the diffusion resistance (r in this text) has the dimensions of the inverse of velocity, $s m^{-1}$ in SI units. This unit may be employed for water vapour, other gases and for heat transfer resistances. However, in building engineering literature the driving potential for diffusion of water vapour is generally expressed in units of partial pressure, the usual unit of measurement. Thus,

Flux = pressure difference/resistance

In this case dimensional analysis leads to a resistance which unfortunately has units which are the *inverse* of those obtained where concentration is considered as the driving potential, i.e. $m s^{-1}$ in SI. Since the latter form is enshrined in the current British Standards for vapour transfer resistances of building materials, it is impossible to avoid the ambiguity of two sets of resistances in this text in reciprocal units: resistances (r) in $s m^{-1}$ are used in the chapters by McArthur and by Clark and Cena; while in his discussion of water vapour transfer through building materials, Wathes uses r^* in $m s^{-1}$.

In addition, despite the range of symbols available when combining the Latin and Greek alphabets, some characters have more than one common usage. A few cases of multiple use have been retained, where the quantities occur only briefly. Such cases may be resolved by reference to the text at the appropriate point, where local usage is defined.

Symbol	Quantity	Unit
A	Surface area	m^2
a	Age	d
a	Width of jet orifice	m
Ar	Archimedes Number	Dimensionless
B	Width of building	m
b	Height of jet orifice	m
b	Rate of emission of pathogens from an infected animal	s^{-1}
C	Convective heat flux subscripts: i interior o exterior	$W\,m^{-2}$
C_0	Concentration of pathogens per unit volume	
c	Carbon dioxide, volumetric concentration	Dimensionless
c	Specific heat	$J\,kg^{-1}\,K^{-1}$
c_p	Specific heat of air	$J\,kg^{-1}\,K^{-1}$
D	Diffusion coefficient for water vapour in air	$m^2\,s^{-1}$
d	Thickness, characteristic dimension or diameter	m

E	Rate of heat loss by evaporation (latent heat flux)
	subscripts: c cutaneous
	           r respiratory
	           s skin
	           t total
		$W\,m^{-2}$

e	Vapour pressure
	subscripts: a ambient air
	           cn at plane of condensation
	           cs at coat surface
	           i interior ⎱ of building
	           o exterior ⎰
	           s at site of respiratory evaporation
	           sk saturation vapour pressure at skin temperature
		Pa

E_g	Egg production	$g\,d^{-1}$
F	Food intake as metabolizable energy subscripts: m maintenance c compensation for low temperatures	$MJ\,d^{-1}$
f	Water vapour conductance subscripts: i interior o exterior	$kg\,N^{-1}\,s^{-1}$
G	Sensible heat loss	$W\,m^{-2}$
g	Acceleration by gravity	$m\,s^{-2}$
H	Height of building	m
H_a	Fuel heat input	W

Symbol	Quantity	Unit
h	Heat transfer coefficient	
	subscripts: c convective	
	E evaporative	
	R radiative	
h	Horizontal length of jet orifice	m
I	Insulation (resistance) to sensible heat transfer	$m^2\,K\,W^{-1}$

subscripts:

a	air	R	radiation
b	body	HR	combined
c	material	i	interior
f	floor	o	exterior
H	convection	s	structure

Symbol	Quantity	Unit
I^*	Insulation cost index	$£\,W\,m^{-4}\,K^{-1}$
J	Rate of change of heat storage, expressed per unit area	$W\,m^{-2}$
K	Wind chill factor	$W\,m^{-2}$
k	Thermal conductivity	$W\,m^{-1}\,K^{-1}$
	subscripts: H total	
	C convective	
	G conductive	
	R radiative	
k	Efficiency of utilization of metabolizable energy	Dimensionless
k	Proportion of infected animals	Dimensionless
l	Coat depth	m
L	Long-wave radiative flux density	$W\,m^{-2}$
	subscripts: d from atmosphere	
	e environment	
	n net	
	s structure	
M	Metabolic rate	$W\,m^{-2}$
	subscript: tn thermoneutral	
M_w	Molecular weight of water	$kg\,mol^{-1}$
m	Mass	kg
N	Number of animals	
P	Pressure	Pa
Q	Heat production or loss totals, per animal or building	W
	subscripts: e latent-heat loss	
	eb latent-heat loss from building	
	s sensible-heat loss	
	sb sensible-heat loss from building	
	tn thermoneutral	
Q_K	Heat loss per degree	$W\,K^{-1}$
q	Rate of emission of a gas into the atmosphere	$kg\,s^{-1}$
q	Rate of production of pathogens	s^{-1}
R^*	Universal Gas Constant	$J\,K^{-1}\,mol^{-1}$

Symbol	*Quantity*	*Unit*
R	Radiative heat flux	$W\,m^{-2}$

subscripts: n   net
              n(i) interior
              n(o) exterior
              ni   net isothermal

Re	Reynolds number	Dimensionless
R_m	Death rate of pathogens	s^{-1}
r	Resistance to sensible heat transfer	$s\,m^{-1}$

subscripts:

a	aerodynamic	f free convection
b	body skin and tissue	H convection
c	coat; structure	R radiative
d	molecular diffusion	tot total
e	environment	v ventilation

r^{*}_{m}	Resistance to vapour transfer of roof materials	$s\,m^{-1}$
r_v	Resistance to vapour transfer	$s\,m^{-1}$

second subscripts: a aerodynamic
                      c coat
                      r respiratory
                      s skin

r_v^{*}	Resistance to vapour transfer	$m\,s^{-1}$

second subscripts: a aerodynamic
                      i interior
                      m structure
                      o exterior

S	Solar radiation flux density	$W\,m^{-2}$

subscripts: e environment
                t total

T	Temperature	°C or K

subscripts:

a	air (dry bulb)	eq equivalent
b	body core	g black globe
c	critical temperature	i interior
cb	critical temperature of building	m mean
cl	lower critical temperature of animal	o exterior
		op operative
cd	upper (dry) critical temperature of animal	R radiant
		s structure
cw	upper (wet) critical temperature of animal	sk skin
		v virtual
cn	at plane of condensation	w wall; wet bulb
cs	at coat surface	
e	environmental or effective	

t	Time	s
t	Wind penetration depth	m
t_d	Thermal time	Degree days
U	Thermal transmittance	$W\,m^{-2}\,K^{-1}$
u	Air velocity	$m\,s^{-1}$
V	Volume (of building)	m^3
$\dot{V}$	Ventilation rate (of building)	$m^3\,s^{-1}$ or $m^3\,h^{-1}$

Symbol	Quantity	Unit
$\dot{V}$	Rate of respiratory ventilation	l min^{-1}
W	Live-weight	kg
W	Rate of mechanical working	W m^{-2}
Z	Thickness	m
z	Resistivity	s cm^{-1}
	subscripts: c coat	
	d diffusion	
	f fleece	
	R radiation	
α	Attenuation factor	Dimensionless
α	Absorptivity for solar radiation	Dimensionless
α	Wind penetration factor	Dimensionless
	subscript: v water vapour	
β	Phase lag	rad
β	Latent heat of fusion of water	J kg^{-1}
γ	Psychrometric constant	kPa K^{-1}
ε	Water vapour flux	kg m^{-2} s^{-1}
	subscripts: cn at plane of condensation	
	s skin	
	sd sweat deposition at skin surface	
	sm evaporation from skin surface	
ε	Emissivity for thermal radiation	Dimensionless
ζ	Heater efficiency	Dimensionless
η	Heating unit efficiency	Dimensionless
Θ	Equivalent temperature	°C or K
$\varkappa$	Thermal diffusivity of dry air	m^2 s^{-1}
λ	Latent heat of evaporation of water	J kg^{-1}
ϱ	Density	kg m^{-3}
ϱ_v	Water vapour resistivity	N s kg^{-1} m^{-1}
σ	Stefan–Boltzmann constant	W m^{-2} K^{-4}
τ	Time constant	s
τ_p	Time period	s
χ	Absolute concentration of water vapour	kg m^{-3}
	subscripts: a air	
	g saturation value	
	sk at skin	
ψ	Gas concentration	kg m^{-3}
Ω	Fuel costs	
ω	Fuel price	
ω	Angular frequency	rad s^{-1}

Reference

MONTEITH, J.L. and MOUNT, L.E. (Eds) (1974). *Heat Loss from Animals and Man*. Butterworths, London

CONTENTS

I

PHYSICAL AND PHYSIOLOGICAL PRINCIPLES

1

THE ENVIRONMENTAL PHYSIOLOGY OF ANIMAL PRODUCTION

DAVID ROBERTSHAW
Department of Physiology and Biophysics, College of Veterinary Medicine and Biomedical Sciences, Colorado State University, Fort Collins, Colorado, USA

Introduction

This chapter reviews the information on the physiological mechanisms whereby the food- or fibre-producing animals maintain a constant body temperature under climatic extremes and the possible impact that these physiological processes may have on productive systems. Thus, it provides the beginnings of a rational basis for an understanding of the empirical observations that productivity is suppressed during climatic extremes. It is only because man has altered the environment that such studies are needed: it can be assumed that through natural selection, an animal has evolved in such a way that it can not only survive but flourish in its own habitat. If, however, man modifies the environment, or utilizes selective breeding systems, then the equilibrium between an animal and its environment is disturbed and productive systems suffer. More than forty years ago this fact was not universally recognized and there were attempts to utilize the grazing lands of countries subject to climatic extremes by the relocation of high-producing animals into these areas. When it was realized that their genetic potential, in its broadest sense, could not be achieved, research programmes were instituted to elucidate the fundamental causes of these failures. This type of research led to an understanding of the need to provide adequate housing for efficient animal production as well as other methods to alleviate the adverse impact of the environment, particularly the thermal environment, on animal production.

Energy balance

Production, whether it be growth, lactation, wool growth or pregnancy, can be understood as comprising the energy transfer processes involved in anabolism and catabolism. In a productive process, the anabolic reactions exceed the catabolic ones. The energy for metabolism is provided in the diet, the gross energy intake being the calorific value of the food. If the energy of the faeces is subtracted from that of the food, we have a measure of the digestibility of the food, sometimes known as the *apparent digestibility*. Not all the digested energy is available to the animal: some energy is lost in the urine and in ruminant animals some is lost as heat and methane in the fermentation process in the gut. If these losses are subtracted from the digestible energy a measure of the energy available for bodily processes is obtained. This is usually referred to as the *metabolizable energy*, and is available to provide energy for maintenance purposes. Maintenance requirements include all the energy needs of a fasting, resting animal (the basal metabolism), plus those associated with physical activity and feeding. Any of the metabolizable energy surplus to maintenance requirements is available for productive processes. If the metabolizable energy is insufficient for maintenance, the animal will make up the deficit by oxidation of its own body tissues. The productive process, whether it be milk, meat, etc., has a calorific value and therefore is a form of stored or retained energy. The gross efficiency of production can be defined as the ratio of the calorific value of the tissue synthesized to the calorific value of the food ingested. The calorific value of weight gain increases with age since juvenile growth generally consists more of water, protein and bone than does later growth, and less of fat. This provides a good example of the limitations of weight gain as a measure of productivity, and emphasizes the value of considering pro-ductivity in terms of the amount of retained energy.

Energy retention can be suppressed in three ways:

(1) By reduction in gross energy intake. This may be in terms of a reduced appetite or in the quality of the food ingested.
(2) By reduction in digestibility. In herbivores, a slow passage through the gut is essential for effective fermentation. Any factors that increase the rate of passage of foodstuff tend to decrease digestibility and vice versa. Kennedy, Young and Christopherson (1977) have demonstrated that an increase in thyroid hormone secretion will reduce retention time and digestibility. Thus, any change in thyroid hormone status, irrespective of its actions on appetite or intermediary metabolism, will be followed by alterations in digestibility.
(3) By any factor that elevates maintenance requirements, which will reduce the proportion of gross energy intake available for productive purposes; that is, efficiency of food conversion, as defined, will be reduced. Appetite may or may not be stimulated to compensate for increased maintenance requirements but this is of little consolation to the producer since he is paying more for less; a form of inflation!

Heat balance

In order to maintain thermal equilibrium the metabolic heat produced by the animal, Q_m, must equal the total heat loss, Q.

$$Q_m = Q \tag{1.1}$$

This equation may be expanded to

$$Q = Q_s + Q_e \tag{1.2}$$

where Q_s is the Newtonian non-evaporative or sensible heat transfer, by conduction, convection and radiation, and Q_e is the evaporative heat loss. Newtonian heat transfer may have either positive or negative sign, since the transfer of heat may be in either direction. If thermal equilibrium is maintained there will, by definition, be no change in heat content. In controlled environmental conditions, thermal equilibrium can be achieved, but under outdoor conditions, heat exchanges are never constant and the thermal balance of an animal is constantly fluctuating.

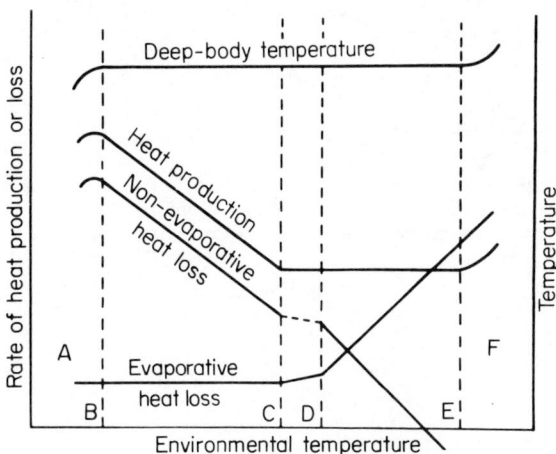

Figure 1.1 Relationship between heat production, evaporative and non-evaporative heat loss and deep-body temperature in a strict homeotherm. A, zone of hypothermia whose border is defined by B; F, zone of hyperthermia whose border is defined by E; C, lower critical temperature; D, temperature of marked increase in evaporative loss (upper critical temperature); CD, zone of minimal thermoregulatory effort; CE, zone of minimal metabolism. (After Mount, 1974)

Figure 1.1 represents equation 1.2 in graphical terms (Ingram and Mount, 1975). The abscissa, labelled 'environmental temperature', incorporates all components of the thermal environment, such as wind, radiation, etc., and is not simply air temperature. The zone CD is the temperature range where there is minimal thermoregulatory effort, usually defined as the *zone of thermal comfort*, and is the thermal environment that an animal would choose for itself (Mount, 1974; Webster, 1974). The *zone of thermal neutrality* is the range (CE) in which metabolic rate is minimal. It is bounded at the lower end by the *lower critical temperature* (C), the temperature below which metabolism increases in order to maintain thermal equilibrium, and at the upper end (E) by the *hyperthermic point*. Since the zone of thermal neutrality is the zone of minimal metabolic rate, it follows that it will be that of maximal energetic efficiency. In order for us to understand the effect of climate on productive processes *Figure 1.1* can be simplified still further (*Figure 1.2*). Here the shaded areas give an indication of the increased maintenance requirements in relation to the environment, and show furthermore how heat loads less than C have a much greater effect on the suppression of production than temperatures in excess of E. The actual values of C and E, and the distance between them, will vary with the species, the breed and the individual. Thus, the width is largely a consequence of the

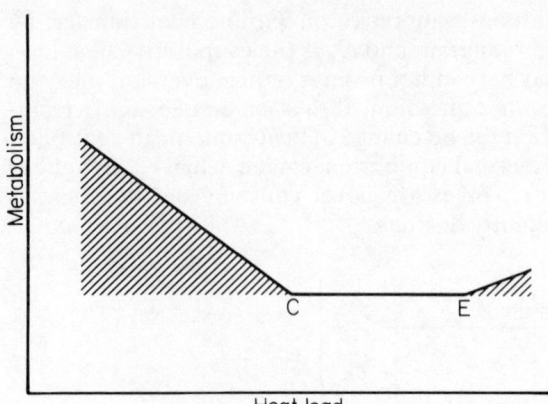

Figure 1.2 Relationship between heat load and metabolism. The shaded areas indicate the increment in maintenance requirement caused by different heat loads. C and E define the heat loads outside which metabolism, and therefore maintenance, is increased. The points C and E correspond to the same points in Figure 1.1

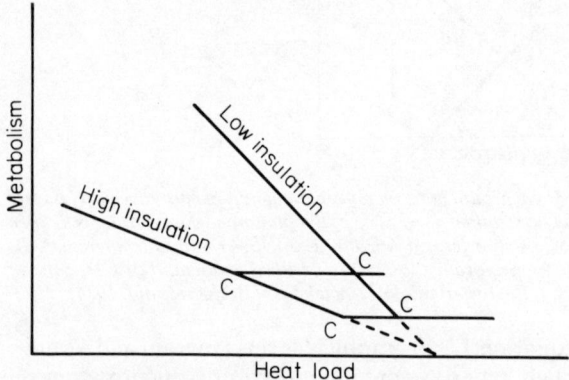

Figure 1.3 Relationship between heat load and metabolism at two levels of insulation and two levels of metabolic rate. The critical temperature (C) varies according to these parameters

ability to regulate evaporative heat loss, whereas C is largely determined by the insulation and heat production. *Figure 1.3* demonstrates how high and low insulation combined with differing metabolic rates can affect the value of C. Species adapted to high environmental heat loads will have a low insulation, well-developed evaporative heat loss mechanisms, and in many cases a lower basal metabolism. Cold-adapted species, on the other hand, tend to have a higher metabolic rate and a greater insulation.

Cold exposure

HEAT PRODUCTION

Heat production increases both with activity and with the level of feeding. The work of Graham *et al.* (1959) demonstrates very clearly that as the level of feeding increases, so does the thermoneutral metabolism (*Figure 1.4*). The increment in metabolism associated with different levels of feeding is a function not only of the level of feeding, but also of the quality of the food: the poorer the quality, the higher the heat increment. Physical activity will also increase heat production. Both sources of heat—feeding and exercise—can substitute for cold thermogenesis.

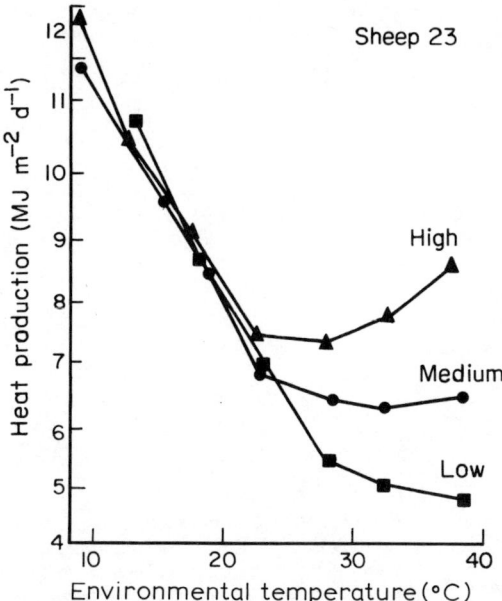

Figure 1.4 Heat production per unit surface area of a sheep at three different levels of feeding in relation to environmental temperature. 1 kcal = 4.18 J. (After Graham et al., 1959)

In the newborn of several species, including man, the oxidation of fat within brown adipose tissue is a source of heat which may or may not be combined with shivering thermogenesis. The proportion of cold thermo-genesis generated from brown-fat catabolism varies with the species, in general being more significant in smaller (e.g. sheep) than in larger species, such as cattle (Alexander, 1979). In the older ruminant, shivering thermo-genesis almost completely replaces brown-fat oxidation. Shivering can be considered as a form of exercise where no external work is done, and is associated with an increase in blood flow to the shivering muscles (Bell *et al.*, 1976). The major substrate for the extra metabolism of shivering is provided by unesterified fatty acids released from white adipose tissue by sympathetic nerves that are activated during cold exposure (Thompson and Clough, 1972). This was demonstrated by Graham *et al.* (1959), who showed that fat is the primary tissue substrate used during cold exposure (*Figure 1.5*). How-ever, it can be seen that the degree of fat utilization depends on the level of feeding and that at high levels of food intake, the amount of fat that is oxidized is much less. Moreover, carbohydrate metabolism has been shown to increase during cold exposure in recently fed ruminants and can provide up to 70 per cent of the metabolism of the tissues of the leg. The question of protein utilization during cold exposure and the level of gluconeogenesis has

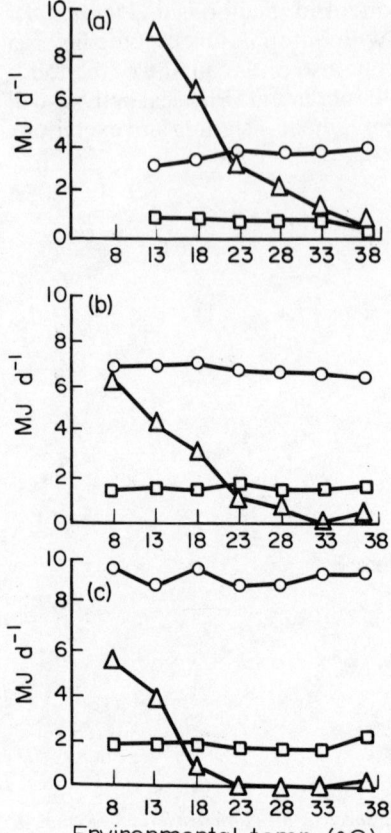

Figure 1.5 Mean heat production arising from the metabolism of (○) carbohydrate, (△) fat and (□) protein by two sheep at (a) low, (b) medium and (c) high levels of feeding in relation to environmental temperature. (After Graham et al., 1959)

been studied by Masaro (1976), who has shown that gluconeogenic enzymes in the rat liver are increased during cold exposure, especially when animals are in negative caloric balance. Stimulation of gluconeogenesis may be related to increased adrenocortical secretion and may thereby be a function of the severity of the cold stress. The question of nitrogen catabolism in the cold, especially severe cold, and its relation to endocrine changes has not been adequately addressed in food-producing animals and would seem an important consideration in understanding the impact of cold exposure on meat production. Glycerol may also be a substrate for hepatic gluconeogenesis, and an increase in its uptake in liver has been demonstrated in cold-exposed sheep (Thompson, Gardner and Bell, 1975).

In summary, there is good evidence that fat provides the main source of energy in cold exposure; it is remarkable that fat catabolism, particularly in animals on a low energy intake, takes place without the development of ketosis (Graham *et al.*, 1959).

The consequence of the elevated metabolism of cold exposure is that less of the metabolizable energy available to an animal is retained for productive purposes. This fact is elegantly demonstrated in *Figure 1.6* (Blaxter and Wainman, 1961), which shows that at a constant food intake the proportion

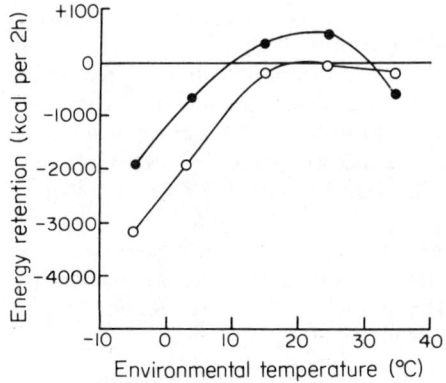

Figure 1.6 Effect of environmental temperature on the energy retention of two steers. (After Blaxter and Wainman, 1961)

of energy retained decreases in an approximately linear manner at environmental temperatures of less than 15 °C.

FOOD INTAKE, RETENTION TIME AND DIGESTIBILITY

Studies carried out mainly at the Department of Animal Science of the University of Alberta in Edmonton, Canada (discussed in more detail by Young in Chapter 10), have shown that the reduction in productive efficiency induced by environmental temperatures below the lower critical temperature is further compounded by a reduction in digestibility. The changes in digestive function are independent of changes in food intake, and are the result of a decrease in retention time associated with an increase in gut motility and of the rate of passage of digesta through the gastrointestinal tract. Maximal digestion in herbivores, and particularly ruminants, occurs only if the passage of food is subject to delay at those sites where microbial action takes place. The mechanism by which cold exposure influences the rate of passage appears to involve the thyroid gland (Westra and

Table 1.1   THE EFFECT OF AMBIENT TEMPERATURE ON THE DIGESTIBILITY, OXYGEN CONSUMPTION, PLASMA LEVELS OF TRIIODOTHYRONINE (T_3) AND THYROXINE (T_4), AND GUT RETENTION TIME OF SHEEP*

	Temperature	
	22–25 °C	*2–5 °C*
Digestibility (%)	57.4	51.8†
Oxygen consumption (ml kg^{-1} h^{-1})	340	545†
T_3 concentration (μg l^{-1})	0.62	1.52†
T_4 concentration (μg l^{-1})	80	135†
Gut retention time (h)	33.4	26.9†

*After Kennedy, Young and Christopherson (1977)
†Figures are significantly different at the 5% level

Christopherson, 1976). Cold exposure reduces retention time, lowers digestibility and stimulates thyroid hormone secretion (*Table 1.1*), although from experiments on thyroidectomized animals, there appear to be additional factors that influence the decreased retention time (Kennedy, Young and Christopherson, 1977). The cold-induced stimulation of appetite is more than able to compensate for the reduced digestibility, so that there is an increase in digestible energy (Kennedy and Milligan, 1978). It has been postulated that the increased rate of passage, particularly through the rumen, allows some nutrients to escape fermentation, resulting in less loss of energy through methane production and thereby increasing the availability of digestible nutrients directly to the tissue of the host (Kennedy and Milligan, 1978).

LACTATION

Cold exposure at temperatures above the lower critical temperature reduces milk yield (MacDonald and Bell, 1958) but the mechanism is unknown, although Holmes (1971) has shown that local cooling of one half of the udder will cause a small drop in milk secretion on the cooled side. With mild cold exposure, sufficient to raise metabolism by 28 per cent, Clarke, Thompson and Thomson (1976) were unable to show a reduction in mammary blood flow, although milk secretion was reduced and composition markedly altered. At more severe levels of cold stress, sufficient to raise metabolism by 50 per cent, a reduction in mammary gland blood flow does occur, and probably impedes substrate supply and further impairs lactation. The suppression of lactation thus appears to result from a combination of two mechanisms, one of which, possibly endocrine, affects secretion while the other brings about a reduction in mammary gland blood flow. The most likely candidate for a hormonal effect would be prolactin, which has been shown to be suppressed by cold exposure (Mills and Robertshaw, 1981).

THE OUTDOOR ENVIRONMENT

The sensible component of the heat balance incorporates such potentially large variables as convective and radiant heat transfer. In all the work cited above, the responses of animals have been defined in terms of air temperature alone, where radiative and convective heat exchanges are constant. Webster (1971) has determined empirically the convective and radiative components of the outdoor environment using an artificial cow. He found that the effect of increasing wind speed, u, is to reduce insulation in proportion to $u^{-0.5}$. Thus, any form of shelter or windbreak will have a significant effect on reducing heat loss, particularly at low ambient temperatures. Radiant heat exchange, whether it be the absorption of heat from the sun or the loss of heat to the clear night sky, represents approximately 9–26 per cent of the total Newtonian heat transfer (Webster, 1971).

On the basis of knowledge of the critical temperature of animals, attempts have been made to predict the impact that climatic variables would have on animal production (Webster, 1970). The broad conclusions that Webster reached were that cold would have a negligible effect on the energetic efficiency of cattle if they were sheltered from the wind, rain and night sky, but given access to the winter sun. In a study in which Milligan and Christison (1974) examined the effects of severe winter conditions on the performance of feedlot steers, it was observed that cold caused marked suppression of productivity. This did not agree with the predictions of Webster (1970). Furthermore, the expected increase in feed intake in the winter months did not materialize; the reason for these differences is not immediately clear, but may be related to the decline in digestibility which accompanies cold exposure. This not only demonstrates the difficulty in predicting environmental influences on productivity but also identifies the considerable intellectual challenge that Webster accepted in making the transition from laboratory to field conditions where other variables, such as feeding behaviour within a group, become apparent.

Effect of heat exposure

The thermoneutral zone (CD, *Figure 1.1*) includes the region for increased evaporative heat loss and suggests that body temperature will remain constant until heat gain exceeds evaporative heat loss (at D); body temperature will then rise until a new thermal equilibrium is established. In reality this rarely happens, and a rise in body temperature (i.e. positive heat storage) often takes place as evaporative heat loss is increasing. An increase in heat storage is a feature of the thermoregulatory response to high temperatures during dehydration: over a 24-hour period, the heat stored during the day can be dissipated at night. However, a rise in rectal temperature will increase metabolism and reduce the amount of energy retained (*Figure 1.2*). This is shown in *Figure 1.6* (Blaxter and Wainman, 1961); in one animal there was a decline in energy retention related to an increase in rectal temperature, whereas the rectal temperature of the other animal remained constant at the higher environmental temperatures, as did energy retention.

The increase in metabolism of animals is mainly accounted for by protein catabolism (Graham *et al.*, 1959; Blaxter and Wainman, 1961). This is demonstrated in *Figure 1.5*, where it will be seen that animals on a high energy intake, which also showed an increment in metabolism at high environmental temperatures, utilized protein as their source of increased calories. Vercoe (1969) and Vercoe and Frisch (1970) have studied the effect of hyperthermia on the energy metabolism of cattle, and verified the protein catabolic effect of hyperthermia. It is generally accepted that the trend towards increased protein catabolism is caused by hormonal factors, in particular cortisol secretion, that accompany either acute (Christison, Mitra and Johnson, 1970) or chronic (Christison and Johnson, 1972) heat exposure.

HEAT EXPOSURE, FOOD INTAKE, RETENTION TIME AND DIGESTIBILITY

The level of food intake is directly related to the level of thermoneutral metabolism (*Figure 1.4*); it is obvious, therefore, that a reduction of food intake is part of the physiological response to high heat loads and that heat-tolerant species will reduce their food intake less than those animals that are adapted to cooler conditions. Animals with a high productivity, whether it be in terms of meat or milk, have a relatively high thermoneutral metabolic rate and are therefore poorly adapted to hot conditions. They will also reduce their food intake to a greater extent than their low-producing, but heat-tolerant, counterparts. The reduction in food intake represents the main mechanism for the suppression of productivity under hot conditions. The actual stimulus to the reduction in food intake has not been identified; for example, it is not known whether heat exposure *per se*, without a rise in body temperature or hyperthermia, is necessary for suppression of food intake. In order for us to be able to predict the effects of heat exposure on the suppression of productivity, this type of information is essential. In some models relating energy intake to heat load, an inverse relationship between energy intake and rectal temperature is assumed (Hahn, 1980).

In both cattle and sheep, heat exposure produces a small increase in digestibility (Graham *et al.*, 1959; Blaxter and Wainman, 1961; Vercoe, 1976). Although there have been no concomitant measurements of retention time, it might be assumed that the rate of passage would be reduced by heat exposure, perhaps as a function of a decline in thyroid secretion rate (Johnson, 1976). All investigators agree that the increase in digestibility is small and sometimes at a low level of significance and does not, therefore, offset the decline in food intake.

HEAT BALANCE IN THE RADIANT ENVIRONMENT

In the hotter countries of the world, the thermal environment cannot be described in terms of air temperature alone; radiant heat exchanges comprise a very significant part of the heat balance equation. The quantitation of these exchanges is difficult and with the minute-by-minute changes in radiant heat load, it represents a much greater challenge for the formulation of a predictive model than is the case for colder environments.

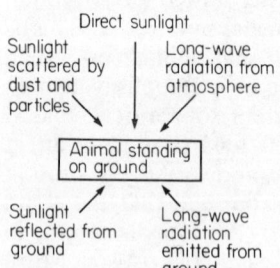

Figure 1.7 *The sources of radiation that impinge on an animal standing in the sun. (After Robertshaw and Finch, 1976)*

Besides the direct solar beam which the animal body intercepts, some solar radiation is diffusely scattered downwards onto the animal by the atmosphere. Furthermore, water vapour, carbon dioxide and ozone, which absorb solar radiation, emit long-wave radiation towards the earth's surface and onto the animal. Short-wave radiation is reflected onto the animal from the ground, the amount depending on the reflectance of the ground, while solar radiation absorbed by the ground is emitted towards the animal as long-wave radiation. The radiation impinging on an animal is shown diagrammatically in *Figure 1.7* (Robertshaw and Finch, 1976). The magnitude of the solar load will depend on the area exposed, the time of day and year, the colour of the animal and the reflectance of the ground. Light-coloured fur will reflect more solar radiation than dark-coloured fur, but solar radiation will also be reflected into the coat and penetrate to a greater depth with light-coloured fur (Hutchinson and Brown, 1969). Whether or not the incoming radiation is absorbed at the surface of the fur (darker coats) or deeper into the fur (lighter coats), some of the heat will be radiated to the surroundings with an emissitivity close to unity, since radiant exchanges in the long-wave part of the spectrum are independent of colour. Some losses by convection take place from the fur, and the difference between the heat absorbed and that lost by convection and re-radiation will be the sensible heat flow into the animal. It is this heat which makes thermoregulatory demands on the animal and which will influence animal production.

Table 1.2 shows the results of calculations of heat balance in an equatorial radiant environment of a *Bos indicus* (zebu) cow at an air temperature of 27 °C and 29–32% relative humidity. The most striking feature of the computation is the magnitude of the heat absorbed, being greater than twelve times the resting metabolism. Of the heat absorbed, 50–60 per cent is re-radiated to the environment and most of the sensible heat flow into the animal is dissipated by cutaneous (sweating) rather than respiratory evaporation (panting); sweating is known to be the major evaporative heat loss mechanism of cattle (McLean, 1963).

If the heat balance of cattle in a radiant environment at a given air temperature is known, it is possible to predict heat exchanges at other air temperatures. The assumption is made that humidity will be at a level that would not suppress evaporative heat loss. Such a prediction is shown diagrammatically in *Figure 1.8* where the heat balances of two species of cattle are given for noon at various air temperatures (Robertshaw and Finch, 1976). Maximal evaporative heat loss for *Bos taurus* and *Bos indicus* was calculated from the data of Hales and Findlay (1968) and Taylor, Robertshaw and Hofmann (1969), respectively. At the air temperatures when evaporative heat loss is maximal the animal stores heat and body temperature rises. Metabolism also increases and the increment is determined using a Q_{10} of two.

As in the case of predictive thermal models for cold-exposed animals, models for heat exposure contain many variables, and several refinements of the model are necessary. Although the magnitude of the solar heat load is constantly changing, the total radiant heat load between 12.00 and 14.00 remains constant, since even though the solar heat load diminishes, the absorbed long-wave radiation from the surroundings increases. The calculation predicts the heat balance of a fasting animal; the energy cost of grazing and the heat increment of feeding must be added to provide information on the productive animal. Such information is becoming available, but measurements of the radiant exchanges to the night sky are needed in order to obtain an integrated 24-hour thermal model.

It is obvious that various management practices can alleviate the heat load

Table 1.2 THE HEAT BALANCE OF *BOS INDICUS* MEASURED IN THE EQUATORIAL NOON SUN (FLUXES IN W m^{-2})

Metabolic heat	Radiant heat	Re-radiated heat	Convective heat loss	Cutaneous heat loss	Respiratory heat loss	Heat storage	Gain	Loss
50.7	638	397.3 (57.7)	63.6 (9.2)	146.2 (21.2)	36.1 (5.2)	7.6 (1.1)	688.7	650.9

The figures in parentheses are the percentage of the total heat dissipated by the various avenues of heat loss
After Robertshaw and Finch (1976)

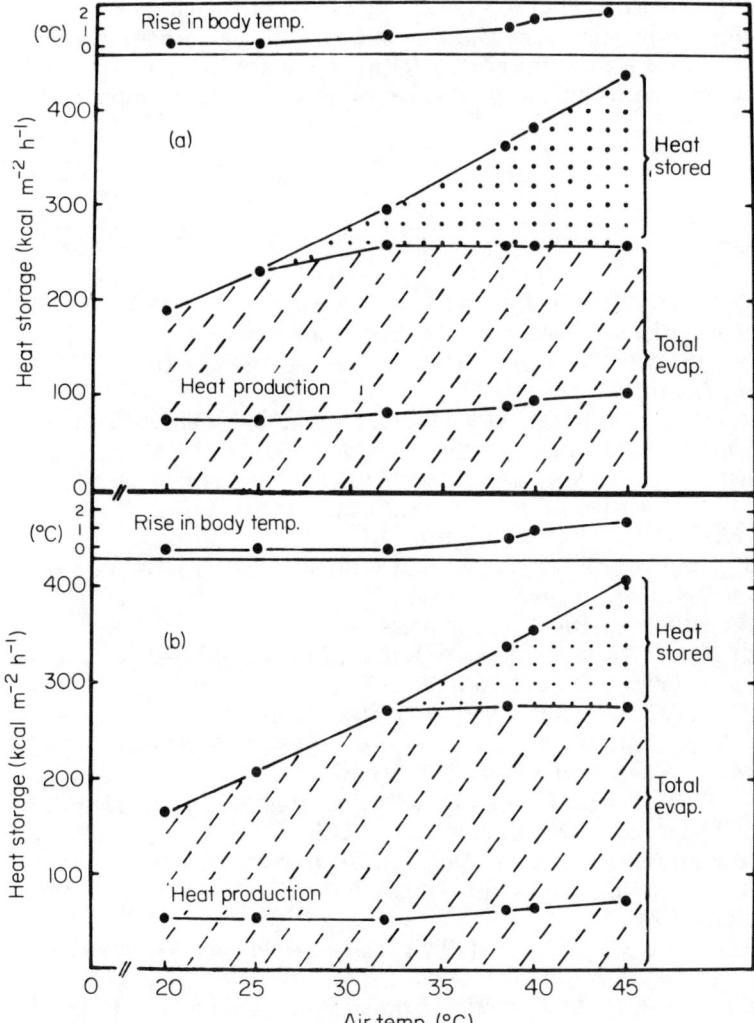

Figure 1.8 Predicted heat balances of (a) Bos taurus *and (b)* Bos indicus *subjected to solar radiation at noon. The graph indicates the effect of different air temperatures on heat storage and thereby predicts a critical air temperature below which a fasting animal can dissipate both the metabolic and solar heat loads. (After Robertshaw and Finch, 1976)*

on an animal; night grazing would allow heat-intolerant animals readily to dissipate the heat increments of feeding and grazing, and the provision of shade will reduce the direct solar heat load. The use of a water spray is a practice which increases the evaporative heat loss and thereby improves heat tolerance (Ansell, Chapter 16).

Conclusions

With greater understanding of the interaction between energy metabolism and heat exchange, we are moving towards the stage where the limiting effects of the environment on productivity can be predicted and improvements in food utilization achieved. An interaction with agronomists, who are working towards the same goals in terms of primary productivity, should allow an approach to animal productivity that is no longer empirical but based on first principles.

References

ALEXANDER, G. (1979). In *Environmental Physiology III*, pp. 43–155. Ed. by D. Robertshaw. University Park Press, Baltimore

BELL, A. W., HILDITCH, T. E., HORTON, P. W. and THOMPSON, G. E. (1976). *J. Physiol., Lond.* **257**, 229–243

BLAXTER, K. L. and WAINMAN, F. W. (1961). *J. agric. Sci., Camb.* **56**, 81–90

CHRISTISON, G. I. and JOHNSON, H. D. (1972). *J. Anim. Sci.* **35**, 1005–1010

CHRISTISON, G. I., MITRA, R. and JOHNSON, H. D. (1970). *J. Anim. Sci.* **31**, 219

CLARKE, P. L., THOMPSON, G. E. and THOMSON, E. M. (1976). *J. Physiol., Lond.* **263**, 176P

GRAHAM, N. McC., WAINMAN, F. W., BLAXTER, K. L. and ARMSTRONG, D. G. (1959). *J. agric. Sci., Camb.* **52**, 1–40

HAHN, G. L. (1981). *J. Anim. Sci.*, in press

HALES, J. R. S. and FINDLAY, J. D. (1968). *Resp. Physiol.* **4**, 333–352

HOLMES, C. W. (1971). *J. Dairy Res.* **38**, 3–7

HUTCHINSON, J. C. D. and BROWN, G. D. (1969). *J. appl. Physiol.* **26**, 454–464

INGRAM, D. L. and MOUNT, L. E. (1975). *Man and Animals in Hot Environments*. Springer Verlag, New York

JOHNSON, H. D. (1976). In *Progress in Biometeorology*, pp. 27–32. Ed. by H. D. Johnson. Swets & Zeitlinger, Amsterdam

KENNEDY, P. M. and MILLIGAN, L. P. (1978). *Br. J. Nutr.* **39**, 105–117

KENNEDY, P. M., YOUNG, B. A. and CHRISTOPHERSON, R. J. (1977). *J. Anim. Sci.* **2**, 1084–1090

MacDONALD, M. A. and BELL, J. M. (1958). *Can. J. Anim. Sci.* **38**, 160–170

McLEAN, J. A. (1963). *J. Physiol., Lond.* **167**, 427–447

MASARO, E. J. (1976). In *Progress in Biometeorology*, pp. 19–26. Ed. by H. D. Johnson. Swets & Zeitlinger, Amsterdam

MILLIGAN, J. D. and CHRISTISON, G. I. (1974). *Can. J. Anim. Sci.* **54**, 605–610

MILLS, D. and ROBERTSHAW, D. (1981). *J. clin. Endoc. Metab.* **52**, 279–283

MOUNT, L. E. (1974). In *Heat Loss from Animals and Man*, pp. 425–439. Ed. by J. L. Monteith and L. E. Mount. Butterworths, London

ROBERTSHAW, D. and FINCH, V. A. (1976). In *Beef Cattle Production in Developing Countries*, pp. 281–293. Ed. by A. J. Smith. University of Edinburgh, Edinburgh

TAYLOR, C. R., ROBERTSHAW, D. and HOFMANN, R. (1969). *Am. J. Physiol.* **217**, 907–910

THOMPSON, G. E. and CLOUGH, D. P. (1972). *Q. Jl exp. Physiol. cogn. med. Sci.* **57**, 192–198

THOMPSON, G. E., GARDNER, J. W. and BELL, A. W. (1975). *Q. Jl exp. Physiol. cogn. med. Sci.* **60**, 107–121

VERCOE, J. E. (1969). *Aust. J. agric. Res.* **20**, 607–612

VERCOE, J. E. (1976). In *Progress in Biometeorology*, pp. 434–441. Ed. by H. D. Johnson. Swets & Zeitlinger, Amsterdam

VERCOE, J. E. and FRISCH, J. E. (1970). *Aust. J. agric. Res.* **21**, 857–863

WEBSTER, A. J. F. (1970). *Can. J. Anim. Sci.* **50**, 563–573

WEBSTER, A. J. F. (1971). *J. appl. Physiol.* **30**, 684–690

WEBSTER, A. J. F. (1974). In *Heat Loss from Animals and Man*, pp. 205–231. Ed. by J. L. Monteith and L. E. Mount. Butterworths, London

WESTRA, R. and CHRISTOPHERSON, R. J. (1976). *Can. J. Anim. Sci.* **56**, 699–708

2

CLIMATE AND THE NEED FOR HOUSING

J. R. STARR
*Meteorological Office, Ministry of Agriculture, Fisheries and Food,
Coley Park, Reading, Berkshire*

The thermal environment—and other reasons for housing

An important impetus for the decision to house livestock comes from the experience that, by controlling the range of an animal's 'thermal environment' (itself an integral part of the interaction between the animal and its environment, *Figure 2.1*), productivity and reproductive efficiency can be increased. It is the aim of this chapter to present various climatic indices that might be used to assess this 'thermal environment', and to comment on them as a means of quantifying the climate experienced by livestock and hence of indicating the need for housing.

In discussing the thermal environment, we must recognize that animals necessarily generate metabolic heat and that this heat has to be dissipated to the environment by the processes of conduction, convection, radiation and evaporation. It is the net effect of these processes which is important, although one process (such as forced convection, in a cool climate) may be dominant. Over a limited range of environments, commonly denoted by the 'thermoneutral zone', a ready balance may be achieved between the animal's heat production and the environmental demand; within this zone the animal is not stressed in achieving a balance. Beyond the 'thermoneutral zone' an animal becomes subject to 'stress' and its productive efficiency decreases (where efficiency is measured in terms of meat, milk or eggs produced per unit of food intake).

There are, of course, management reasons for undertaking housing that are not necessarily related to thermal comfort. For example, in an extensive animal production system the need for prophylactic measures at certain times may lead to the group's being brought together in order to protect either the animals (e.g. when dipping or lambing) or the soil (e.g. avoidance of poaching).

The acceptable environment will, then, be defined by the type of animal, the purpose for which it is kept and the proposed management system. The housing of ruminants, for example, is normally rather different from that of poultry and pigs. For the latter, positive attempts may be made to meet precisely defined environmental/feed energy/production relationships. For young and fattening ruminant stock, also, it is often desirable to control the environment for optimum health, growth rate, food conversion rate, etc., while for adult and breeding stock simple protection from environmental

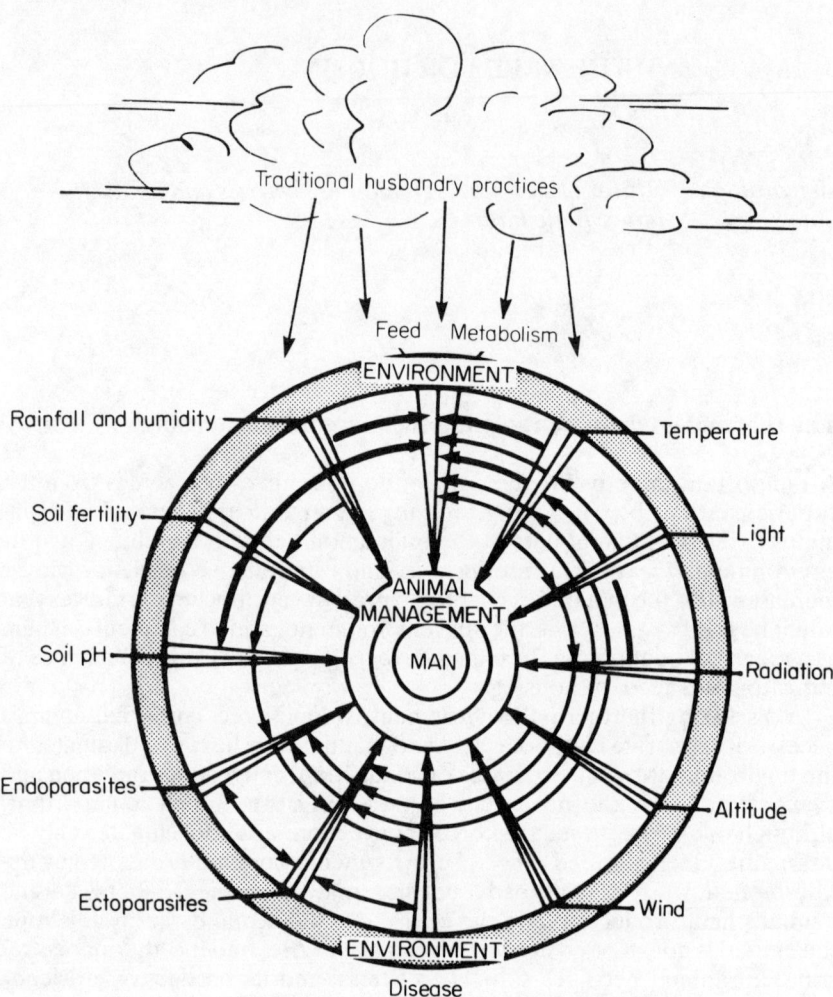

Figure 2.1 Elements of the environment which directly or indirectly influence the performance of animals. (From McDowell, 1967, Figure 1, p. 279; reproduced with the permission of R. E. McDowell and the American Association for the Advancement of Science)

extremes may suffice (Webster, 1974). Whatever the species of animal, it is also likely that local breeds will have different needs for housing compared with exotic varieties.

Specifying the environment

Since an important aim in housing livestock is to reduce the environmental extremes, which on the one hand lead to a reduction of energy intake and on the other lead to an increase in 'non-productive' energy expenditure (Smith, 1971), more precise relationships between environmental factors and the response of livestock production are needed. So also is a rational specification of the environmental factors leading to 'stress'. Apart from specifying these environmental factors, two other questions that must be answered are: 'What range of thermal environments is desirable for a given age and species of stock?' and, this having been decided, 'How can structures be specified that will provide them?' (a question considered in other contributions to this volume). It is relatively easy to measure the physical environment, but there must also be an awareness of the need to translate this understanding into economic terms and investment decisions by asking: 'By how much and how often is energy intake depressed and what is the energy loss?'.

The search for a rational specification of environment led Monteith (1974) to the general principles that such a specification should be applicable to all homeotherms, to all climates, and based on standard meteorological records. A caveat (Findlay, 1972) is that, even for an environment specified in this way, the extent of the climate's influence will depend on such factors as the age, breed, posture and the nutritional state of the animal. Furthermore, although any stress experienced by the animal will be expressed in physiologically demonstrable factors such as raised respiration or rectal temperature (McDowell, 1965), it will not necessarily result in lower productivity, especially if there is a long 'rest' period under thermoneutral conditions between stresses. The outcome of environmental stress may also depend on the degree of acclimatization, the duration of the stress and on the animal's response time (Smith, 1970). Fluctuations in environmental elements imposing stress necessitate corresponding readjustments by the animal's physiology. Even where it is possible to adjust management practices to such fluctuations this may not be advisable; the husbandry of poultry and swine might be adjustable frequently, but lactating cattle might not respond well to such disturbances (McDowell, 1972). (In contrast, diurnal variations of non-stressful environments have been shown to be productively beneficial.)

Since some of the physiological effects of a 'stressful' environment may be short-lived, while others develop over a longer period, measurements of both the intensity and duration of significant ranges of values of climatic elements are therefore necessary. The value of 'microenvironment' measurements of some variables at animal height is clear, especially if they can be related to measurements at a nearby 'standard' meteorological station.

The assessment of climate

Various attempts have been made to 'specify the environment'; that is, to classify biometeorologically important factors which would enable mapping of the geographical distribution of areas suitable for various agricultural enterprises. For example, an early classification of world climate by Köppen and Geiger (1936) is appropriate to plant growth but not necessarily so for defining animal husbandry practices. Duckham and Masefield (1970) concentrated on the distribution of farming systems in terms of the crop 'thermal growing season' and 'hydroneutrality', but remarked on the plethora of crop and livestock diseases which limit livestock stature, production and distribution in the wet tropics, and also on the general environmental needs of specific species when transferred to other zones.

Many other classifications and climate parameters have been developed. These will be discussed after a brief consideration of the effects and implications of the meteorological elements which commonly form the basis of these 'indices'.

Temperature is probably the most important single bioclimatic factor. The temperature of both the air and the surroundings should be considered and may exercise indirect influences via food supply and disease incidence. Wind speed and direction influence the effects of temperature and their covariance should be taken into account, since the use of daily or monthly averages will mask the true range of fluctuations, as emphasized by Mumford (1979) in discussing 'wind chill'.

The effects of altitude on temperature are extremely important: at sea level the annual range of monthly mean temperature in the British Isles is from 6 to 16 °C in the extreme south-west, and from 3 to 12.5 °C in north-east Scotland. However, since the mean air temperature falls by about 0.6 K per 100 m increase in height, the higher parts of Dartmoor experience temperatures similar to those at sea level in the Wick area, five hundred miles further north.

In some parts of the hot, wet zone close to the equator the annual range of monthly air temperature is only 1 K; even in savanna regions the difference in mean temperature between 'cool' and 'hot' is rarely more than 7 K. While the temperatures in subtropical zones will fluctuate from day to day in response to the synoptic weather patterns, these changes are generally less frequent and abrupt than in temperate zones.

Outside the temperature zone within which thermoneutrality may easily be maintained, other climatic elements, such as *humidity,* may become important. While an animal can lose adequate heat by radiation and convection, relative humidity (RH) is of little consequence; however, its ability to sweat will be an important influence on the animal's ability to respond to high temperature. The relationship established for the rate of heat loss from the bovine respiratory tract indicates, for example, that at 29 °C and 60% RH a cow loses 3.4 W m^{-2} while at 90% RH this heat loss is reduced by 25 per cent (McDowell, 1972). Heat balance can become a problem at 20 °C and above when RH is in excess of 60%. High humidity also affects animal health indirectly, by encouraging conditions conducive to the propagation of various disease organisms and skin fungi. Hot, very dry environments may also be hazardous.

Wetting of the skin or hair coat by *precipitation* can significantly reduce coat insulation and increase heat dissipation through evaporation. The intensity and duration of the precipitation and the nature of the coat determine the extent of wetting. In temperate and cool climates, this added heat loss may lead to cold stress (Alexander, 1964) and, if sustained, to the cold/starvation syndrome in young lambs described by Slee (1977). In both cold and hot climates the onset of precipitation may cause stock to cease grazing and seek shelter and may, therefore, reduce food intake. Precipitation also has important indirect effects through its influence on the type and quality of vegetation, on the incidence of certain diseases (Smith and Ollerenshaw, 1967) and on the degree of pasture 'poaching', an important limitation on grazing stock under otherwise benign climatic conditions.

Air motion affects the rate of convective heat exchange from the animal and reduces coat insulation, and in this the characteristics of the 'surface' of the animal are also important. In tropical zones the 'advection' of hot air over the animal may increase heat stress; in cold locations air motion may cause or enhance cold stress. In climatic zones between these extremes air motion generally mitigates against climatic stress. A figure of 2.25 m s⁻¹ is quoted by McDowell (1972) as ideal in hot, dry, daytime environments; after sunset the restoration of the heat balance is encouraged by wind speeds of the order 2.25–4.5 m s⁻¹. In other climates, where external air temperature is commonly well below skin temperature, any shelter provided aims to make use of metabolic heat to raise ambient temperatures. Control of the flow of air and attention to air circulation patterns are important to avoid environmental stress within the building and to reduce the risk of the build-up of toxic waste gases and airborne infection (Smith, 1971). Air motion is an important vector of certain animal diseases (Wright, 1979).

Outdoors, an important influence on animals fed for high performance is the incident short-wave *solar radiation,* both direct and diffuse. Insolation will increase the heat load on an animal in a complex way that depends on the animal's orientation, coat type and coat colour. Long-wave radiation is also important in the energy balance. In the tropics the intensity of solar radiation under clear skies is fairly uniform from place to place, and the duration of sunshine is a useful indicator of radiation load. There is evidence for significant increases in performance efficiency in these warm (dry) areas when shades or shelter are introduced into dry-lot feeding schemes. The benefits of shade provision may well be questionable, however, in hot, humid environments that are associated with low air movement, if the congregation of stock further restricts that air movement and hence evaporation and convection.

Other practical considerations may influence the choice of orientation for housing or shelter. A north–south orientation may be preferred for the housing of dairy and beef cattle in warm climates with high solar radiation, because the relatively short period of shading of a particular ground area during the daily radiation cycle will allow the ground to dry out, minimizing sanitation problems. However, a shelter orientated east–west may intercept more solar radiation and so provide a 'cooler' environment in which ground temperature will also be lower.

The direct effects on animal physiology of normal *atmospheric pressure* variations at any given site have yet to be established, though difficulties may be encountered when transferring livestock between enterprises at different altitudes (Hyslop, 1980). Air pressure is one element that conventional housing does not normally set out to modify (though fan-driven ventilation systems may be operated to provide a 'pressurized' house to counteract unwanted air infiltration).

Climatic indices

Many indices have been developed to characterize the thermal comfort zones of clothed and unclothed human beings. Some of these may be more applicable to specific ranges and types of environment than others and their use will need some caution when applied to other homeotherms. The object, though, with both animals and man, is to achieve some readily comprehensible index for comparing environments and to present, in a single variable, factors that characterize or imply both the thermal environment and the stress it imposes on particular organisms. This search has led physiologists in two directions: empirical indices have been derived from laboratory and 'field' experiments, and instruments have been developed to simulate the heat loss.

An understanding of the physical processes involved in animal heat balance leads to the suggestions that an 'environmental demand' index should include measurements of the meteorological elements discussed, and that weight should be given to the various elements to reflect their relative importance to the animal. There are over thirty such indices that have been found useful over specific ranges of (mainly human) environments. Many of the more common ones have been reviewed, for example, by Smith (1970), Kerslake (1972), Landsberg (1972), Mather (1974), Morgan and Baskett (1974) and Mount (1979) and the ranges and limitations of some will now be discussed.

The *wind chill factor,* **K** (Siple and Passel, 1945), is based on the effect imposed on a clothed person by cold environmental conditions. This factor is intended to indicate when there is a danger of exposed flesh freezing, and represents the environmental cooling power in shade and without regard to evaporation. It has units of $W\,m^{-2}$ and is given by

$$\mathbf{K} = 1.163(10u^{0.5} + 10.45 - u)(33 - T_a) \tag{2.1}$$

where T_a is the air temperature in °C and u is the air speed in $m\,s^{-1}$. The index was developed from the observed chilling under laboratory conditions of an apparently arbitrary volume of water in a glass phial; although it has many shortcomings it is still useful as a rough measure of heat loss under conditions of forced convection.

The likelihood that **K** is invalid for animals with 'external insulation' has led workers such as Ames (1974) to develop prediction expressions for the rate of heat loss from dry sheep and cattle, based on calorimetic experiments. Controlled-climate chambers have also been used by Kibler and Brody (1954) to prepare 'comfort diagrams' for cattle describing the effect of different combinations of temperature, humidity and wind speed on cattle's wellbeing. The 'zone of normal rectal temperature' was determined using a wind speed of $0.2\,m\,s^{-1}$, although it was shown that normal rectal temperatures could also be maintained at higher air temperature and humidity if the wind speed exceeded $2.7\,m\,s^{-1}$ to increase heat dissipation. It is interesting to observe that humidity seems to have little effect on cattle when in the field, but that some reaction is found under these climate-chamber conditions.

One of the most important sources of heat stress for livestock and man can be solar radiation. An index of this stress is that devised by Minard, Belding and Kingston (1957): the *wet-bulb globe temperature* (WBGT). This is based on measurement of the temperature of a standard black globe (T_g), the wet-bulb temperature (T_w) and shaded dry-bulb temperature (T_a) such that

$$\text{WBGT} = 0.7T_w + 0.2T_g + 0.1T_a \tag{2.2}$$

If a shaded, adequately ventilated wet bulb of temperature T_w is substituted, then

$$\text{WBGT} = 0.7T_w + 0.3T_g \tag{2.3}$$

No direct measurement of air movement is required. In the absence of a radiant heat load, air movement would not affect the index value at all. The WBGT is unlikely to be applicable to all types of hot environment and the direct application to animals with coat absorptivity differing greatly from the characteristic black-globe value must be questionable. Furthermore, the size of the globe will determine the sensitivity to wind speed. Kerslake (1972) has, however, suggested that the WBGT offers a simply measured index applicable to humans under a wide range of stressful environments, from cold store to blast furnace. Further work might be justified in adapting the WBGT index as a standard for comparing livestock environments where heat stress is a primary consideration.

The *effective temperature* (T_e) (Yaglou, 1927) is defined as the temperature of a 'still-air' environment, saturated with water vapour, which produces the same sensations of warmth for human subjects as the actual environment. It emphasizes and combines the effects of heat loss or gain by convection and evaporation and, obtained from a nomogram, is widely used as a 'steady-state' index. The effect of radiation exchanges is introduced through the concept of the *corrected effective temperature*, in which globe thermometer readings replace those of the dry bulb. There is, again, the problem of assigning the correct 'weighting' to the radiation term, since its importance will be a function both of the radiation wavelength and the nature of the animal's coat.

Another parameter biased towards the radiative and convective elements of an environment is the *equivalent temperature**, T_{eq}. It is defined as the temperature of a uniform enclosure in which, in still air, a sizeable black body at 24 °C would lose heat at the same rate as in the given environment. Experiments suggest that

$$T_{eq} = 0.522T_a + 0.478T_w - 0.207\,(37.8 - T_a)u^{0.5} \tag{2.4}$$

T_a, T_w and u having been defined earlier. T_{eq} was originally derived for work on human comfort. The index is considered inappropriate when humidity effects may be important, particularly when T_a is above about 20 °C.

The *operative temperature* (T_{op}) of an environment presents an evaluation of the combined physical effects of the radiation temperature T_R of the surroundings, the ambient air temperature and the air movement. It identifies an environment of equal wall and air temperature, T_a, and of a standard air speed, in which the heat loss of an organism by radiation and convection is the same as in the original environment. The equation defining T_{op} includes both air temperature and velocity. Inclusion of the ratio of the transfer coefficients for convection and radiation enables 'shape factors' to

---

*Not to be confused with the meteorologist's term ⊖, defined for a 'wet system' as $\ominus = T_a + e/\gamma$, where T_a is the air temperature, e the vapour pressure and γ the psychrometric constant. *See* p. 31.

be taken into account and also accommodates animals of different surface area.

When the temperature differences are small it is possible to estimate radiant exchange with only a small error from a simple temperature difference, instead of from the difference between the fourth powers of the absolute temperatures. Then T_{op} can be regarded as the weighted average of air and radiant temperatures:

$$T_{op} = \frac{h_R \bar{T}_R + h_c T_a}{h_R + h_c} \tag{2.5}$$

where h_c and h_R are respectively the convective and radiative heat transfer coefficients, $\bar{T}_R$ the mean radiant temperature and T_a the air temperature. Since h_c is approximately proportional to the square root of the wind speed, u, variations in air movement can be incorporated (Gagge, 1965):

$$T_{op} = \frac{1}{1 + K}\left[K(17.7 + \bar{T}_R) + \frac{u}{u_o}(17.7 + T_a) - \left(\frac{u}{u_o}\right)^{0.5}\right.$$
$$\left. (17.7 + T_{sk}) - 0.55 \right] - 17.7 \tag{2.6}$$

where T_{sk} is the skin temperature, $K = h_r/h_c$ and u_o is the standard air velocity.

Humidity effects are ignored in this index. The operative temperature, then, is another simple variable, calorimetrically derived and biased towards the radiative and convective elements of heat loss (generally the dominant components under the climatic conditions of the British Isles), which will enable comparison of internal climates to be quickly and simply made, as demonstrated by Smith (1965) in his assessment of pig housing environments. Evaporative heat loss must be assessed separately. A practical disadvantage in using this scale is that the mean skin temperature, T_s, must be known.

The *temperature–humidity index* (THI) (*Figure 2.2*), was developed by the US Weather Bureau as a warm-weather discomfort index for the evaluation of livestock stress. Its merits have not been extensively investigated, although Johnson *et al.* (1963) found a relationship with milk production for lactating cows. The THI has units of temperature and is given in °C by

$$THI = 0.72(T_a + T_{dp}) + 40.6 \tag{2.7}$$

where T_a is the air temperature and T_{dp} is the dewpoint temperature. Monthly mean THI values can be used to compare the stress of areas which might display diverse (mean) temperature and humidity ranges.

Many other climatic indices have been proposed. For example, Cena and Slomka (1966) designated five categories of cooling power based on monthly mean temperature and wind speed. The sensations associated with the (unlikely) simultaneous occurrence of these mean quantities will obviously differ from those associated with the combinations of wind and temperature which actually occur and which, when integrated, lead to the monthly mean

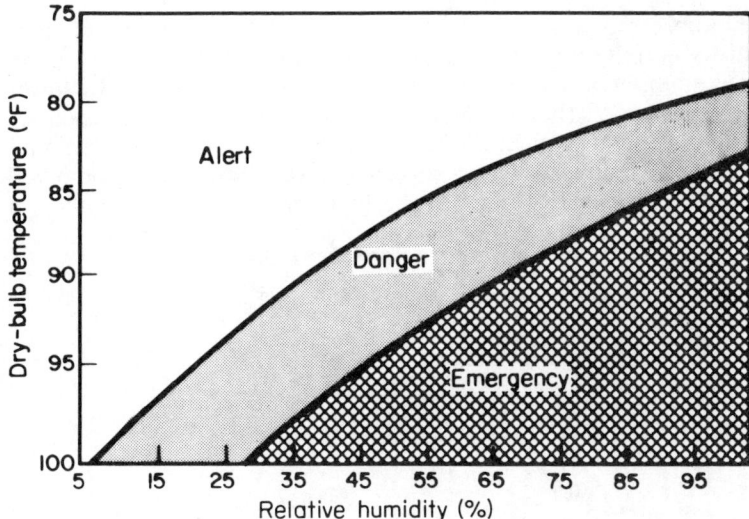

Figure 2.2 Temperature–humidity index (developed by the US Weather Bureau) as presented in a 'Livestock Safety' booklet distributed by Livestock Conservation Inc., USA, and reprinted with permission

values. This system is not specific enough for closer analytical work in bioclimatology. Terjung (1966) categorized *effective temperatures* modified by 'wind chill' and corrected for solar radiation. The effect was, again, largely vitiated by the use of mean monthly values for certain variables. Bultot (1962) produced a methodologically more satisfying approach, using a temperature and vapour pressure diagram with simultaneous hourly values. In it isolines of Lee's (1958) *relative climate strain* were shown. Lee's emphasis was on the 'hot' end of the spectrum, since man can protect himself relatively easily against excessive heat loss under cold conditions. Both temperature and vapour pressure were given on a probabilistic scale: for example, 95 per cent of observed combinations fell within the plotted ellipse. If data are in machine-processable form the frequency distribution of variables may be established and probability levels in a joint frequency distribution found. Landsberg (1972) stated that 'it would be desirable if future bioclimatic classification attempts were based on probability levels of the observed elements' (as in Smith's (1974) discussion of the temporal variability in 'cold stress'). He went on to say: 'The enormous complexity and variety of bodily functions and atmospheric variations defy simple solutions . . . one can only cope with limited objectives'.

A few climatic indices have been developed specifically for domestic and non-domestic ruminants. Picton (1979) used a 'climatic fluctuations' index due to Lamb (1963) to assess the impact of climate upon deer populations. Such an index 'allows effective reduction of standard weather data to a form relevant to the study of population biology'. Øritsland (1974) presented a 'wind chill' and solar radiation index for fur or plumaged creatures based on the heat balance relationship. The index is for average heat loss as a function of body weight and insulative cover under all naturally occurring dry weather conditions; it is not a complete expression for animal heat loss, but a reference index for the response to cold. Smith (1974) produced predictions of cold stress in determining the need for the housing of steers. Data on steers fed at maintenance level in Ireland were used, and he defined a 'lower critical temperature' T_c (below which a raised metabolic rate is induced) for the conditions of the investigation. He produced a map of average accumulated temperature below T_c (*Figure 2.3*) enabling cold stress to be compared in various parts of the UK, an approach that could usefully be applied to

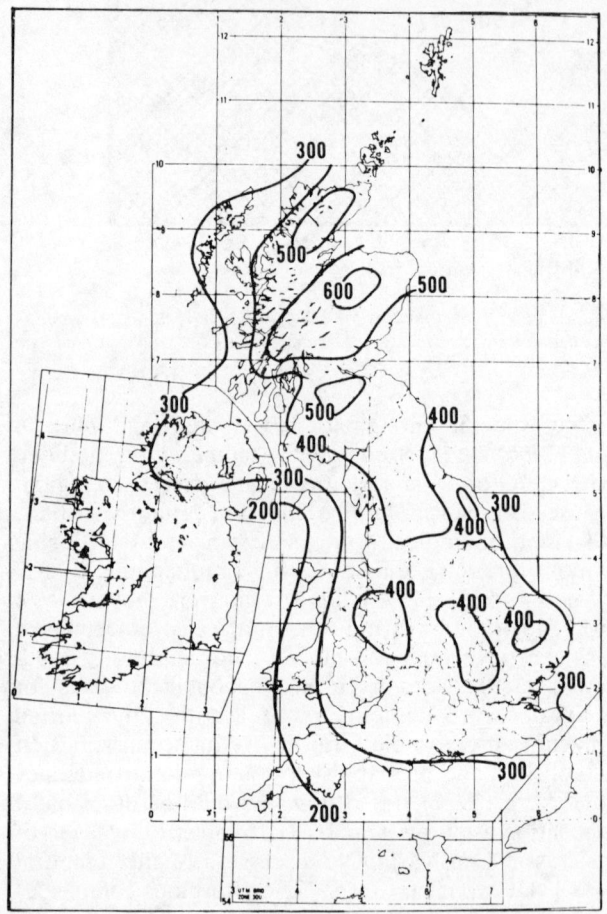

Figure 2.3 Average accumulated temperature in day-degree Celsius below 5.6°C (at mean county height below 300 m) indicative of the degree of cold stress on steers. (Reproduced from Smith, 1974)

other classes of livestock on a world basis. The variation of cold stress was demonstrated for a sample decade and Smith showed that the associated weight loss in the more extreme seasons could be double that for the average winter—a translation of environmental stress into economic terms.

Experimental data on the influence of weather on lowland lamb mortality, presented by Starr (1980, 1981) were analysed using the 'wind chill' relationships for young lambs due to Alexander (1964). The 'critical daily heat loss'

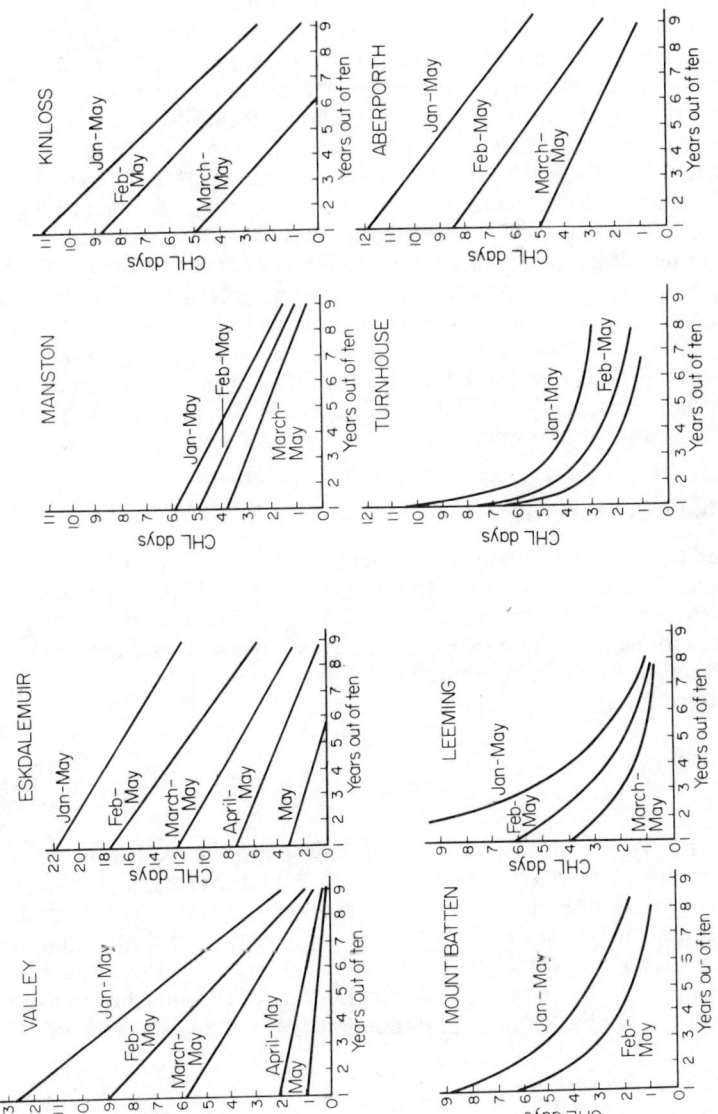

Figure 2.4 Probability of days with a critical daily heat loss (CHL) of more than 80 times the basal metabolic rate for lambs in different parts of the UK for periods between January and May

(CHL) for lambs weighing less than 3 kg was estimated from hourly meteorological measurements at lamb height in the light of lamb mortality over the preceding 24 hours. This CHL of 80 times the basal metabolic rate was interpreted in terms of various combinations of the environmental variables using Alexander's (1964) data for lambs with fine birth coats. *Figure 2.4* presents the number of times the CHL was reached for various periods (in terms of years in 10) at particular stations in the UK. Experimental daily forecasts of daily heat loss, based on this work, have been made by Benson Meteorological Office in the spring of 1980 for a farm near Reading, Berkshire, enabling the farm manager to decide whether to retain newborn lambs in housing for a period or put them out on pasture.

Any equivalence between environments with different radiative–convective combinations, established on the basis of indices such as those discussed, would strictly apply to a particular body only, since 'shape factors' might introduce different constants into the heat transfer process. The alternative is to estimate separately heat transfer through each of the four exchange channels and then to combine the results to give the total thermal effect of the environment. Experimental and theoretical modelling techniques using this principle provide a powerful approach to the assessment of housing needs, through the evaluation of the thermal interaction between the animal and its environment.

The heat balance equation

By appealing to the steady-state heat balance relationship for a homeotherm, information may be combined about the physiological state of the organism, the physiological and physical conditions of the organism/environment interfaces, the state of the environment and the extent to which all are thermally coupled (Monteith, 1974). The conventional form for the steady-state heat balance of a homeotherm is

$$\text{Gains} = \text{losses}$$
$$\mathbf{M} - \mathbf{W} = \mathbf{C} + \mathbf{E} + \mathbf{R} \tag{2.8}$$

where $\mathbf{M}$ and $\mathbf{W}$ are the rates of metabolic energy production and of mechanical working respectively and $\mathbf{C}$, $\mathbf{E}$ and $\mathbf{R}$ are the rates of heat loss by convection, evaporation of water and radiation respectively, all expressed in watts per square metre ($\mathrm{W\,m^{-2}}$) of the external surface of the animal. By straightforward algebraic manipulation, Monteith arranged this equation in a form which helps to distinguish the conditions of the interface and the environment and their coupling. The radiative loss $\mathbf{R}$ may be expressed as

$$\mathbf{R} = h_{\mathrm{R}}\,(T_0 - T_\mathrm{a}) - \mathbf{R}_{\mathrm{ni}} \tag{2.9}$$

and

$$\sigma(T_0{}^4 - T_\mathrm{a}{}^4) = h_{\mathrm{R}}\,(T_0 - T_\mathrm{a}) \tag{2.10}$$

where h_{R} is the linearized radiative transfer coefficient ($h_{\mathrm{R}} \simeq 4\sigma T_\mathrm{a}{}^3$), T_0 is the interface temperature, T_a the wall (air) temperature and σ is Stefan's constant. The surface emissivity is assumed to be unity. $\mathbf{R}_{\mathrm{ni}}$ is the additional radiation gained from surfaces not at T_a, the 'isothermal net radiation'; it represents the radiative energy that the interface would absorb if its temperature were equal to T_a, and can be estimated from the distribution of short- and long-wave radiation over the interface. The convective heat exchange may be written

$$\mathbf{C} = h_c(T_0 - T_a) \qquad (2.11)$$

where h_c is the heat transfer coefficient for convection. The latent heat loss has a similar form:

$$\mathbf{E} = h_E(e_0 - e_a)/\gamma \qquad (2.12)$$

where h_E is the heat transfer coefficient for evaporation, e_0 is the mean vapour pressure of the interface in millibars and e_a is the vapour pressure of the surrounding air in the same units. γ is the psychrometric constant, at sea level approximately 66 kPa K^{-1}. Adding equations 2.9, 2.11 and 2.12 we obtain

$$\mathbf{M} - \mathbf{W} = \mathbf{C} + \mathbf{E} + \mathbf{R} = (h_c + h_R)(T_0 - T_a) + h_E\frac{(e_0 - e_a)}{\gamma} - \mathbf{R}_{ni} \qquad (2.13)$$

It is often convenient to simplify this equation by amalgamating the sensible and latent heat terms. For this purpose an 'apparent equivalent temperature', Θ^*, can be defined:

$$\Theta^* = T + e/\gamma^* \qquad (2.14)$$

where

$$\gamma^* = \frac{(h_c + h_R)}{h_E} = \frac{h_{cR}}{h_E} \qquad (2.15)$$

(*Note:* the quantity $\Theta^* - T$ can be regarded as a humidity increment proportional to vapour pressure and dependent on wind speed, albeit weakly, when this is greater than 5 m s^{-1}.) Then, substituting in equation 2.13,

$$\mathbf{C} + \mathbf{E} + \mathbf{R} = h_{cR}(\Theta_0^* - \Theta_a^*) - \mathbf{R}_{ni} \qquad (2.16)$$

where Θ_0^* and Θ_a^* are the apparent equivalent temperatures of the interface and the surrounding air, respectively.

Finally, the thermal radiation increment is introduced, this quantity being the increase in air temperature that would be needed to compensate for removing an external source of radiant energy:

$$\Theta_e^* - \Theta_a^* = \mathbf{R}_{ni}/h_{cR} \qquad (2.17)$$

where Θ_e^* is an apparent equivalent temperature of the environment. Substitution in equation 2.8 now gives

$$\mathbf{M} - \mathbf{W} = h_{cR}(\Theta_0^* - \Theta_e^*) \qquad (2.18)$$

The state of the environment is represented by values of h_{cR} and $\Theta_e{}^*$ whereas that of the interface is represented by $\Theta_o{}^*$. However, h_{cR} and h_E both depend on the size and shape of the organism and $\Theta_o{}^*$ depends on wind speed (through the value of γ^*). An additional term is needed in equation 2.18 to account for the heat loss through the respiratory tract of a panting animal, which may constitute a significant fraction of the latent heat loss of an animal exposed to heat stress.

In this analysis Monteith avoided the problem of describing the physical behaviour of the integument by referring to the state of an interface. However, the heat balance equation can be written in a form directly applicable to an animal with a hair coat by making arbitrary assumptions based on our existing knowledge of the relationship between insulation, wind speed and intercepted radiation and about the distribution of evaporating sweat. The main value of the heat balance approach is its generality. Many facets of thermal physiology are not described explicitly but new experimental results can readily be incorporated within the framework provided by this fundamental approach to the specification of the environment.

Even though an environment may be identified by the heat demand or heat stress it imposes, the fact remains that there is no simple, single measurement with standard meteorological equipment which will summarize or integrate the effects of the environment. For this reason Blaxter (1959) and McArthur (1977) with 'sheep', Mount (1964) with 'pigs' and Webster, Chlumelky and Young (1970) with 'cattle' have approached this problem by physical simulation. In each case heat loss was measured from artificially heated bodies with dimensions, form and insulation that approximated to those of the animals. The environmental demand was then estimated under a full range of (non-evaporative) conditions from the electrical energy needed to maintain a constant 'deep-body' temperature.

Since the heat balance equation can incorporate all the environmental and physiological variables it appears to be the strongest candidate for the basis of an index; McArthur will be considering more detailed applications of this heat balance approach in the next chapter.

Summary

The climatic assessment of housing need obviously requires more than simply a consideration of climatic averages or extremes. The extent to which essential climate characteristics can be assessed for a region or a given location, furthermore, will depend on the availability of appropriately sited observing stations.

Climate classification maps may be helpful provided the specific purposes for which such maps were prepared and the range of relevance of the 'indices' used are borne in mind and the importance of local climate appreciated. No sharp boundaries exist between the various climate zones: there are usually many 'meso-' and 'micro-climatic' subzones in each class where certain critical, relatively rare values of the combined climatic elements may be of greater importance in identifying the need for housing than more frequently encountered conditions. Surveys of the local livestock types and their diseases, and crop and soil types and condition offer a useful means of identifying local environmental problems, whilst a knowledge of the elevation, latitude and factors such as the distance from the sea is valuable in extrapolating information from other areas that have already been assessed. McDowell (1972) suggested that, after mapping with a suitable 'index', additional climate variables that might assist in establishing and refining an assessment of housing need should include the following:

(1) mean monthly temperatures and rainfall; which can be used to help determine whether the climate is cool, temperate or tropical, the availability of forage, the prediction of disease levels and the likelihood of soils to favour 'poaching';
(2) mean monthly maximum and minimum temperatures; which characterize the extremes and the diurnal heating and cooling rates. Information on degree days below the appropriate 'critical temperature' will help to quantify the economic penalties of not housing;
(3) air movement and relative humidity; knowledge of these variables and the duration of their extremes will help in the choice of housing structure and location, while calculations of THI values provide statements of the joint occurrence of these elements helpful in decision-making on ventilation;
(4) mean monthly radiation levels; measurements of which can be of aid in defining shelter; and
(5) local pollution.

More detailed data allow estimates of average daily durations and intensity of the climatic elements, of their covariance over shorter time scales and the period for which acceptable limits are exceeded or not attained. For example, Bonsma and Joubert (1957) developed climate summaries for South Africa that enabled the potential for livestock production to be established, whilst McDowell (1967) quoted an example of the use of climate summaries in northern Columbia, based on surveys of local climate variables, which provided the basis for the development of calf housing (as well as feeding and parasite control). The end result was a substantial reduction in calf mortality, since the (optimum) practices suggested by attention to climate data proved quite different from the traditional husbandry of the area.

In conclusion, one can suggest that livestock housing has different emphasis depending on whether the primary concept is shelter for optimum production or the avoidance of problems when animals are brought together for other management reasons. When it has been established, by the techniques discussed, that climate is likely to limit productivity, the use and cost of housing to mediate climate effects must be weighed against the management and economic penalties if housing measures are not taken and the requirement for a higher level of management implicit in the adoption of intensive, controlled-environment production systems. Intensive production systems simply present alternative environmental hazards if such management is not forthcoming.

Acknowledgements

This paper is published by permission of the Director General of the Meteorological Office. Mr J. Gloster helped with the calculation of CHI values.

References

ALEXANDER, G. (1964). *Proc. Aust. Soc. Anim. Prod.* **5**, 113–122

AMES, D. R. (1974). *Proc. Int. Livestock Environment Symposium, 1974.* American Society of Agricultural Engineers, St Joseph, Michigan

BLAXTER, K. L. (1959). *J. agric. Sci., Camb.* **52**, 25–40

BONSMA, F. N. and JOUBERT, D. M. (1957). *Dept. agric. Sci. Bull.* 380, Pretoria

BULTOT, F. (1962). *Revue belge statistq. Rech. oplle* 3, 73–78

CENA, K. M. and SLOMKA, J. (1966). *Roczn. Nauk roln.* **119D**, 33–88

DUCKHAM, A. N. and MASEFIELD, G. B. (1970). *Farming Systems of the World.* Chatto & Windus, London

FINDLAY, J. D. (1972). *Wld Rev. Anim. Prod.* **8**, 35–44

GAGGE, A. P. (1965). *Proc. Int. Physiol. Congr., Tokyo*

HYSLOP, N. St G. (1980). *Int. J. Biomet.* **24**, 57–58

JOHNSON, H. D., RAGSDALE, A. C., BERRY, I. L. and SHANKLIN, M. D. (1963). *Research Bulletin of the University of Missouri Agricultural Experiment Station,* No. 846

KERSLAKE, D. McK. (1972). *The Stress of Hot Environments.* Cambridge University Press, Cambridge

KIBLER, H. H. and BRODY, S. (1954). *Research Bulletin of the University of Missouri Agricultural Experiment Station,* No. 552

KÖPPEN, W. and GEIGER, R. (1936). *Handbuch der Klimatologie,* Vol. 1, Part C. Bornträger, Berlin

LAMB, H. H. (1963). *Symp. Changes Climate, Arid Zone Res. (Paris)* **20**, 125–150

LANDSBERG, H. E. (1972). 'An assessment of human bioclimate.' *WMO Tech. Note* 123. World Meteorological Organization, Geneva

LEE, D. H. K. (1958). *Arid Zone* **10**, 102–124

LIVESTOCK CONSERVATION INC. (undated). Booklet: *Livestock Safety,* pp. 11–12

McARTHUR, A. (1977). *PhD Thesis,* University of Nottingham

McDOWELL, R. E. (1965). *132nd Meeting of AAAS, Berkeley, Calif.* (US Agric. Res. Service paper 44–182, Aug. 1966)

McDOWELL, R. E. (1967). *Ground Level Climatology.* Pub. No. 86, American Association for the Advancement of Science

McDOWELL, R. E. (1972). *Improvements of Livestock in Warm Climates.* W. H. Freeman, San Francisco

MATHER, J. R. (1974). *Climatology: Fundamentals and Applications.* McGraw-Hill, New York

MINARD, D., BELDING, H. S. and KINGSTON, J. R. (1957). *J. Am. med. Ass.* **165**, 1813–1818

MONTEITH, J. L. (1974). In *Heat Loss from Animals and Man,* pp. 1–18. Ed. by J. L. Monteith and L. E. Mount. Butterworths, London

MORGAN, D. and BASKETT, R. (1974). *Int. J. Biomet.* **18**, 184–198

MOUNT, L. E. (1964). *J. agric. Sci., Camb.* **63**, 335–339

MOUNT, L. E. (1979). *Adaptation to Thermal Environment: Man and his Productive Animals.* Arnold, London

MUMFORD, A. M. (1979). *Weather, Lond.* **34**, 424–429

ØRITSLAND, N. A. (1974). *J. theor. Biol.* **47**, 413–420

PICTON, A. D. (1979). *Int. J. Biomet.* **23**, 115–122

SIPLE, P. A. and PASSEL, C. F. (1945). *Proc. Am. phil. Soc.* **89**, 177–199

SLEE, J. (1977). *Anim. Breeding Res. Org. Rep.*, Edinburgh

SMITH, C. V. (1965). Unpublished agricultural memorandum No. 138 (available from Meteorological Office, Bracknell, Berkshire)

SMITH, C. V. (1970). 'Meteorological observations in animal experiments.' *WMO Tech. Note* 107. World Meteorological Organization, Geneva

SMITH, C. V. (1971). 'Some environmental problems of livestock housing.' *WMO Tech. Note* 122. World Meteorological Organization, Geneva

SMITH, C. V. (1974). In *Heat Loss from Animals and Man*, pp. 345–366. Ed. by J. L. Monteith and L. E. Mount. Butterworths, London

SMITH, L. P. and OLLERENSHAW, C. B. (1967). *Agriculture, Lond.* **74**, 256–260

STARR, J. R. (1980). *Int. J. Biomet.* **24**, 224–229

STARR. J. R. (1981). *Agric. Met.*, in press

TERJUNG, W. H. (1966). *Ann. Ass. Am. Geogr.* **56**, 141–179

WEBSTER, A. J. F. (1974). In *Heat Loss from Animals and Man*, pp. 205–232. Ed. by J. L. Monteith and L. E. Mount. Butterworths, London

WEBSTER, A. J. F., CHLUMELKY, J. and YOUNG, B. A. (1970). *Can. J. Anim. Sci.* **50**, 89–100

WRIGHT, P. B. (1969). *Weather* **24**, 204–213

YAGLOU, C. P. (1927). *J. ind. Hyg. Toxicol.* **9**, 297–309

3

THERMAL INSULATION AND HEAT LOSS FROM ANIMALS

A. J. McARTHUR
Department of Physiology and Environmental Science,
University of Nottingham

Introduction

A homeothermic animal attempting to maintain body core temperature at a relatively steady value despite changes in its thermal environment must balance the rates at which heat is gained by and dissipated from its body. The heat gained originates primarily from metabolic conversion of the chemical energy stored in food. The rate of metabolic heat production in a thermo-neutral environment, M_{tn}, ranges from about 50 to 200 W m^{-2}, depending on species and the level of production (Graham *et al.*, 1959; Webster, 1974). In strong sunshine outdoors, the heat gained by absorption of solar radiation can exceed M_{tn} by a factor of 3 or 4 and is a major component in an animal's heat balance, but indoors this gain is usually negligible. Heat loss to the environment occurs by two routes, sensible (i.e. 'dry' convective and thermal radiation) and latent heat (i.e. evaporative) transfer.

As outlined by Robertshaw in Chapter 1, at low environmental tempera-tures the loss is mainly as sensible heat, and if the rate of heat loss exceeds M_{tn} then the rate of metabolic heat production must be raised to prevent a drop in body temperature. Metabolic rate can reach 500 W m^{-2} in response to cold (at summit metabolism), but this level cannot be maintained for more than a few hours. As heat production is achieved by the oxidation of food intake or body reserves (e.g. fat), an increase in metabolic rate above M_{tn} results in lowered productivity, because a higher proportion of the food eaten is employed in thermoregulation. Indeed, if metabolic rate exceeds the rate of supply of metabolizable energy (as feed intake), then body mass will decrease.

At high environmental temperatures, the rate of sensible heat loss may be lower than M_{tn} and an animal relies on the evaporation of water to dissipate its excess heat, either from the skin surface as a result of sweating or from the respiratory system by panting. This is necessary to prevent a rise in its body temperature due to storage of thermal energy. Although heat dissipation by evaporation may prevent an increase in body core temperature, elevated skin temperatures, high respiratory rates and the associated thermal dis-comfort reduce an animal's appetite and therefore lower its productivity because of the reduced food intake. High respiratory rates can also result in an increase in metabolic rate above M_{tn}, because of the increased muscular activity associated with panting.

The range of environmental temperatures outside which thermal strain causes a loss of productivity depends on the metabolic rate and on the thermal resistances to heat and mass transfer between the body core and the surroundings. This chapter is therefore concerned with the factors which determine thermal strain, with the components of thermal resistance for livestock indoors and with the estimation of heat loss in relation to environmental temperature, wind speed and humidity. It also shows that when changes in local rather than whole-body average resistances are considered, different conclusions are reached concerning the responses of animals to their thermal environment. Heat loss from non-sweating species (e.g. poultry) and those which sweat (e.g. cattle) will be considered separately.

Units

The rate of sensible heat flow, G, across unit area of an insulating layer with thermal resistance r can be expressed as

$$G = \varrho c_p \, \Delta T / r \tag{3.1}$$

where ΔT (K) is the temperature difference across the layer and ϱc_p ($=1.29 \times 10^3$ J m^{-3} K^{-1}) is the volumetric specific heat of air at an arbitrary temperature, conveniently 0 °C (Monteith, 1975; Cena and Clark, 1978).

The corresponding relation for the evaporative heat transfer E (W m^{-2}) from a wet surface is given by

$$\lambda \mathcal{E} = E = \varrho c_p \, \Delta e / \gamma r_v \tag{3.2}$$

where Δe (kPa) is the difference in vapour pressure between the surface and the surrounding air, r_v is the resistance to mass transfer, and γ ($= 0.066$ kPa K^{-1}) is the psychrometric constant. The quantity $\lambda (\simeq 2400$ J g^{-1}) is the latent heat of vaporization of water and $\mathcal{E}$ (g m^{-2} s^{-1}) is the rate of water loss per unit area by evaporation. In equations 3.1 and 3.2 the thermal resistance values (r and r_v) are both in units of s m^{-1}, where 1 s cm^{-1} is equivalent to an insulation of 0.078 m^2 K W^{-1}.

Non-sweating animals

In a simple model describing the sensible heat flow from a non-sweating animal to its surroundings (indoors), the total thermal resistance $\bar{r}_{tot}$ between the body core and environment comprises three resistances in series, provided by the body tissue ($\bar{r}_b$), the coat ($\bar{r}_c$) and the environment ($\bar{r}_e$), such that

$$\bar{r}_{tot} = \bar{r}_e + \bar{r}_c + \bar{r}_b \tag{3.3a}$$

The bars indicate an average value for the whole body. The rate of sensible heat loss G is then given by

$$G = \varrho c_p (T_b - T_e) / \bar{r}_{tot} \tag{3.3b}$$

where T_b and T_e are the temperature of the body core and the environment respectively. The resistances in equation 3.3a will now be considered separately.

ENVIRONMENTAL RESISTANCE, $\bar{r}_e$

Sensible heat transfer between the outer surface of an animal's coat and the environment takes place by thermal radiation exchange (L_n) with its surroundings, in a building primarily to the roof, floor, walls and other animals, and by convection (C) to the air. The resistances to these heat

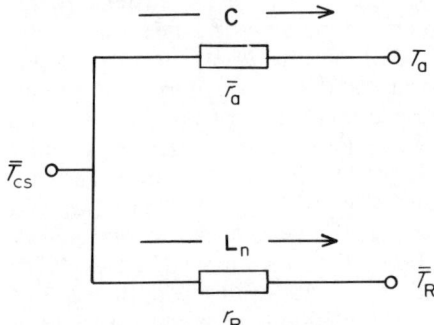

Figure 3.1 Resistance analogue describing sensible heat loss from the coat surface of an animal to its indoor environment. ($\bar{T}_{cs}$, mean radiative temperature of the coat surface; T_a, air temperature; $\bar{T}_R$, mean radiative temperature of surroundings; $\bar{r}_a$, mean boundary-layer resistance; r_R, resistance to thermal radiation exchange; C, heat flux density by convection; L_n, net flux density of heat by thermal radiation exchange)

transfer processes, represented by r_R and $\bar{r}_a$ respectively in *Figure 3.1*, are defined by

$$L_n = \varrho c_p(\bar{T}_{cs} - \bar{T}_R)/r_R \tag{3.4}$$

and

$$C = \varrho c_p(\bar{T}_{cs} - T_a)/\bar{r}_a \tag{3.5}$$

where $\bar{T}_{cs}$ is the mean radiative temperature of the coat surface, T_a is the air temperature and $\bar{T}_R$ is the mean radiative temperature of the surroundings. The total flux density of sensible heat G ($= L_n + C$) from the body surface is therefore given by

$$G = \varrho c_p(\bar{T}_{cs} - T_a)/\bar{r}_e + \varrho c_p(T_a - \bar{T}_R)/r_R \tag{3.6}$$

where $\bar{r}_e$, equal to $\bar{r}_a r_R/(\bar{r}_a + r_R)$, is the combined resistance to convection and thermal radiation transfer.

Radiative resistance, r_R

The thermal radiation flux from a surface behaving as a black-body emitter is proportional to the fourth power of its absolute temperature. The rate of heat exchange by thermal radiation, L_n, is given by

$$L_n = \sigma(\bar{T}_{cs}{}^4 - \bar{T}_R{}^4) \tag{3.7}$$

where σ (= 5.67×10^{-8} W m^{-2} K^{-4}) is the Stefan–Boltzmann constant and the temperatures are in kelvins. For small temperature differences this equation can be linearized and expressed in a form equivalent to equation 3.1. The linearized form is given by equation 3.4. The resistance to radiative transfer, r_R, is approximately equal to $\varrho c_p/(4\sigma \bar{T}^3)$, where $\bar{T}$ (K) is the average of $\bar{T}_{cs}$ and $\bar{T}_R$. For values of $\bar{T}$ between 273 and 303 K, $2.8 > r_R > 2.0$ s cm^{-1}.

Boundary-layer resistance, $\bar{r}_a$

The boundary-layer resistance $\bar{r}_a$ governing convective heat transfer depends on the rate of air movement, the geometry of the body and the nature of the interface (Mitchell, 1974; McArthur and Monteith, 1980a). In still or slowly moving air, buoyancy forces are important and, because of *free* convection, $\bar{r}_a$ decreases with increasing temperature difference $(\bar{T}_{cs} - T_a)$. At wind speeds above about 0.5 m s^{-1}, the effects of buoyancy can usually be ignored and $\bar{r}_a$ decreases with increasing wind speed in *forced* convection. The rate of air movement in an animal house varies with the type of stock and environmental temperature. For example, the wind speed within poultry houses in the UK is usually between about 0.1 and 0.3 m s^{-1} (Wathes, 1978) whereas in hotter climates forced ventilation, with wind speeds in excess of 0.5 m s^{-1}, may be used to increase heat dissipation from housed stock (e.g. Wiersma and Stott, 1966). Either free- or forced-convection regimes can therefore exist for housed animals, and in slowly moving air (wind speed $\simeq$ 0.2 m s^{-1}) both free and forced convection can occur simultaneously (*mixed* convection). Wathes concluded that mixed convection was the usual condition for housed laying hens in the UK.

For simplicity in estimation of convective heat loss, animal bodies are usually treated as standard geometrical shapes: quadrupeds as cylinders and birds as spheres. In *Figure 3.2*, boundary-layer resistance values for cylindrical bodies are plotted against wind speed u (m s^{-1}). These relationships between $\bar{r}_a$ and u have been derived from more general equations relating Nusselt and Reynolds numbers. Line a is the standard relation for a smooth isothermal cylinder (McAdams, 1954), given by

$$\bar{r}_a = 2.7d^{0.40}/u^{0.60} \tag{3.8a}$$

where d (= 0.39 m) is the cylinder's diameter. For animals, the characteristic dimension d is usually taken to be the trunk diameter. Line b is the relation

$$\bar{r}_a = 2.2d^{0.47}/u^{0.53} \tag{3.8b}$$

established by Wiersma and Nelson (1967) for cattle. Line c is the relation

$$\bar{r}_a = 3.0d^{0.50}/u^{0.50} \tag{3.8c}$$

calculated by Monteith (1975) from the data for sheep reported by Joyce, Blaxter and Park (1966), and line d is the relation

$$\bar{r}_a = 2.4d^{0.20}/u^{0.80} \tag{3.8d}$$

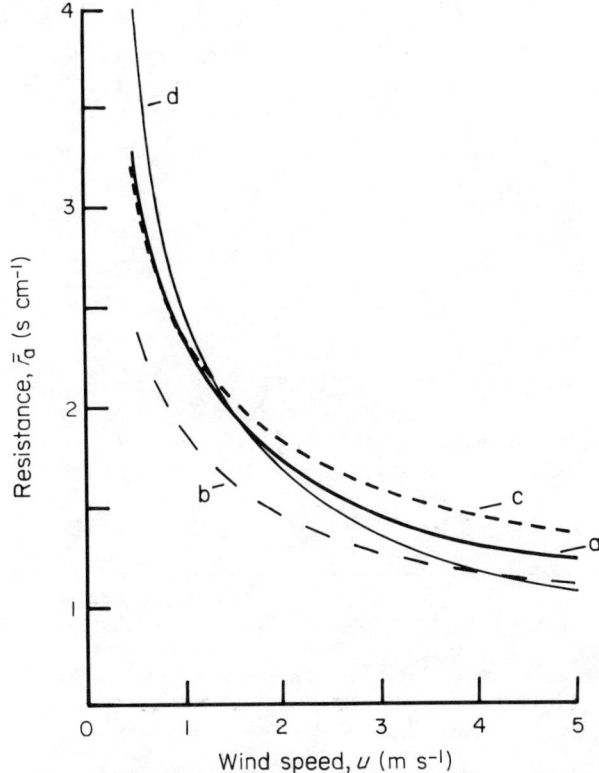

Figure 3.2 Boundary-layer resistance $\bar{r}_a$ v. wind speed u for cylindrical bodies. Line a, smooth isothermal cylinder (McAdams, 1954); line b, cattle (Wiersma and Nelson, 1967); line c, sheep (Monteith, 1975); line d, sheep (McArthur and Monteith, 1980a)

established for sheep by McArthur and Monteith (1980a). The value $d = 0.39$ m was used in each case.

The high values of $\bar{r}_a$ at low wind speeds predicted by equation 3.8d can be attributed to the use of the *radiative* surface temperature of the coat, rather than the temperature of the actual hair tips, in the calculation of $\bar{r}_a$. Insulation is provided by a layer of air, a few millimetres in depth, between the effective surface for radiation exchange and the actual coat surface. However, it is convenient in heat balance analysis to describe the interface between the animal and its environment by the radiative temperature of the coat surface (e.g. equation 3.6), as this quantity can be easily measured (e.g. Cena and Clark, 1973; Clark, Cena and Monteith, 1973).

Figure 3.3 shows boundary-layer resistance values in still air for a smooth horizontal cylinder (line a; $d = 0.39$ m) and a horizontal flat plate (line b; $d = 0.28$) plotted against the temperature difference $(T_{cs} - T_a)$ between the surfaces and the surrounding air. Both surfaces are assumed to be isothermal. The values of $\bar{r}_a$ were calculated using

$$\bar{r}_a = 9.2d^{0.25}(T_{cs} - T_a)^{-0.25} \tag{3.9a}$$

and

$$\bar{r}_a = 8.8d^{0.25}(T_{cs} - T_a)^{-0.25} \tag{3.9b}$$

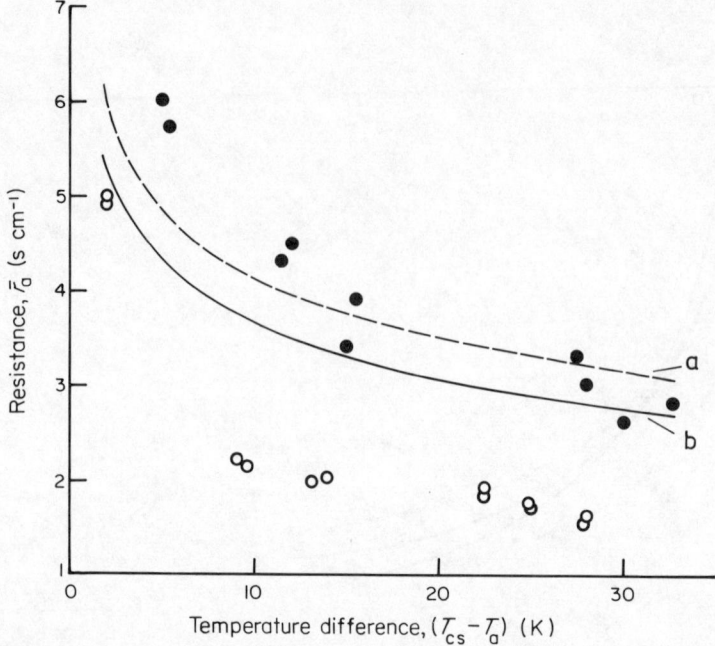

Figure 3.3 Boundary-layer resistance $\bar{r}_a$ for cylinder (line a) and horizontal flat plate (line b) plotted against the temperature difference $(T_{cs} - T_a)$ between the coat surface and the surrounding air. (●) Rabbit fur; (○) cow hide. (From data of Cunningham, 1979)

respectively, derived from standard relations for free convection recommended by Ede (1967). In each case values of $\bar{r}_a$ decrease from about 5.5 to 3.0 s cm⁻¹ when $(T_{cs} - T_a)$ increases from 2 to 30 K.

The experimental points in *Figure 3.3* are boundary-layer resistance values for flat (horizontal) coat samples in still air reported by Cunningham (1979). The values of $\bar{r}_a$ were calculated using the mean radiative temperature of the coat surface, $\bar{T}_{cs}$ measured with an infrared radiometer. The closed points for rabbit fur ($d = 0.28$ m), which had guard hairs extending about 3 mm beyond the main surface of the coat, lie close to the values for a flat plate predicted by equation 3.9b. However, the open points for cow hide ($d = 0.28$ m), which did not have guard hairs, are up to 40 per cent lower than values predicted by equation 3.9b. These measurements suggest that heat transfer by free convection from a hairy surface may not be accurately described by relationships derived for solid surfaces, possibly as a result of the upward movement of warm air through the fibrous interface.

In a mixed-convection regime, the dominant type of convection depends on the ratio of the buoyancy to inertial forces (e.g. Oosthuizen and Madan, 1970). *Figure 3.4* is a diagrammatic representation of the relationship between $\bar{r}_a$ and u (line a) for a sphere ($d = 0.165$ m) representing a chicken subjected to low rates of air movement and with $(\bar{T}_{cs} - T_a) = 20$ K (constant). Line b is the standard relation between $\bar{r}_a$ and u for a sphere in forced convection, given by

$$\bar{r}_{a1} = 248d/(1 + 63u^{0.5}\,d^{0.5}) \tag{3.10}$$

Line c is the boundary-layer resistance for free convection, given by

$$\overline{r}_{a2} = 248d/[1 + 30d^{0.75}(\overline{T}_{cs} - T_a)^{0.25}] \tag{3.11}$$

At high wind speed, forced convection is dominant, but at low wind speed the heat loss occurs mainly by free convection. At $u = 0.46$ m s^{-1}, $\overline{r}_{a1} = \overline{r}_{a2} = 2.3$ s cm^{-1} and the resistance $\overline{r}_a$ for mixed convection (line a) is less than $\overline{r}_{a1}$ and $\overline{r}_{a2}$ by an amount Δr_0. Above $u = 0.46$ m s^{-1}, $\overline{r}_a$ is reduced (by Δr_1)

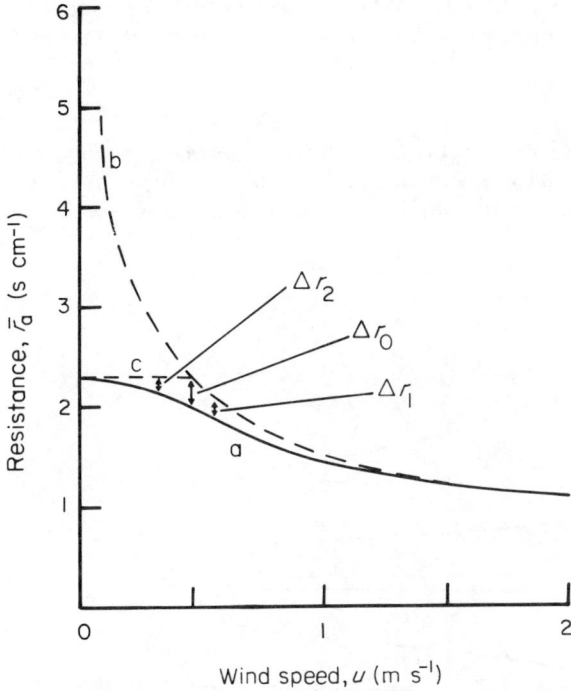

Figure 3.4 Diagrammatic representation of relationship between boundary-layer resistance $\overline{r}_a$ and wind speed u for a sphere. Line a, subjected to low rates of air movement and with a surface temperature 20 K above the air temperature, mixed convection; line b, standard relation for forced convection (equation 3.10); line c, standard relation for free convection (equation 3.11). Δr_0, Δr_1, and Δr_2 are the differences between the actual value of $\overline{r}_a$ and those predicted by equations 3.10 and 3.11, as indicated

below the value predicted by equation 3.10, because of buoyancy effects; while below $u = 0.46$ m s^{-1}, $\overline{r}_a$ is reduced (by Δr_2) below the value predicted by equation 3.11, because of the effects of forced convection. Empirical formulae for evaluating Δr_0, Δr_1, and Δr_2 are given by Yuge (1960), but the common procedure of calculating both $\overline{r}_{a1}$ and $\overline{r}_{a2}$ and using the smaller value to determine convective heat loss will underestimate the flux. Conversely, the procedure of treating $\overline{r}_{a1}$ and $\overline{r}_{a2}$ as parallel resistances will overestimate the heat loss by convection. However, these errors may be small in comparison with that introduced by assuming that the boundary-layer resistance for a hairy surface is the same as that for a solid interface of similar shape (*see Figure 3.3*).

Combined radiation and convection

When the air temperature T_a and the mean radiative temperature $\bar{T}_R$ are identical, equation 3.6 reduces to

$$\mathbf{G} = \varrho c_p(\bar{T}_{cs} - T_e)/\bar{r}_e \qquad (3.12)$$

where the environmental temperature $T_e = T_a = \bar{T}_R$. In animal houses, the air and radiative temperatures may not be the same and the environmental temperature lies between T_a and $\bar{T}_R$. When T_a differs from $\bar{T}_R$, equation 3.12 can still be used to estimate sensible heat loss if T_e is evaluated using

$$T_e = T_a + \Delta T \qquad (3.13)$$

where $\Delta T = (\bar{T}_R - T_a)/(1 + r_R/\bar{r}_a)$ is the difference between the environmental temperature sensed by an animal and the air temperature. *Figure 3.5* shows values of ΔT plotted against wind speed for environments in which T_a

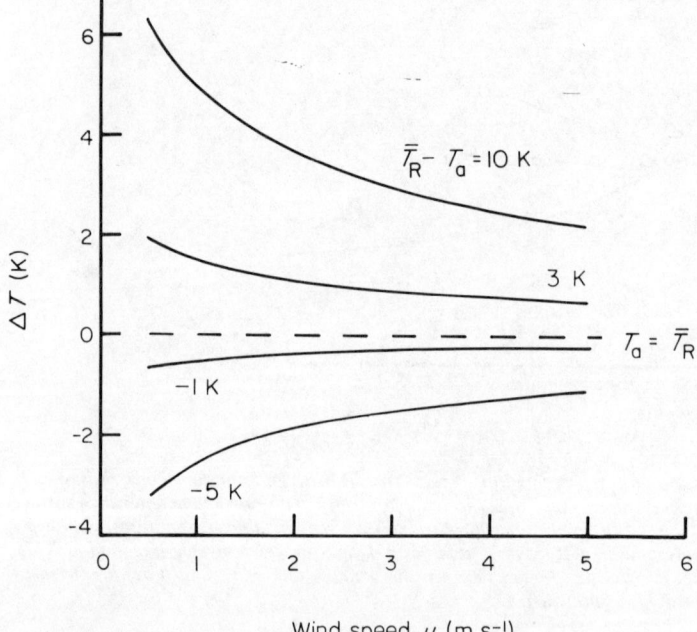

Wind speed, u (m s^{-1})

Figure 3.5 The difference ΔT between the environmental temperature sensed by an animal ($d = 0.39$ m) and the air temperature plotted as a function of wind speed u. (T_a = air temperature; $\bar{T}_R$ = mean radiative temperature of surroundings)

and $\bar{T}_R$ differ. In this example the resistance r_R was assumed to be $2.0\,\mathrm{s\,cm^{-1}}$, and values of $\bar{r}_a$ were calculated from equation 3.8d with $d = 0.39$ m. The lines on the graph correspond to $(\bar{T}_R - T_a) = 10, 3, -1$ and -5 K as indicated. The graph illustrates the importance of the thermal radiation environment when housed animals are subjected to low rates of air movement. For example, the rate of air movement should be minimized in the vicinity of young stock (e.g. piglets), using infrared heaters if $\bar{T}_R > T_a$ and T_e is to be maximized, because as the wind speed increases and boundary-layer resistance is reduced, the environmental temperature approaches the air temperature.

COAT RESISTANCE, $\bar{r}_c$

The mean thermal resistance of an animal's coat, $\bar{r}_c$, is defined by

$$G = \varrho c_p (\bar{T}_{sk} - \bar{T}_{cs})/\bar{r}_c \tag{3.14}$$

where $\bar{T}_{sk}$ is the mean temperature of the skin surface. The resistivity $\bar{z}_c$ of the coat, defined as the coat resistance per unit depth of hair and expressed in s cm^{-2}, is given by

$$\bar{z}_c = \bar{r}_c/\bar{l} \tag{3.15}$$

where $\bar{l}$ (cm) is the mean coat depth.

Sensible heat transfer through a coat occurs simultaneously by conduction, radiation and convection (Cena and Monteith, 1975a). The relative magnitudes of these processes, and the resultant coat resistance, depend on the coat structure and depth, wind speed and the mean temperature difference across the hair layer.

Coat resistance in 'still' air

Figure 3.6 is a resistance analogue describing the sensible heat transfer through a coat in still or slowly moving air. We can regard the resistances to molecular diffusion (r_d), thermal radiation (r'_R) and free convection (r_f)as acting in parallel. Conduction along the fibres themselves is usually a small component of the total heat flow and has been neglected.

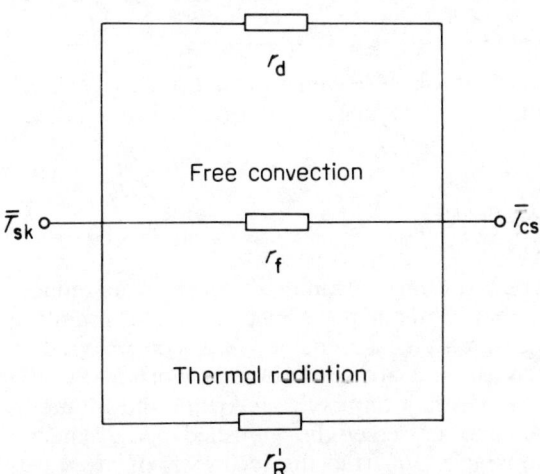

Figure 3.6 Resistance analogue describing sensible heat transfer through a coat in still or slowly moving air. ($\bar{T}_{sk}$, mean skin temperature; $\bar{T}_{cs}$, mean radiative temperature of coat surface; r_d, r_f and r'_R, resistances to molecular conduction, free convection and thermal radiation respectively)

The values of r_d and r'_R can be established theoretically: the resistivity z_d ($= r_d/\bar{l}$) to heat transfer by molecular diffusion, which depends on the mean temperature of the air trapped within the coat, is about 5.0 s cm^{-2}, while the resistivity z'_R ($= r'_R/\bar{l}$) to radiant exchange from fibre to fibre depends on the size, density and orientation of the fibres. For a cow's hide $z'_R \simeq 30$ s cm^{-2}, whereas for fleece $z'_R \simeq 6.5$ s cm^{-2} (Cena, 1974). The writer's own measurements on a Clun Forest ewe show that the resistivity of fleece, z_f ($= r_f/l$), to heat transfer by free convection can be described by

$$z_f = 14.5(\bar{T}_{sk} - \bar{T}_{cs})^{-0.53} \qquad (3.16)$$

Equation 3.16 indicates that z_f for fleece decreases from about 6 to 2.5 s cm^{-2} when the temperature difference across the coat increases from 5 to 30 K, and that free convection can often be the major mechanism of heat transfer through the coat. The dependence of coat resistance on $(\bar{T}_{sk} - \bar{T}_{cs})$ has not been established for other species.

The overall resistivity $\bar{z}_c$ of a coat can therefore be evaluated from

$$\bar{z}_c = (z_d^{-1} + z'^{-1}_R + z_f^{-1})^{-1} \qquad (3.17)$$

Table 3.1 presents coat resistivity for a number of species measured in slowly moving air ($u \simeq 0.3$ m s^{-1}). Values of $\bar{z}_c$ range from about 1 to 3 s cm^{-2}. Although heat transfer by free convection must be occurring in addition to

Table 3.1   COAT RESISTIVITY VALUES FOR DIFFERENT SPECIES

Species	Coat resistance per unit depth, $\bar{z}_c$ (s cm^{-2})	Source
Cattle	1.1	Webster (1974)
Cheviot sheep	1.5	Joyce, Blaxter and Park (1966)
Blackface sheep	1.5	Joyce and Blaxter (1964)
Merino sheep	2.8	Bennett and Hutchinson (1964)
White Leghorn chicken (housed)	2.0–3.0	Barnes (1978)

diffusion and radiation, to account for these values of coat resistivity (Cena and Monteith, 1975a), the effects of buoyancy forces on coat insulation are rarely considered.

Coat resistance in wind

It has long been known that penetration of an animal's coat by wind reduces its effectiveness as a barrier to heat flow and there is a considerable literature of work on this topic. In the past most measurements have been analysed on the assumption that the decrease in coat resistance is proportional to the square root of wind speed. However, Campbell, McArthur and Monteith (1980) have recently reviewed and reassessed the published measurements. They suggest that coat conductance $1/\bar{r}_c$ (i.e. the reciprocal of the resistance) usually increases linearly with wind speed, u, according to

$$1/\bar{r}_{c(u)} = 1/\bar{r}_{c(0)} + \alpha u \qquad (3.18)$$

where $1/\bar{r}_{c(0)}$ is the conductance (cm s^{-1}) in still air and α is a constant of proportionality dependent on coat type. For example, the minimum conductance of a 3.5 cm depth of fleece is about 0.14 cm s^{-1} and $\alpha \simeq 0.025$.

A relationship of the type expressed in equation 3.18 could arise in several ways, but the simplest postulate is that wind destroys the insulation of a layer of coat, the thickness of which depends on the velocity. *Figure 3.7* is an equivalent resistance analogue describing the sensible heat flow through

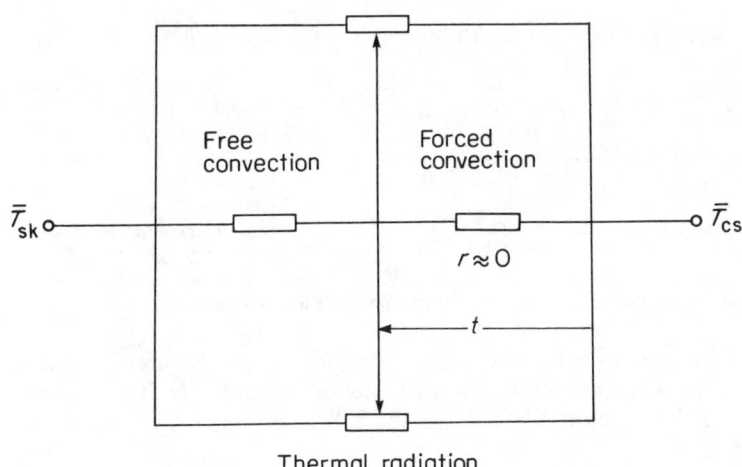

Figure 3.7 Resistance analogue describing sensible heat transfer through a coat in wind. ($\bar{T}_{sk}$, mean skin temperature; $\bar{T}_{cs}$, mean radiative temperature of coat surface; t, wind penetration depth)

fleece in wind (McArthur and Monteith, 1980b). Both free and forced convection occur within the exposed fleece: close to the skin, heat transfer occurs by molecular diffusion, thermal radiation and free convection; but between a wind penetration depth t and the outer surface of the coat the heat transfer is dominated by forced convection and the coat resistance in this outer section is close to zero. The penetration depth t is equal to $\bar{l}\alpha u/[\alpha u + 1/\bar{r}_{c(0)}]$, and increases with wind speed. A wind speed of 5 m s^{-1} is sufficient to destroy about half the insulation provided by fleece.

The orientation of a sheep to the wind has little effect on the mean fleece resistance on its trunk (Bennett and Hutchinson, 1964; McArthur and Monteith, 1980b). However, for birds or for animals whose pelage has a directional pile the orientation to the wind is likely to have a substantial effect on the wind penetration and therefore on coat resistance.

TISSUE RESISTANCE, $\bar{r}_b$

The last of the resistance components to be considered is the tissue resistance. At first sight this is the smallest and least important resistance, at least in coated animals, but we will show later that changes in local tissue resistance can have unexpected effects on the total resistance to heat loss.

The mean thermal resistance $\bar{r}_b$ between the body core and the skin surface is defined by

$$G_b = \varrho c_p (T_b - \bar{T}_{sk})/\bar{r}_b \qquad (3.19)$$

where G_b is the total heat flux density through the body tissue. For a non-sweating animal, $G_b \approx G$. *Table 3.2* presents values of $\bar{r}_b$ for a number

Table 3.2   RANGE OF MEAN TISSUE RESISTANCE VALUES FOR DIFFERENT SPECIES

Species	Tissue resistance, $\bar{r}_b$ (s cm⁻¹)	Source
Calf	0.5–1.2	Blaxter (1967)
Steer	0.5–1.8	Blaxter (1967)
Pig (1–2 months)	0.7–1.2	Stombaugh and Roller (1977)
Sheep	1.0–2.5	McArthur (1980)

The higher value corresponds to vasoconstriction, the lower to vasodilation

of species. Two values of $\bar{r}_b$ are given, corresponding to the maximum and minimum, i.e. vasoconstriction and vasodilation respectively. The value of $\bar{r}_b$ may change by a factor of between 2 and 3 as a consequence of vasomotor action.

Physiological control of blood flow gives an animal some ability to regulate its heat loss and changes in blood flow are most pronounced on the

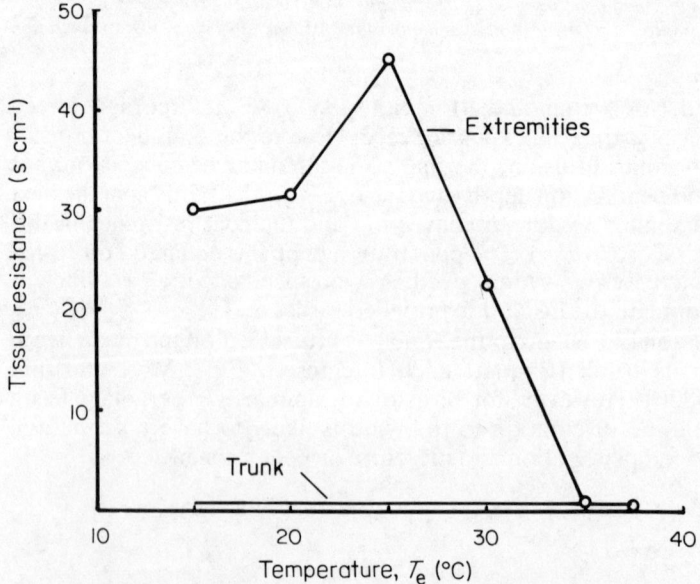

Figure 3.8 Tissue resistance versus environmental temperature for young pigs. (From data of Stombaugh and Roller, 1977)

extremities. For example, *Figure 3.8* shows the local tissue resistance for the extremities ($\bar{r}_{bx}$) and trunk ($\bar{r}_{bt}$) of young pigs (aged about 50 days) plotted against environmental temperature, T_e, calculated from the data of Stombaugh and Roller (1977). Though $\bar{r}_{bt}$ was constant ($\simeq 0.62$ s cm^{-1}) over the range of T_e investigated, $\bar{r}_{bx}$ increased from about 0.7 to 45 s cm^{-1} when T_e fell from 35 to 25 °C. The corresponding increase in the mean tissue resistance for the whole body ($\bar{r}_b$) was from about 0.7 to 1.2 s cm^{-1}. Measurements on sheep and poultry indicate similar changes in tissue resistance on the extremities, which are usually poorly insulated in comparison with the trunk, but changes in tissue resistance have not been observed on the trunk (e.g. Blaxter *et al.*, 1959; Richards, 1974). When expressed as an average for the whole of an animal's body, changes in mean tissue resistance therefore conceal the large local variations on its extremities.

In animals adapted to sub-zero temperatures, a decrease in tissue resistance (cold-induced vasodilation) will occur on the extremities to maintain skin temperatures above 0 °C and prevent damage to the body tissue in the cold (e.g. Webster and Blaxter, 1966). The author has estimated that the local tissue resistance for the legs of a sheep standing in the cold must decrease from about 8 to 3 s cm^{-1} if environmental temperature falls from 0 to –10 °C ($u \simeq 0.4$ m s^{-1}), resulting in a threefold increase in heat loss from this region (McArthur, 1980).

SENSIBLE HEAT LOSS IN RELATION TO ENVIRONMENTAL TEMPERATURE

The rate of sensible heat loss from each region of an animal's body depends on both the environmental temperature and the local thermal resistance provided by the body tissue, coat and environment. The sensible heat loss from the trunk of an animal with a pelage is therefore determined largely by the resistance of the coat, but the sensible heat loss from the extremities depends mainly on the local tissue resistance. As environmental temperature changes the proportion of the total heat loss which occurs from the extremities will change. For example, at an environmental temperature of 10 °C only a small percentage of the total heat loss from birds occurs through the legs, but at 35 °C most of their metabolic heat can be dissipated by this route (Steen and Steen, 1965). The physical characteristics of the coat may remain unchanged but, when environmental temperature changes, causing a vasomotor response, *apparent* changes in the mean coat resistance $\bar{r}_c$ occur owing to the inhomogeneity of the body (McArthur, 1981). Surprisingly, these apparent changes in coat resistance can exceed the corresponding changes in that of the body ($\bar{r}_b$). They will be most pronounced on animals with a thick pelage and with extremities which are poorly insulated with hair—not only birds but animals such as sheep that are usually regarded as well insulated. For example, *Figure 3.9* shows values of the total resistance ($\bar{r}_{tot}$) for four sheep (Cheviot and Suffolk) plotted against environmental temperature. The values of $\bar{r}_{tot}$ were calculated from the data of Webster and Blaxter (1966) using equation 3.3b and assuming a respiratory heat loss of 15 W m^{-2}. The decrease in $\bar{r}_{tot}$ at sub-zero temperatures is about 0.1 s cm^{-1} K^{-1}. This figure is about twice that attributable to a decrease in tissue resistance as a consequence of cold-induced vasodilation, indicating that the coat resistance must also have decreased, owing to a preferential heat loss from the extremities. It is inappropriate, therefore, to use a constant value of $\bar{r}_c$ for an animal exposed to different environmental temperatures.

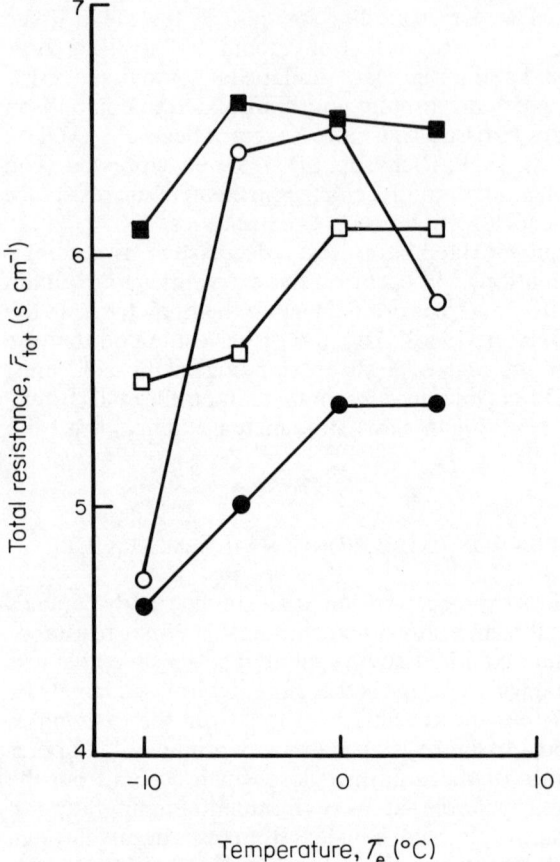

Figure 3.9 Total resistance, $\bar{r}_{tot}$, for four sheep plotted against environmental temperature T_e. (From data of Webster and Blaxter, 1966)

Sensible heat loss in the cold, and the metabolic costs of cold exposure, may therefore be underestimated if a linear relation between G and T_e is assumed. Further, the rate at which an animal can dissipate sensible heat at high environmental temperatures will also be underestimated if the apparent decrease in the mean coat resistance due to vasodilation is ignored.

CASE STUDY—POULTRY

Figure 3.10 shows changes with temperature of the total resistance $\bar{r}_{tot}$ for poultry, calculated from measurements reported by Richards (1977). The closed and open points correspond to well and poorly feathered birds respectively. The graph illustrates the dependence on feather insulation of

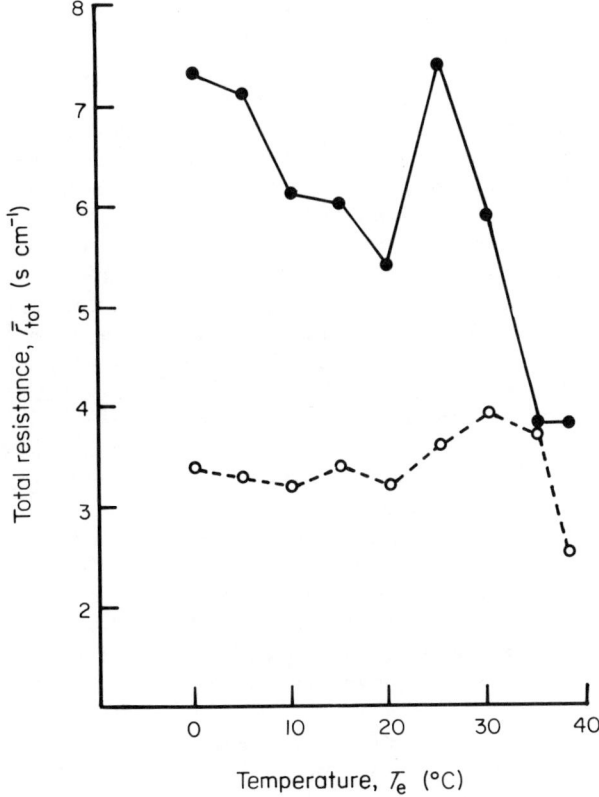

Figure 3.10 Total resistance, $\bar{r}_{tot}$, for poultry plotted against environmental temperature T_e. (●) Well feathered birds; (○) poorly feathered birds. (From data of Richards, 1977)

the sensible heat loss from poultry at low T_e. For example, at 10 °C the respective rates of sensible heat loss were about 65 and 125 W m⁻², the latter value being more than double the minimum metabolic rate, $\mathbf{M_{tn}}$.

For the well-feathered birds, the increase in $\bar{r}_{tot}$ from about 3.8 s cm⁻¹ at $T_e = 35$ °C to 7.3 s cm⁻¹ at $T_e = 0$ °C cannot be accounted for by a change in tissue resistance alone. Three other factors may contribute to this increase in resistance:

(1) a real increase in $\bar{r}_c$ due to fluffing of the feathers;
(2) an apparent increase in $\bar{r}_c$ as a consequence of vasomotor action and the inhomogeneity of the body;
(3) an apparent increase in $\bar{r}_c$ due to a postural adjustment (e.g. protection of the legs beneath the feathers).

Clearly, the relationship between sensible heat loss and environmental temperature is non-linear for the well feathered birds. In contrast, the increase in $\bar{r}_{tot}$ with decreasing T_e for the poorly feathered birds is comparatively small and can be accounted for by an increase in tissue resistance alone. Unlike well feathered birds, poorly feathered birds show little change in $\bar{r}_c$ due to behavioural responses or inhomogeneity of heat flow because the actual coat insulation is small. For the poorly feathered bird, sensible heat loss increases almost linearly with decreasing temperature between 35 and 0 °C, as in this range $\bar{r}_{tot}$ is approximately constant.

At high environmental temperatures the rate of sensible heat loss from a bird is almost independent of its feather status. For example, at $T_e = 35$ °C (*Figure 3.10*) the losses are about 23 W m⁻² for both sets of birds, equivalent to a heat flux of about 3.5 W. The metabolic rate (M_{tn}) is about 60 W m⁻². As the thermal resistance provided by the feathers on the 'trunk' will be about 6 s cm⁻¹ (Wathes, 1978), a significant proportion of the sensible heat loss from the well feathered birds must be occurring through the extremities (legs and comb) rather than through the plumage. The surface area of the extremities is at least 0.02 m², about 15 per cent of the bird's total surface area. Assuming that 30 per cent of the total sensible heat loss occurs through the extremities at $T_e = 35$ °C and that the environmental resistance $\bar{r}_{ex}$ for the extremities is 1.0 s cm⁻¹, then the local tissue resistance $\bar{r}_{bx}$ must have decreased to about 0.5 s cm⁻¹ owing to vasodilation. The rate of heat loss from the extremities at high environmental temperature is therefore strongly dependent on the environmental resistance, $\bar{r}_{ex}$. A decrease in $\bar{r}_{ex}$, caused for example by an increase in the rate of air movement over the legs and head, could enable poultry to dissipate more of their metabolic heat production through the extremities. Faster wind speeds could relieve thermal strain at high environmental temperatures, by lowering the rate of respiratory evaporation necessary to prevent an increase in body core temperature.

When the sensible heat loss is insufficient for balance, poultry respond by panting. The rate of heat dissipation by evaporation from the respiratory system, E_r (W m⁻²), can be expressed as

$$E_r = \varrho c_p [e_s - e_a]/\gamma r_{vr} \tag{3.20}$$

where e_s is the saturation vapour pressure at the site of evaporation, e_a is the vapour pressure of the air in the house and r_{vr} is the resistance to water vapour loss from the respiratory system. It is convenient to treat e_s as a constant determined by the deep-body temperature (= 8 kPa for poultry). The resistance r_{vr} can be evaluated using

$$r_{vr} = 600A/\dot{V} \tag{3.21}$$

where $\dot{V}$ (l min⁻¹) is the rate of respiratory flow and A (m²) is the surface area of the body.

Figure 3.11 shows values of r_{vr} for poultry (Brown Leghorns) plotted against respiratory rate, RR (min⁻¹), calculated from the evaporation rates measured by Hutchinson (1954). The resistance r_{vr} decreases from about 100 to 25 s cm⁻¹ when respiratory rate increases from 12 to 350 min⁻¹. These results indicate, for example, that if a bird has to dissipate 40 W m⁻² by respiratory evaporation (e.g. at $T_e = 35$ °C), the respiratory rate that must be achieved is about 250 min⁻¹ ($r_{vr} \simeq 35$ s cm⁻¹) at $e_a = 1$ kPa and about 400 min⁻¹ ($r_{vr} \simeq 20$ s cm⁻¹) at $e_a = 4$ kPa.

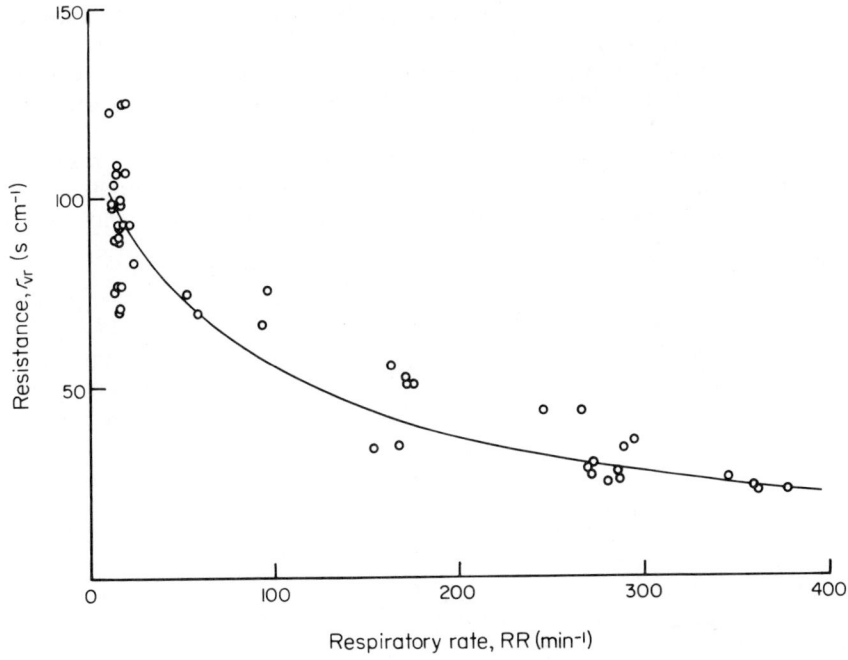

Figure 3.11 Resistance to water vapour loss from the respiratory system of poultry, r_{vr}, plotted against respiratory rate RR. (From data of Hutchinson, 1954)

With a knowledge of the relationship between r_{vr} and RR, the evaporative heat loss from birds can be estimated using equation 3.20 by measuring respiratory rate and ambient vapour pressure. The solid line fitted to the experimental points (RR > 10) in *Figure 3.11* is the relation

$$r_{vr}^{-1} = (9.1 \times 10^{-5}\,RR) + 9.0 \times 10^{-3} \tag{3.22}$$

Sweating animals

Figure 3.12 is a resistance analogue describing the sensible (**G**) and evaporative (**E**ₛ) heat losses from the body surface of a sweating animal to an indoor environment specified by an air temperature T_a, mean radiative temperature $\bar{T}_R$ and vapour pressure e_a. The total heat loss per unit surface area, G_b, is given by

$$G_b = G + E_s \tag{3.23}$$

where **G** (= **C** + **L**ₙ) is the sensible heat flow through the coat. The flux density of latent heat, **E**ₛ (= $\lambda \varepsilon_s$), from the skin surface is given by

$$E_s = \omega \varrho c_p (e_{sk} - e_a)/\gamma \bar{r}_{vs} \tag{3.24}$$

where e_{sk} is the saturation vapour pressure at skin temperature and $\bar{r}_{vs}$ is the resistance to water vapour transfer provided by the coat ($\bar{r}_{vc}$) and boundary layer ($\bar{r}_{va}$) in series (*Figure 3.12*).

The quantity ω is the proportion of skin surface area which is covered with sweat (Gagge, 1937). Sweat rate increases with increasing skin temperature (e.g. Berman, 1971), but the rate of sweat secretion varies with species. For example, cattle can achieve a sweat rate of $0.08 \text{ g m}^{-2}\text{s}^{-1}$ whereas for sheep the maximum rate is less than $0.02 \text{ g m}^{-2}\text{s}^{-1}$ (Hales, 1974). The maximum rate of evaporation, ε_{sm}, from the skin surface of an animal when fully

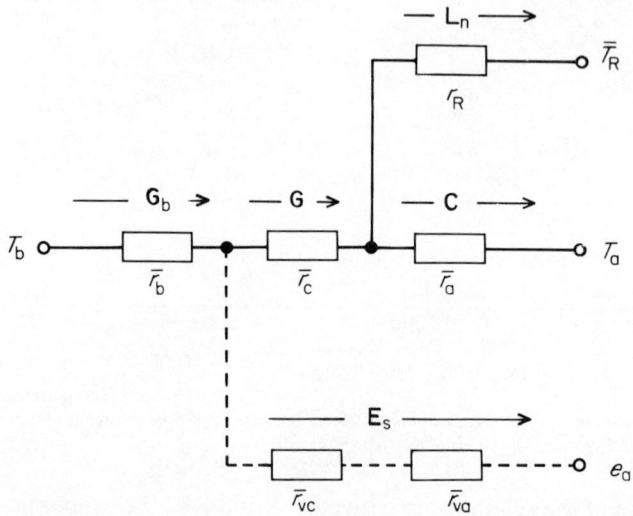

Figure 3.12 *Resistance analogue describing sensible and evaporative heat losses from the body surface of a sweating animal to an indoor environment. (T_b, body core temperature; $\bar{T}_R$, mean radiative temperature of surroundings; T_a, air temperature; e_a, vapour pressure; G_b, total heat loss per unit surface area; G, sensible heat flux density through coat; C, heat flux density by convection; L_n, net flux density of heat by thermal radiation exchange; E_s, flux density of latent heat from skin surface; $\bar{r}_b$, $\bar{r}_c$ and $\bar{r}_a$, mean resistances of body tissue, coat and boundary layer respectively; r_R, resistance to thermal radiation exchange; $\bar{r}_{vc}$ and $\bar{r}_{va}$, mean resistances to water vapour transfer provided by the coat and boundary layer respectively)*

wetted is determined by two factors: the resistance $\bar{r}_{vs}$ and the vapour pressure difference ($e_{sk} - e_a$). If ε_{sm} is less than the rate of sweat deposition at the skin surface, ε_{sd}, then $\varepsilon_s = \varepsilon_{sm}$ and $\omega = 1.0$. However, if $\varepsilon_{sm} > \varepsilon_{sd}$ then the actual rate of evaporation is less than ε_{sm}, $\omega < 1.0$, and the rate of evaporative heat loss is determined physiologically rather than directly by the physical processes of diffusion.

COAT RESISTANCE, $\bar{r}_{vc}$

Few measurements of the resistance of animal coats to water vapour transfer have been made. Cena and Monteith (1975b) reported that in still or slowly moving air the transfer of water vapour through sheep's fleece takes place by diffusion, enhanced by free convection induced by gradients of both temperature and pressure. The molecular diffusion resistivity for water vapour in air is about 4.2 s cm^{-2}. When free convection occurs the resistance $\bar{r}_{vc}$ depends on the difference in virtual temperature between the skin and coat surface rather than air temperature. The virtual temperature $\bar{T}_v$ (K) of air at temperature T_a (K) and vapour pressure e is the temperature at which dry air has the same density, and is given by

$$T_v = T_a(1 + 0.38e/P) \qquad (3.25)$$

where P is the total pressure in the same units as e (Monteith, 1975).

Figure 3.13 shows values of $\bar{r}_{vc}$ for fleece ($\bar{l} = 2.0$ cm) plotted against the mean temperature difference ($\bar{T}_{sk} - \bar{T}_{cs}$) across the coat. The values of $\bar{r}_{vc}$ were calculated for $\bar{T}_{sk} = 39$ °C ($e_{sk} = 7$ kPa) using equations 3.16 and 3.25 and with the assumption that the ratio of the resistances to heat and water vapour transfer is 1.1 (Cena and Monteith, 1975b). The lines in *Figure 3.13* correspond to vapour pressures at the coat surface, e_{cs}, of 0 (the

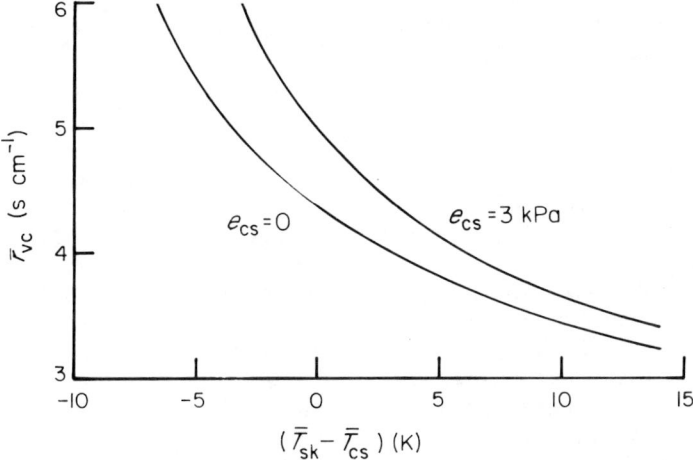

Figure 3.13 Resistance $\bar{r}_{vc}$ to water vapour transfer through fleece (depth 2 cm) plotted against the mean temperature difference across the coat ($\bar{T}_{sk} - \bar{T}_{cs}$), assuming $\bar{T}_{sk} = 39$°C. (e_{cs} is the assumed vapour pressure at the coat surface)

hypothetical minimum) and 3 kPa respectively, as indicated. The graph shows that a sheep with 2 cm of fleece could by sweating dissipate heat at a rate which exceeds its metabolic heat production, even when the surface temperature of the coat exceeds skin temperature. For example, at $e_{cs} = 3$ kPa if $\omega = 1.0$ and ($\bar{T}_{sk} - \bar{T}_{cs}$) = –1 K the potential latent heat loss, E_{sm}, is 150 W m^{-2}. Poor sweating ability rather than the physical processes of diffusion therefore appears to limit the rate at which sheep can dissipate heat by evaporation from their skin surface, although Waites and Voglmayr (1962) reported that sweat glands in sheep are more numerous on the legs than on the trunk, implying that cutaneous moisture loss from these extremities may be an important pathway for dissipation of surplus heat at high environmental temperatures.

Forced air movement is used extensively to increase the sensible and evaporative heat loss from the body surface of livestock at high environmental temperatures. Mixing of the air within a coat by wind will reduce the vapour transfer resistance as it does that for sensible heat exchange. Campbell, McArthur and Monteith (1980) reported that the vapour conductance $1/\bar{r}_{vc}$ (cm s^{-1}) of clothing increased linearly with wind speed, as did that for sensible heat (equation 3.18), according to the relation

$$1/\bar{r}_{vc(u)} = 1/\bar{r}_{vc(0)} + \alpha_\omega u \qquad (3.26)$$

where $1/\bar{r}_{vc(0)}$ is the water vapour conductance (cm s^{-1}) in still air and α_ω is a constant, the value of which ranged from 0.20 to 2.1 depending on the penetrability of the clothing. Measurements are still required to establish the values of α_v for animal coats, although the value of $1/\bar{r}_{vc(u)}$ can be estimated from analysis of the sensible heat transfer through the coat, correcting for radiation transfer between the fibres, and assuming that the processes of heat and water vapour transfer are similar.

The resistance to mass transfer $\bar{r}_{va}$ provided by the boundary layer can also be estimated from equation 3.8 by applying the similarity between heat and mass transfer, i.e. for forced convection from a cylinder in wind the ratio $\bar{r}_v/\bar{r}_a = 0.80$ (Monteith, 1975).

CASE STUDY—CATTLE

Figure 3.14 shows values of $\bar{r}_b$, $\bar{r}_c$ and $\bar{r}_e$ for Jersey cows plotted against environmental temperature. These resistance values were calculated from the data of Worstell and Brody (1953), using the appropriate heat flux density and the mean skin and coat surface temperatures of the trunk. The increase in tissue resistance $\bar{r}_b$ with decreasing T_e is consistent with vaso-constriction. However, as the cows were kept at each temperature setting for about 2 weeks, the accompanying increase in coat resistance $\bar{r}_c$ can be attributed largely to an increase in the length and thickness of the coat. There was no evidence of cold-induced vasodilation at $T_e < 0\,°C$. The environmental resistance $\bar{r}_e$ remains almost constant at about 1.8 s cm^{-1} ($u \simeq 0.25$ m s^{-1}).

The extremities comprise about 25 per cent of the total surface area of a cow, and changes in blood flow to these parts are used by an animal to increase or decrease heat loss to the environment (Whittow, 1962). For animals with a pelage and a poor sweating capacity (such as sheep and poultry), a decrease in the tissue resistance on the trunk would be of little benefit as a mechanism to increase heat dissipation in a hot environment. However, a decrease in tissue resistance on the trunk would be of thermal advantage to a sweating animal subjected to high environmental temperatures. Berman (1971) reported that the local tissue resistance on the trunk of Holstein dairy cows fell from about 0.28 to 0.12 s cm^{-1} when the corresponding skin temperature rose from about 36.0 to 39.0 °C. The accompanying rise in body core temperature to about 39.7 °C indicates that heat could be dissipated through the body tissue on the trunk at a rate of about 75 W m^{-2} when the mean skin temperature on the trunk ($\bar{T}_{st}$) reached 39 °C. However, few published measurements allow similar analysis, and more information is needed on the local changes in tissue resistance of cattle

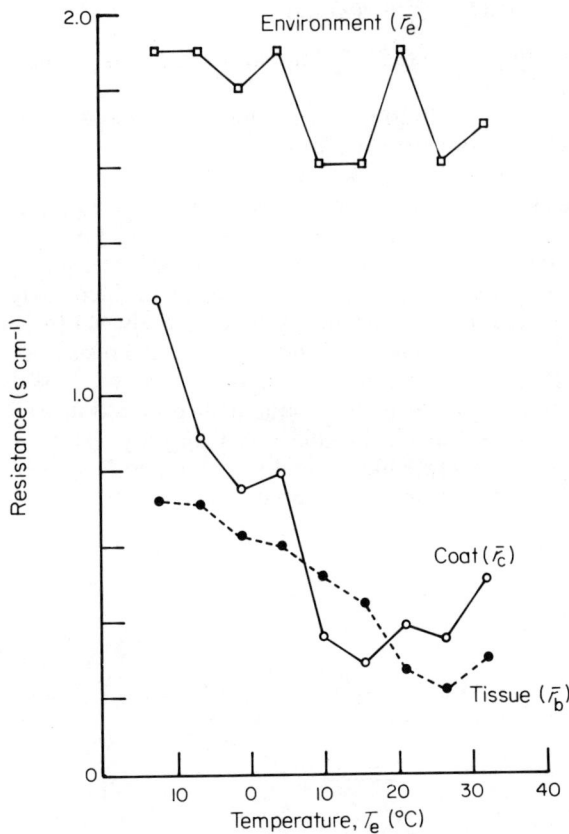

Figure 3.14 Tissue ($\bar{r}_b$), coat ($\bar{r}_c$) and environmental ($\bar{r}_e$) resistances for Jersey cows plotted against environmental temperature T_e. (From data of Worstell and Brody, 1953)

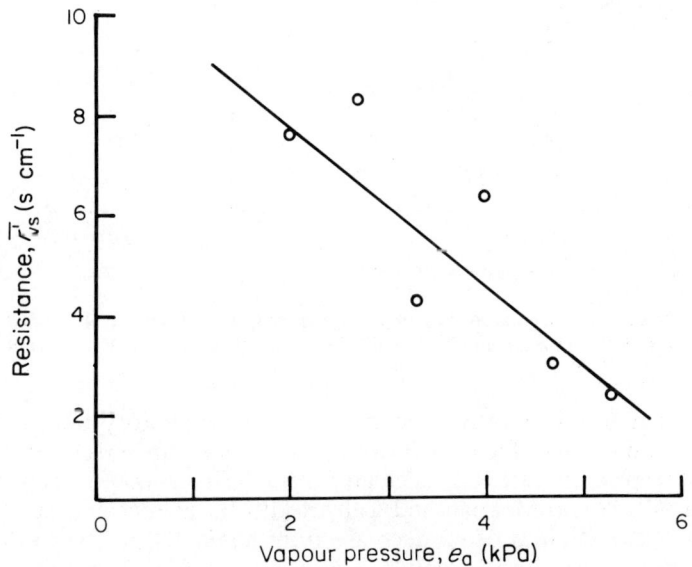

Figure 3.15 Resistance $\bar{r}'_{vs}$ ($= \bar{r}_{vs}/w$) to water vapour loss from skin surface of Jersey cow plotted against pressure e_a. (From data of Knapp and Robinson, 1954)

in hot environments and on the role of the different body regions as avenues for heat dissipation.

Knapp and Robinson (1954) measured the variations of cutaneous water loss from a Jersey cow with vapour pressure in a controlled environment chamber, with $T_e = 40\,°C$. *Figure 3.15* shows the resistance $\bar{r}'_{vs} (= \bar{r}_{vs}/\omega)$, calculated from their results, plotted against vapour pressure e_a. The decrease in $\bar{r}'_{vs}$ from about $8\,s\,cm^{-1}$ at $e_a = 2\,kPa$ to about $2\,s\,cm^{-1}$ at $e_a = 5.5\,kPa$ can be attributed to an increase in ω, due to an increase in sweat rate with increasing skin temperature and to a decrease in ε_{sm} with increasing vapour pressure e_a. Assuming that the skin surface is completely wet ($\omega = 1.0$) at $e_a = 5.5\,kPa$ and $T_e = 40\,°C$, *Figure 3.15* indicates that the resistance $\bar{r}_{vs}$ of the coat and boundary layer is $2.0\,s\,cm^{-1}$, and that in dry air ($e_a = 0$) at $T_e = 40\,°C$ the animal could dissipate its metabolic heat production (say $140\,W\,m^{-2}$) by evaporation of sweat alone with $\omega \simeq 0.20$ only. At $e_a = 5.3\,kPa$ the cutaneous evaporation rate measured by Knapp and Robinson was about $140\,W\,m^{-2}$ and at $2\,kPa$ the rate was about $120\,W\,m^{-2}$.

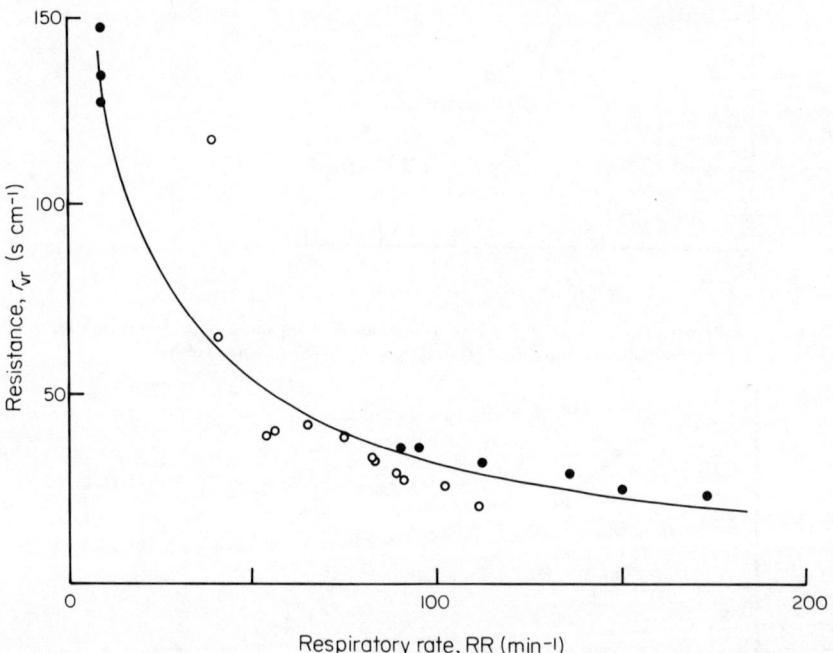

Figure 3.16 Resistance r_{vr} to water vapour loss from respiratory system of Jersey cows plotted against respiratory rate RR. (From data of Knapp and Robinson, 1954 (O) and McLean and Calvert, 1972 (●))

Cattle lose heat by evaporative cooling from the respiratory tract in addition to the skin surface. *Figure 3.16* shows values of r_{vr} for Jersey cows plotted against respiratory rate RR, calculated from the data of both Knapp and Robinson (1954) and McLean and Calvert (1972), as indicated, and assuming $e_s = 7\,kPa$. Values of r_{vr} decrease from about 130 to $25\,s\,cm^{-1}$ when RR increases from 10 to $150\,min^{-1}$. These resistance values are substantially higher than the resistance to water vapour transfer between the skin surface and the environment (e.g. *Figure 3.15*). The graph indicates that a cow with a respiratory rate of 100 min^{-1} can dissipate about 40 W m^{-2} by respiratory evaporation when $e_a = 1$ kPa, but only about 20 W m^{-2} when $e_a = 4$ kPa. The relation between r_{vr} and RR is given by

$$r_{vr}^{-1} = (2.7 \times 10^{-4} \, RR) + 5.0 \times 10^{-3} \qquad (3.27)$$

Figures 3.11 and *3.16* indicate that panting cattle can dissipate heat by respiratory evaporation (expressed per unit area of body surface) faster than poultry with the same respiratory rate.

Summary

This chapter has described the dependence of the resistances to heat and water vapour loss from livestock on physical factors, such as coat depth and wind speed, and on physiological factors, such as vasomotor action and respiratory rate. The next stage in this sort of analysis must be the development of a general model which incorporates all the relevant factors to predict the thermal status of animals in specified environments. Only then can the relative importance of the different environmental variables be identified for each type of stock and more exact recommendations concerning the control of these variables be forwarded for the improvement of animal productivity. The limited assessment presented here suggests that estimates of heat dissipation using simple application of whole-body averages of coat or tissue resistance, measured over small temperature intervals, may be misleading. Because of local resistance changes, such an approach may seriously underestimate the metabolic costs of cold exposure and could exaggerate sensitivity to heat. A complete analysis should also include the effects of postural adjustments (e.g. lying or standing) which may be invoked by animals to maximize or minimize the changes in their thermal insulation.

References

BARNES, A.P. (1978). *BSc Dissertation*, University of Nottingham

BENNETT, J.W. and HUTCHINSON, J.C.D. (1964). *Aust. J. agric. Res.* **15**, 427–445

BERMAN, A. (1971). *J. Physiol., Lond.* **215**, 477–489

BLAXTER, K.L. (1967). *The Energy Metabolism of Ruminants*, 2nd edn. Hutchinson, London

BLAXTER, K.L., GRAHAM, N.McC., WAINMAN, F.W. and ARMSTRONG, D.G. (1959). *J. agric. Sci., Camb.* **52**, 25–39

CAMPBELL, G.S., McARTHUR, A.J. and MONTEITH, J.L. (1980). *Boundary-Layer Met.* **18**, 485–493

CENA, K. (1974). In *Heat Loss from Animals and Man*, pp. 33–58. Ed. by J.L. Monteith and L.E. Mount. Butterworths, London

CENA, K. and CLARK, J.A. (1973). *J. Mammal.* **54**, 1003–1007

CENA, K. and CLARK, J.A. (1978). *J. thermal Biol.* **3**, 173–174

CENA, K. and MONTEITH, J.L. (1975a). *Proc. R. Soc., Ser. B* **188**, 377–393

CENA, K. and MONTEITH, J.L. (1975b). *Proc. R. Soc., Ser. B* **188**, 413–423

CLARK, J.A., CENA, K. and MONTEITH, J.L. (1973). *J. appl. Physiol.* **35**, 751–754

CUNNINGHAM, I.R. (1979). *BSc Dissertation*, University of Nottingham

EDE, A.J. (1967). *An Introduction to Heat Transfer Principles and Calculations*. Pergamon Press, Oxford

GAGGE, A.P. (1937). *Am. J. Physiol.* **120**, 277–287

GRAHAM, N.McC., BLAXTER, K.L., WAINMAN, F.W. and ARMSTRONG, D.G. (1959). *J. agric. Sci., Camb.* **52**, 13–24

HALES, J.R.S. (1974). In *Environmental Physiology*, pp. 107–162. Ed. by D. Robertshaw. Butterworths, London

HUTCHINSON, J.C.D. (1954). *J. agric. Sci., Camb.* **45**, 48–59

JOYCE, J.P. and BLAXTER, K.L. (1964). *Br. J. Nutr.* **18**, 5–27

JOYCE, J.P., BLAXTER, K.L. and PARK, C. (1966). *Res. vet. Sci.* **7**, 342–359

KNAPP, B.J. and ROBINSON, K.W. (1954). *Aust. J. agric. Res.* **5**, 568–577

McADAMS, W.H. (1954). *Heat Transmission*, 3rd edn. McGraw-Hill, New York

McARTHUR, A.J. (1980). *Proc. R. Soc., Ser. B* **209**, 219–237

McARTHUR, A.J. (1981). *J. thermal Biol.* **6**, 43–47

McARTHUR, A.J. and MONTEITH, J.L. (1980a). *Proc. R. Soc., Ser. B* **209**, 187–208

McARTHUR, A.J. and MONTEITH, J.L. (1980b). *Proc. R. Soc., Ser. B* **209**, 209–217

McLEAN, J.A. and CALVERT, D.T. (1972). *J. agric. Sci., Camb.* **78**, 303–307

MITCHELL, D. (1974). In *Heat Loss from Animals and Man*, pp. 59–76. Ed. by J.L. Monteith and L.E. Mount. Butterworths, London

MONTEITH, J.L. (1975). *Principles of Environmental Physics*. Arnold, London

OOSTHUIZEN, P.H. and MADAN, S. (1970). *J. Heat Transfer* **92**, 194–196

RICHARDS, S.A. (1974). In *Heat Loss from Animals and Man*, pp. 255–276. Ed. by J.L. Monteith and L.E. Mount. Butterworths, London

RICHARDS, S.A. (1977). *J. agric. Sci., Camb.* **89**, 393–398

STEEN, I. and STEEN, J.B. (1965). *Acta physiol. scand.* **63**, 285–291

STOMBAUGH, D.P. and ROLLER, W.L. (1977). *Trans. Am. Soc. agric. Engrs* **20**, 1110–1118

WAITES, G.H.H. and VOGLMAYR, J.K. (1962). *Nature, Lond.* **196**, 965–967

WATHES, C.M. (1978). *PhD Thesis*, University of Nottingham

WEBSTER, A.J.F. (1974). In *Heat Loss from Animals and Man*, pp. 205–232. Ed. by J.L. Monteith and L.E. Mount. Butterworths, London

WEBSTER, A.J.F. and BLAXTER, K.L. (1966). *Res. vet. Sci.* **7**, 466–479

WHITTOW, G.C. (1962). *J. agric. Sci., Camb.* **58**, 109–120

WIERSMA, F. and NELSON, G.L. (1967). *Trans. Am. Soc. agric. Engrs* **10**, 733–737

WIERSMA, F. and STOTT, G.H. (1966). *Trans. Am. Soc. agric. Engrs* **9**, 309–311

WORSTELL, D.M. and BRODY, S. (1953). *Missouri agric. exp. Sta. Res. Bull.* **515**

YUGE, T. (1960). *J. Heat Transfer* **82**, 214–220

II

ENVIRONMENTAL INFLUENCES ON REPRODUCTION

4

THE ENVIRONMENT AND REPRODUCTION

N.B. HAYNES
C.M. HOWLES
Department of Physiology and Environmental Studies, University of Nottingham

Introduction

It is a feature of animal reproduction that for optimum success as a species, animals should produce young when the latter stand the best chance of survival. Prime climatic conditions for the young animal in the wild occur only at certain times during the year in most parts of the world and, as a consequence, animals have developed a breeding season so that young are born only at such times. With the exception of animals which can show delayed implantation, the period of gestation is of a fixed length for a given species and hence, if an animal is to control the time of parturition it must regulate the initial events, namely the time of ovulation in the female and optimal testicular function in the male, by using proximate environmental cues (Baker, 1938). In theory, the nature of the cue is unimportant. However, the solar cycle has marked consequences for the climatic environment. The energy from the sun varies in a fairly regular manner in terms of intensity and photoperiod for particular latitudes, and it is not surprising that animals have adapted to use temperature and photoperiod as cues for triggering reproductive activity at the best time. In addition, reproductive activity is often 'fine-tuned' for efficiency within the broad-based cues of photoperiod and temperature by adaptive mechanisms in which animals of one sex utilize cues emanating from the other.

In some domesticated species, the need for breeding seasonality has been selected against and such animals will produce young at any time of the year. Notwithstanding this, the elements responsible for a waxing and waning of breeding activity concomitant with climate seem to be retained and under certain environmental conditions the animals will revert to seasonal breeding, or at least show periods of seasonal infertility.

The purpose of this chapter is to review some of the literature regarding the effects of these environmental variables, namely temperature, photoperiod and the social environment, upon the reproductive process in mammals. The review is relevant to animal housing only in so far as those designing, building and operating such housing can, if they wish, exert considerable control over these variables.

General physiological principles

Some of the effects of the environment on reproduction no doubt take place through direct action upon the reproductive tissues. For example, elevated temperature acts upon the testis to impair spermatogenesis (Van Demark and Free, 1970) and on the uterus to cause death or damage to the early embryo (Edwards, 1978). However, abundant evidence exists showing that environmental cues may modify the reproductive process in an indirect and more subtle manner. To achieve this, the cue is perceived by the animal and an assessment of its nature made. This information is then transferred to the hypothalamic–pituitary axis, which affects gonadal function via the pituitary trophic hormones. Some alteration in activity of this axis must then ensue (*Figure 4.1*). To take a specific example, Lincoln and co-workers have

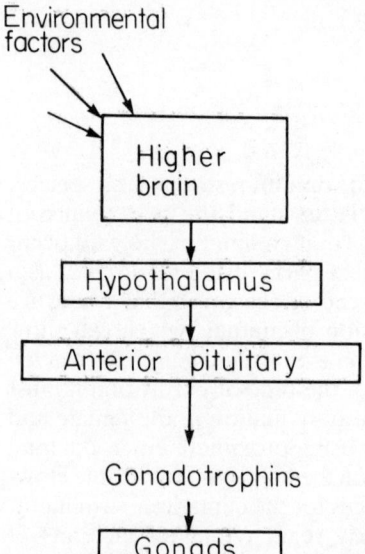

Environmental factors

Higher brain

Hypothalamus

Anterior   pituitary

Gonadotrophins

Gonads

Figure 4.1 The hypothalamo–pituitary–gonad axis—site of action of many environmental factors

documented in detail the sequential changes which occur in gonadotrophin secretion, testicular function and sexual behaviour in the ram as a result of photoperiodic change (Lincoln, 1976a, b; Lincoln and Peet, 1977; Lincoln and Davidson, 1977). These are summarized in *Figure 4.2*. That such changes occur as a result of altered hypothalamic activity has been demon-strated by the fact that pulsatile infusion of the hypothalamic hormone GnRH into seasonally quiescent rams results in full seasonal reproductive activity despite a depressive photoperiod (Lincoln, 1979). Part of the environmentally induced change in the hypothalamic–pituitary axis may result from a change in its sensitivity to negative feedback (via steroid hormones) in both the ram (Pelletier and Ortavant, 1975a, b; *Figure 4.3*) and the ewe (Karsch *et al.*, 1978; Legan and Karsch, 1979). Whilst most of the evidence that the environment modifies reproductive function through the brain–pituitary axis has accrued from experiments involving manipu-lation of photoperiod, it is reasonable to assume that other environmental variables such as temperature and the social conditions in which animals are

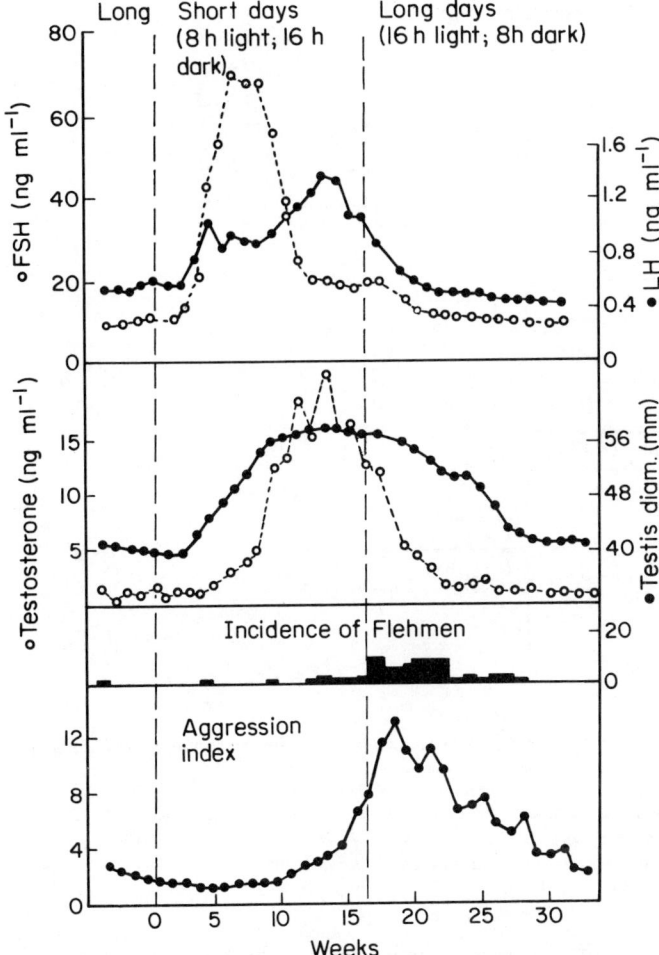

Figure 4.2 Plasma FSH, LH and testosterone levels, testicular diameters and aggression scores (weekly mean) for six Soay rams housed under artificial lighting conditions. The total weekly observations of Flehmen are also shown. (From Lincoln and Davidson, 1977)

reared work in the same way, namely through brain receptors which modify the hypothalamic–pituitary axis and thus turn the secretion of gonadotrophins on or off.

Temperature effects on reproduction

Much of the literature concerned with the effects of temperature on the reproductive system should be treated with caution, for a number of reasons. First, temperature and photoperiod are positively correlated in many parts of the world, and where studies have been carried out under natural conditions it is often difficult to distinguish what proportion of the effect is due to temperature or to photoperiod. Secondly, changes in

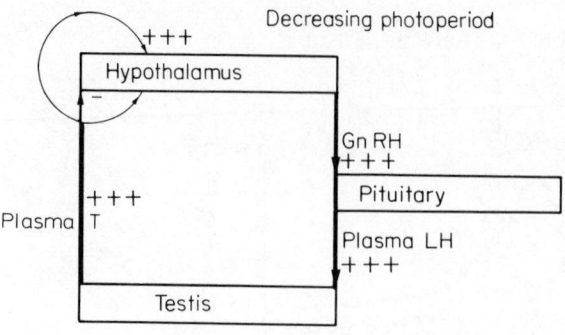

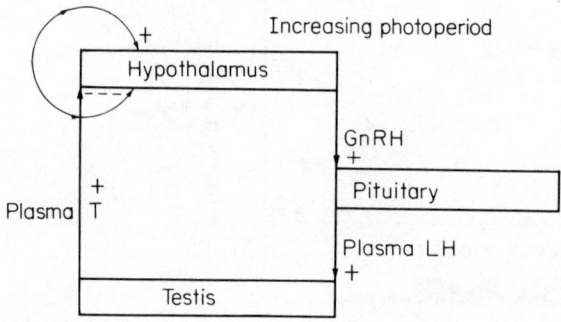

Figure 4.3 Diagram representing the influence of photoperiod on hypothalamo–pituitary activity in the ram. The number of plus signs indicates the intensity of hormone release into the blood and minus signs the intensity of negative feedback. Decreasing photoperiod may act by: (1) stimulating the hypothalamo–pituitary system; and (2) decreasing the negative feedback effect of testosterone. Conversely, increasing photoperiod: (1) produces less stimulation of hypo-thalamo–pituitary activity; (2) increases the negative feedback effect of testosterone. (Adapted from Pelletier and Ortavant, 1975b)

temperature often modify food intake and changes ascribed to temperature may really be due to altered nutrition. Thirdly, in experiments carried out under laboratory conditions, animals have sometimes been subjected to temperature changes more abrupt or greater than those experienced naturally. Despite these limitations, there is evidence that temperature can affect the reproductive process at a number of points ranging from pubertal development through conception and embryonic mortality. In rodents, there have been definitive studies showing that low and high environmental temperatures cause a delay in the onset of puberty in both males and females (Barnett, 1962; Nazion and Piacsek, 1976, 1977). There is some evidence for this in domestic species also: for instance, Brahman and Shorthorn cows reached puberty some five months later at 26 °C than at 10 °C (Dale, Ragsdale and Cheng, 1959). Few studies have been carried out to determine the physiological cause of such delayed puberty; this is not surprising since the hormonal events leading to normal puberty are poorly understood (Foxcroft, 1978). However, the temperature-induced delay in puberty in rats appeared to be mediated through the endocrine system, since it was associated with a delay in the time of occurrence of the prepubertal luteinizing hormone (LH) surge. Since the pattern of androgen secretion found in these animals was normal, the interesting suggestion was made that secondary sex organ sensitivity to testosterone was lowered, because of a delay in the development of LH-induced receptor activity in those organs (Nazion and Piacsek, 1977, 1978).

There are similarities between the onset of puberty and the re-establishment of reproductive activity after seasonal quiescence in seasonally breeding animals (Illius, 1976), and the onset of the breeding season in sheep can also be markedly affected by temperature (Dutt and Bush, 1955; Godley, Wilson and Hurst, 1966). In the studies of Dutt and Bush (1955), ewes maintained at an environmental temperature of 7–9 °C from May to October had a mean date of 10 July for first ovulation, compared with 26 September for control animals at a maximum daily temperature of 32 °C. Prolactin concentrations in the plasma of sheep are high during long and depressed by short photoperiod (Pelletier, 1973; Forbes *et al.*, 1975; Lincoln, McNeilly and Cameron ,1978), and on this basis it has been suggested that the return to sexual cyclicity at the start of the breeding season may be brought about by the removal of an anti-gonadotrophic effect exerted by high concentrations of prolactin in blood (Walton *et al.*, 1977). Prolactin in some species is also depressed by low ambient temperatures (*see* p. 72) and it is possible that an accelerated lowering of prolactin levels by cooling could have been responsible for the early cyclicity of sheep in the above experiments.

In the adult female the problem of low fertility during the hot summer months has been widely recognized for a number of domestic species (Stott and Williams, 1962; Edwards *et al.*, 1968; Thwaites, 1968; Plasse, Warnick and Koger, 1970; Dunlap and Vincent, 1971; Ingraham, Gillette and Wagner, 1974; Thatcher *et al.*, 1974; Rosenberg *et al.*, 1977; Love, 1978). An example of the findings is shown in *Table 4.1*. There would appear to be a

Table 4.1 CONCEPTION RATE OF DAIRY COWS (MULTIPAROUS AND PRIMIPAROUS) INSEMINATED IN SUMMER AND WINTER

		Summer	*Winter*
All inseminations	No. of inseminations	37	44
	Conception rate	57%*	75%
First inseminations	No. of inseminations	26	26
	Conception rate	46%*	77%

Data from Rosenberg *et al.* (1977)
*Significantly different from conception rate in winter (at $p < 0.05$; χ^2 test)

variety of causes: predominantly the low fertility arises from a failure of conception, or from early embryonic mortality resulting from elevated temperature around this time. This is borne out by controlled experiments, in which animals have been subjected to elevated temperatures shortly after mating or artificial insemination and the fertility of that particular mating recorded. The severity of infertility which can ensue is illustrated in the studies of Dunlap and Vincent (1971) in which short-term exposure of cattle to heat (32 °C for 72 hours) immediately following breeding resulted in no conceptions in 23 animals compared with 14 conceptions in 25 animals from a control group maintained at 21 °C. In a number of similar studies, rectal temperatures were shown to be increased during elevated ambient temperature and furthermore, uterine temperature and average ambient temperature on the day of insemination were inversely related to fertility in animals maintained under natural conditions (Gwazdauskas, Thatcher and Wilcox, 1973). On the basis of such evidence and the findings of Alliston, Howarth and Ulberg (1965) that fertilized rabbit eggs in culture were killed by small rises in temperature, it was suggested that a hot uterine environment may have a direct lethal effect on young embryos (Wolff and Monty, 1974). On the other hand, oestrous periods are known to be shortened by elevated temperatures (Teague, Roller and Grifo, 1968; Plasse, Warnick and Koger, 1970; Madan and Johnson, 1973; Abilay, Johnson and Madan, 1975) as illustrated in *Table 4.2*. This implies a response through the hypothalamo–pituitary–gonad axis resulting in lowered oestrogen secretion; the effect of this on the status of the oviduct and uterus could also result in increased embryonic mortality. It is not possible on the basis of currently available evidence to distinguish between such mechanisms.

Table 4.2   MEAN VALUES (±SEM 6 animals/group) FOR OESTROUS CYCLE LENGTH, DURATION OF OESTRUS, RECTAL TEMPERATURE, FEED AND WATER INTAKE OF GUERNSEY HEIFERS KEPT AT ENVIRONMENTAL TEMPERATURES OF 18.2 AND 33.5 °C

	18.2 °C	*33.5 °C*
Oestrous cycle length (d)	$19.5^a \pm 0.2$	$21.4^b \pm 0.3$
Duration of oestrus (h)	$17.0^a \pm 0.7$	$12.5^b \pm 1.0$
Rectal temperature (°C)	$38.3^a \pm 0.1$	$39.6^b \pm 0.2$
Feed intake (kg/d)	$9.4^a \pm 0.3$	$6.2^b \pm 0.3$
Water intake (l/d)	$29.0^a \pm 1.4$	$34.6^b \pm 1.1$

Data from Abilay, Johnson and Madan (1975)
Values with different superscripts are statistically different (at $p < 0.05$)

Furthermore, poor conception rates and early embryonic mortality may not occur solely through the female. Reports of deleterious effects upon spermatogenesis through high ambient temperatures acting upon the testis are numerous (cf. Van Demark and Free, 1970; Blackshaw, 1977), and reduced pregnancy rates have been recorded in sows at normal temperatures mated with boars previously kept at high ambient temperatures (Wetteman *et al.*, 1976). Also, behaviour of the male may be altered to the detriment of fertility. For example, Knecht, Wright and Toraason (1978) have shown that male rats exposed daily to a temperature of 38 °C for 55 minutes had a lowered frequency of copulation with females kept at temperatures around 20 °C. Conception rates were lower and fetal survival was reduced also.

To return to the female, several studies have attempted to define endocrinological mechanisms which may affect fertility in situations of elevated temperature and have been concerned essentially with measurements of progesterone and LH. These data are confusing. Relatively high environmental temperatures on the day of insemination were correlated with progesterone concentrations in blood higher than those usually observed around the day of oestrus (Gwazdauskas, Thatcher and Wilcox, 1973). In support of this, Mills *et al.* (1972) found a significant increase in plasma progesterone in heifers which had been exposed to continuous heat stress for three days and elevated progesterone levels in gilts associated with increased embryonic mortality (Kreider and Wettemann, 1975). Chronic exposure to heat has also been reported to cause elevated progesterone, since levels were higher in lactating cows in summer than in winter (Vaught, Monty and Foote, 1977) and exposure under laboratory conditions to 33 °C for two oestrous cycles, an ambient temperature sufficient to cause an elevation of rectal temperature to 1.5 K above normal, resulted in increased plasma progesterone concentrations during most of the first cycle and part of the second (Abilay, Johnson and Madan, 1975). On the other hand, the average monthly plasma progesterone concentrations of cows exposed to normal summer conditions in Arizona were reported to be lower than the levels in cows kept under shaded conditions in the same location (Stott and Wiersma, 1973). These results were confirmed in a detailed study by Rosenberg *et al.* (1977), which revealed that progesterone levels during the cycle were significantly lower in the summer than in winter in Friesian dairy cows in Israel. Such cows show a significant depression in fertility when inseminated during late summer and autumn (Heiman, 1972). Moreover, in animals which did not conceive in the summer, progesterone concentrations in the cycles preceding insemination began to decrease much earlier than in the second cycle. This accords with other observations that the level of blood progesterone during the luteal phase of the cycle preceding the first insemination may be positively related to conception (Folman *et al.*, 1973; Corah *et al.*, 1974). Examples of conflicting data in regard to plasma progesterone levels resulting from elevated temperature are detailed in *Figure 4.4*. The reasons for such marked differences are hard to find, although the experimental protocols may differ between laboratories. One possibility is suggested, namely that if animals are suddenly, rather than gradually, subjected to temperature change, they may have no time to make physiological adjustments. As a result, there may be a significant stress-induced adrenal contribution to plasma progesterone resulting in the elevated progesterone reported in some of the above studies. It should be noted, however, that Vaught, Monty and Foote (1977) present arguments against this being a factor in their experiments and corticoid levels were actually depressed in the studies of Abilay, Johnson and Madan (1975). The few studies carried out on the important reproductive hormone LH are no clearer in terms of a unified picture. In the work of Vaught, Monty and Foote (1977) described above, LH levels were similar in high-temperature and control groups, and in other experiments heat stress did not alter blood LH levels in ewes (Hooley, Findley and Stevenson, 1979). Conversely, Riggs, Alliston and Wilson (1974) reported that LH levels in gilts were increased at the time of ovulation in conditions of high ambient temperature, and there are other reports that both basal and peak blood levels of LH are lower in cattle subjected to elevated temperature (Madan and Johnson, 1973; Miller and Alliston, 1974). The former authors were careful to point out that feed intake was lower under high temperatures and depressed LH may be a direct consequence of this (*Figure 4.5*).

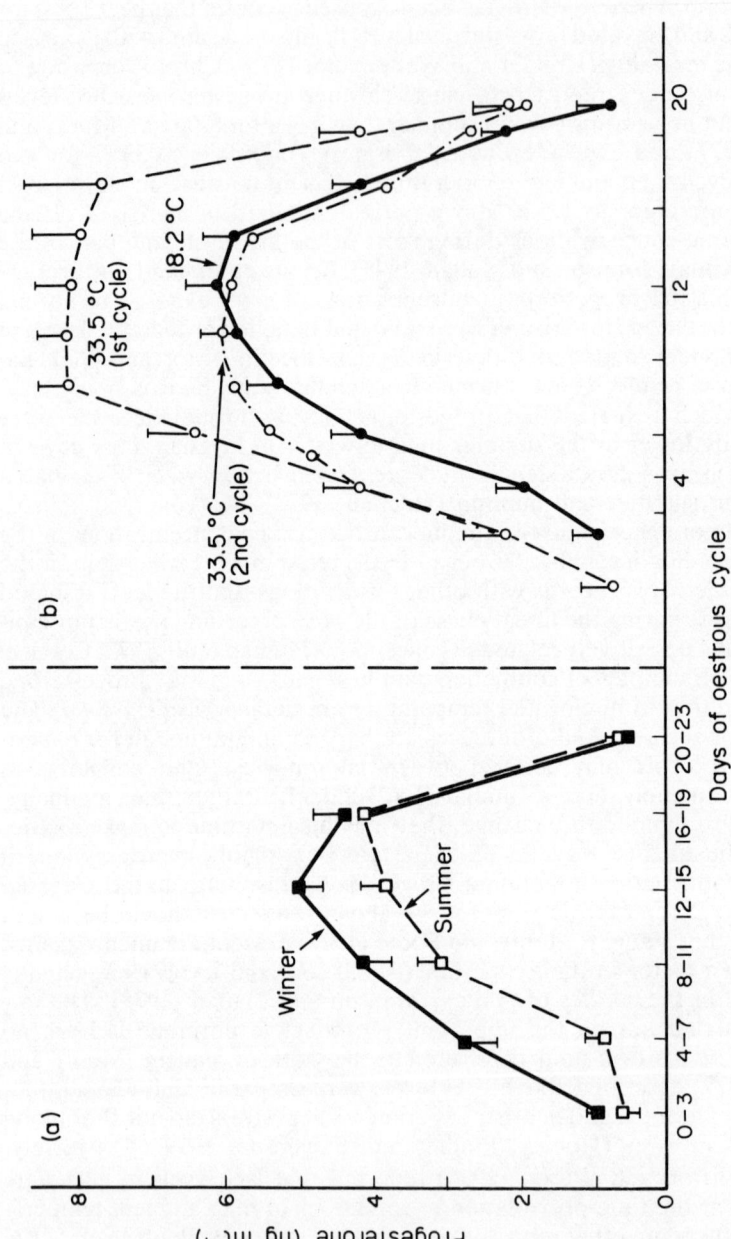

Figure 4.4 (a) Plasma progesterone concentrations (mean ± SEM) during the cycles of multiparous cows in summer (□, 14 cycles of 10 cows) and winter (■, 21 cycles of 16 cows). (From Rosenberg et al., 1977). (b) Plasma progesterone concentrations (mean ± SEM) during the second oestrous cycle of six Guernsey heifers at 18.2 °C (●——●) and the first (○——○) and second (○––○) cycle at 33.5°C. (From Abilay, Johnson and Madan, 1975)

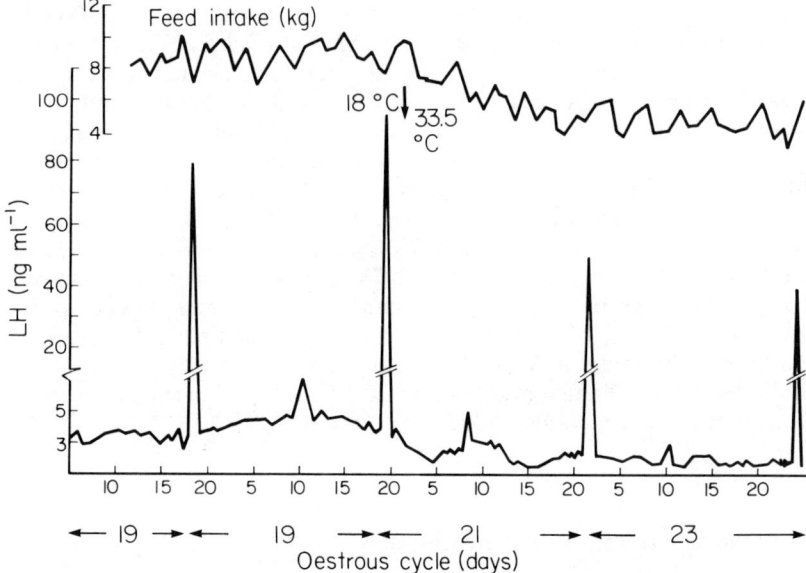

Figure 4.5 Plasma LH in a single animal measured through consecutive cycles at two tempera-ture conditions. The figure shows the relation between the onset of oestrus, the LH peak and the feed consumed by the animal. The vertical arrow designates the change in chamber temperature from 18.2 to 33.5 °C. (From Madan and Johnson, 1973)

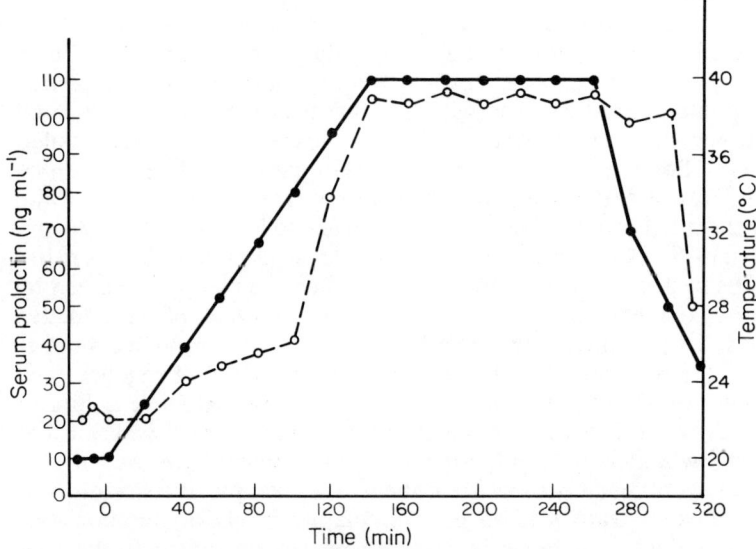

Figure 4.6 Effect of environmental temperature changes on serum prolactin concentrations in steers. (●) Temperature; (○) prolactin. (From Smith, Hacker and Brown, 1977)

Finally, in regard to temperature effects upon reproductive hormones, the only consistent response seems to be with prolactin. Blood concentrations of this hormone were elevated by temperatures above the norm and depressed by temperatures below it in both male and female cattle and sheep (Wettemann and Tucker, 1974; Smith, Hacker and Brown, 1977; Hooley, Findley and Stevenson, 1979). The data of Smith, Hacker and Brown (1977) are illustrated in *Figure 4.6*. The studies referred to above were not carried out to obtain information on fertility, but it is known that hyperprolactinaemia is associated with lack of cyclicity in women (Butt, 1978) and in ewes (Kann, Martinet and Schirar, 1978) and with lack of libido in the males of a number of species (Hartmann, Endröczi and Lissák, 1966; Thorner and Besser, 1977). Moreover, depressed prolactin results in lack of testicular development in rats, mice and hamsters, and the decline in testicular function in hamsters kept under short photoperiod can be reversed by treatment with exogenous prolactin (Bartke *et al.*, 1978).

Photoperiodic effects upon reproduction

Photoperiod is no doubt the most important agent used as a proximate cue by animals which exhibit seasonal reproduction. This has been the subject of a number of excellent reviews dealing with both mammals and birds (cf. Follett, 1978; Turek and Campbell, 1979) and is covered by another participant in this Easter School (Morris, Chapter 5). Therefore, only one particular aspect is discussed here, namely what happens when animals are maintained under constant photoperiod. Some of the results are surprising, bizarre and could have important connotations for the lighting regimes used in animal housing.

It is generally accepted that 'photoperiodic' animals respond to changes from short to long or long to short days, depending upon the species, with an ordered sequence of endocrine events leading ultimately to the animal showing optimal reproductive activity. This has been referred to earlier and illustrated for the ram in *Figure 4.2*. This figure merits more detailed examination. Some of the responses to photoperiodic change, namely those of FSH, LH and testosterone, begin to wane towards the end of the period of short light, but *before* the reversal back to long light. Lincoln and Davidson (1977) point out that this tends to suggest that the ram has endogenous cycles of reproductive activity which are merely entrained by photoperiodic change and not caused by it. Indeed, cycles of reproductive activity have been shown to occur under constant photoperiod in a number of species (Goss and Rosen, 1973; Gwinner, 1975; Pengelley and Asmundson, 1974; Michael and Bonsall, 1977). The situation is further complicated by the fact that the length of light periods may be critical in determining whether the endocrine changes actually give rise to the behavioural events usually ascribed to them. For example, if testosterone is implanted in a castrate stag to give circulating levels of this hormone sufficient to maintain the secondary sex organs, rutting behaviour does not ensue unless the photoperiodic conditions are those at the time rutting normally occurs, namely autumn (Lincoln, Guinness and Short, 1972). Thus photoperiod may modify the sensitivity of behaviour centres in the brain to circulating hormones.

These points are further illustrated in the following experiment. Three groups of ram lambs were reared under constant photoperiod, one under 16 hours light and 8 hours dark, a second under 8 hours light and 16 hours dark while the control group experienced the natural photoperiod. Both constant-photoperiod groups showed a cycle of testis growth and increased testosterone production followed by regression, the cycle approximating in length to that shown in rams under natural photoperiod but out of phase with it. Testis volume data are shown in *Figure 4.7*. On the other hand, the development of sexual behaviour was much slower in the long-light than the short-light group (*Figure 4.8*), despite the fact that testosterone levels were equivalent. Subsequent study revealed a further complicating feature. The alteration of prolactin secretion by photoperiod in sheep, and the possibility

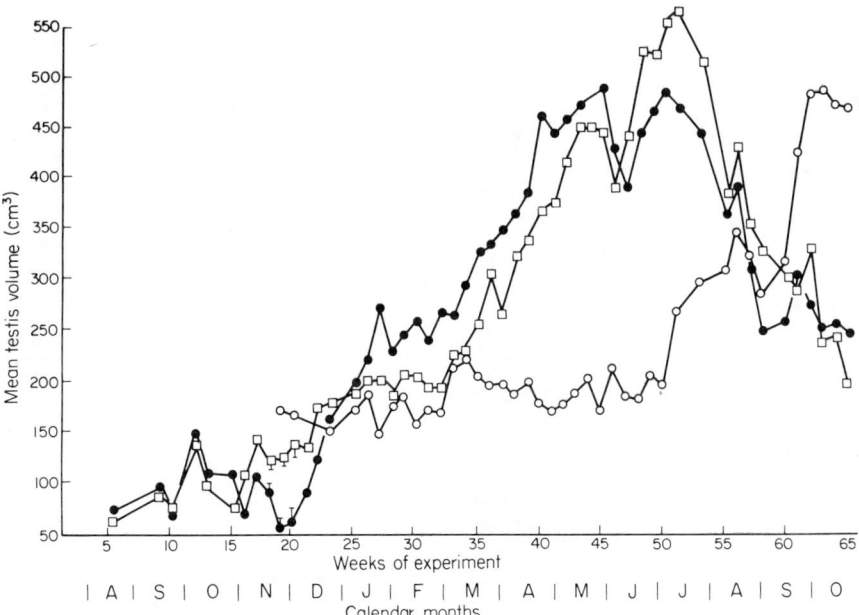

Figure 4.7 Mean testis volume for rams reared under short (●), long (◻) or normal (○) photoperiod. (From Howles, Webster and Haynes, 1980)

that this relates to breeding activity, has been referred to earlier, and prolactin levels in cattle are positively correlated with photoperiod (Bourne and Tucker, 1975; Leining, Bourne and Tucker, 1979). Prolactin levels were also markedly affected in the ram rearing experiments described above and are shown in *Figure 4.9*. Prolactin levels were much higher (and more variable) in the long- as compared with the short-light group. Hence, rams reared in constant photoperiods have different developmental patterns in terms of reproduction compared with animals under natural photoperiod. Therefore, on the one hand, constant but different photoperiods are not

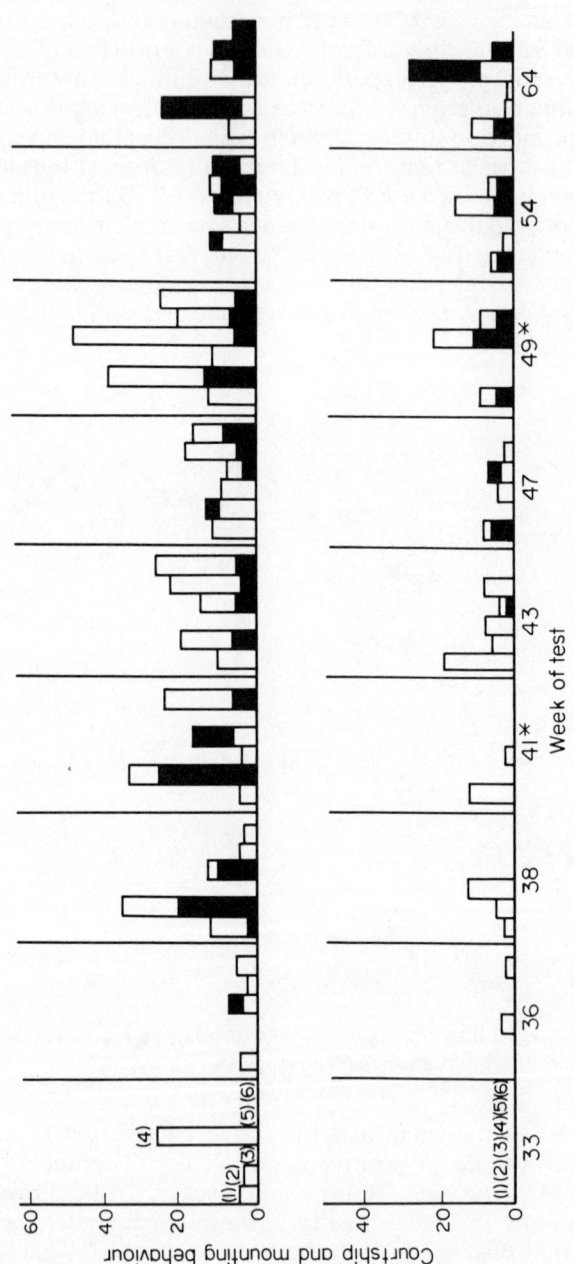

Figure 4.8 Numbers of courtship behaviours (unshaded areas) and mounts (shaded areas) shown by each ram towards a ewe, during one 20-minute behavioural test at the week indicated. The total for two tests carried out on the same day is indicated by an asterisk. Ram identification numbers are shown in parentheses. Top, short-light group; bottom, long-light group (see text). (From Howles, Webster and Haynes, 1980)

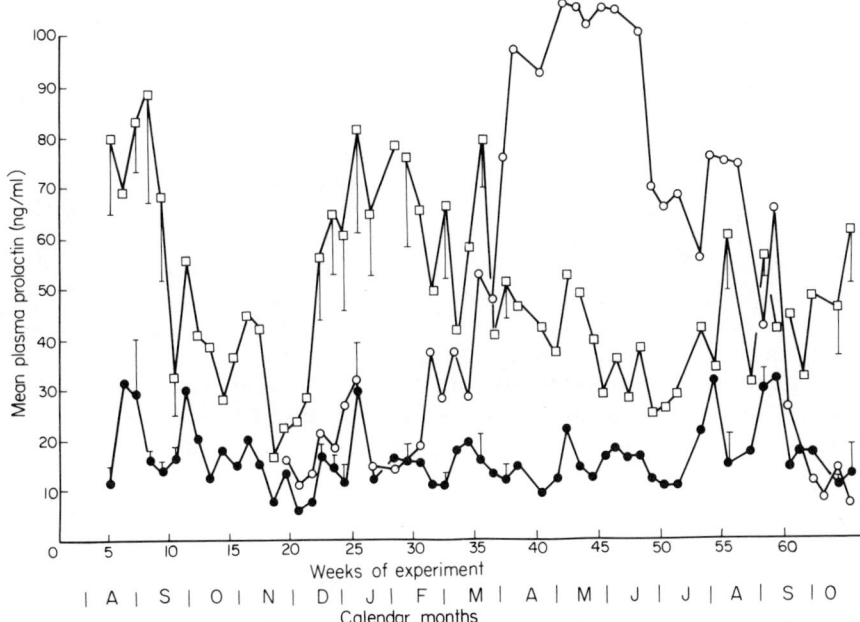

Figure 4.9 Mean plasma prolactin concentrations for rams reared under short (●), long (□) or natural (○) photoperiod. The SEM value is shown for short- and long-photoperiod animals when the means were significantly different. (From Howles, Webster and Haynes, 1980)

recognized by the ram in terms of testicular growth and regression; while on the other, they are recognized in relation to development of sexual behaviour and prolactin production.

Social factors and reproduction

The type of social relationship of an animal with others of its species during its life can have both profound and subtle effects on reproductive activity. The effect of too high a population density is well established for many species, and results in a large spectrum of reproductive abnormalities ultimately leading to sterility (Christian, Lloyd and Davis, 1965; Christian, 1975). However, it is not necessary to use such extremes of rearing conditions to demonstrate effects on the reproductive process. Many effects of a more subtle nature have been demonstrated with laboratory species, namely rodents, mostly depending upon whether animals are kept as single- or mixed-sex groups, and undoubtedly some of these apply to domestic species also.

The most striking effect is the temporal regulation of ovulation, which is brought about by male proximity. These actions are particularly dramatic in relation to the first pubertal ovulation. In mice, urinary cues, presumably acting through priming pheromones from the male, accelerate the process, whilst cues from other females decelerate it (Vandenbergh, 1975). In fact, in mice, male presence is crucial for the proper organization of puberty.

Females isolated from males from 21 days of age show prolonged and disorganized vaginal and uterine cycles and the cycles are often anovulatory (Stiff, Bronson and Stetson, 1974). Cyclicity in such animals often does not start until 50 days of age, whereas pairing with a male can give rise to puberty as early as 25 days after birth (Vandenbergh, Drickamer and Colby, 1972; Bronson and Stetson, 1973). Moreover, Colby and Vandenbergh (1974) have shown that the accelerating action of the male can be mimicked by male urine, indicating the pheromonal nature of the cue. Bronson and Desjardins (1974) have produced detailed information about the hormonal pathways involved in induction of puberty by male presence. Exposing females to a male resulted in a stimulus of the hypothalamo–pituitary axis, giving rise to an immediate elevation in LH followed by a dramatic increase in serum oestradiol. FSH was depressed and rose again on the third day of exposure, accompanied by an ovulatory surge of LH and ovulation some 72 hours after the initial male exposure (*Figure 4.10*). This effect of the male on the induction of puberty is certainly not limited to rodents, and the use of the

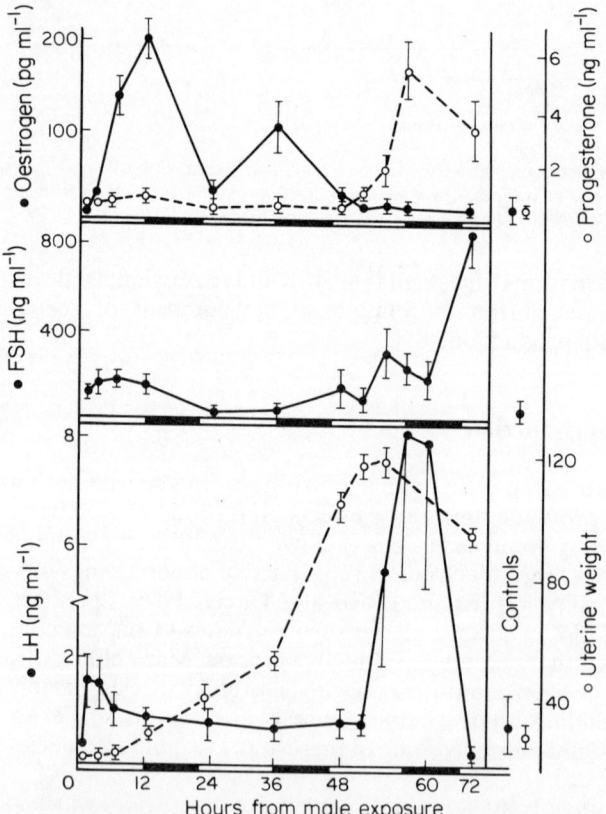

Figure 4.10 Changes in serum LH, FSH, oestradiol and progesterone concentrations during experimental induction of the pubertal ovulation in young female mice by cohabitation with adult males. Ovulation occurred late in the third dark phase, as shown by light and dark bars at the bottom of the graph. Control females, housed in groups in a room free of male odour, were killed at 72 hours. (From Bronson and Desjardins, 1974)

boar to induce early puberty in gilts kept in intensive housing has been applied successfully (Brooks and Cole, 1970; Thompson and Savage, 1978).

Ovulation in the adult female is also regulated by male presence in a similar manner to that seen in pubertal animals. Female mice exhibit a short oestrous cycle when housed with males, a longer cycle when housed in isolation and the longest cycle of all when housed in groups in the absence of the odour of a male (Whitten, 1956; van der Lee and Boot, 1956; Bronson, 1968). Similarly, if rats are given a restricted feed intake, the oestrous cycles become lengthened and irregular, but if a male is placed adjacent to the female, the effect of underfeeding on cyclicity is much less (Cooper and Haynes, 1967). Also, the novelty stimulus of regularly changing the male is much more effective than using the same male (Cooper, Purvis and Haynes, 1972). Again, these effects of male presence on the cycles of the adult female are not confined to rodents. It is a well known fact that sheep will enter the breeding season earlier (and with synchronized cycles) if a ram is introduced into the flock as the season approaches. This has been applied as a husbandry technique (Schinckel, 1954a, b).

The male as a reproductive entity is also affected by social rearing conditions. There is much evidence that social experience during rearing has a marked effect on the sexual behaviour of the male of several species. In the dog, chimpanzee, guinea pig, rat and rhesus monkey the copulatory performance of the mature male is reduced when the environment allows only a little social experience (Gerall, 1965; Gerall, Ward and Gerall, 1967; Beach, 1968; Young, Goy and Phoenix, 1964). This has been investigated recently with specific regard to the problem of low levels of sexual activity in breeding boars which occurs in intensively managed piggeries in Australia (Hemsworth, Beilharz and Galloway, 1977). In these studies boars were reared from 20 days to 30 weeks of age in isolation, as all-male groups or as a mixed-sex group respectively. Courting behaviour and copulation were markedly depressed in the group reared out of visual and physical contact with other pigs (*Figures 4.11* and *4.12*). Whilst not being researched in a controlled manner, this situation has been stated to occur in other species where males are reared with a view to use in artificial insemination. Furthermore, the situation did not appear to be reversible in the boar experiments referred to above. A similar state can exist in adult animals, as also demonstrated by the fact that boars 15 months old and kept in isolation, or in physical contact with other boars only, showed a progressive decline in sexual behaviour over a three-month period compared with animals housed near sexually receptive sows. This effect was not so severe as that for immature boars since it was reversible. These data are shown in *Figure 4.13* (Hemsworth *et al.*, 1977). The results led to the following conclusions:

(1) that the practice of rearing boars from three weeks of age in the absence of visual and physical contact with other pigs has an undesirable effect on sexual behaviour;
(2) that in order to maintain high levels of sexual behaviour, pig producers and technicians collecting semen should not isolate breeding boars from oestrous females.

In the pig it is not established whether hormonal changes are involved in depression of behaviour as a result of aberrant social conditions. However,

78

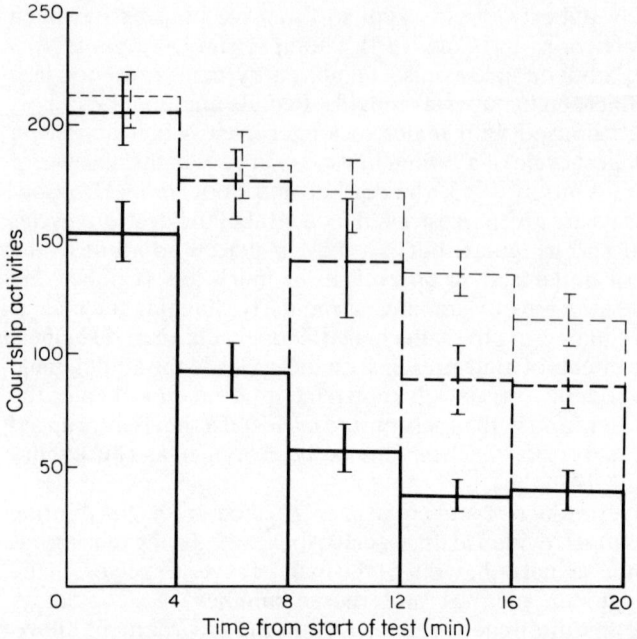

Figure 4.11 Mean number of times a courting behaviour activity (±SEM) was exhibited in the mating test by each group of boars. (———) Socially restricted animals; (– – –) all-male group; (- - -) mixed-sex group. (From Hemsworth, Beilharz and Galloway, 1977)

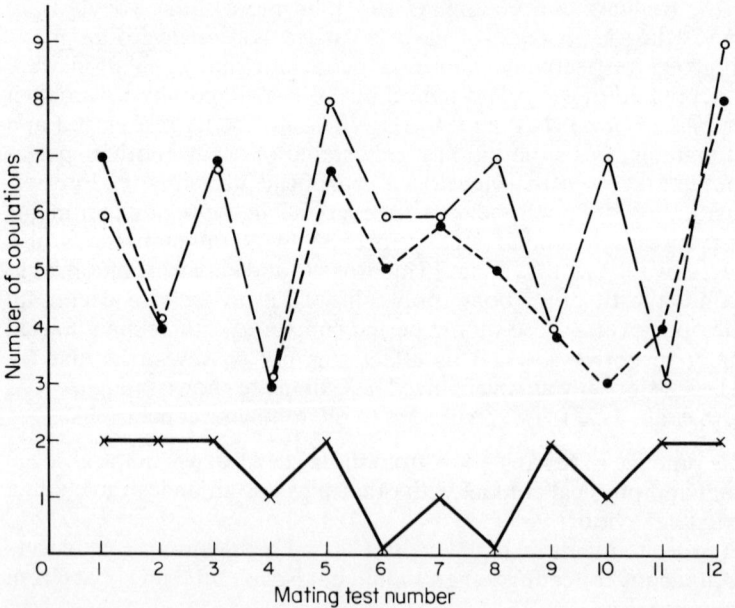

Figure 4.12 Total number of copulations achieved by each group of boars in each of 12 mating tests. (×) Socially restricted animals; (O) all-male group; (●) mixed-sex group. (From Hemsworth, Beilharz and Galloway, 1977)

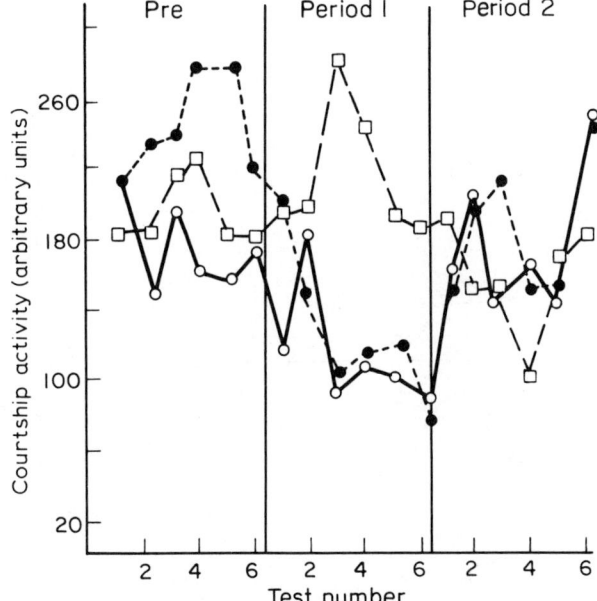

Figure 4.13 Sum of the courting behaviour activities exhibited by each group of mature boars in each mating test. (O) Social restriction in period 1; (●) semi-social restriction in period 1; (□) near-cyclic females in period 1. (From Hemsworth et al., 1977)

there is evidence that this may be so from studies on other species. As early as 1936, Steinach stated that atrophy of the reproductive system occurred in rats kept in the absence of females or female odour, and this was confirmed in studies by Drori and Folman (1964), Thomas and Neiman (1968) and Purvis and Haynes (1972). On the other hand, the rearing situation in which rams were kept before puberty, i.e. in isolation, as a homosocial or a heterosocial group did not affect plasma testosterone levels. However, yearling rams reared adjacent to, but not in physical contact with ewes had higher testosterone levels than rams reared in isolation from ewes (*Figure 4.14*). The rams reared next to females also showed marked elevations in plasma LH and testosterone as a result of copulation, whereas the animals reared away from females did not (Illius, Haynes and Lamming, 1976; Illius et al., 1976). The significance of these findings in relation to behaviour was not firmly established, but the heterosocial group was more aggressive and had higher libido than the homosocial group as assessed in a limited series of tests.

It seems, therefore, that the social situation in which animals are reared may be important for reproductive efficiency since both hormone levels and sexual behaviour can be depressed under some conditions, particularly by the relatively common practice of rearing animals as single-sex groups.

Summary and conclusions

It is a feature of animal reproduction that, for optimum success as a species, animals should produce young when the latter stand the best chance of

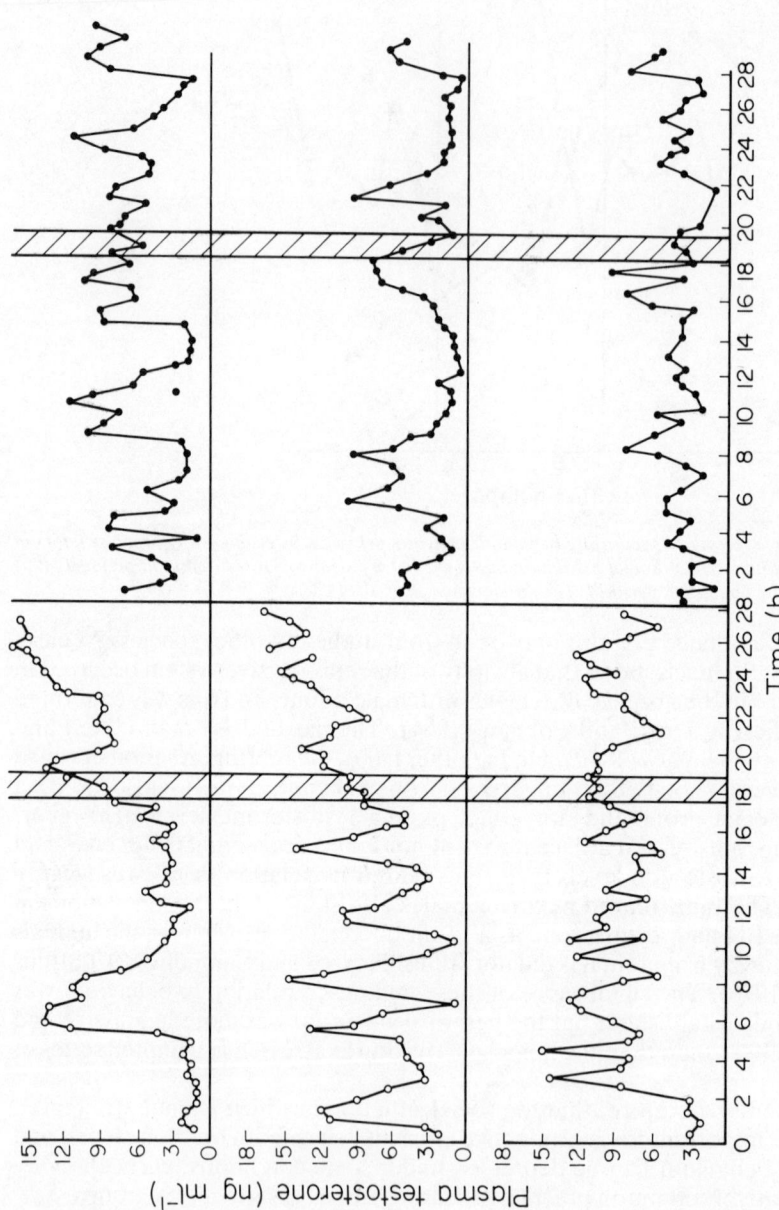

Figure 4.14 Plasma testosterone profiles for individual rams reared adjacent to (O) or in isolation from (●) ewes before and after copulation with a female. The period of copulation is shown by the hatched area. (From Illius et al., 1976)

survival. To achieve this, some species have evolved the facility to control the time of ovulation in the female and optimal testicular function in the male by using proximate cues, such as photoperiod and temperature. In addition, reproductive activity is 'fine-tuned' for efficiency by animals of one sex utilizing cues emanating from the other. In some domesticated animals, the need to produce young at a specific time has been selected against and they will breed all the year round. However, the mechanisms for establishing seasonal breeding seem to be retained and under particular circumstances domestic stock will revert to seasonality, or at least show periods of sub-fertility. Therefore, for optimum reproductive activity under controlled environmental conditions, some knowledge of how temperature, photoperiod and social factors may affect reproduction is essential. As a generalization, some exceptions being concerned with temperature, these environmental factors act on reproduction through the hypothalamo–pituitary–gonad axis, ultimately either stimulating or regressing gonadal activity.

Whilst the literature is somewhat conflicting, there is evidence that temperature can affect reproduction at a number of points ranging through pubertal development, conception and embryonic mortality. In particular, elevated temperature delays the onset of puberty, lowers conception rate and increases embryo loss. Some of the effects are through direct action upon reproductive tissues, namely the testis and uterus. On the other hand, temperature may exert its action through hormones, since the length of oestrous periods, sexual behaviour, progesterone and LH concentrations are sometimes abnormal in animals showing 'summer infertility'.

That photoperiod has profound effects on reproduction is well known. Some of the responses to photoperiodic change, particularly those related to sexual behaviour, are complex and merit more investigation.

The social situation in which animals are kept is also important. There is much evidence that rearing animals of one sex in the absence of the other can affect subsequent reproductive activity. Puberty is delayed in single-sex groups, oestrous cycles of females may be disrupted and subsequent sexual behaviour of males is not normal. It should be stressed that most of the data described in this chapter were not collected with animal housing specifically in mind. How much reproductive potential is lost through inefficient housing is an unknown quantity, but on the basis of circumstantial evidence it could be considerable. The situation obviously merits more detailed study.

References

ABILAY, T. A., JOHNSON, H. D. and MADAN, M. (1975). *J. Dairy Sci.* **58**, 1836–1840

ALLISTON, C. W., HOWARTH, B., Jr and ULBERG, L. C. (1965). *J. Reprod. Fert.* **9**, 337–341

BAKER, J. R. (1938). In *Evolution: Essays on Aspects of Evolutionary Biology*, pp. 79–104. Ed. by G. R. de Beer. Clarendon Press, Oxford

BARNETT, S. A. (1962). *J. Reprod. Fert.* **4**, 327–335

BARTKE, A., HAFIEZ, A. A., BEX, F. J. and DALTERIO, S. (1978). *Biol. Reprod.* **18**, 44–54

BEACH, F. A. (1968). *Behaviour* **30**, 218–238

BLACKSHAW, A. W. (1977). In *The Testis*, vol. 4, pp. 517–545. Ed. by A. D. Johnson and W. R. Gomes. Academic Press, New York

BOURNE, R. A. and TUCKER, H. A. (1975). *Endocrinology* **97**, 473–475

BRONSON, F. H. (1968). In *Reproduction and Sexual Behaviour*, pp. 341–361. Ed. by M. Diamond. Indiana University Press, Bloomington

BRONSON, F. H. and DESJARDINS, C. (1974). *Endocrinology* **94**, 1658–1668

BRONSON, F. H. and STETSON, M. H. (1973). *Biol. Reprod.* **9**, 449–459

BROOKS, P. H. and COLE, D. J. A. (1970). *J. Reprod. Fert.* **23**, 435–440

BUTT, W. R. (1978). In *Control of Ovulation*, pp. 357–371. Ed. by D. B. Crighton, N. B. Haynes, G. R. Foxcroft and G. E. Lamming. Butterworths, London

CHRISTIAN, J. J. (1975). In *Hormonal Correlates of Behaviour*, pp. 205–274. Ed. by B. E. Eleftheriou and R. L. Sprott. Plenum, New York

CHRISTIAN, J. J., LLOYD, J. A. and DAVIS, D. E. (1965). *Recent Prog. Horm. Res.* **21**, 507–578

COLBY, D. R. and VANDENBERGH, J. G. (1974). *Biol. Reprod.* **11**, 268–279

COOPER, K. J. and HAYNES, N. B. (1967). *J. Reprod. Fert.* **14**, 317–320

COOPER, K. J., PURVIS, K. and HAYNES, N. B. (1972). *J. Reprod. Fert.* **28**, 473–475

CORAH, L. R., QUEALY, A. P., DUNN, T. G. and KALTENBACH, C. C. (1974). *J. Anim. Sci.* **39**, 380–385

DALE, H. E., RAGSDALE, A. C. and CHENG, S. C. (1959). *J. Anim. Sci.* **18**, 1363–1366

DRORI, D. and FOLMAN, Y. (1964). *J. Reprod. Fert.* **8**, 351–359

DUNLAP, S. E. and VINCENT, C. K. (1971). *J. Anim. Sci.* **32**, 1216–1218

DUTT, R. H. and BUSH, L. F. (1955). *J. Anim. Sci.* **14**, 885–896

EDWARDS, M. J. (1978). *Adv. vet. Sci. comp. Med.* **22**, 1–28

EDWARDS, R. L., ORMTVEDT, E. J., TURMAN, E. J., STEPHENS, D. F. and MAHONEY, G. W. A. (1968). *J. Anim. Sci.* **27**, 1634–1637

FOLLETT, B. K. (1978). In *Control of Ovulation*, pp. 267–293. Ed. by D. B. Crighton, N. B. Haynes, G. R. Foxcroft and G. E. Lamming. Butterworths, London

FOLMAN, Y., ROSENBERG, M., HERZ, Z. and DAVIDSON, M. (1973). *J. Reprod. Fert.* **34**, 267–278

FORBES, J. M., DRIVER, P. M., ELSHAHAT, A. A., BOAZ, T. G. and SCANES, C. G. (1975). *J. Endocr.* **64**, 549–554

FOXCROFT, G. R. (1978). In *Control of Ovulation*, pp. 117–138. Ed. by D. B. Crighton, N. B. Haynes, G. R. Foxcroft and G. E. Lamming. Butterworths, London

GERALL, H. D. (1965). *J. Personality soc. Psychol.* **2**, 460–464

GERALL, H. D., WARD, I. L. and GERALL, A. A. (1967). *Anim. Behav.* **15**, 54–58

GODLEY, W. C., WILSON, R. L. and HURST, V. (1966). *J. Anim. Sci.* **25**, 212–216

GOSS, R. J. and ROSEN, J. K. (1973). *J. Reprod. Fert.* **19**, 111–118

GWAZDAUSKAS, F. C., THATCHER, W. W. and WILCOX, C. J. (1973). *J. Dairy Sci.* **56**, 873–877

GWINNER, E. (1975). In *Avian Biology*, pp. 221–285. Ed. by D. S. Farner and J. R. King. Academic Press, New York

HARTMANN, G., ENDRÖCZI, E. and LISSÁK, K. (1966). *Acta physiol. hung.* **30**, 53–59

HEIMAN, M. M. (1972). *Proc. 7th Int. Congr. on Animal Reproduction and Artificial Insemination, Munich*, Vol 3, pp. 2007–2010

HEMSWORTH, P. H., BEILHARZ, R. G. and GALLOWAY, D. B. (1977). *Anim. Prod.* **24**, 245–252

HEMSWORTH, P. H., WINFIELD, C. G., BEILHARZ, R. G. and GALLOWAY, D. B. (1977). *Anim. Prod.* **25**, 305–310

HOOLEY, R. D., FINDLEY, J. K. and STEVENSON, R. G. (1979). *Aust. J. biol. Sci.* **32**, 231–235

HOWLES, C. M., WEBSTER, G. M. and HAYNES, N. B. (1980). *J. Reprod. Fert.* **60**, 434–437

ILLIUS, A. W. (1976). *PhD Thesis,* University of Nottingham

ILLIUS, A. W., HAYNES, N. B. and LAMMING, G. E. (1976). *J. Reprod. Fert.* **48**, 25–32

ILLIUS, A. W., HAYNES, N. B., PURVIS, K. and LAMMING, G. E. (1976). *J. Reprod. Fert.* **48**, 17–24

INGRAHAM, R. H., GILLETTE, D. D. and WAGNER, W. D. (1974). *J. Dairy Sci.* **57**, 476–481

KANN, G., MARTINET, J. and SCHIRAR, A. (1978). In *Control of Ovulation*, pp. 319–335. Ed. by D. B. Crighton, N. B. Haynes, G. R. Foxcroft and G. E. Lamming. Butterworths, London

KARSCH, F. J., LEGAN, S. J., RYAN, K. D., and FOSTER, D. L. (1978). In *Control of Ovulation*, pp. 29–48. Ed. by D. B. Crighton, N. B. Haynes, G. R. Foxcroft and G. E. Lamming. Butterworths, London

KNECHT, E. A., WRIGHT, G. L. and TORAASON, M. A. (1978). *Can. J. Physiol. Pharmac.* **56**, 747–753

KREIDER, D. L. and WETTEMANN, R. P. (1975). *Res. Rep. agric. exp. Sta., Oklahoma State Univ.*, pp. 191–195

LEGAN, S. J. and KARSCH, F. J. (1979). *Biol. Reprod.* **20**, 74–85

LEINING, K. B., BOURNE, R. A. and TUCKER, H. A. (1979). *Endocrinology* **104**, 289–294

LINCOLN, G. A. (1976a). *J. Endocr.* **69**, 213–226

LINCOLN, G. A. (1976b). *J. Reprod. Fert.* **47**, 351–353

LINCOLN, G. A. (1979). *J. Endocr.* **83**, 251–260

LINCOLN, G. A. and DAVIDSON, W. (1977). *J. Reprod. Fert.* **49**, 267–276

LINCOLN, G. A., GUINNESS, F. and SHORT, R. V. (1972). *Horm. Behav.* **3**, 375–396

LINCOLN, G. A., McNEILLY, A. S. and CAMERON, C. L. (1978). *J. Reprod. Fert.* **52**, 305–311

LINCOLN, G. A. and PEET, M. J. (1977). *J. Endocr.* **74**, 355–367

LOVE, R. J. (1978). *Vet. Rec.* **103**, 443–446

MADAN, M. L. and JOHNSON, H. D. (1973). *J. Dairy Sci.* **56**, 1420–1423

MICHAEL, R. P. and BONSALL, R. W. (1977). *J. Reprod. Fert.* **49**, 129–131

MILLER, H. L. and ALLISTON, C. W. (1974). *Biol. Reprod.* **11**, 187–190

MILLS, A. G., THATCHER, W. W., DUNLAP, S. E. and VINCENT, C. K. (1972). *J. Dairy Sci.* **55**, 400–401

NAZION, S. J. and PIACSEK, B. E. (1976). *J. exp. Zool.* **198**, 13–16

NAZION, S. J. and PIACSEK, B. E. (1977). *Biol. Reprod.* **17**, 668–675

NAZION, S. J. and PIACSEK, B. E. (1978). *Biol. Reprod.* **19**, 256–260

PELLETIER, J. (1973). *J. Reprod. Fert.* **35**, 143–147

PELLETIER, J. and ORTAVANT, R. (1975a). *Acta endocr., Copenh.* **78**, 435–441

PELLETIER, J. and ORTAVANT, R. (1975b). *Acta endocr., Copenh.* **78**, 442–450

PENGELLEY, E. T. and ASMUNDSON, S. J. (1974). In *Circannual Clocks*, pp. 95–160. Ed. by E. T. Pengelley. Academic Press, New York

PLASSE, D., WARNICK, A. C. and KOGER, J. (1970). *J. Anim. Sci.* **30**, 63–72

PURVIS, K. and HAYNES, N. B. (1972). *Physiol. Behav.* **9**, 401–407

RIGGS, B. L., ALLISTON, C. W. and WILSON, S. P. (1974). *J. Anim. Sci.* **39**, 159–160

ROSENBERG, M., HERZ, Z., DAVIDSON, M. and FOLMAN, Y. (1977). *J. Reprod. Fert.* **51**, 363–367

SCHINCKEL, P. G. (1954a). *Aust. J. agric. Res.* **5**, 465–469

SCHINCKEL, P. G. (1954b). *Aust. vet. J.* **30**, 189–195

SMITH, V. G., HACKER, R. R. and BROWN, R. G. (1977). *J. Anim. Sci.* **44**, 645–649

STEINACH, E. (1936). *Wien. klin. Wschr.* **49**, 164–172

STIFF, M. E., BRONSON, F. H. and STETSON, M. H. (1974). *Endocrinology* **94**, 492–496

STOTT, G. H. and WIERSMA, F. (1973). *Int. J. Biomet.* **17**, 115–122

STOTT, G. H. and WILLIAMS, R. J. (1962). *J. Dairy Sci.* **45**, 1369–1375

TEAGUE, H. S., ROLLER, W. L. and GRIFO, A. P. (1968). *J. Anim. Sci.* **27**, 408–411

THATCHER, W. W., GWAZDAUSKAS, F. C., WILCOX, C. J., TOMS, J., HEAD, H. H., BUFFINGTON, D. E. and FREDRIKSSON, W. B. (1974). *J. Dairy Sci.* **57**, 304–307

THOMAS, T. R. and NEIMAN, C. N. (1968). *Endocrinology* **83**, 633–635

THOMPSON, L. H. and SAVAGE, J. S. (1978). *J. Anim. Sci.* **47**, 1141–1144

THORNER, M. O. and BESSER, G. M. (1977). In *Prolactin and Human Reproduction*, pp. 285–302. Ed. by P. G. Crosignani and C. Robyn. Academic Press, London

THWAITES, C. J. (1968). *Int. J. Biomet.* **12**, 29–34

TUREK, F. W. and CAMPBELL, C. S. (1979). *Biol. Reprod.* **20**, 32–50

VAN DEMARK, N. L. and FREE, M. J. (1970). In *The Testis*, vol. 3, pp. 233–312. Ed. by A. D. Johnson, W. R. Gomes and N. L. Van Demark. Academic Press, New York

VANDENBERGH, J. G. (1975). In *Hormonal Correlates of Behaviour*, pp. 551–584. Ed. by B. E. Eleftheriou and R. L. Sprott. Plenum, New York

VANDENBERGH, J. G., DRICKAMER, L. C. and COLBY, D. R. (1972). *J. Reprod. Fert.* **28**, 397–405

VAN DER LEE, S. and BOOT, L. M. (1956). *Acta physiol. pharmac. néerl.* **5**, 213–214

VAUGHT, L. W., MONTY, D. E., Jr and FOOTE, W. C. (1977). *Am. J. vet. Res.* **38**, 1027–1030

WALTON, J. S., McNEILLY, J. R., McNEILLY, A. S. and CUNNINGHAM, F. J. (1977). *J. Endocr.* **75**, 127–136

WETTEMANN, R. P. and TUCKER, H. A. (1974). *Proc. Soc. exp. Biol. Med.* **146**, 908–911

WETTEMANN, R. P., WELLS, M. E., ORMTVEDT, E. J., POPE, C. E. and TURMAN, E. J. (1976). *J. Anim. Sci.* **42**, 664–669

WHITTEN, W. K. (1956). *J. Endocr.* **18**, 102–107

WOLFF, L. K. and MONTY, D. E., Jr (1974). *Am. J. vet. Res.* **35**, 187–192

YOUNG, W. C., GOY, R. W. and PHOENIX, C. H. (1964). *Science, NY* **143**, 212–218

5

THE INFLUENCE OF PHOTOPERIOD ON REPRODUCTION IN FARM ANIMALS

T.R. MORRIS
Department of Agriculture and Horticulture, University of Reading

Introduction

The majority of plants and animals that are adapted to temperate latitudes show a seasonal variation in reproductive activity which is timed to ensure that progeny are produced when there is maximum likelihood of their survival. Amongst the homeothermic animals, photoperiod is almost always the primary external regulator controlling seasonal breeding. For the animal in its natural environment, day length measurement provides a calendar, and natural selection has arranged matters so that the breeding cycle begins at an appropriate date to ensure that the young are born in favourable conditions. Photoperiod is not, of course, the only environmental factor influencing reproduction: as discussed by Haynes and Howles in the previous chapter, criteria such as temperature, food supply or courtship may have important effects on the timing and success of reproduction; but seasonal gonadal development, which precedes other preparations for breeding, is usually controlled by changes in photoperiod.

Species affected

The farm animals whose natural breeding season is restricted to a part of the year are the sheep, the goat, the turkey and the goose. The duck and the fowl were seasonal breeders sixty years ago but the current domesticated varieties of these species will breed all the year round even when maintained in natural daylight at high latitudes. The fact that twenty or thirty generations of selective breeding can change a population from seasonal breeding to year-round breeding implies that there is substantial genetic variation in sensitivity to photoperiod. The same conclusion may be drawn from evidence that breeds of sheep which have developed at different latitudes in comparatively recent times show breeding seasons that are clearly adapted to their environmental origin (*Figure 5.1*).

Modern fowls (and modern ducks) show the remnants of a seasonal breeding mechanism. When they are reared in natural daylight, the age at which they reach sexual maturity is influenced by hatch date (*Figure 5.2*) and their adult reproductive rate is affected by changes in photoperiod (Morris, Fox and Jennings, 1964). Seasonal responses are also observed in the horse (Ginther, 1974, 1979), the cow (Hawk, Tyler and Casida, 1954; Menge *et al.*,

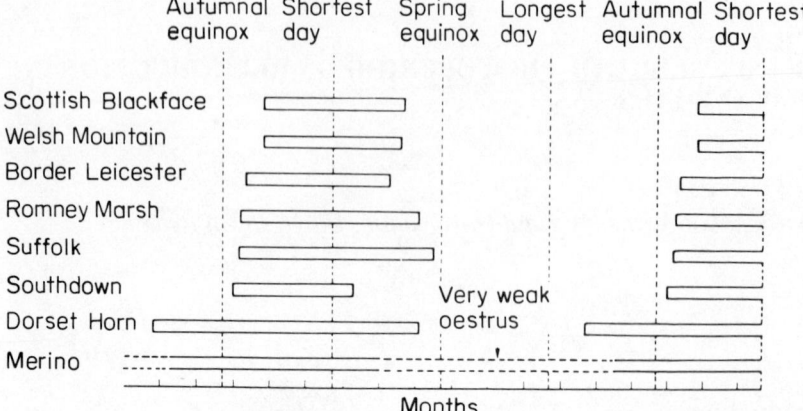

Figure 5.1 *Duration of the breeding season in eight breeds of sheep. (From Yeates, 1965)*

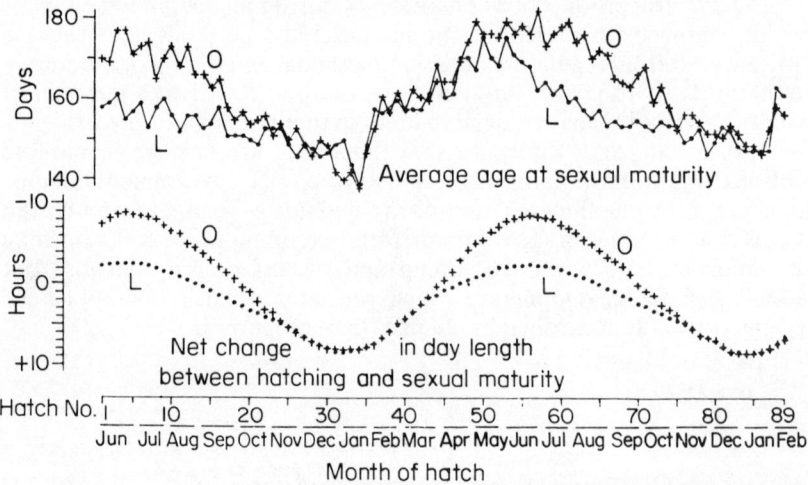

Figure 5.2 *The effect of season of hatch on mean age at first egg for pullets raised in natural daylight at 51 °N (treatment 0), or given natural daylight to 16 weeks of age and a constant 14 h day thereafter (treatment L). (From Morris and Fox, 1958)*

1960; Roy, Gillies and Shotton, 1975) and the rabbit (Sittmann *et al.*, 1964). Age at puberty and adult fertility in these species show seasonal effects which are certainly (in the case of the horse) or probably (in the case of the heifer) due to changes in photoperiod. Evidence of seasonal effects in a species does not of itself prove the influence of photoperiod: control by temperature or seasonal feed supply must also be considered. However, in cases that have been thoroughly explored it turns out that, given adequate nutrition, responses to artificially manipulated photoperiods are sufficient to explain almost all the regular seasonal breeding patterns observed in temperate-zone homeotherms.

The evidence of seasonal differences in fertility in the pig is slight and there is little response to using artificially long or short photoperiods

(Dufour and Bernard, 1968; Waddill, Chaney and Dutt, 1968; Hacker, King and Bearss, 1974).

Photoperiodic responses can be demonstrated in both males and females of the seasonally breeding species, although the duration of the breeding season is not necessarily the same in both sexes. For example, the ram may be potentially fertile at times when ewes of the same breed are anoestrus, but circulating gonadotrophin levels and testis size increase as the breeding season for females approaches (Ortavant, 1977). In laboratory birds, males have been used much more than females in photoperiodic experiments, chiefly because testis size is a very sensitive measure of response. In domestic poultry, most of the evidence relates to the egg-laying females. Cockerels do respond to photoperiodic treatment (Parker and McSpadden, 1943; Parker and McCluskey, 1964, 1965; Parker, 1972) but, as with the ram, probably remain fertile to some extent in conditions which cause a cessation of ovulation in the female. The turkey cock, on the other hand, is very sensitive to short photoperiods and it is normal practice to increase day length for turkey males two or three weeks before lighting the females, so as to obtain good semen yields for artificial insemination.

General properties of photoperiodism

The characteristic feature of a photoperiodic response, whether it be reproduction or coat shedding in a farm animal, diapause in an insect or flower initiation in a plant, is that the response is related to the *period* of illumination and not to the *amount* of light energy received. For photoperiodic effects (as distinct, for example, from photosynthesis) there is no scope for offsetting short periods of illumination by increasing light intensity. This is the essential feature which makes the mechanism reliable in determining the date.

When considering light intensity, there appears to be a threshold value of about 10 lx, which is similar (or perhaps identical) for all vertebrate species: above this value the animal reckons it is daylight and below this value it reckons that night has fallen. The detailed evidence for the chicken, which is considered below, indicates that the threshold may not be sharply defined.

Although the essential feature of a photoperiodic response is that it depends upon measurement of day length, the responses observed in birds and mammals (as opposed to plants) are rarely a simple function of the photoperiod to which the animal is currently exposed. Previous photoperiod is an important element affecting the response, and this is logically indistinguishable from a statement that the animal responds to changes in photoperiod. Much confusion has arisen over this issue since zoologists first demonstrated in the 1930s that *gradual* changes in photoperiod, imitating natural days, were not required to produce gonadal responses. It was found to be sufficient to put a bird or a ferret in a constant long day to induce gonadal growth or to expose it to short days to cause gonadal regression (Burger, 1939a, b; 1940; Hammond, 1952). It was not then fashionable to say that each of these procedures involved *changing* the photoperiod from some 'pre-experimental' value to a new value (Hammond, 1954). However, evidence was available in the 1950s that a 12 h photoperiod, which acts as a 'long day' for a migratory bird (*Junco hyemalis*) transferred from winter day lengths (Wolfson and Winn, 1948), can also act as a 'short day' for a bird transferred from longer photoperiods (Wolfson, 1958). Such experiments show that change in photoperiod is important in the migratory species that have been commonly used in laboratory studies, but the significance of day length change has mostly been ignored or denied in the literature on photoperiodism. The principle has, however, recently been reintroduced in connection with the Japanese quail (Robinson and Follett, 1979).

In the case of farm animals, it was argued by Yeates (1949) that the sheep responds to a specified change in photoperiod rather than to a threshold value, and the same thesis was advanced for the chicken by Morris and Fox (1958). Satisfactory evidence to discriminate between the alternative hypotheses had to await the availability of multiple light-proof chambers, in which constant photoperiods and changing photoperiods could be applied to groups of animals that were of comparable age and physiological status at the start of the experiment. Such chambers were built at Reading in 1960 for chickens and in 1965 for sheep and soon yielded unequivocal evidence that for these two species, at least, change in photoperiod is more important than the absolute length of the photoperiod (Morris, 1962; Morris, Fox and

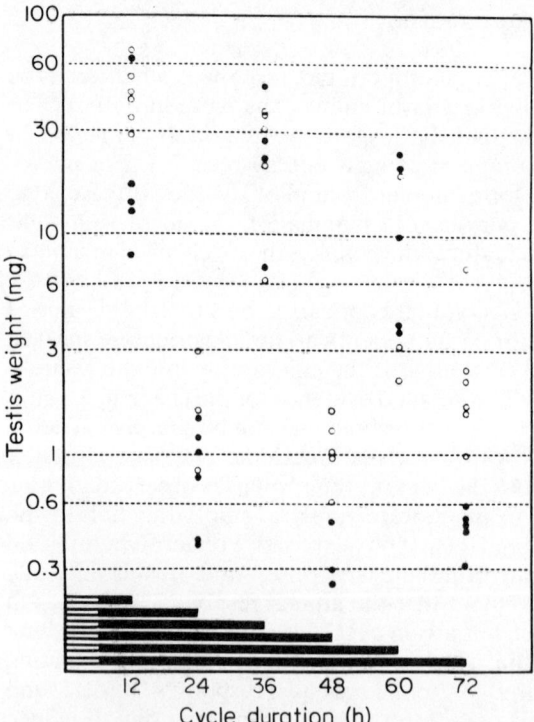

Figure 5.3 The effect on testis weight of exposing house finches (Carpodacus mexicanus) to repeated light–dark cycles of 12, 24, 36, 48, 60 or 72 hours' duration with 6 hours of light in each cycle. When the short light period recurs at 24, 48 or 72 hour intervals it is non-stimulatory but when it recurs at intervals of 12, 36 or 60 hours there is a marked response. (●) Experiment 1, in which treatments were applied for 33 days; (○) experiment 2, in which treatments were repeated for 22 cycles. (From Hamner, 1963)

Jennings, 1964 for the fowl: Ducker, Thwaites and Bowman, 1970a, b; Ducker and Bowman, 1970 for the sheep).

There is now no reason to doubt that, in all vertebrates studied, the photoperiodic mechanism is complex and cannot be satisfactorily modelled by supposing that a given physiological response, such as a rate of hormone release, is a simple function either of current photoperiod or of change in photoperiod. The physiological status of the animal (time since onset or termination of last breeding season, pregnancy, lactation, broodiness, etc.) and the whole pattern of photoperiods previously experienced are relevant matters affecting the response.

Another issue which was at one time hotly debated was whether photoperiodic responses depend upon day length, night length or the ratio of light to dark. It is now clear from 'resonance' experiments (*Figure 5.3*) and from many other trials using artificial patterns of light and dark (*see* Saunders, 1977; Follett, 1978) that neither day length nor night length nor light : dark ratio is critical. To explain results such as those in *Figure 5.3* one is obliged to postulate an endogenous circadian rhythm and to say that a response is obtained when light coincides with a particular portion of the endogenous cycle, called the photo-inducible phase (*Figure 5.4*). The location of this photo-inducible phase is itself determined by light (because the biological clock is entrained by environmental signals, especially light) and so it is not possible to induce the photoperiodic response by exposing the animal to a

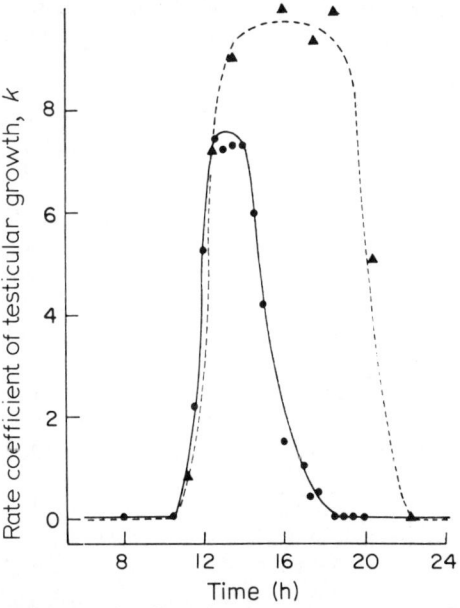

*Figure 5.4 The effect of exposing Japanese quail (*Coturnix coturnix japonica*) to repeated light schedules with a base period of 6 hours of light (from hours 0 to 6 on the time scale shown) and an additional pulse of light of 15 minutes' duration (●) or 2 hours' duration (▲) given at the time indicated on the abscissa. The photo-inducible phase appears to be longer when using the 2 h light pulse but this may be because the bird shifts its subjective dawn and, for example, reads 6L : 10D : 2L : 6D as 2L : 6D : 6L : 10D. k is the fitted rate coefficient of testicular growth in an equation of the form $y = at^k$, where y is testis weight, a is a constant and t is the trial duration in days. (From Follett and Sharp, 1969)*

succession of short photoperiods at 24 h intervals, timed to coincide with the photo-inducible phase.

Photo-induction can be achieved by quite short exposure to light and this leads to the notion of 'skeleton' photoperiods, in which a short photoperiod (e.g. 6 h or 8 h light) is followed by one or more pulses of light at the appropriate time. This phenomenon is well known in plants and has been repeatedly demonstrated in birds but has not, so far, been exploited in animal production systems. The application of intermittent lighting in poultry production will be considered below.

The pathways by which light acts to cause the overt reproductive response have not been completely identified. It was postulated many years ago that light affects the output of gonadotrophins from the anterior pituitary gland and this has been amply confirmed in recent years as sensitive assays for circulating hormones have become available (*see* Follett, 1978). The problem is to explain how light falling on the head of an animal induces changes in the secretion of gonadotrophin-releasing factor(s) from the hypothalamus. It has been shown in laboratory experiments with birds that neither the eyes nor the pineal gland are necessary to obtain a response (Benoit and Ott, 1944; Harrison, 1972, 1974; Oishi and Lauber, 1973). Testicular responses in the Japanese quail can even be induced by implanting beads of luminous paint directly into specific areas of the hypothalamus (Oliver and Baylé, 1976). However, these experiments do not show that the retina is not involved in the primary reception and transmission of the light signal in the case of intact animals maintained in low-intensity artificial lighting.

Effects of constant photoperiods

It is useful to know how animals respond when maintained for long periods on constant photoperiods, but there are few experiments that provide satisfactory evidence. Some species, notably the migratory birds that have been studied in the laboratory and also the turkey (probably the goose and the guinea fowl too, but evidence is lacking) show no gonadal development if kept permanently on short days. These same species breed when transferred to long days, but if such a bird is maintained on constant long days the gonads eventually regress (a phenomenon which has been called 'refractoriness to light') and remain quiescent until a period of short days is provided to 'prepare' the animal for the next breeding cycle.

The sheep, the chicken and probably the domestic duck do not respond in this way. These species will reach sexual maturity whether reared on constant short days or on constant long days (*Figure 5.5* and *5.6*). They will then have a breeding cycle (not necessarily of normal length), pass through the characteristic non-breeding phase and begin a new cycle in due course. For the modern domestic fowl, the time of onset of moult and start of the second period of egg-laying show wide individual variation, with the result that a flock held for two or more years under constant lighting conditions reaches a low, constant average rate of lay because the number of birds returning to production is in equilibrium with the number going out of lay (*Figure 5.7*).

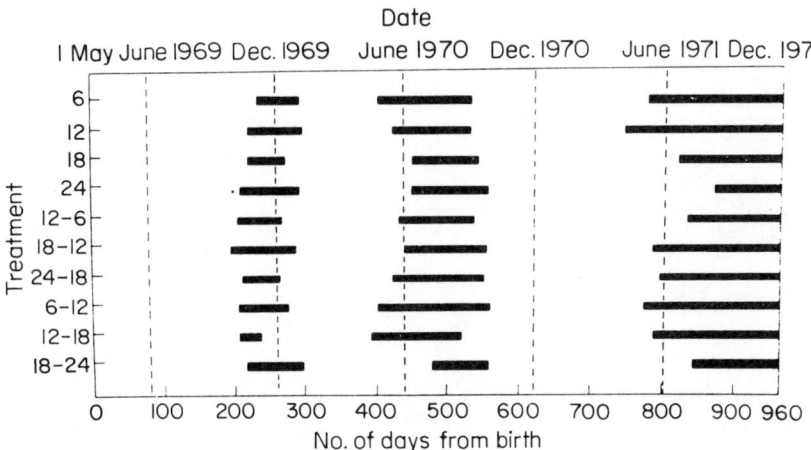

Figure 5.5 Mean duration of oestrus for ewes born in April and May 1969 and maintained throughout their lives on constant photoperiods of 6, 12, 18 or 24 hours (first four treatments) or transferred in September at approximately 5 months of age from one constant photoperiod to another. (From Ducker, Bowman and Temple, 1973)

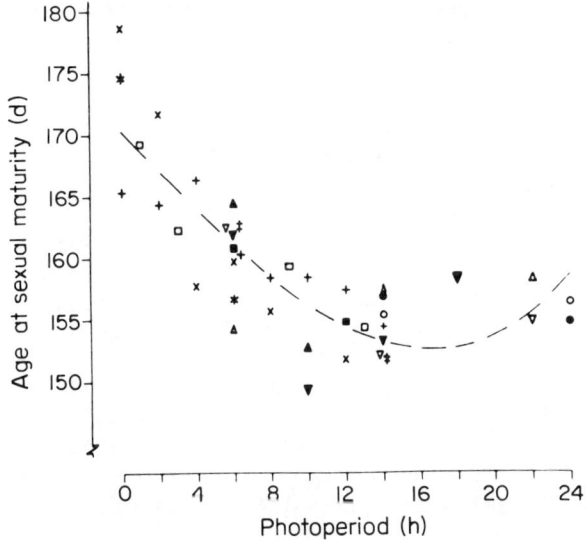

Figure 5.6 The effect of photoperiod on age at sexual maturity when photoperiod is maintained at a constant level from hatching onwards. Different symbols represent different sources. (From Morris, 1968)

The data for ewes (*Figure 5.5*) have been interpreted as showing evidence of an endogenous circennial rhythm controlling the onset of successive breeding seasons (Ducker, Bowman and Temple, 1973) but it is possible that the return to breeding is influenced by seasonal environmental factors which were not excluded in this experiment. The critical test would involve maintaining two sets of animals, born six months apart, under constant photoperiods for three or four years. If, in that case, breeding seasons

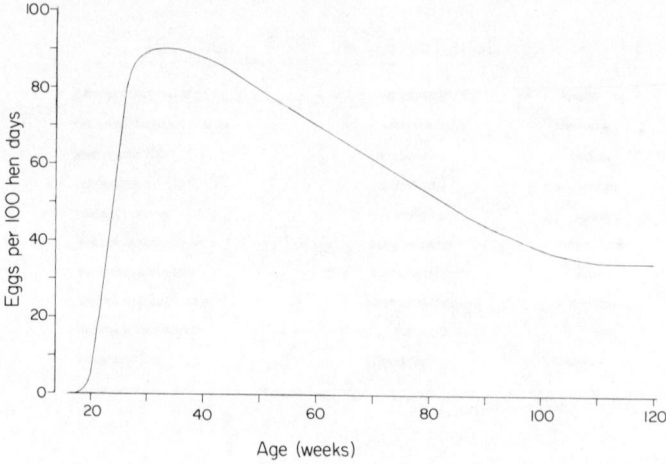

Figure 5.7 Egg production of hens maintained under a constant photoperiod (14L : 10D) for two years. (Based on data of Hansen, 1966)

recurred regularly at approximately annual intervals, but six months out of phase in the two groups, the hypothesis of external control would be eliminated, leaving only the alternative hypothesis of an endogenous circennial rhythm.

Manipulation of breeding seasons by changing photoperiods

The simplest form of using artificial light to adjust a breeding season is to supplement winter day length to advance the onset of breeding in species such as the turkey and the goose that are normally reproductively inactive under short-day conditions. This practice does not require light-proof housing and the responses are prompt and dramatic (Asmundson and Moses, 1950; Kinney, Burger and Shoffner, 1959). To advance the breeding season in a sheep or a goat requires a reduction in photoperiod, and this can only be achieved if animals are housed in light-proof accommodation for at least part of the day (Hart, 1950) or are pretreated for several months with artificial light, maintaining a long photoperiod so that switching off the lights provides an appropriate reduction in photoperiod (Ducker and Bowman, 1972).

If animals are housed in light-proof buildings, it is easy to control the breeding season, so that the animal breeds annually, but out of phase with the normal season (*Figure 5.8*), or breeds at intervals of less than twelve months (Robinson, 1974; Vesely, 1975, 1978; *Figure 5.9*). More difficult, but of greater potential advantage in the case of sheep, is the use of artificial light to supplement natural daylight in a pattern designed to induce, for example, three breeding seasons in two years (*Figure 5.10*). Since sheep do not benefit from being housed in an elaborate building allowing control of temperature, it may be of advantage to house them in cheap structures which admit fresh air and daylight but allow photoperiod to be manipulated to some extent.

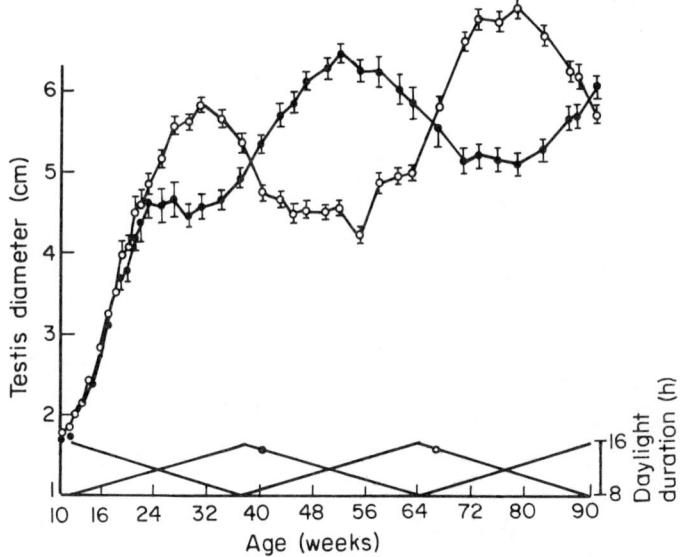

Figure 5.8 Variations in mean testis diameter for two groups of rams given light schedules six months out of phase. (From Alberio, 1976)

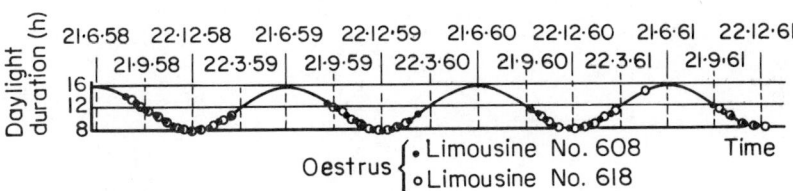

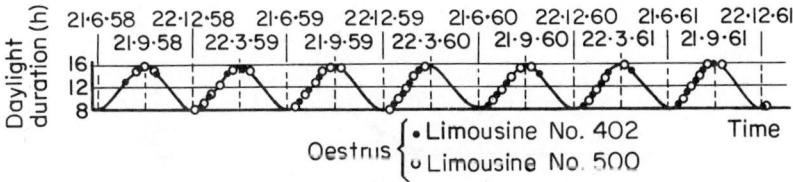

Figure 5.9 Records of oestrus in two sheep maintained on natural daylight (top) and two sheep given increasing and decreasing light at twice the natural rate to form a series of six-month cycles (bottom). Notice that, because the ewe has a long latency period between onset of decreasing days and onset of breeding, oestrus actually occurs during the phase of increasing light in the second case. (From Mauleon and Rougeot, 1962)

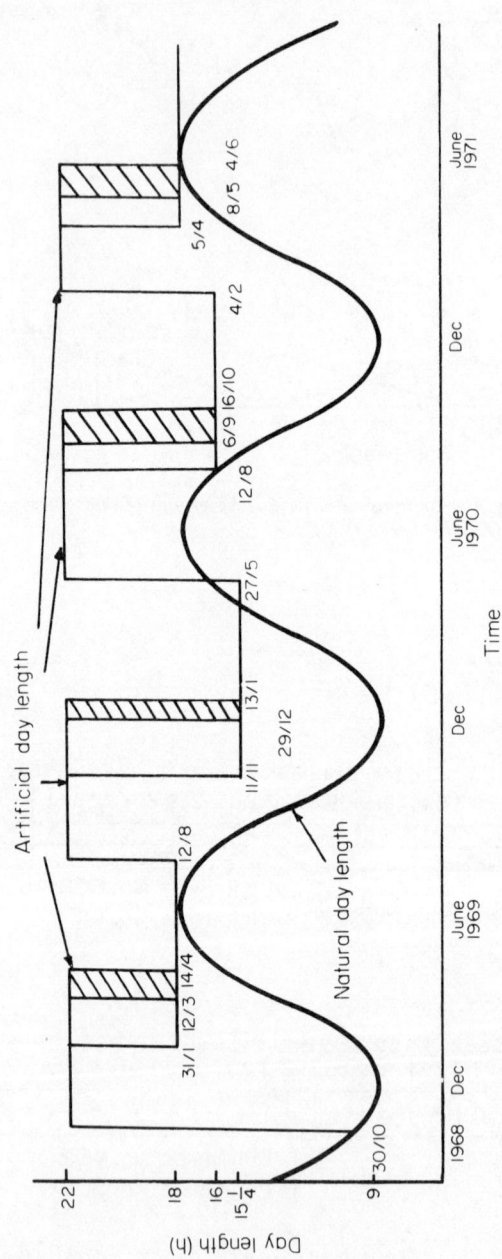

Figure 5.10 A lighting schedule designed to induce three lambings in two years without requiring light-proof accommodation. Ewes were given abrupt decreases in day length either on a fixed date before lambing or at lambing (the shaded areas in the figure). (From Ducker and Bowman, 1972)

Manipulation of reproductive rate by changing photoperiods

In poultry, artificial light has long been used as a means of increasing rate of lay. There is now some evidence that milk yield in dairy cows can also be increased (Peters *et al.*, 1978), perhaps owing to an effect on prolactin secretion (Peters, Chapin and Tucker, 1979).

It is interesting that the early statements about supplementing winter day length for chickens (Curtis, 1920) describe the response as an increase in rate of lay, although it seems likely that the dramatic changes in mean rate of lay reported were the result of bringing non-laying members of the flock back into production. The effect was primarily on duration of the breeding season and not on rate of lay of those birds already in production. Experiments with more modern fowls, which are capable of sustained egg production under constant short-day conditions, show that rate of lay of those birds which are already in production can be enhanced by increasing photoperiod (Morris, Fox and Jennings, 1964). However, the effects of photoperiod on annual egg production in modern fowls are smaller than is sometimes supposed. For example, the data in *Table 5.1* show that pullets maintained throughout their

Table 5.1 EGG NUMBERS PER BIRD HOUSED FOR PULLETS MAINTAINED ON CONSTANT PHOTOPERIODS FROM 0 TO 500 DAYS OF AGE, OR GIVEN A 6 h PHOTOPERIOD FROM 0 TO 126 DAYS WITH INCREMENTS THEREAFTER

	Eggs per pullet housed, to 500 d of age
1. 18L : 6D (0–500 d)	235 ⎫
2. 14L : 10D (0–500 d)	234 ⎬ mean 236 ± 2.8
3. 10L : 14D (0–500 d)	239 ⎭
4. 6L : 18D (0–500 d)	211 ± 5.3
5. 6L : 18D (0–18 weeks; photoperiod increased by 20 min/week from 18 weeks to reach 17L : 7D at 50 weeks)	245 ± 1.9

From data given in Morris (1979). Data are taken from two experiments conducted at Reading University. Treatments 1, 2, 3 and 4 were compared in the first trial and treatments 1, 2, 3 and 5 in the second. The means given are adjusted for differences between trials. The standard errors are approximate. In each trial, four treatments were applied to three different stocks, but there were no significant stock × treatment interactions. A total of 3456 pullets was used in each trial. L, light; D, dark

lives on a photoperiod of 6 h gave an egg yield only 10% less than that of birds maintained on 14 h photoperiods throughout, and an 'optimal' lighting programme exploiting the maximum useful increase in photoperiod during the laying year improved yield by only 4%, compared with constant photoperiods of 10 h, 14 h or 18 h. Of course, a 4% increase in gross output could represent a doubling of net income in a competitive industry that has already taken the step of housing its laying hens in light-proof buildings. There is an additional feed cost associated with the extra yield, but the cost of extra feed in this context is generally less than half the revenue from the extra production. Given the windowless house and provision of artificial lighting, the running cost for lights, using an optimal programme of short-day rearing followed by increasing light in the laying stage, is less than the running cost for a constant 14 h photoperiod throughout.

Another question that may be asked is whether the improved egg production to be expected from use of an optimal light pattern in a light-proof house is sufficient to justify the cost of the housing involved. The answer is that it all depends what alternatives are being considered. In a temperate climate with winter temperatures below the optimum for efficient egg production and with little or no summer heat stress there are large benefits to be obtained from well-insulated buildings. In such buildings windows are a disadvantage, and so are best substituted by artificial light. Having justified a well-insulated, windowless building there is no difficulty in showing that the extra cost of making that building light-proof will be readily recouped by the small increase in productivity attributable to having total control of photoperiod. In a warmer climate, where the main problem is to deal with summer heat stress, the advantage probably lies with open-sided housing, and in this case one accepts the small loss in productivity associated with the use of constant long photoperiods (equal to midsummer day length at the latitude in question) as the only practicable lighting programme.

Lighting intensity requirements in poultry houses

Three questions arise in designing lighting systems for light-proof poultry houses:

(1) What is the minimum intensity of artificial lighting required?
(2) What is the effect of using lamps giving different spectral emission curves?
(3) What is the maximum intensity of infiltering daylight that can be accepted?

Logically, the first two questions are linked, but there is no satisfactory evidence showing how colour and intensity interact. Perhaps with monochromatic light sources there would be no interaction: there might be no response to illumination outside a narrow waveband. But with commercial lamps there is a wide range of spectral emission so that 'blue' lights, for example, give some radiation in the red part of the spectrum and if it is found that blue lighting requires a greater intensity than red lighting to achieve a given response (McGinnis *et al.*, 1966), this may indicate that blue light is relatively less effective than red or it might mean that blue light is totally ineffective, the responses observed being due to the red light output of the 'blue' lamp. In principle, the matter can be resolved by the use of appropriate filters, but it is difficult in practice to achieve sufficient monochromatic illumination over a wide enough area to test the productive responses of replicated groups of chickens. The best available evidence relates to testicular growth in the drake, where orange and red light (600–750 nm) were found to be effective and blue, green and yellow light were not (Benoit, Walter and Assenmacher, 1950).

No advantage has so far been demonstrated from using one particular type of commercial lamp rather than another, but this is an area where more evidence is needed. In particular, there is need for a comparison of tungsten filament and fluorescent lamps each providing a range of intensities, so that the relationship between intensity (and power input) and productive response can be measured. Fluorescent lamps are being used successfully in practice, but there is no evidence to show whether the full advantage of their higher light output/power input ratio can be exploited or whether some of this advantage is lost because of the lower red and orange component of the fluorescent lamp output.

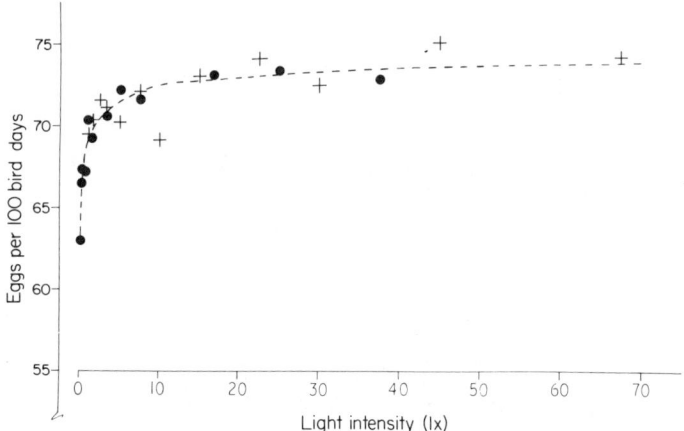

Figure 5.11 Data from two unpublished experiments (Reading University) in which laying pullets were exposed from 20 to 72 weeks of age to incandescent lamps giving light intensities, measured at the cage front, ranging from 0.13 lx to 37.5 lx in Expt. 1 (●) and from 1.1 lx to 67.5 lx in Expt. 2 (+). The means shown are each based on 288 pullets housed (12 lots of 8 pullets for each of three laying stocks). The equation of the fitted curve is

$$y = (68.95 + k) + 4.192x - 0.8162x^2$$

where y is the rate of lay (eggs per 100 bird days), x is the logarithm of the light intensity in lux, and k is a constant to adjust for the mean difference between experiments

The principal evidence about light intensity, based on work with tungsten filament lamps, is summarized in *Figure 5.11*. Although the responses to intensities in the range 20–40 lx are not significant, it does seem that a curvilinear model is appropriate. Hence the idea of a 'threshold' requirement value has to be discarded in favour of the concept of an 'optimum dose', which is a function of costs and returns. It should be noted that the data in *Figure 5.11* were obtained using large groups of chickens in battery cages and it is possible that the curvilinear response is related to the variation in exposure to light which inevitably occurs under these conditions. Perhaps, as average light intensity is increased, so the proportion of the flock receiving a sufficient intensity approaches unity. Assuming an electricity cost of 4p/kWh, a power input of 0.1 W per bird for each 1 lx intensity achieved at bird level and a marginal return (egg income minus marginal feed cost) of 2p/egg, the optimum light intensity calculated from the curve shown in *Figure 5.11* is about 5 lx.

The maximum stray light that can be tolerated in a poultry house without vitiating the lighting programme being applied is probably about 0.4 lx (*Figure 5.12*).

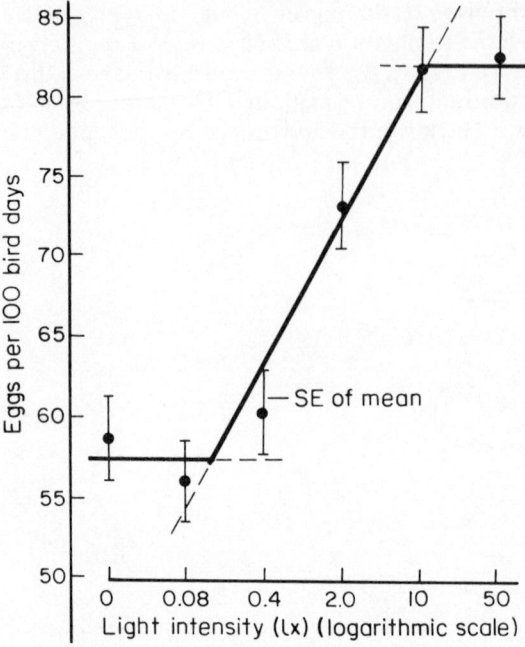

Figure 5.12 The effect of light intensity on egg production in a short-term experiment (Morris and Owen, 1966) in which hens were given 6 hours' light each day at 50 lx and the remainder of the photoperiod (8 h for some birds, 16 h for others) at intensities ranging from 0 to 50 lx. The data indicate that 0.4 lx gives a response that is indistinguishable from total darkness

Intermittent lighting and ahemeral cycles

It may be appropriate to conclude this chapter by mentioning two recent developments in poultry lighting practices. As described above, 'skeleton' photoperiods have been used in experiments with many species and it is clear that a period of illumination (e.g. a '14 h day') can be repeatedly interrupted without altering the response obtained. In other words, the animal has a system for identifying 'dawn' and 'sunset', computing the interval between these events and storing this information. This leads to the question: how little light can be used to make up an effective skeleton photoperiod? Curiously, this question has not been systematically studied in any farmed species of animal and so the complete rules for operating intermittent lighting successfully are not known. Nevertheless, some successful intermittent lighting schemes have been described (van Tienhoven and Ostrander, 1973; Purina, 1978) and it is likely that work in this field will expand rapidly, leading to the adoption of intermittent lighting as the normal practice for laying flocks. An example of a successful intermittent lighting treatment, taken from a recent short-term experiment at Reading, is given in *Table 5.2*.

Ahemeral cycles are those in which the light and dark periods do not add up to 24 h (ahemeral, 'not a day'). For example, a cycle of 14L : 14D gives a 28 h cycle and six such cycles fit neatly into a week. Ahemeral light–dark cycles with periods between 20 h and 30 h have the effect of entraining the

Table 5.2 THE EFFECTS OF AN INTERMITTENT-LIGHTING TREATMENT AND TWO CONTROL TREATMENTS APPLIED FROM 17 WEEKS OF AGE TO ROSS TINT PULLETS (4 groups of 24 birds per treatment)

	Mean age at first egg (d)	Eggs per 100 bird d from 27 to 29 weeks of age	Feed consumption in 28th week (g/bird)
Negative control, 8L :16D	151.8	91.0	106.5
Positive control, 16L : 8D	146.2	96.5	111.3
Intermittent light (8 h light; then 1 min light pulses at hourly intervals for 8 h; then 8 h dark)	147.8	97.0	106.9
s.e. of each mean	0.72	1.05	1.88

Unpublished experiment, Reading University

biological rhythms of the hen (Bhatti and Morris, 1978) and, in particular, the rhythm that determines the permitted hours for the surge in the circulating level of luteinizing hormone (LH) which causes ovulation. Ovulatory LH releases are limited to a maximum of one per cycle and so the use of a cycle longer than 24 h will reduce rate of lay (compared with 24 h controls) for the most prolific birds in the flock (Morris, 1973). The change in cycle length does not alter the rate of yolk accumulation in the ovary (except perhaps with cycles as long as 30 h) and so there is an increase in yolk size, which leads to an increase in egg size inversely proportional to the reduction in rate of lay. Also, because of the increased time which each egg spends in the shell gland, there is an increase in the thickness (and strength) of shells laid under cycles longer than 24 h (Leeson, Summers and Etches, 1979; Yannakopoulos and Morris, 1979).

Some poultrymen have used ahemeral cycles as a means of increasing early egg weight or of improving shell thickness late in the laying year. The benefits are small and depend upon the market prices for eggs of different sizes when the flock is young, or the extent of the cracked-egg problem in the older flock. It is possible to switch backwards and forwards between 24 h and ahemeral cycles provided that certain special rules are observed (Morris, 1978) and so the poultryman who chooses to increase early egg size at the expense of rate of lay is not obliged to accept increased mean egg weight for the remainder of the year. This is in marked contrast to other techniques for increasing early egg size, such as delaying sexual maturity with a 'step down' pattern of lighting (Morris and Fox, 1960; Morris, 1968), when egg size is permanently and irrevocably increased throughout the laying year. The inconvenience of having lights in the poultry house going on and off at odd times can be avoided by using alternating bright and dim light. The rules for entrainment with bright and dim lighting have been worked out (Morris and Bhatti, 1978) and are quite distinct from the rules about photoperiodic stimulus given in *Figures 5.11* and *5.12*.

It does not seem likely that ahemeral cycles will find universal application in the poultry industry but, as a special tool for special circumstances, they have a part to play and they illustrate the elaborate nature of the control now exercised over productive responses in poultry by the artificial manipulation of the environment.

References

ALBERIO, R. (1976). *PhD Thesis*, 3e Cycle University, Paris

ASMUNDSON, V. S. and MOSES, B. D. (1950). *Poult. Sci.* **29**, 34–41

BENOIT, J. and OTT, I. (1944). *Yale J. Biol. Med.* **17**, 27–43

BENOIT, J., WALTER, F. X. and ASSENMACHER, I. (1950). *C. r. Séanc. Soc. Biol.* **144**, 1206–1211

BHATTI, B. M. and MORRIS, T. R. (1978). *Br. Poult. Sci.* **19**, 333–340

BURGER, J. W. (1939a). *J. exp. Zool.* **80**, 249–257

BURGER, J. W. (1939b). *J. exp. Zool.* **81**, 333–341

BURGER, J. W. (1940). *J. exp. Zool.* **84**, 351–361

CURTIS, G. M. (1920). *Use of Artificial Light to Increase Winter Egg Production*. Reliable Poultry Journal Publishing Company, Quincy, Illinois

DUCKER, M. J. and BOWMAN, J. C. (1970). *Anim. Prod.* **12**, 465–471

DUCKER, M. J. and BOWMAN, J. C. (1972). *Anim. Prod.* **14**, 323–334

DUCKER, M. J., BOWMAN, J. C. and TEMPLE, A. (1973). *J. Reprod. Fert., Suppl.* **19**, 143–150

DUCKER, M. J., THWAITES, C. J. and BOWMAN, J. C. (1970a). *Anim. Prod.* **12**, 107–113

DUCKER, M. J., THWAITES, C. J. and BOWMAN, J. C. (1970b). *Anim. Prod.* **12**, 115–123

DUFOUR, J. and BERNARD, C. (1968). *Can. J. Anim. Sci.* **48**, 425–430

FOLLETT, B. K. (1978). In *Control of Ovulation*, pp. 267–293. Ed. by D. B. Crighton, G. R. Foxcroft, N. B. Haynes and G. E. Lamming. Butterworths, London

FOLLETT, B. K. and SHARP, P. J. (1969). *Nature, Lond.* **223**, 968–971

GINTHER, O. J. (1974). *Am. J. vet. Res.* **35**, 1173–1179

GINTHER, O. J. (1979). In *Animal Reproduction, 3rd Beltsville Symposium in Agricultural Research*, pp. 291–305. John Wiley, New York

HACKER, R. R., KING, G. J. and BEARSS, W. H. (1974). *J. Anim. Sci.* **39**, 155

HAMMOND, J., Jr (1952). *J. agric. Sci., Camb.* **42**, 293–303

HAMMOND, J., Jr (1954). *Vitams Horm.* **12**, 157–206

HAMNER, W. M. (1963). *Science, NY* **142**, 1294–1295

HANSEN, R. S. (1966). *Proc. 8th Annual Washington State University Poultrymen's Inst. Conf.*, pp. 17–24

HARRISON, P. C. (1972). *Poult. Sci.* **51**, 2060–2064

HARRISON, P. C. (1974). *Poult. Sci.* **53**, 560–564

HART, D. S. (1950). *J. agric. Sci., Camb.* **40**, 143–149

HAWK, H. W., TYLER, W. J. and CASIDA, L. E. (1954). *J. Dairy Sci.* **37**, 252–287

KINNEY, T., BURGER, R. E. and SHOFFNER, N. N. (1959). *Poult. Sci.* **38**, 1469–1470

LEESON, S., SUMMERS, J. D. and ETCHES, R. J. (1979). *Poult. Sci.* **58**, 285–287

McGINNIS, J., RAMIREZ, H., BOYD, J. and LAUBER, J. K. (1966). *Poult. Sci.* **45**, 1104

MAULEON, P. and ROUGEOT, J. (1962). *Annls Biol. anim. Biochim. Biophys.* **2**, 209–222

MENGE, A. C., MARES, S. E., TAYLOR, W. J. and CASIDA, L. E. (1960). *J. Dairy Sci.* **43**, 1099–1107

MORRIS, T. R. (1962). *Proc. 12th World's Poultry Congress. Symposia*, pp. 115–124

MORRIS, T. R. (1968). In *Environmental Control in Poultry Production*, pp. 15–39. Ed. by T. C. Carter. Oliver & Boyd, Edinburgh

MORRIS, T. R. (1973). *Poult. Sci.* **52**, 423–445

MORRIS, T. R. (1978). *Br. Poult. Sci.* **19**, 207–212

MORRIS, T. R. (1979). In *Animal Reproduction, 3rd Beltsville Symposium in Agricultural Research*, pp. 307–322. John Wiley, New York

MORRIS, T. R. and BHATTI, B. M. (1978). *Br. Poult. Sci.* **19**, 341–348

MORRIS, T. R. and FOX, S. (1958). *Nature, Lond.* **181**, 1453–1454

MORRIS, T. R. and FOX, S. (1960). *Br. Poult. Sci.* **1**, 25–36

MORRIS, T. R., FOX, S. and JENNINGS, R. C. (1964). *Br. Poult. Sci.* **5**, 133–147

MORRIS, T. R. and OWEN, V. M. (1966). *Proc. 13th World's Poultry Congress*, pp. 458–461

OISHI, T. and LAUBER, J. K. (1973). *Am. J. Physiol.* **225**, 155–158

OLIVER, J. and BAYLÉ, J. D. (1976). *J. Physiol., Paris* **72**, 627–637

ORTAVANT, R. (1977). *Symp. Management of Reproduction in Sheep and Goats*, pp. 58–71. University of Wisconsin, Madison

PARKER, J. E. (1972). *Poult. Sci.* **51**, 1848

PARKER, J. E. and McCLUSKEY, W. H. (1964). *Poult. Sci.* **43**, 1401–1405

PARKER, J. E. and McCLUSKEY, W. H. (1965). *Poult. Sci.* **44**, 23–27

PARKER, J. E. and McSPADDEN, B. J. (1943). *Poult. Sci.* **22**, 142–147

PETERS, R. R., CHAPIN, L. T., LEINING, K. B. and TUCKER, H. A. (1978). *Science, NY* **199**, 911–912

PETERS, R. R., CHAPIN, L. T. and TUCKER, H. A. (1979). *J. Dairy Sci.* **62** (Suppl. 1), 113–114

PURINA (1978). Mimeographed report from Ralston Purina, Checkerboard Square, St Louis, Missouri

ROBINSON, J. E. and FOLLETT, B. K. (1979). Paper presented at a meeting of the Society for the Study of Fertility, Glasgow, 1979

ROBINSON, J. J. (1974). *Proc. Br. Soc. Anim. Prod.* **3**, 31–40

ROY, J. H. B., GILLIES, C. M. and SHOTTON, S. M. (1975). *Proc. Seminar on Early Calving of Heifers and its Impact on Beef Production*, pp. 128–142. Commission of the European Communities, Copenhagen

SAUNDERS, D. S. (1977). *Biological Rhythms.* Blackie, London

SITTMANN, D. B., ROLLINGS, W. C., SITTMANN, K. and CASADY, R. B. (1964). *J. Reprod. Fert.* **8**, 29–37

VAN TIENHOVEN, A. and OSTRANDER, C. E. (1973). *Poult. Sci.* **52**, 998–1001

VESELY, J A, (1975). *Anim. Prod.* **21**, 165–174

VESELY, J. A. (1978). *Anim. Prod.* **26**, 169–176

WADDILL, D. G., CHANEY, C. H. and DUTT, R. H. (1968). *J. Reprod. Fert.* **15**, 123–125

WOLFSON, A. (1958). *J. exp. Zool.* **139**, 349–379

WOLFSON, A. and WINN, W. S. (1948). *Anat. Rec.* **101**, 720–721

YANNAKOPOULOS, A. L. and MORRIS, T. R. (1979). *Br. Poult. Sci.* **20**, 337–342

YEATES, N. T. M. (1949). *J. agric. Sci., Camb.* **39**, 1–43

YEATES, N. T. M. (1965). *Modern Aspects of Animal Production*, p. 29. Butterworths, London

6

THE INCUBATION OF EGGS

K. F. LAUGHLIN†
Agricultural Research Council's Poultry Research Centre, Roslin, Midlothian

Introduction

The artificial incubation of their eggs is an essential prerequisite for the exploitation of the high reproductive potential of domesticated birds and therefore for the maintenance of the modern commercial poultry industry. However, artificial incubation is not new and has been practised in China and Egypt for more than two thousand years. The techniques of artificial incubation are well documented in the monograph of Landauer (1967), which also provides detailed information on the factors of environment and heredity which influence hatchability in modern chicken eggs. Data for the European Economic Community indicate the current scale of artificial incubation. The annual production of meat chickens in the EEC exceeds 2000 million birds. Turkeys, guinea fowl and ducks add another 230 million birds and a further 230 million female egg-laying chicks are placed on farms annually. Assuming that the chickens placed on farms represent 75 per cent of the eggs incubated, then approximately 3000 million chicken eggs and 300 million eggs of other species are placed in incubators in the EEC annually, the majority in about 1600 large commercial hatcheries.

For the purposes of this review 'incubation' is extended to include consideration of the environmental influences on the hatching egg from the time it is collected from the breeder house to the time of placement of the day-old chick in the brooding house. Thus, the storage of eggs and transportation of hatched chicks are included although normally, the term 'incubation' is applied only to the process during which the eggs are artificially heated in a machine. However, for the purposes of commercial practice and research it probably is sensible to consider the whole process from the laying of the egg by the breeder hen to the placement of the chick on the growing farm as a continuum, because factors outside the incubation machine can have a significant effect on the performance of the eggs when they are in it and on the health of the chicken after hatching. In effect this adds the transport and storage of the eggs and chick transport to the true incubation period. Two major reviews cover the two former aspects in domestic poultry. The storage and handling of hatching eggs were examined by Proudfoot (1969) and the effects on hatchability of physical factors of the incubator environment were considered by Lundy (1969). These will be considered in turn.

†Now with D. B. Marshall (Newbridge) Ltd, Newbridge, Midlothian

The major factors influencing the egg, and the development of the embryo which it contains, are temperature and the gaseous environment, particularly humidity. There have been few recent developments regarding the temperature of incubation but in the past decade there has been a significant advance in our understanding of the processes of gas exchange across the eggshell. This permits a reappraisal of the effects of humidity and the control of water loss from the egg during incubation. Problems of incubation are examined and discussed in relation to information on practical commercial experience. Recent work on the influence of light on embryo development is also discussed.

Storage

There is general agreement that the optimum temperature for the storage of hatching eggs depends on the duration of the storage period (Proudfoot, 1969). Thus 15–16 °C is preferred for storage periods up to one week, but beyond this time a temperature of 11–12 °C is recommended. The interaction between storage time and temperature has been examined further (Kaltofen and El Jack, 1972; El Jack and Kaltofen, 1975). These authors were concerned with egg storage in hot climates in which cooling of egg stores was a problem and shortening storage time might be advantageous. They showed that for periods of storage of less than 3–4 d, temperatures of 20–25 °C were preferable.

However, when ambient temperatures exceeded 30 °C it was beneficial to cool eggs to 20 °C for any storage period of more than one day. Thus, the inverse relationship between temperature and storage time probably extends back to one day of storage although no continuous series of experiments has been carried out.

In the nests of wild birds, eggs are periodically rewarmed when the female lays each additional egg of the clutch. This fact has prompted an interest in the use of fluctuating temperature during storage of commercial eggs. In one reported experiment on this subject eggs of Japanese quail were warmed to 37.7 °C for 30 min during each day of storage. The eggs subjected to rewarming maintained better hatchability than controls held at a constant 20 °C for 10 d (Kraszewska-Domanska and Pawluczuk, 1977).

Despite the importance of work on the storage of hatching eggs, the ultimate solution to the problems of egg storage is improvement in production scheduling to remove the need for storage. Although this effectively occurs now in large integrated broiler companies, there will, however, be a continuing need to store eggs from breeding stock prior to incubation, and these eggs are particularly valuable.

Incubation environment

There are only two areas in which significant additions can be made to the review of Lundy (1969) on the physical environment during incubation. These concern gas exchange across the eggshell and the effects of light stimulation during incubation.

During the past ten years there have been considerable advances in our understanding of the principles of gas exchange across the eggshell. For example, Wangensteen, Wilson and Rahn (1970/71) measured the permeability of eggshells to oxygen and predicted from this the permeability to carbon dioxide and water vapour. They concluded that the porosity of the shell is set when the egg is laid and is of prime importance in determining oxygen uptake and carbon dioxide and water loss from the developing embryo. These workers were the first to include explicitly a term for the vapour pressure difference across the shell in describing its porosity. Following this work Ar et al. (1974) defined the term *water vapour conductance* to describe measurements of the mass of water evaporated from the whole egg[†], instead of estimation of the volume of vapour lost per unit area of eggshell. Thus according to Ar et al. (1974)

$$\eta = \frac{\dot{w}}{(e_s - e_a)}$$

where η is the water vapour conductance in $mg\,d^{-1}\,kPa^{-1}$, $\dot{w}$ is the rate of weight loss due to evaporation, in $mg\,d^{-1}$, e_s is the saturation vapour pressure of water at the temperature of the egg and e_a is the vapour pressure in the surrounding air. $e_s - e_a$ is therefore the vapour pressure difference across the eggshell. Ar et al. (1974) studied eggs from many species of wild birds and found that the water vapour conductance was proportional to (fresh weight)$^{0.78}$. They further showed that water vapour conductance was proportional to the quotient of the functional pore area (pore number × average pore size) of the shell and its thickness.

Using the techniques described by these workers it can be shown that the water loss from eggs is simply a function of time, the porosity of the shell (measured as water vapour conductance) and the vapour pressure difference across the egg shell. Since water loss from eggs appears to be a simple passive process, it is possible to predict from measurements of the average water vapour conductance of a batch of eggs either the conditions necessary to obtain a desired percentage weight loss or, alternatively, the weight losses which will result when eggs are kept in a specified environment. Collection of detailed data on eggs of domestic poultry produced from various genotypes, bird ages and husbandry systems should in future allow specification of the humidity requirements necessary for the optimal weight loss during incubation from the point of view of hatchability and chick survival.

The stimulation of embryo development by light treatments has received considerable attention during the past twenty years. For example, a reduction of incubation time by 16 h resulted from stimulation with incandescent light (Shutze et al., 1962), while Lauber and Shutze (1964) reported a reduction of 20 h, Siegel et al. (1969) 30 h and Walter and Voitle (1973) 9 h. Coleman (1978) has reviewed her own extensive work on photoacceleration of the development of avian embryos and the publications of other workers. It is clear that light stimulation can affect not only the rate of embryo growth but also chick viability and hatched chick weight. The biochemical (or other) mechanisms which mediate these effects are as yet

(†Editor's note. Strictly, conductances should be expressed per unit area.)

not fully explained but it is important that the techniques are developed to allow treatment on a commercial scale to fully evaluate their potential.

Summary

Large-scale commercial incubators have changed little over the last two decades but developments in modern electronic monitoring and control systems are beginning to be introduced. Further research and development is needed on the whole of the handling system for hatching eggs from the production farm through to the incubator. There is also a need to develop energy-saving features in the machines themselves. Increasingly, the emphasis will be on chick quality and subsequent performance and not simply hatchability of eggs as a measure of incubation success.

References

AR, A., PAGANELLI, C. V., REEVES, R. B., GREENE, D. G. and RAHN, H. (1974). *Condor* **76**, 153–158

COLEMAN, M. A. (1978). *Proc. 16th World's Poultry Congress*, Vol. I, 14–19

EL JACK, M. H. and KALTOFEN, R. S. (1975). *Arch. Geflügelk.* **39**, 198–202

KALTOFEN, R. S. and EL JACK, M. H. (1972). *Arch. Geflügelk.* **36**, 116–120

KRASZEWSKA-DOMANSKA, B. and PAWLUCZUK, B. (1977). *Br. Poult. Sci.* **18**, 531–533

LAUBER, J. K. and SHUTZE, J. V. (1964). *Growth* **28**, 179–190

LANDAUER, W. (1967). Storrs Agricultural Experimental Station. *Monograph 1*

LUNDY, H. (1969). In *The Fertility and Hatchability of the Hen's Egg*, pp. 143–176. Ed. by T. C. Carter and B. M. Freeman. Oliver & Boyd, Edinburgh

PROUDFOOT, F. G. (1969). In *The Fertility and Hatchability of the Hen's Egg*, pp. 127–141. Ed. by T. C. Carter and B. M. Freeman. Oliver & Boyd, Edinburgh

SHUTZE, J. V., LAUBER, J. K., KATO, M. AND WILSON, W. (1962). *Nature, Lond.* **96**, 594–595

SIEGEL, P. B., ISAKSON, S. T., COLEMAN, F. N. and HUFFMAN, B. J. (1969). *Comp. Biochem. Physiol.* **28**, 753–758

WALTER, J. H. and VOITLE, R. A. (1973). *Br. Poult. Sci.* **14**, 533–540

WANGENSTEEN, O. D., WILSON, D. and RAHN, H. (1970/71). *Resp. Physiol.* **11**, 16–30

III

ENVIRONMENTAL INFLUENCES ON THE PRODUCTIVITY OF FARM ANIMALS

7

THE ENVIRONMENTAL PHYSIOLOGY OF JUVENILE ANIMALS

P. POCZOPKO

Institute of Animal Physiology and Nutrition, Polish Academy of Sciences, Jabłonna, near Warsaw, Poland

Introduction

This chapter reviews the relationship between young animals and their thermal environment. It is intended to have a broad biological outline and therefore will not be limited to animals of agricultural importance.

Among the approximately one million known animal species, only about twenty thousand are homeotherms. However, the majority of animals bred by man for economic reasons belong to this group. The homeotherms are usually divided into *altrical* and *precocial* animals. Altrical animals are born or hatched blind, with bare skin and ineffective locomotor mechanisms while, by contrast, precocial animals are born with open eyes, their skin covered with pelage or down and they are able to move about soon after birth. These characteristics of altrical and precocial animals fit only the extreme cases, and many intermediate forms can be distinguished (King and Farner, 1961; Davydov and Keskpajk, 1967; Dawson and Hudson, 1970). Even among precocial species, a category that includes almost all farm animals, the degree of independence of the newborn varies considerably.

A common feature of all juvenile animals, both altrical and precocial, is that their lower critical temperature is high and the zone of thermal neutrality narrow in comparison to those for adult specimens of the same species. This means that young animals are more sensitive to cold than adult ones. The reasons for this sensitivity and the diverse means used by young animals in order to survive cold conditions will be discussed in subsequent sections.

Physical factors

EFFECT OF SIZE AND BODY COVERING

Many textbooks of animal physiology suggest that newborn homeotherms do not possess fully developed mechanisms of thermoregulation. However, Hull (1973) concluded that 'Although many newborn mammals are relatively helpless, the physiological mechanisms which control their internal environment are neither poorly developed nor ineffectual, though they are often different from those which operate in adults'. Hull emphasized in his review the importance of an animal's size, its surface to body mass ratio and insulation, all of which change during growth. The larger an animal the smaller is its relative surface area. For instance, in two species of passerine birds (*Spizella pusilla* and *S. passerina*) a 50 per cent decrease in the relative surface area occurred within 6 days after hatch (Dawson and Evans, 1957, 1960). A similar change may be observed in rabbits within 17 days after birth (Poczopko, 1969a). Relatively efficient thermoregulation in these animals is also acquired within 6 and 17 days, respectively. There is no doubt that this decrease in the relative surface area contributes to the increase in cold resistance, but the development of body covering, which takes place at the same time, seems to be more important.

Altrical animals are completely or nearly completely naked at the beginning of their post-embryonic life. For instance, the pelage begins to play a role as a barrier for heat flow from the body at the age of 18 days in rats (Poczopko, 1961) and 8 days in rabbits (Poczopko, 1969b). Among altrical birds an interesting example is the snow bunting (*Plectrophenax nivalis*). This species incubates its eggs when the ambient temperature is far below freezing point. The newly hatched chicks are almost naked, but a dense down develops during the first week, and at the same time considerable thermal independence is established (Blix and Steen, 1979).

In newborn precocial animals the skin is covered with pelage or down, but their ability to maintain a constant body temperature varies from species to species. Koskimies and Lahti (1964) have studied the cold hardiness of ducklings belonging to 10 European species, in the first day after hatching. The ducklings were exposed to temperatures of 20, 8–10 or 0–2 °C. In all the ducklings examined, a 20 minute exposure to either 20 °C or 8–10 °C had little effect on their body temperature. This finding contrasted with the data the same authors obtained in the gallinaceous birds and gulls, which at similar ambient temperatures quickly become hypothermic. The ducklings of one species (*Mergus mergasner*) maintained a constant body temperature for 15 hours even on exposure to 0–2 °C, whereas the body temperature of *Anas platyrhynchos* fell immediately at this exposure. According to Koskimies and Lahti (1964), differences in the susceptibility to cold among young precocial birds are due not only to differences in their thermal insulation, but also to differences in heat production. The rate of heat production, which will be discussed in more detail later in this chapter, is indeed of considerable importance. Low heat production probably explains why even the snowy owl (*Nyctea scandiaca*), which after hatch looks like a ball of down, attains thermal independence only at the age of 3 to 4 weeks (Blix and Steen, 1979). Among domesticated animals, sheep are born with considerable coat insulation and are more resistant to cold than animals such as piglets (Slee, 1978).

Subcutaneous fat plays a considerable role in the thermal insulation of many adult animals, but not in the newborn; in newborn piglets, for instance, the total fat content amounts only to approximately 1 per cent of body weight (Mount, 1968), so the amount under the skin is negligible. Since newborn pigs also have very sparse hair, they have little thermal insulation. This explains why the lower critical temperature is high, 34–35 °C (Mount, 1968). The pups of pinnipeds (seals, sealions and walruses) also have surprisingly poor thermal insulation. At birth their body is covered only by a lanugo pelt of low insulative quality and they lack (with a single known exception) subcutaneous blubber. However, in contrast to piglets they are quite resistant to cold. The pups of many species of pinnipeds are born on ice during a season when air temperature frequently falls below –20 °C and strong winds are common (Blix and Steen, 1979). Weighing from 5 to 10 kg at birth, they have smaller relative surface area than piglets, but neither this difference nor the difference in thermal insulation explains their superior cold tolerance. This must depend on other factors.

PHYSICAL THERMOREGULATION

An animal's insulation is under the control of two physiological mechanisms: vasomotor and pilo- or pteromotor. For instance, because of these mechanisms the overall thermal insulation (tissue plus coat) in the newborn pig increases 1.73 times when ambient temperature decreases from 30 to 5 °C (Mount, 1963) and, in 2- to 5-day-old goslings, 1.55 times with a fall in ambient temperature from 25 to 17 °C (Poczopko, 1972). These examples show that even very young animals of some species can utilize their means of physical thermoregulation, but these mechanisms cannot operate equally in all juvenile animals. Those which are born without a coat certainly cannot utilize the mechanism of piloerection. However, even in those which are born with a coat the efficiency of this mechanism is inferior to that in adults, because the coat is less developed. For instance, in piglets the increase in coat insulation evoked by cold is not measurable up to the age of 2 weeks, whereas in older ones it may amount to about 15 per cent of overall insulation (Mount, 1964).

Vasomotor regulation of heat loss does not operate in some small animals during the first few days of post-embryonic life. In rats this may be inferred from the fact that body and skin temperature remain nearly equal when ambient temperature decreases from 30 to 25 °C, and that the local cooling of the skin (by an ice-filled test tube) does not produce any changes in the colour of the cooled region (Poczopko, 1961). The results obtained by Lagerspetz (1964, 1966) and Hissa (1964) suggest that the vasomotor response is also absent in newborn hamsters, mice and lemmings. In rabbits

Table 7.1 BODY WEIGHT, ZONE OF THERMAL NEUTRALITY AND OXYGEN CONSUMPTION IN RHODE ISLAND RED CHICKENS OF DIFFERENT AGES

Age	Body weight (g)	Zone of thermal neutrality (°C)	Thermoneutral O_2 consumption (ml O_2 g^{-1} h^{-1})	Cold exposure Temperature (°C)	O_2 consumption (ml O_2 g^{-1} h^{-1})	Metabolic quotient
2–6 d	36	34–36	1.25	21	2.85*	2.28
2 w	90	33–35	1.40	21	2.45*	1.75
5 w	260	30–34	1.35	10	2.45*	1.85
8 w	590	29–33	1.10	0	2.30*	2.09
12 w	1030	27–32	0.90	–15	1.90*	2.11
18 w	1610	23–31	0.77	–12	1.30	–
23 w	1960	20–29	0.65	–12	1.05	–
1 y	2430	16–27	0.62	–12	0.90	–

Recalculated from the data of Barott and Pringle (1946). The values marked by asterisks may be considered as summit metabolism. Age: d, days; w, weeks; y, year exposed to cold, vasoconstriction of blood vessels in the skin takes place even for animals only 1 day old, but it produces little increase in thermal insulation (Poczopko, 1969). On the other hand, vasoconstriction is considerable in newborn pigs exposed to 10 °C, so that the skin temperature of the trunk may drop as much as 12 K below rectal temperature (Ingram, 1964).

The better the thermal insulation and its regulation, the lower is the critical temperature for an animal and the wider the thermal neutrality zone. Since a gradual improvement in thermal insulation takes place during ontogeny, corresponding changes in the zone of thermal neutrality occur. This is shown in *Table 7.1*, which is based on measurements published by Barott and Pringle (1946).

Behavioural thermoregulation

Behavioural control of body temperature can be regarded as an extension of physical thermoregulation, because it changes the physical conditions of heat exchange between an animal and its surroundings. We must remember, however, that entirely different neural centres are involved in the two types of thermoregulation.

In behavioural thermoregulation one can distinguish between the behaviour of parents and that of offspring. Parental behaviour includes the preparation of nests, burrows or dens and brooding, which may be constant or intermittent.

A spectacular example of constant brooding is found in polar bears. The naked and blind cubs are delivered (one to three in a litter) in December or January when air temperature often falls to –40 °C and the temperature inside the dens is probably close to freezing. The female curls up and presses the cubs to the nipples with her furred legs. After 3 months of continuous brooding the cubs attain a weight of 10 kg and their fur becomes quite thick. The lower critical temperature is then about –30 °C (Kostyan, 1954; Blix and Lentfer, 1979; Blix and Steen, 1979). Similar brooding probably occurs in the wolverine (*Gulo gulo*), because the early development of the naked and blind pups also takes place in snow dens (Blix and Steen, 1979). In contrast, among pinnipeds from very cold environments brooding is known only in the walrus (*Odobenus rosmarus*), whose calves weigh about 45 kg at birth but are almost lacking in thermal insulation (Fay and Ray, 1968).

The snowy owl (*Nyctea scandiaca*) broods her young continuously for the first 3 days (Blix and Steen, 1979) and the snow bunting (*Plectrophenax nivalis*) for 1 week (Maher, 1964). However, birds which incubate their eggs in warmer seasons are typically intermittent brooders. To these belong the ancestors of the domestic hen, goose, duck and turkey.

In many rodents both parents care for the progeny. An example is the bank vole (*Cletrionomys glareolus*), which broods the offspring from the 1st to the 14th day after birth. The female spends on average a little over 20 hours per day in the nest and less than 4 hours outside it. The male leaves the nest far less frequently and for shorter intervals. Only rarely are the male and female outside the nest simultaneously (Gębczyński, 1975). In contrast, the female rabbit feeds her young between one and three times a day, but at other times stays outside the nest, which is usually covered. However, the parent still provides behavioural thermoregulation. The author's (unpublished) observations on rabbits kept in an open shed showed that during warm sunny days the nests remained uncovered, whereas they were carefully covered during relatively cold May nights.

When juvenile animals are subjected to cold they tend to huddle. This behaviour is also present in adult animals of small body size, and occasionally even in quite large animals, such as pigs. The studies by Mount (1960) and Mount and Holmes (1969) showed that in 1-day-old piglets kept singly the metabolic rate doubles when ambient temperature decreases from 38 to 18 °C, whereas in those kept in groups of four to six it increases only by 37 per cent. The values collected in *Table 7.2* show results for bank voles which

Table 7.2   LITTER SIZE AND OXYGEN CONSUMPTION BY 1- TO 9-DAY-OLD BANK VOLES AT DIFFERENT AMBIENT TEMPERATURES

Litter size	Oxygen consumption (ml O_2 g^{-1} h^{-1})				
	15 °C	20 °C	25 °C	30 °C	35 °C
1	3.99	4.19	4.13	3.59	2.38
2	3.95	3.74	3.55	3.15	2.37
3	3.29	2.18	2.01	2.04	1.72
4	2.83	2.24	1.84	1.51	1.30
5	2.91	2.03	1.87	1.49	1.42
6	2.92	1.84	1.62	1.32	1.33

Based on data by Gębczyński, 1975, reproduced by courtesy of the editor of *Acta theriologica*

illustrate the importance of group size; the larger the group of animals, the lower is the metabolic rate of its members. Earlier measurements by the same authors showed that at an ambient temperature of 5 °C, two voles kept together had metabolic rates 22 per cent lower than single individuals, while rates for the groups of three and five specimens were lower by approximately 31 and 35 per cent respectively (Gębczyński, 1969). Similar economies due to group size may be observed in young domestic animals (Mount, 1968).

Metabolic factors

CHANGES IN THE STANDARD METABOLIC RATE WITH AGE

A high metabolic rate obviously reduces the danger of hypothermia even in animals with relatively poor thermal insulation. It follows from this statement that criteria are needed to define what is meant by a high or low metabolic rate. One such criterion may be derived from mathematical analysis of the relationship between body size and metabolic rate. Where adult homeotherms are concerned the relationship between these variables is of the form

$$\mathbf{M} = aW^b \tag{7.1}$$

where **M** is the standard metabolic rate (SMR, in W kg^{-1}) determined with the animal at rest, in a post-absorptive state and in thermoneutral surroundings, W is the body weight in kilograms and a and b are constants.

The values of a and b may be determined by plotting **M** against W on a log–log graph, since

$$\log \mathbf{M} = \log a + b \log W \qquad (7.2)$$

Equation 7.1 shows that the metabolic rate of adult homeotherms is proportional to a power of the body weight. Calculations made by nine different authors, and based on quite different sets of data, give remarkably similar values for the constant b (Poczopko, 1979). The figure proposed by Kleiber (1947, 1961) and Hemmingsen (1960), namely 0.75, is commonly accepted. Kleiber (1961) suggested that the body weight in kilograms raised to the power of ¾ should be called the 'metabolic body size' and the daily heat production divided by the metabolic body size the 'metabolic level'. Analysing data concerning SMR in adults of 12 species of mammals, from a 21 g mouse to a 600 kg steer, Kleiber came to the conclusion that the average metabolic level for homeotherms is $70 \, \text{kcal} \, \text{kg}^{-0.75} \, \text{d}^{-1}$; which equals $293 \, \text{kJ} \, \text{kg}^{-0.75} \, \text{d}^{-1}$, or approximately $3.4 \, \text{W} \, \text{kg}^{-0.75}$. More recently it has been found that other metabolic levels can be distinguished as characteristic for particular groups of animals (Poczopko, 1971, 1979). Nevertheless, Kleiber's level $(3.4 \, \text{W} \, \text{kg}^{-0.75})$ can be used as a convenient reference standard. This level is shown by the horizontal line in *Figure 7.1*.

Changes of metabolic rate with size within a single species are more complex. The intraspecific relation between body size and SMR is not

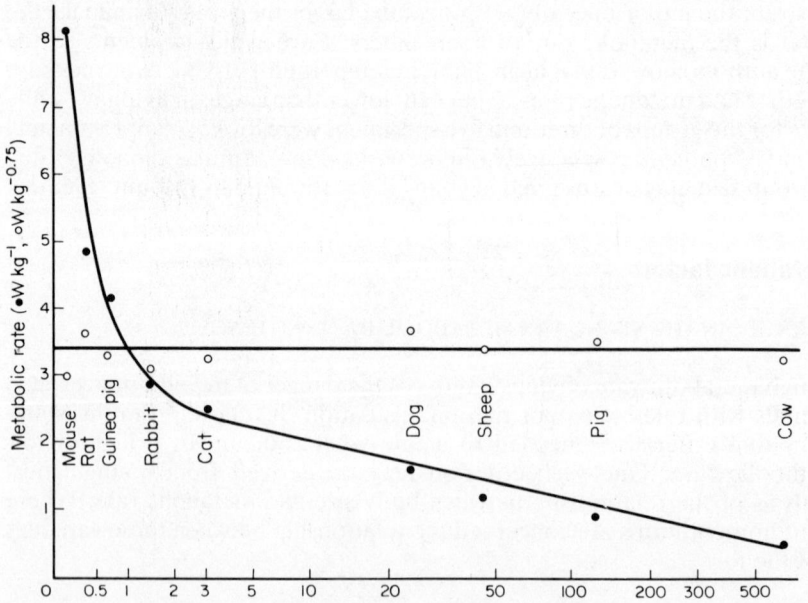

Figure 7.1 Relationships between standard metabolic rate and body size in adult mammals, expressed per kilogram of body weight (●) and relative to metabolic body size ($kg^{-0.75}$) (○). (From Poczopko, 1979, courtesy of the editor of Acta theriologica)

represented by a straight line even on a log–log plot (Brody, 1945; Poczopko, 1979), but by a curve, divided by some authors into several straight sections (Freeman, 1964; Hissa, 1968; Piekarzewska, 1977). According to Freeman (1964), these represent different metabolic phases. The data compiled by Poczopko (1979) suggest that such a pattern for the metabolic rate changes is a common phenomenon in developing animals. However, many authors have calculated the exponent *b* in equation 7.2 for particular animal species, using the results obtained during a part of the lifespan only. Consequently the values of *b* reported vary from 0.25 to almost 5 (Poczopko, 1979).

The evidence quoted above shows that it is impossible to establish a single metabolic body size for growing animals. However, accepting the metabolic body size as found in interspecific comparisons enables us to decide whether the metabolic rate of an animal is high or low at a particular stage of its development. *Figure 7.2* shows that the metabolic rate of growing hens and

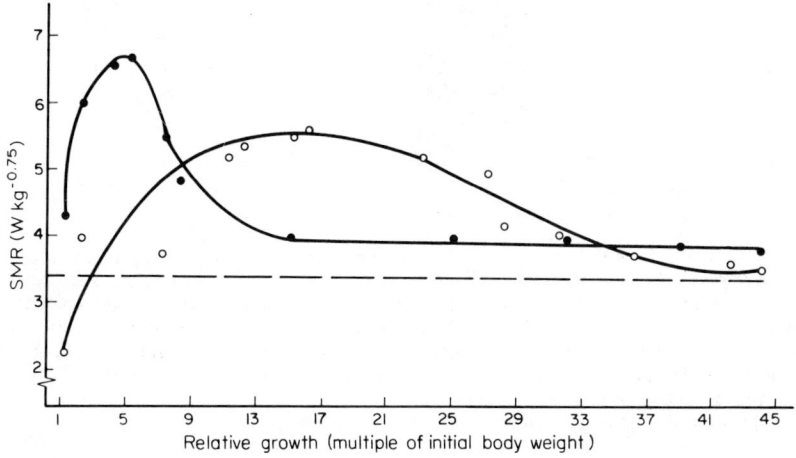

Figure 7.2 Relationship between standard metabolic rate, expressed per kg$^{0.75}$, and relative size as a multiple of mean initial body weight, in growing rats and chickens. (●) Hens, body wt 43–2080 g; (○) rats, body wt 6.5–280 g; (- - -) mean interspecific level. (Adapted from Poczopko, 1979, courtesy of the editor of Acta theriologica*)*

rats, expressed per kg$^{0.75}$, changes in a particular manner. At the beginning of post-embryonic life the metabolic rate is relatively low; it subsequently increases to reach a level approximately twice as high as that in adults, and then gradually decreases again. More evidence of the changes in SMR with age is presented in *Table 7.3*. The level of metabolism during the first few days of post-embryonic life is in most cases similar to that noted in adult animals. Since thermal insulation in small newborn animals is inferior to that in adults, one may conclude that the production of metabolic heat is less adjusted to the possibility of heat loss than in fully grown specimens.

The rate of post-natal increases in SMR depends on species (*Table 7.3*). For instance, in pigs the maximum level is attained within 7 days, in rats within 36 days and in turkeys within 60 days. The cause and mechanism of this post-natal increase in SMR are not fully understood. According to

Table 7.3   THE STANDARD METABOLIC RATE (SMR) IN YOUNG ANIMALS AS COMPARED WITH THAT IN ADULTS

Species	Age (d)	SMR ($W kg^{-0.75}$)	Age (d)	SMR ($W kg^{-0.75}$)	SMR in adults ($W kg^{-0.75}$)
Mammals					
Rat[1,2]	1	2.19	36	5.57	3.53
Rabbit[3]	1–2	2.69	14	6.20	3.20
Guinea pig[3]	1	3.65	17	4.45	3.87
Pig[4]	1	3.14	7	5.96	2.90
Sheep[5]	7	4.68	21	6.81	3.20
Cow[6]	1–3	5.62	?	6.29	3.00
Birds					
Tree sparrow[7]	1	2.26	14	10.65	?
Domestic sparrow[7]	1	2.42	15	11.82	6.24
Japanese quail[8]	1	3.79	21	9.20	6.40
Domestic pigeon[9]	3	5.15*	11	7.42	3.50
Hen (Rhode Island Red)[6]	1–6	4.26	30	7.02	4.60
Goose (White Italian)[11]	2–4	2.81	15	6.83	4.20
Turkey[10]	1–3	3.25	60	6.31	3.34
Domestic duck (Peking)[12]	1–2	3.92	27	6.53	4.70
Wild ducks[13]	0–1				
Anas platyrhynchos	0–1	2.77	–	–	–
Mergus serrator	0–1	3.12	–	–	–
Bucephala clangula	0–1	4.26	–	–	–

Sources of data: 1—Taylor, 1960; 2—for adult rat only, Kleiber, Smith and Chernikoff, 1965; 3—Piekarzewska, 1977; 4—Mount and Rowell, 1960; Brody, 1945; 5—Graham, Searle and Griffiths, 1974; 6—Brody, 1945; 7—Myrcha, Pinowski and Tomek, 1973; 8—Freeman, 1967; 9—Riddle, Nussman and Benedict, 1932; 10—Kotrbáček, 1977; 11—Poczopko, 1969b; Poczopko, Kaciuba-Uścilko and Jusiak, 1973; 12—Kotrbáček, 1973; 13—Koskimies and Lahti, 1964 (among 10 species listed by these last authors three examples were selected with the lowest, intermediate and the highest SMR)
*Probably obtained at an ambient temperature below the critical temperature

Mount (1969), the oxygen consumption of newborn pigs is about 8–10 ml O_2 kg^{-1} min^{-1} for the first few hours after birth, and by the second day it reaches 15 ml O_2 kg^{-1} min^{-1}. However, this rise in metabolic rate in the first day *post partum* does not take place in piglets fasted from birth. Misson (1977) suspects that the energetic cost of growth, the transition from poikilothermia to homeothermia and stabilization of endocrine systems are involved in this phenomenon. Among endocrine factors the thyroid hormones probably play an important role. This supposition is based on the fact that there is a remarkable parallelism between the changes in thyroid gland activity with age and those of SMR in several animal species, such as pigs (Kaciuba-Uścilko, 1971; Kaciuba-Uścilko, Mount and Legge, 1970), rabbits and guinea pigs (Piekarzewska, 1977), geese (Poczopko, Kaciuba-Uścilko and Jusiak, 1973), chicken (Jastrzębski, Pietras and Gajewska, 1977) and ducks (Poczopko, Witkowska and Uliasz-Poniewierska, 1979). Whatever the reason, the high metabolic rates of juvenile animals, which are relatively small and have poor thermal insulation, allow them to maintain thermal homeostasis in a cold environment and lower their critical temperatures.

COLD-INDUCED INCREASES IN THERMOGENESIS

When ambient temperature drops below the critical value for an animal, the animal's ability to increase heat production is essential for maintaining homeothermy. Two mechanisms for increasing thermogenesis in cold-exposed animals are usually recognized: shivering and non-shivering thermogenesis. However, Ivanov (1962) and Ivanov and Alimuchamedov (1963) have also suggested that 'thermoregulatory tonus', i.e. an increase in the muscular tone (detectable with sensitive electromyographic methods), may be accompanied by an increase in heat production. Voluntary muscular activity also plays some role in an animal's defence against hypothermia.

Reports of studies on shivering in the newborn animal are scarce, and the majority concern visual observations only. Nevertheless, it is known that many altrical mammals are unable to shiver for some time after birth. In rats, for instance, visible shivering appears only on the 8th or 9th day after birth (Gulick, 1937) and in mice on the 10th day (Lagerspetz, 1966). Hissa (1968) studied the development of shivering thermogenesis in golden hamsters and Norwegian lemmings, and found that in young animals, in contrast to the adult specimens, a decrease in body temperature is necessary to produce shivering. Moreover, the maximal increase in electrical muscle activity changes with age. In hamsters no increase in this activity could be detected up to the 9th day after birth. In 12-day-old specimens it increased by a factor of four between the normal body temperature and a body temperature of 21–23 °C, whereas in 34-day-old animals activity increased by a factor of 9 at a body temperature of 25–26 °C. In lemmings a small increase in the electrical activity of muscles in response to cooling was detectable on the 6th day after birth, but the increase in this activity with age was not as pronounced as in hamsters.

Hissa (1964) has shown that the effector muscles are already well developed at birth, even in animals that do not shiver in response to cooling, and that the muscles can be stimulated by appropriate motor nerves. He concluded therefore that the lack of response of these muscles to cooling is due to immaturity of the integrating systems. Since many species of altrical mammals and birds are unable to shiver for some time after birth, they can be regarded as 'poorly developed'. In contrast, in precocial animals the ability to shiver starts rather early. The author's own observations (Poczopko, 1967) revealed that newly hatched and still wet goslings shiver vigorously when removed from the incubator. However, 2- to 3-day-old goslings (with dry down cover) exposed to 5 °C showed no visible shivering although their heat production was twice as high as at thermoneutrality (Poczopko and Uliasz, 1975a). Since non-shivering thermogenesis has not been found in this species (Poczopko and Uliasz, 1977; Poczopko and Poniewierski, 1977), invisible shivering or Ivanov's thermoregulatory tonus is probably responsible for the observed increase in heat production. Similar situations may also exist in some other precocial animals. It is possible, however, that in young precocial animals shivering is not as efficient as in adults. This problem certainly deserves further comparative studies.

Non-shivering thermogenesis (NST) has been commonly recognized for some 15 years. In his valuable review, Janský (1973) defined NST as 'heat production mechanisms liberating chemical energy due to processes which do not involve muscular contraction'. This author distinguished 'obligatory' or basal NST, which occurs under conditions of basal metabolism, and 'regulatory' NST, occurring at ambient temperatures below the thermo-neutral zone. Age changes in obligatory NST, which modify the range of the lower critical temperature, were discussed earlier in this chapter. The sequence of events taking place during regulatory NST may be summarized as follows: a cold environment stimulates the adrenergic system and, because of this stimulation, noradrenaline (NA) release from the sympathetic nerve endings increases. Owing to the lipolytic action of NA, free fatty acids are mobilized and then oxidized.

The brown adipose tissue (BAT) present in the body of many newborn mammals has been postulated as a principal site of NA action during non-shivering thermogenesis (Hull, 1966; Hull and Hardman, 1970; Smith and Horwitz, 1969). In newborn rabbits the temperature and oxygen consumption of this tissue increased both on cold exposure and after NA injection (Heim and Hull, 1966). Moreover, when a proportion of BAT was excised the increase in heat production after an injection of NA was smaller than in sham-operated controls (Leduc and Rivest, 1969), though more recent experiments have shown that the depression of the calorigenic effect of NA by the removal of BAT is transient: it disappears after 4 days at most (Foster, 1974). The latter finding suggests that the role of BAT is smaller than has often been supposed; indeed, Janský (1973) estimated that in non-shivering thermogenesis by small animals only 6–8 per cent of the heat is generated in brown adipose tissue, compared with up to 50 per cent in skeletal muscles and 25 per cent in the liver.

'Loosely coupled' mitochondria have been discovered in BAT, and these play an important role in the liberation of heat from the chemical energy (Christiansen, 1977). Himms-Hagen *et al.* (1976) found that the mito-chondria are smaller and more numerous in the muscles of cold-acclimated rats than in controls, although the mitochondrial mass per gram of the tissue is the same. These authors did not mention whether or not they found loosely coupled mitochondria in the muscles of cold-acclimated rats, but such mitochondria have been discovered recently in the muscles of the fur seal (Grav and Blix, 1979).

Haldmaier (1971) pointed out that the larger an animal the smaller is its calorigenic response to exogenous NA. For instance, in a species of bat (*Myotis lucifugus*) the heat production after NA injection increases to over 3 times the SMR, whereas in animals weighing more than 10 kg the increase is minimal. This suggests that non-shivering thermogenesis is less important in large animals than in small ones. However, the tendency shown by Haldmaier is not applicable to seal pups adapted to the extreme cold occurring in their birthplace: the pups of harp seals (*Pagophilus grenlandicus*) weigh about 10 kg at birth and can live on ice at an air temperature of –20 °C and in strong winds, owing to intensive NST and the abundant supply of energy in milk containing approximately 52 per cent fat (Blix, Grav and Ronald, 1975; Grav and Blix, 1979). Blix, Grav and Ronald (1975) suggested that no decrease in capacity for NST occurs in the pups of harp seals during their terrestrial stage of life. On the contrary, this capacity probably reaches its maximum at the time of transition to aquatic life; that is, a month after birth. A similar capacity for NST also exists in other pinniped species, which also seem to have rather high SMR: over 5 W kg$^{-0.75}$.

The above observations show that the capacity for intensive NST has also developed, during the course of evolution, in large animals whose offspring are compelled to face severe cold. Among domestic animals, lambs and calves seem well adapted to face a cold environment. *Table 7.3* shows that shortly after birth they have a high SMR. In both species the capacity for NST also seems to be higher than could be predicted on the basis of Haldmaier's curve. In newborn lambs the heat production increases greatly during cold exposure (Alexander, 1962), and according to the estimate made by Alexander and Williams (1968) approximately 40 per cent of the maximal increase is derived from NST. The ability to produce heat by NST declines rapidly during the month following birth, and at the same time the BAT is lost. However, when partly shorn lambs are kept for several weeks at an ambient temperature of 3 °C, a significant retention of BAT is observed, and the thermogenic effect of infusion of NA is retained (Alexander, Bell and Williams, 1970). Studies by Alexander, Bennet and Gemmel (1975) showed that in newborn calves approximately 2 per cent of the body weight is BAT. The sensitivity of calves to NA is also consistent with this finding. Infusion of this amine into newborn calves exposed to a thermoneutral environment, at the rates of 1 and 5 µg kg^{-1} min^{-1}, increased the metabolic rate two- to threefold. This response is comparable with that found in bats by Haldmaier (1971).

Newborn pigs have no brown fat (LeBlanc and Mount, 1968; Rowlatt, Mrosovsky and English, 1971) but a definite thermogenic effect of exogenous NA has been shown in 2-week-old pigs by LeBlanc and Mount (1968). This finding was confirmed by Kaciuba-Uściłko and Poczopko (1973, 1975), who found NA injection (2 mg kg^{-1}, s.c.) to have no calorigenic effect in piglets aged 4–6 days, but to cause an increase in heat production of about 18 per cent in those aged 11–13 days. Kaciuba-Uściłko and Poczopko (1975) and Kaciuba-Uściłko and Ingram (1977) also showed that propranolol injection depresses the cold-induced elevation of the metabolic rate in piglets at both 4–6 and 11–13 days. Since no BAT was found in piglets, one may presume that loosely coupled mitochondria are present in the muscles of these animals, as in the muscles of fur seals, and that they play the same role in NST as does BAT in other animal species. However, the relatively small metabolic response of piglets to NA suggests that NST in this species is considerably less marked than in lambs or calves.

Brown adipose tissue is also absent in birds (Freeman, 1971), but Wekstein and Zolman (1969) reported that NA plays some role in thermogenesis in the newly hatched chicken, because propranolol injection reduced their heat production. However, the results obtained by other authors did not confirm this, although some metabolic changes induced by cold exposure or NA injection were similar to those found in mammals during NST. For instance, cold exposure increased plasma free fatty acid (FFA) levels in the chicken (Davison, 1973; Freeman, 1970, 1977) and in goslings (Poczopko and Uliasz, 1975a). Increases in the plasma NA level on exposure to cold have been observed in blackheaded gulls (Palokangas and Hissa, 1971), in pigeons (Saarela and Hissa, 1976) and goslings (Poczopko and Uliasz, 1975a). However, NA injection had no effect on the plasma FFA concentration in the titmouse (Hissa and Palokangas, 1970) or hen (Carlson *et al.*, 1964). An injection or infusion of NA also failed to stimulate heat production in the chicken (Freeman, 1970) and in blackheaded gulls (Palokangas and Hissa, 1971).

Freeman (1971) suggested that in birds NST, if it exists, may be mediated by the thyroid hormones. Glucagon has also been proposed as a hormone involved in NST in birds. It has been shown, however, that while glucagon stimulates lipolysis it does not stimulate heat production (Freeman, 1975; Freeman and Manning, 1977). In 17- to 21-day-old goslings (but not in younger ones) a small but significant increase in heat production was noted after injection of NA (Poczopko and Uliasz, 1975a). Exogenous T_4 also exerted a calorigenic effect (Poczopko and Uliasz, 1975b). However, while thiouracil treatment (intraperitoneal injections for 3 consecutive days) diminished cold-induced thermogenesis, the injection of propranolol had no such effect (Poczopko and Uliasz, 1977). Gallamine also caused a rapid fall in the metabolic rate and body temperature of goslings and chickens (Poczopko and Poniewierski, 1977).

The data briefly summarized above suggest that NST is absent in birds. Presumably, therefore, the FFA mobilized by cold exposure are instead used in shivering thermogenesis.

Summit metabolism

The maximal rate of heat production that is achieved in response to cold and can be sustained for some time without hypothermia, was defined by Giaja (1938) as the *summit metabolism*. According to that author, the summit metabolism in adult animals is 3–4 times higher than SMR, but in some rodents it may be as much as 8 times higher (Hart, 1971).

The data presented in *Table 7.3* show that SMR in newborn animals is close to or even lower than that in adults of the same species, and also often lower than the Kleiber's interspecific mean, viz. 3.4 W $kg^{-0.75}$. Since the relative surface area is large and thermal insulation is poor in the newborn, such a metabolic rate should be accepted as low, especially in the case of altrical animals. Under cold conditions these animals need to raise their heat production more than do adults, in order to keep their body temperature constant. Yet in many newborn animals the metabolic quotient, i.e. the ratio of summit to standard metabolism, is considerably lower than those listed by Giaja (1938) and Hart (1971).

Small altrical birds are essentially poikilothermic at hatching. In the 1-day-old tree sparrow and house sparrow the SMR are only 2.26 and 2.42 W $kg^{-0.75}$. The SMR in three species of small passerine birds examined by Lukina and Makarova (1967) are probably of the same magnitude. These authors showed that in 1-day-old sparrow chicks, body temperature decreased immediately even if the environmental temperature fell by as little as from 36 to 33 °C, and no increase in oxygen consumption could be detected. A small rise of oxygen consumption in response to cold was noted, depending on species, between the 2nd and 5th day after hatch.

Among mammals, even 1-day-old (5 g) rats increase their heat production in response to cooling, but the summit metabolism, which occurs at an ambient temperature of 34 °C, is only twice as high as the SMR. Since the SMR in the newborn rat is low (2.19 W kg$^{-0.75}$), even at summit metabolism the heat production of this animal is not high. In 21-day-old rats (30–40 g) the SMR is approximately 3.18 W kg$^{-0.75}$, and a fourfold increase in this value may be evoked by cold (Taylor, 1960). Newborn rabbits (60 g) also have a low SMR (*Table 7.3*), and only a twofold increase in the metabolic rate occurs when ambient temperature drops from 35 to 28 °C. A similar increase was also noted in rabbits aged 9–10 days (approximately 200 g) by Hull (1965), but the SMR at this age is almost twice as high as in the newborn rabbit (Piekarzewska, 1977). Newborn kittens (120 g) are blind and have poorly developed locomotor ability, but they are covered with pelage and their SMR (recalculated from data given by Hull, 1965) is relatively high (3.5 W kg$^{-0.75}$). It increases 2.5 times with a drop in ambient temperature from 33 to 18 °C.

In all the above-mentioned altrical animals, the ambient temperature at which the metabolic rate reaches a summit value is high (ranging from 34 °C in rats to 18 °C in kittens) and the maximal rate of heat production achieved in response to cold is considerably lower than that usually found in adults. This justifies the opinion that the newborn of altrical animals have poorly developed mechanisms of thermogenesis. However, this is probably not true of most precocial animals.

In the domestic fowl, a typical precocial bird, the changes in SMR with age and cold-induced heat production were studied carefully by Barott and Pringle (1946). In the youngest chicks examined (1–6 days), the summit metabolism, which occurred at an ambient temperature of 21 °C, was only 2.2 times higher than the SMR (*Table 7.1*), but the basal value in the chicken of that age is approximately 4.3 W kg$^{-0.75}$, already quite high (*Table 7.3*).

Among small precocial mammals the guinea pig is of interest. Its size at birth is approximately the same as that of the rabbit, but besides a dense pelage and well-developed locomotor activity it has a considerably higher SMR (3.65 as compared with 2.69 W kg$^{-0.75}$; *see Table 7.3*). The ability of the guinea pig to increase heat production in response to cold is also superior to that of the rabbit, so its summit metabolism (at 8 °C) is 3.5 times higher than its SMR (Brück and Wünnenberg, 1965).

The pig has a low SMR, but even at 1 day old it can increase its metabolic rate greatly: the summit metabolism (at 0 °C) is 4.5 times as high as the SMR. In piglets aged 5–6 days the summit metabolism is only 3.3 times as high as SMR, but the latter is almost twice as high as that found in newborns (Mount and Stephens, 1970).

A very high SMR has been found in newborn lambs (Graham, Searle and Griffiths, 1974), calves (Brody, 1945; Roy, Huffman and Reineke, 1957) and caribou calves (Hart, Heroux and Cottle, 1961). When there is no rain and only moderate wind these animals can maintain a constant body temperature at an ambient temperature a few degrees below freezing, increasing their heat production to about twice that found at thermoneutrality. However, during rainy and windy weather they increase their metabolic rate fivefold even at air temperature well above freezing, but many nevertheless become hypothermic. In lambs deep hypothermia and death have been observed at +13 °C (Alexander, 1961, 1962, 1964). As noted earlier, a high SMR and remarkable ability to increase heat production in response to cold have also been observed in calves and the pups of pinnipeds.

Changes in deep-body temperature with age

Most newborn animals can maintain a constant body temperature within a limited range of ambient temperatures. Usually, however, the body temperature of newborn animals is regulated at a somewhat lower level than that of adults and it increases gradually (*Figure 7.3*). The difference between

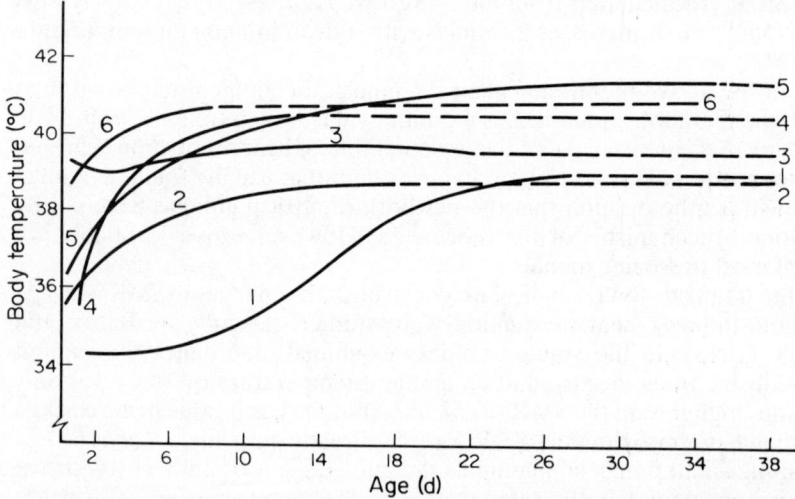

Figure 7.3 Changes in the normal body temperature of animals during postnatal development. (1) Bank vole (Gębczyński, 1975); (2) pig (Newland, McMillen and Reineke, 1952); (3) sheep South Down (Simms, 1971); (4) canary (Lukina and Makarova, 1967); (5) hen (Simkova, 1960); (6) goose (Poczopko, 1967)

the body temperature of 1-day-old animals and that of the adults is generally larger in altrical animals than in precocial. The time necessary for stabilization of body temperature to the adult level varies from several hours, in lambs (Simms, 1971), to approximately one month, e.g. in rabbits (Poczopko, 1969b). At the same ambient temperature with similar thermal insulation, animals with a lower body temperature dissipate less heat than those with high body temperatures. From this point of view the low body temperature of newborns may be regarded as favourable although the advantage deriving from being colder by only 1–4 K is not great.

Hypothermia and its significance

In most natural habitats juvenile animals are endangered by cold. This holds especially true for altrical animals with their poor thermal insulation and deficient physical and chemical thermoregulation. It often happens that these animals must remain for some time without the protection of their parents. For instance, swifts, which live exclusively on airborne insects, are deprived of food during cold and rainy weather. It was observed that during such weather adult swifts inhabiting Great Britain may leave for France and in consequence their offspring must remain unattended for a few days in their nests. Although they soon become completely torpid, brooding by their returning parents brings them back to normal life (Lack, 1956). Similarly, the chicks of the parrot crossbill (*Loxia pytyopsittacus*) hatch in February and March, when the air temperature falls far below 0 °C. The naked chicks are brooded almost continuously during the first days after hatch. Olson (1964) observed that when a chick aged 6 days was abandoned for 90 minutes at an ambient temperature of 0 °C it soon lay motionless, and no breathing could be observed. However, when its mother returned the chick regained normal vitality after 7 minutes' brooding. High tolerance to deep hypothermia has also been observed in young gulls (Koskimies and Lahti, 1964; Dawson and Hudson, 1970).

Altrical mammals also seem to be more tolerant to hypothermia soon after birth than at a more advanced age. Newborn lemmings (*Lemmus lemmus*) were cooled to deep-body temperatures of 2–5 °C. The cooled animals were almost completely stiff and breathing could not be detected, but when they were returned to their nests they regained normal behaviour very soon and no ill effects of the cooling could be observed (Østbye, 1965). The changes with time of the lower lethal body temperature in the developing mouse are well illustrated in *Figure 7.4*, based on data of Lagerspetz (1962).

The question of whether or not newborns of precocial animals are more tolerant to hypothermia than adults has not been definitely answered. In adult bears moderate (about 7 K decrease) but prolonged hypothermia

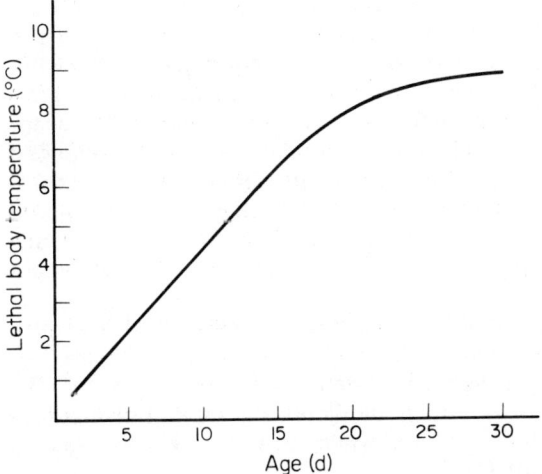

Figure 7.4 Changes with age of the lower lethal body temperature in the mouse. The curve is based on the data of Lagerspetz (1962). (From Poczopko, 1973, courtesy of the editor of Acta physiologica polonica*)*

during hibernation is a normal phenomenon (Folk, 1974). On the other hand, the lethal body temperature for the newborn guinea pig is similar to that for the adult, values being 14–16 °C and 17 °C respectively (Prosser and Brown, 1961).

Domestic animals such as piglets, lambs and calves are born with well-developed mechanisms of thermoregulation. Nevertheless in bad climatic conditions high mortality rates are observed in piglets (Curtis, 1970) and lambs (Alexander, 1961, 1962, 1964). This suggests that for these precocial animals hypothermia is fatal.

Summary and conclusions

GENERAL REMARKS

The thermal insulation of newborn animals varies from negligible to being well developed. It is poor in all altrical animals, from mice and lemmings to the wolverine and polar bear and from small passerine birds to storks. In newborn precocial animals thermal insulation is generally better, and in lambs and caribou for instance it is high. However, in piglets it is rather poor and it is unexpectedly poor in the pups of seals, which after birth are exposed to very cold climatic conditions.

The mechanisms of physical thermoregulation are entirely lacking in some altrical animals at birth (e.g. rats) and in others (e.g. rabbits) they are poorly developed. The mechanisms of chemical thermoregulation are also either not developed or poorly developed in many newborn altrical animals. The SMR at the beginning of post-embryonic life in small passerine birds and rodents, and even in some relatively large animals such as geese and pigs, is lower than the interspecific mean for adult animals ($3.4\,W\,kg^{-0.75}$).

Small altrical animals usually have a poorly developed mechanism of shivering thermogenesis. Although young mammals (but not birds) are usually able to increase their heat production in other ways, in many altrical animals this increase satisfies the thermostatic heat requirements only within a very limited range of environmental temperatures. During the first few days of post-embryonic life for small passerine birds (and probably for shrews, mice etc.), the increase in heat production in response to cooling, if it exists at all, has no practical significance. In altrical animals such as rats and rabbits the metabolic response to cooling is clear, but the summit metabolism is only twice that under standard conditions. In contrast, in such precocial animals as guinea pigs, pigs and lambs, SMR and summit metabolism may differ by a factor of 3–5.

The ambient temperature at which the maximal increase of heat production is achieved (that of summit metabolism) varies considerably: for instance in newborn rats it is 34 °C, in kittens 18 °C, for chickens 21 °C, guinea pigs 8 °C and piglets 0 °C. In day-old lambs the summit metabolism probably occurs at –10 °C (but at +13 °C when the coat is wet and there is wind) and in some seals at –40 °C.

When ambient temperature falls below that evoking the summit metabolism, hypothermia seems unavoidable. Nevertheless, even some altrical animals are born or hatched in the polar regions and, in spite of all their deficiencies in regard to heat production and conservation, they live and grow there. The remarkable tolerance towards hypothermia found in some young altrical animals enables them to withstand short absences by their parents, but the crucial conditions that must be fulfilled are proper nesting and brooding as well as a sufficient supply of food.

The food supply is of major importance. Insectivorous birds certainly have no chance to raise their offspring during the winter, but some granivorous birds do. According to Olson's estimate (1964), each crossbill chick consumes about 65 000 pine seeds before fledging. This kind of food is abundant in the taiga forests. The female of the polar bear certainly has ample milk, which contains 33% fat (Jenness, 1974), to feed its young for three months. Among precocial animals, the pups of seals are extremely dependent on food supply, because their survival in an adverse environment is possible almost exclusively owing to a very high rate of heat production. But the milk they receive, containing from 40 to 53% of fat (depending on species: Jenness, 1974) is used not only as a substrate for heat production but also as the material for a rapidly developing subcutaneous blubber. Koskimies and Lahti (1964) found that ducklings of some wild species, when less than 1 day old, can maintain a constant body temperature on exposure to 0–2 °C for up to 15 hours. However they suddenly become hypothermic when the energy stores in the body are exhausted. It is also worth recalling that the rise in SMR that usually occurs during the first days of post-embryonic life, does not take place in piglets which are fasted from birth (Mount, 1969), and that the metabolic response in rats which have not yet suckled is much smaller than in those which have already filled their stomachs with milk (Taylor, 1960). These examples suggest that food ingestion by newborn animals acts as a kind of starter of the metabolic machinery.

THE EFFECTS OF COLD ON JUVENILE FARM ANIMALS

All the important species of farm animals are precocial and therefore relatively resistant to cold. Nevertheless cold may be harmful for animal production. Cold exposure may entail an increase in feed expenditure, retardation of growth and an increase in the mortality rate.

Curtis (1970) estimated that in the USA approximately 25 per cent of piglets born alive die before weaning. Most of these losses take place during the first three days after birth, and are due to the direct effects of cold. Lambs, as has been mentioned earlier, can survive exposure to temperatures somewhat below freezing when their coat is dry. However, during rainy and windy weather, losses may occur at air temperatures as high as +13 °C. Alexander (1961) also estimated that one in every five lambs died soon after birth in Australia, where they are usually delivered and kept on pasture. Among farm mammals calves are the most resistant to cold (Roy, Huffman and Reineke, 1957; Roy *et al.*, 1971; Webster, 1974; Webster, Gordon and McGregor, 1978). However, many cattle producers maintain that frequent wetting of their coat may predispose young calves to disease and retard their growth. Among domestic birds turkeys are the least cold-resistant and ducks and geese the most, but there is one common feature in the husbandry of young birds. During the first few weeks after hatching birds are usually kept in heated incubators, and incubator failure may produce considerable losses, in part because of natural behavioural thermoregulation. Huddling is beneficial for small groups of birds in their natural habitats, but may result in deaths from suffocation when hundreds of chicks huddle in response to cold in the high stock densities of an intensive production unit.

It is therefore obvious that proper housing, which ensures an optimal thermal environment, can reduce losses due to cold. Less obvious, but also important, is early feeding of newborn animals, as discussed in the previous section.

MEASUREMENTS OF METABOLIC RATE

There is one further matter which warrants special mention in this chapter. This concerns the presentation and assessment of measurements of metabolic rate in young animals and in productive adult animals such as the dairy cow and laying hen.

Kleiber's relation (Kleiber, 1961) is the recognized basis for the comparison of metabolic rates in animals. For a very wide range of species and sizes the assumption that the standard metabolic rate is inversely proportional to body weight to the power of 0.75 enables remarkably accurate prediction of metabolic rates for post-absorptive adult animals. However, it is unlikely that Kleiber himself envisaged that this could be applied either to young, growing, well-fed animals or to productive adults. Indeed, comparison of the values of standard metabolic rate (*Table 7.3*) with Kleiber's reference value, $3.4 \, W \, kg^{-0.75}$, shows that markedly different values are obtained for the majority of young animals. These values also change with age, as shown in *Figure 7.2*. Selective breeding for high productivity is likely to alter the metabolic rate, even without allowing for the increments of growth and digestion. If the standard metabolic rate changes with age, and therefore with weight, there is a strong probability that the exponent of 0.75 is to say the least of doubtful validity for intraspecific comparisons. Serious consideration should therefore be given to the ultimate validity of the standard practice, used in this and other chapters, of expressing the feed consumption and metabolic rates of farm animals as '(per $kg^{0.75}$)', although it does lead to remarkably good predictions of the energy requirements for production.

References

ALEXANDER, G. (1961). *Proc. 4th Int. Congr. on Reproduction*, vol. 3, pp. 630–637. The Hague

ALEXANDER, G. (1962). *J. agric. Res.* **13**, 100–121

ALEXANDER, G. (1964). *Proc. Aust. Soc. Anim. Prod.* **5**, 113–122

ALEXANDER, G., BELL, A. W. and WILLIAMS, D. (1970). *Biologia Neonat.* **15**, 198–210

ALEXANDER, G., BENNET, J. W. and GEMMEL, R. T. (1975). *J. Physiol., Lond.* **244**, 223–234

ALEXANDER, G. and WILLIAMS, D. (1968). *J. Physiol., Lond.* **198**, 251–276

BAROTT, H. G. and PRINGLE, E. M. (1946). *J. Nutr.* **31**, 35–50

BLIX, A. S., GRAV, H. J. and RONALD, K. (1975). *Acta physiol. scand.* **94**, 133–135

BLIX, A. S. and LENTFER, J. W. (1979). *Am. J. Physiol.* **236**, R67–R74

BLIX, A. S. and STEEN, J. B. (1979). *Physiol. Rev.* **59**, 285–304

BRODY, S. (1945). *Bioenergetics and Growth.* Reinhold, New York

BRÜCK, K. and WÜNNENBERG, B. (1965). *Pflügers Arch. ges. Physiol.* **282**, 362–375

CARLSON, A. L., LILIEDAHL, S. O., VERDY, M. and WIRSEN, C. (1964). *Metabolism* **13**, 227–231

CHRISTIANSEN, E. N. (1977). *Comp. Biochem. Physiol.* **56B**, 19–24

CURTIS, S. E. (1970). *J. Anim. Sci.* **31**, 576–587

DAVISON, T. F. (1973). *Comp. Biochem. Physiol.* **44A**, 979–989

DAVYDOV, A. F. and KESKPAJK, Ju. E. (1967). In *Fizjologia Ptic*, pp. 147–152. Akademia Nauk Estonskoj SSSR, Tallinn

DAWSON, W. R. and EVANS, F. C. (1957). *Physiol. Zoöl.* **30**, 315–328

DAWSON, W. R. and EVANS, F. C. (1960). *Condor* **62**, 329–336

DAWSON, W. R. and HUDSON, J. W. (1970). In *Comparative Physiology of Thermoregulation*, vol. 1, pp. 223–310. Ed. by G. C. Whittow. Academic Press, New York and London

FAY, F. H. and RAY, C. (1968). *Zoologica, NY* **53**, 1–18

FOLK, G. E., Jr (1974). In *Progress in Biometeorology*, pp. 581–584. Ed. by S. W. Tromp. Swets & Zeitlinger, Amsterdam

FOSTER, D. O. (1974). *Can. J. Physiol. Pharmac.* **52**, 1051–1061

FREEMAN, B. M. (1964). *Br. Poult. Sci.* **5**, 263–267

FREEMAN, B. M. (1967). *Br. Poult. Sci.* **8**, 147–152

FREEMAN, B. M. (1970). *Comp. Biochem. Physiol.* **33**, 219–230

FREEMAN, B. M. (1971). In *Non-shivering Thermogenesis*, pp. 83–96. Ed. by L. Janský. Academia, Prague

FREEMAN, B. M. (1975). *J. thermal Biol.* **1**, 59–60

FREEMAN, B. M. (1977). *J. thermal Biol.* **2**, 145–149

FREEMAN, B. M. and MANNING, A. C. C. (1977). *Comp. Biochem. Physiol.* **57A**, 211–214

GĘBCZYŃSKI, M. (1969). *Acta theriol.* **14**, 427–440

GĘBCZYŃSKI, M. (1975). *Acta theriol.* **20**, 379–437

GIAJA, G. (1938). *Homeothermie et Thermoregulation.* Hardman, Paris

GRAHAM, N. McC., SEARLE, T. W. and GRIFFITHS, A. D. (1974). *Aust. J. agric. Res.* **25**, 957–971

GRAV, H. J. and BLIX, A. S. (1976). *Can. J. Physiol.* **54**, 409–412

GRAV, H. J. and BLIX, A. S. (1979). Cited by Blix and Steen (1979), *q.v.*

GULICK, A. (1937). *Am. J. Physiol.* **119**, 323

HALDMAIER, G. (1971). In *Non-shivering Thermogenesis*, pp. 73–80. Ed. by L. Janský. Academia, Prague

HART, J. S. (1971). In *Comparative Physiology of Thermoregulation*, vol. 2, pp. 2–149. Ed. by G.C. Whittow. Academic Press, New York and London

HART, J. S., HEROUX, O. C. and COTTLE, W. H. (1961). *Can. J. Zool.* **39**, 845–856

HEIM, T. and HULL, D. (1966). *J. Physiol., Lond.* **186**, 42–55

HEMMINGSEN, A. (1960). *Rep. Steno meml Hosp.* **9**, 1–110

HIMMS-HAGEN, J., BEHRENS, W., HBOUS, A. and GREENWAY, D. (1976). In *Regulation of Depressed Metabolism and Thermogenesis*, pp. 243–260. Ed. by L. Janský, X. J. Musacchia and G. Wilber. Charles C. Thomas, Springfield, Illinois

HISSA, R. (1964). *Experientia* **20**, 326–327

HISSA, R. (1968). *Annls zool. fenn.* **5**, 345–385

HISSA, R. and PALOKANGAS, R. (1970). *Comp. Biochem. Physiol.* **33**, 941–953

HULL, D. (1965). *J. Physiol., Lond.* **177**, 192–202

HULL, D. (1966). *Br. med. Bull.* **22**, 92–96

HULL, D. (1973). In *Comparative Physiology of Thermoregulation*, vol. 3, pp. 167–200. Ed. by G. C. Whittow. Academic Press, New York and London

HULL, D. and HARDMAN, M. J. (1970). In *Brown Adipose Tissue*, pp. 97–115. Ed. by O. Lindberg. Elsevier, New York

INGRAM, D. L. (1964). *Res. vet. Sci.* **5**, 357–364

IVANOV, K. P. (1962). *Fiziol. Zh. SSSR* **48**, 436–443

IVANOV, K. P. and ALIMUCHAMEDOV, A. (1963). *Fiziol. Zh. SSSR* **49**, 484–488

JANSKÝ, L. (1973). *Biol. Rev.* **48**, 85–132

JASTRZĘBSKI, M., PIETRAS, M. and GAJEWSKA. M. (1977). *Roczn. Nauk roln., Ser. B* **42**, 55–62

JENNESS, R. (1974). In *Lactation*, vol. 3, pp. 3–107. Ed. by B. R. Larson and V. R. Smith. Academic Press, New York

KACIUBA-UŚCILKO, H. (1971). *Bull. Acad. pol. Sci. Sér. Sci. biol.* **19**, 145–151

KACIUBA-UŚCILKO, H. and INGRAM, D. L. (1977). *Comp. Biochem. Physiol.* **56C**, 53–55

KACIUBA-UŚCILKO, H., MOUNT, L. E. and LEGGE, F. (1970). *J. Physiol., Lond.* **206**, 229–241

KACIUBA-UŚCILKO, H. and POCZOPKO, P. (1973). *Experientia* **29**, 108–109

KACIUBA-UŚCILKO, H. and POCZOPKO, P. (1975). In *Temperature Regulation and Drug Action*, pp. 202–208. Ed. by J. Lomax, E. Schönbaum and J. Jacob. Karger, Basel

KING, J. R. and FARNER, D.S. (1961). In *Biology and Comparative Physiology of Birds*, vol. 2, pp. 215–288. Ed. by A. J. Marshall. Academic Press, New York and London

KLEIBER, M. (1947). *Physiol. Rev.*, **27**, 511–541

KLEIBER, M. (1961). *The Fire of Life*. John Wiley, New York and London

KLEIBER, M., SMITH, A. H. and CHERNIKOFF, T. (1965). *Am. J. Physiol.* **186**, 9–12

KOSKIMIES, J. and LAHTI, L. (1964). *Auk* **81**, 281–307

KOSTYAN, E. Ya. (1954). *Zool. Zh.* **33**, 207–215

KOTRBÁCEK, V. (1973). *Acta vet., Brno* **42**, 15–21

KOTRBÁCEK, V. (1977). *Acta vet., Brno* **46**, 71–80

LACK, D. (1956). *Swifts in a Tower*. Methuen, London

LAGERSPETZ, K. Y. H. (1962). *Experientia* **18**, 282–284

LAGERSPETZ, K. Y. H. (1964). *Annls Med. exp. Biol. fenn.* **42**, 43–48

LAGERSPETZ, K. Y. H. (1966). *Annls Med. exp. Biol. fenn.* **44**, 71–73

LeBLANC, J. and MOUNT, L. E. (1968). *Nature, Lond.* **217**, 77–78

LEDUC, J. and RIVEST, P. (1969). *Revue can. Biol.* **28**, 49–66

LUKINA, E. V. and MAKAROVA, A. R. (1967). In *Fizjologia Ptic*, pp. 153–157. Akademia Nauk Estonskoj SSSR, Tallinn

MAHER, W. J. (1964). *Ecology* **45**, 520–528

MISSON, B. H. (1977). *J. thermal Biol.* **2**, 107–110

MOUNT, L. E. (1960). *J. agric. Sci., Camb.* **55**, 101–105

MOUNT, L. E. (1963). *J. Physiol., Lond.* **168**, 698–705

MOUNT, L. E. (1964). *J. Physiol., Lond.* **170**, 286–295

MOUNT, L. E. (1968). *The Climatic Physiology of the Pig.* Arnold, London

MOUNT, L. E. (1969). *Proc. Nutr. Soc.* **28**, 52–56

MOUNT, L. E. and HOLMES, W. C. (1969). In *Energy Metabolism of Farm Animals*, pp. 311–318. Ed. by K. L. Blaxter, J. Kielanowski and G. Thorbek. Oriel Press, Newcastle upon Tyne

MOUNT, L. E. and ROWELL, J. G. (1960). *Nature, Lond.* **186**, 1054–1055

MOUNT, L. E. and STEPHENS, D. B. (1970). *J. Physiol., Lond.* **207**, 417–427

MYRCHA, A., PINOWSKI, J. and TOMEK, T. (1973). In *Productivity, Population Dynamics and Systematics of Granivorous Birds*, pp. 59–83. Ed. by S. C. Kendeigh and J. Pinowski. PWN, Warsaw

NEWLAND, H. W., McMILLEN, W. N. and REINEKE, E. P. (1952). *J. Anim. Sci.* **11**, 118–133

OLSON, V. (1964). *Br. Birds* **57**, 118–123

ØSTBYE, E. (1965). *Nytt Mag. Zool.* **12**, 65–75

PALOKANGAS, R. and HISSA, R. (1971). *Comp. Biochem. Physiol.* **38A**, 743–750

PIEKARZEWSKA, A. B. (1977). *Acta theriol.* **22**, 159–180

POCZOPKO, P. (1961). *J. cell. comp. Physiol.* **57**, 175–184

POCZOPKO, P. (1967). *Acta physiol. pol.* **18**, 425–434

POCZOPKO, P. (1969a). *Acta theriol.* **14**, 449–462

POCZOPKO, P. (1969b). In *Energy Metabolism of Farm Animals*, pp. 361–367. Ed. by K. L. Blaxter, J. Kielanowski and G. Thorbek. Oriel Press, Newcastle upon Tyne

POCZOPKO, P. (1971). *Acta theriol.* **16**, 1–21

POCZOPKO, P. (1972). *Acta physiol. pol.* **23**, 843–851

POCZOPKO, P. (1973). *Acta physiol. pol.* **24** (Suppl. 6), 101–115

POCZOPKO, P. (1979). *Acta theriol.* **24**, 125–136

POCZOPKO, P., KACIUBA-UŚCILKO, H. and JUSIAK, R. (1973). *Roczn. Nauk roln., Ser. B* **94**, 135–145

POCZOPKO, P. and PONIEWIERSKI, W. (1977). *Bull. Acad. pol. Sci. Sér. Sci. biol.* **25**, 483–486

POCZOPKO, P. and ULIASZ, M. (1975a). *Acta physiol. pol.* **26**, 149–158

POCZOPKO, P. and ULIASZ, M. (1975b). *Acta physiol. pol.* **26**, 241–248

POCZOPKO, P. and ULIASZ, M. (1977). *Acta physiol. pol.* **28**, 51–60

POCZOPKO, P., WITKOWSKA, H. and ULIASZ-PONIEWIERSKA, M. (1979). *Roczn. Nauk roln., Ser. B* **99**, 55–64

PROSSER, C. L. and BROWN, F. A. J. (1961). *Comparative Animal Physiology.* W.B. Saunders, Philadelphia and London

RIDDLE, O., NUSSMANN, T. C. and BENEDICT, E. G. (1932). *Am. J. Physiol.* **101**, 251–259

ROWLATT, U., MROSOVSKY, N. and ENGLISH, A. (1971). *Biologia Neonat.* **17**, 53–83

ROY, J. H. B., HUFFMAN, C. F. and REINEKE, E. P. (1957). *Br. J. Nutr.* **11**, 373–381

ROY, J. H. B., STOBO, I. J. F., GASTON, H. J., GANDERTON, P. and SHOTTON, S. M. (1971). *Br. J. Nutr.* **26**, 363–381

SAARELA, S. and HISSA, R. (1977). *Comp. Biochem. Physiol.* **56C**, 25–30

ŠIMKOVA, A. (1960). *Živočišna Výroba* **5** (33), 449–460

SIMMS, R. H. (1971). *J. Anim. Sci.* **32**, 296–300

SLEE, J. (1978). *Anim. Prod.* **27**, 43–119

SMITH, R. E. and HORWITZ, B. A. (1969). *Physiol. Rev.* **49**, 330–425

TAYLOR, P. M. (1960). *J. Physiol., Lond.* **154**, 153–168

WEBSTER, A. J. F. (1974). In *Heat Loss from Animals and Man,* pp. 205–231. Ed. by J. L. Monteith and L. E. Mount. Butterworths, London

WEBSTER, A. J. F., GORDON, J. G. and McGREGOR, R. (1978). *Anim. Prod.* **26**, 85–92

WEKSTEIN, D. R. and ZOLMAN, J. P. (1969). *Fedn Proc. Fedn Am. Socs exp. Biol.* **22**, 1023–1028

8

THERMAL INFLUENCES ON POULTRY

M. VAN KAMPEN
Laboratory for Veterinary Physiology, Utrecht, The Netherlands

Introduction

This chapter discusses the physiological responses of laying hens to various environmental factors. Knowledge of these responses should allow definition of an environment in which maximum food conversion will occur with the highest production of egg material: that is, those conditions which allow the most efficient production of eggs. When a hen has equal energy intake and output, then the intake of metabolizable energy (ME) is equal to the sum of heat production and net energy for egg production. By changing climatic factors such as dry- and wet-bulb temperatures and wind speed, ME intake and heat production may change differently, hence net energy for egg production will change, unless the hen is losing or gaining body mass or changing in body composition. Exploitable body reserve is limited and if, after a change in temperature, ME intake was initially insufficient, it has to be corrected within an acclimation period by a lower maintenance requirement concurrent with the lower body weight, by an increase in ME intake or by reduced egg production.

Changes in food intake and temperature may also cause a change in water intake. Especially at high temperatures the evaporative water loss may be considerable, although this depends on the humidity level. This water loss must be compensated by an increased water intake or a decreased water loss via excreta. In general, drinking-water temperature follows ambient temperature. Water is usually supplied *ad libitum*. Controlling the amount and temperature of drinking water may have several beneficial effects.

Ambient temperature

INFLUENCE ON FOOD INTAKE

The effect of temperature on food intake has been reviewed by Sykes (1977), Balnave, Farrell and Cumming (1978) and Byerly (1979). These authors derived equations for ME intake from the published data. Balnave, Farrell and Cumming (1978) suggested the equation

$$\text{ME}_m = 388W^{0.75}\, e^{0.027(22 - T)} + 8.67E_g \tag{8.1}$$

where ME_m is the minimum ME requirement, expressed in kJ d^{-1}; W is the body weight in kg, and E_g is the egg product in g d^{-1}. Byerly (1979) proposed the relation

$$ME = (558 - 5.58T)W^{0.75} + 33.5\,\Delta W + 2.3E_g \qquad (8.2)$$

where ME, per bird, is expressed in kJ d^{-1}; W is the body weight in kg; ΔW is the weight gain in g d^{-1} and E_g is the egg product in g d^{-1}. For a bird of 1.5 kg Sykes (1977) gives the simplified relation

$$ME = 1690 - 20.1T \qquad (8.3)$$

A temperature factor is not included in many of the published equations for the prediction of food requirements because the experiments were done at a fixed temperature (usually between 20 and 25 °C), which is assumed to be within the zone of thermoneutrality. However, while van Es *et al.* (1973) found a lower critical temperature of 10 °C in laying hens, Arieli, Meltzer and Berman (1979) reported values of 15.8 °C in winter and 24.2 °C in summer. Furthermore, many data suggest that there is no clear zone of thermoneutrality in the fowl, but only a narrow range in which metabolism is minimal. This range is between 30 and 40 °C, and depends on factors such as strain, degree of acclimation, food intake, production level and physical activity.

Comparisons of the predicted ME intakes (*Table 8.1*) reveal that in the range 20–25 °C the differences are comparatively small, while at low and high ambient temperatures the equations predict values higher than the measured values. From the data of Wilson, McNally and Ota (1957), Zimmerman and Snetsinger (1975), Lillie *et al.* (1976), Vo, Boone and Johnston (1978) and *Figure 8.4* (M. van Kampen, unpublished; *see* p. 140), it may be calculated that the change in food intake for each degree Celsius change in temperature is less than 1 g d^{-1} when the ambient temperature is below 20 °C, and in the ranges 20–25, 25–30 and 30–35 °C it is 1.3, 2.3 and 4 g d^{-1} respectively. These values are in accord with those suggested by

Table 8.1   INTAKE OF METABOLIZABLE ENERGY BY LAYING HENS (kJ kg$^{-0.75}$ d^{-1}) AT SPECIFIED AMBIENT TEMPERATURES

T_a (°C)	Balnave, Farrell and Cumming (1978)	Byerly (1979)	Sykes (1977)	Davis, Hassan and Sykes (1973)
10	1013*	924†	1099†	802‡
15	946	896	1025	778
20	886	869	951	755
25	835	841	876	731
30	789	813	802	708
35	750	785	728	684
Compound % change per °C	1.21	0.65	1.66	0.64

Values are measured or predicted from equations proposed by the authors, as indicated
*Minimum ME requirement   †Calculated ME intake   ‡Actual ME intake
The calculated values in columns 2 and 3 are for a laying hen, 25–45 weeks of age with a body weight of 1.7 kg, growth of 3 g d^{-1} and an egg production of 55 g d^{-1}

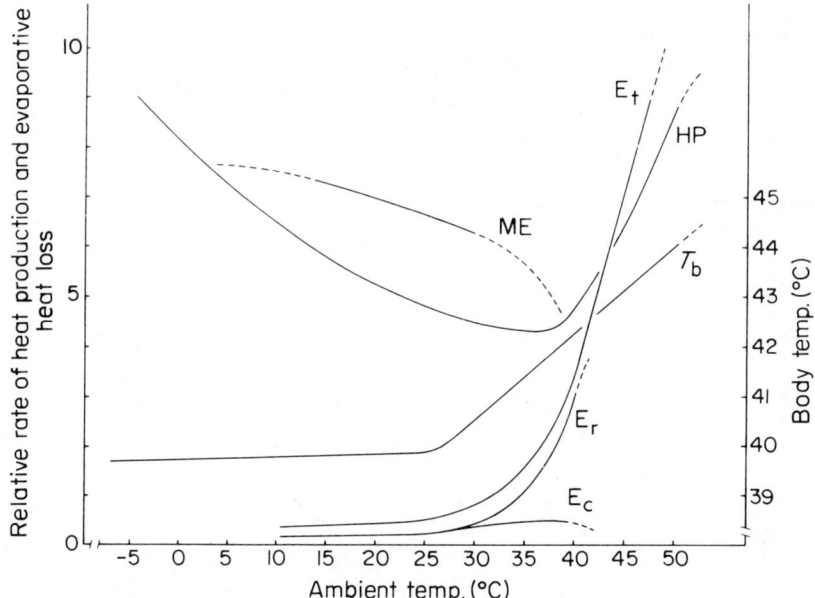

Figure 8.1 Schematic diagram showing the relationship between ME intake, heat production, evaporative heat loss and body temperature and ambient temperature in the fowl. (T_b, body temp; HP, heat production; E_t, E_r and E_c, total, respiratory and cutaneous evaporative heat loss. (After van Kampen, 1976)

Emmans (1974). This indicates that if the temperature range is wide enough the prediction equation will be a hyperbola, as in *Figure 8.1.*

INFLUENCE ON HEAT PRODUCTION

Heat production decreases with increasing ambient temperature up to about 35 °C (*Figure 8.1*); over limited parts of this range the relationship between heat production and ambient temperature can be described by a linear equation, but above this critical temperature the metabolic rate rises. In the temperature range which includes the critical temperature (where metabolic rate is minimal), the relationship is either quadratic (Nichelmann, Thomas and Lyhs, 1974; van Kampen, 1974) or may be described by two linear equations which intersect at the critical temperature. In short-term experiments with unacclimatized hens the critical temperature is about 33 °C (van Kampen, 1974; Richards, 1977; *Figure 8.2*).

Calculated values of the heat production at various ambient temperatures are presented in *Table 8.2*. The difference in sensitivity to temperature between columns 2 and 3 is probably a result of differences in behavioural temperature regulation between singly and group-housed birds, as illustrated in *Figure 8.6* (*see* p. 144). The slope for the group is lower than for the single bird. The differences in slope may also be caused by differences in feather cover and in the uninsulated areas of the fowl. The difference in heat production between poorly and well feathered layers increases with

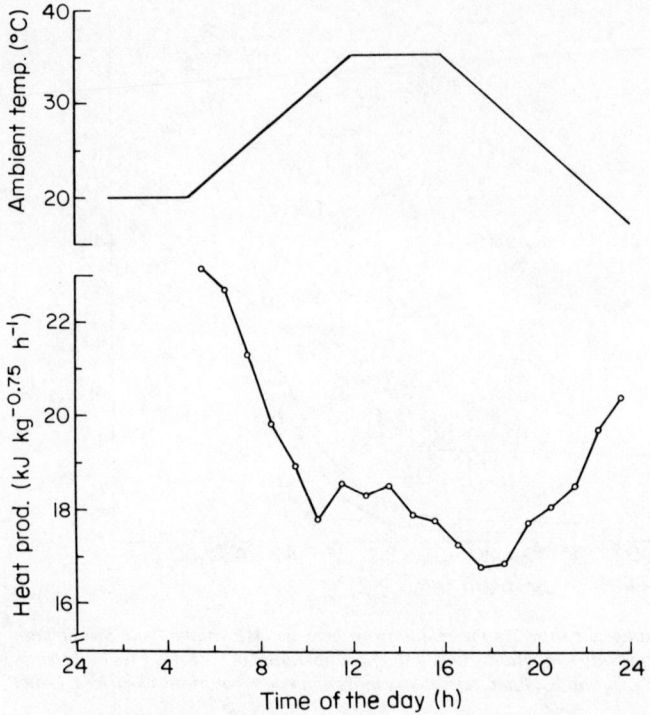

Figure 8.2 Mean heat production of six single White Leghorn hens during a simulated 'tropical' day. (Adapted from van Kampen, 1977)

Table 8.2 HEAT PRODUCTION OF FOWLS (kJ kg $^{-0.75}$ d^{-1}) AT SPECIFIED AMBIENT TEMPERATURES

T_a (°C)	O'Neill, Balnave and Jackson (1971)*	Davis, Hassan and Sykes (1973)†	Van Kampen (1974)‡
10	573	586	639
15	526	562	575
20	479	539	523
25	433	516	484
30	386	493	457
35	339	470	442
Compound % change per °C	2.12	0.89	1.49

*Relation for cockerels, body weight 2.5–2.8 kg; temperature range 15–34 °C:

$$M = 667 - 9.37T$$

†Relation for layers, body weight 1.65–1.90 kg; temperature range 7.2–35 °C:

$$M = 632 - 4.64T$$

‡Relation for layers, body weight 1.5–1.9 kg; temperature range –5 to +40 °C:

$$M = 803 - 18.9T + 0.245T^2$$

decreasing temperature and reaches a factor of two at about 0 °C (Romijn and Lokhorst, 1961; Richards, 1977). Some areas of the fowl, e.g. comb and wattles, are highly vascularized. In White Leghorn hens the surface area of the head appendages is relatively large, some 7–8 per cent of the total body surface, and through them up to 25 per cent of the sensible heat can be dissipated (van Kampen, 1974). This may explain the high maintenance ME requirement of the White Leghorns, although Leeson and Morrison (1978) could not correlate efficiency of food use with the surface area of the comb and wattles. The reduction in heat production per 1 K rise is greater for cockerels than for layers. If this difference is the result of egg production, as has been suggested by Balnave, Farrell and Cumming (1978), then the egg production at lower ambient temperatures must be very low, because the regression lines for heat production against ambient temperature for layers and cockerels intersect in the low-temperature range. Consistent with this, the curves of ME intake and heat production for laying hens, shown in *Figure 8.1,* are not parallel in the lower and upper ambient temperature range, indicating that egg production cannot be maintained unless stored energy is used. Indeed, birds subjected to low temperatures lose weight (Campos, Wilcox and Shaffner, 1962; Davis, Hassan and Sykes, 1973). The relatively lower ME intake may be also a consequence of the maximum filling capacity of the intestinal tract and the length of the light period.

INFLUENCE ON EGG PRODUCTION

Egg production can be maintained in hens allowed a limited intake of food, though accompanied by a substantial loss of body weight (Sykes, 1972), and in hens subjected to high ambient temperatures (Davis, Hassan and Sykes, 1973). Therefore, provided that the diet is suitably formulated, production is relatively constant (*Table 8.3*) in the temperature range 10–30 °C. The decrease in ME intake and heat production with increasing ambient temperatures, while egg production is maintained, results in a continuous improvement in the food requirement per egg. However, the food required per kilogram of egg is at a minimum at about 25 °C (*Table 8.4*) because of changes in egg size with temperature. It is evident from *Tables 8.3* and *8.4* that egg size varies inversely with temperature. At temperatures above 30 °C there is a tendency towards both reduced egg production and a progressive decline in egg size. The decreased egg production may be a result of decreased food intake and a direct effect of high temperature. By restricting the energy intake, MacLeod, Tullett and Jewitt (1979) found that there is loss of body weight and egg weight. Consequently, the smaller eggs may have been a reflection of the smaller body mass.

It is also possible that the bird may be less able to synthesize lipid when it has a smaller amount of body fat. Thus chicks reared at high ambient temperatures have a lower mature body weight, contain less fat, and produce smaller eggs (Payne, 1966a; van Kampen, 1980). Similarly, the loss of body weight in hens on restricted energy intake or housed at high temperatures is principally due to the loss of body fat (Davis, Hassan and Sykes, 1973). In the egg, the weight of the yolk is decreased more than the weight of the albumen in energy-restricted hens (Simon, 1973) and in hens at

Table 8.3 RESULTS OF TRIAL MEASUREMENTS OF EGG SIZE AND EGG PRODUCTION BY WHITE LEGHORNS AT VARIOUS AMBIENT TEMPERATURES

Reference	Temperature (°C)*	Egg size (g)	Laying %†
Al-Soudi and Al-Jebouri (1979)	± 10	56.0	44.0
	± 45	51.7	45.0
Bray and Gesell (1961)	5.6	58.0	88.3
	24.4	56.9	86.5
	30.0	52.9	87.8
Davis, Hassan and Sykes (1972)	7	54.7	92.0
	35	53.7	89.5
Davis, Hassan and Sykes (1973)	7.2	64.9	76.2
	15.6	59.3	86.3
	23.9	59.6	85.1
	29.4	60.1	82.1
	35.0	58.5	79.2
Mowbray and Sykes (1971)	± 12.5 (10–15)	56.0	88
	± 22.2 (13–35)	53.8	85
	± 23.0 (18–30)	54.5	88
	30.0	54.0	87
Payne (1966b)	18	57.3	78.1
	± 20.3 (15–24)	56.3	81.9
	± 22.5 (18–30)	56.0	82.6
	24	55.9	81.4
	± 25.5 (18–30)	55.6	84.4
	30	55.8	84.2
Wilson, Itoh and Siopes (1972)	10	61.3	70.2
	23	60.0	65.8
	36	48.3	62.0

*A plus-or-minus sign indicates the mean of a fluctuating temperature; temperature limits within brackets.
†Laying percentage is the number of eggs laid per day by 100 birds.

Table 8.4 THE INFLUENCE OF AMBIENT TEMPERATURE ON LAYER PERFORMANCE

Mean temp.* (°C)	Relative production (%)	Relative egg size (%)	Relative food required per egg (%)	Relative food required per unit of egg mass (%)
16	100	100	100	100
18	100	100	95	95
21	100	100	91	91
24	100	99	88	89
27	99–100	96	86	89
29	97–100	93	85	91
32	94–100	86	84	98

After Zimmerman and Snetsinger (1975)
*The daily temperature range was 4.7–7.1 K

38 °C (Smith and Oliver, 1972). This point could be proved if a positive correlation could be shown between body fat mass and egg weight in a group of chickens of similar body weight. However, the variation in body fat is very large and there is little or no such correlation.

Smith and Oliver (1972) found that hens kept at 21 °C with the same food intake as hens at 32 °C produced heavier eggs. These results suggest a direct temperature effect, but the body weight of the hens at 32 °C was lower. However, night heat stress has been found to cause a significant decrease in egg production (Wolfenson *et al.*, 1979). The decrease in egg weight may be the result of a decreased blood flow to the ovarian follicles (Wolfenson *et al.*, 1978).

The time interval between ovipositions is also increased at high ambient temperatures (Nordstrom, 1973; Wolfenson *et al.*, 1979), with a concurrent decrease in clutch size and egg production. This may be a reason to extend the light period at high temperatures. A change in temperature from 38 to 21 °C resulted in an immediate change in egg shell thickness but not in egg size (Smith and Oliver, 1972), which indicated that the reduction in egg size was not caused by a direct temperature effect.

Egg production is also decreased by low temperatures, particularly if the temperature drop is sudden (Campos, Wilcox and Shaffner, 1962) or prolonged (Bruckner, 1936). However, low temperature has little effect on egg weight (Campos, Wilcox and Shaffner, 1962; Wilson, McNally and Ota, 1957).

A COMPARISON BETWEEN CONSTANT AND FLUCTUATING AMBIENT TEMPERATURES

It has been suggested that hens exposed to varying temperatures lay more eggs than those in a constant thermal environment (Mueller, 1961; Payne, 1966b; Wilson, Itoh and Siopes, 1972). However, the range of fluctuation is important. Trials in which the constant temperature was equal to the mean of the cycling temperature (18 or 13–23 °C, 29 or 24–34 °C and 31 or 26–36 °C did not reveal differences in food intake, egg production and egg weight (Zimmerman and Snetsinger, 1975). There is no reason for a difference in heat production and ME intake between cycling and constant temperature if they are linearly related to ambient temperature. This means that only when the temperature fluctuates outside the range 10–30 °C, where ME intake and heat production are in no way parallel to each other, will egg production be worse than that at the equivalent constant temperature. However, while egg production is constant in the temperature range 10–30 °C, egg size and food intake vary: at 30 °C both will be minimal and at 10 °C maximal.

Acclimation

Responses to climatic changes can be divided into three phases: neuronal, hormonal and morphological. The neuronal phase is the first and is seen immediately, while the morphological response is manifest only in the long term. Harrison and Biellier (1969) found that fowls exposed to a change in temperature from 35 to 5 °C produced an increase in metabolic rate for 2 hours (neuronal response) and a second rise (hormonal response) after 2 days. Chicken reared at high ambient temperatures developed enlarged combs and wattles (Lamoreux, 1943), they contained less fat and their feather cover was less than that of controls (van Kampen, 1980). Respiratory rate, blood pressure, pulse rate (Harrison and Biellier, 1969) and egg shell thickness (Smith and Oliver, 1972) approached new values within one day of a temperature change in either direction. Rectal temperature is adjusted after 2–5 days (Hillerman and Wilson, 1955; Harrison and Biellier, 1969) and heat production after 3–12 days (Harrison and Biellier, 1969; Shannon and Brown, 1969). Davis, Hassan and Sykes (1973) found an acclimation period of 1–2 weeks for food intake, egg production and egg size. They claimed that acclimation to high temperatures took longer than to low temperatures and suggested that about six weeks is needed for body weight to become stabilized in a hot environment. Generally speaking, once body weight has stabilized after a temperature change, the bird will be well acclimatized.

Water

Although water has been described as 'the forgotten nutrient', some work has been done in recent years (*see* Leeson, Summers and Moran, 1976; Hill, Powell and Charles, 1979). High-quality drinking water is becoming scarcer and more expensive. Intermittent or restricted watering programmes may result in a saving of water, in a decreased production of wet excreta (Maxwell and Lyle, 1957), in improved food efficiency (Maxwell and Lyle, 1957; Hill and Richards, 1975; Spiller, Dorminey and Arscott, 1976) and in a reduction in food consumption (Savory, 1978). Controlling the temperature of the drinking water may have also beneficial effects at low and high ambient temperatures.

WATER INTAKE

Water intake is affected by such factors as food intake, composition of the diet, growth, egg production, housing and watering system, water temperature and climatic conditions. It increases with increasing ambient temperature, as shown in *Figure 8.3*. Up to 27 °C, water intake increases by about 3 g per day per °C, but above 29 °C the increase may be as great as 11 g per day per °C.

Figure 8.3 clearly shows the large differences in water intake observed in the trials. The differences are a consequence not only of the above-mentioned factors but also of breed, strain and individual differences. Bordas, Obeidah and Mérat (1978) showed, for two strains, that one hen in six drank about twice as much as other hens. If the water intake can be reduced by 200 g by selection, and the drinking-water temperature is 20 °C, then the excreta production will be about 200 g less and the hen has to warm up to body temperature 200 g less of water. Less excreta results in smaller drying costs and/or transportation costs and a prevention of a 20 K rise in temperature of 200 g of water saves 17 kJ or 1.4 g of food per bird per day.

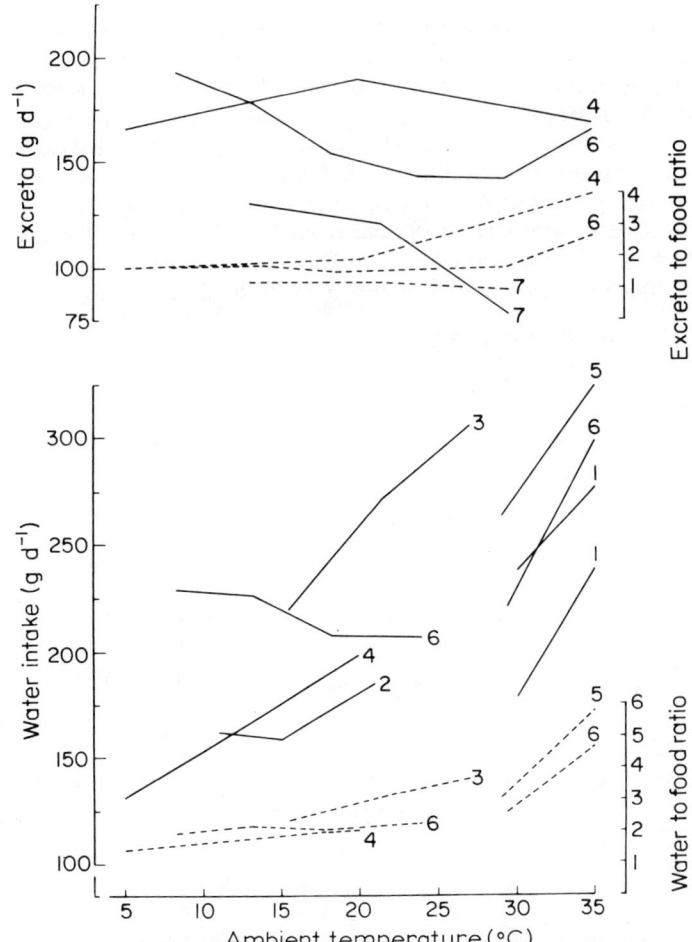

Figure 8.3 Variation with ambient temperature of the water intake and excreta output (————) and water and excreta to food ratios (- - - -). Data for layers with body weights of 1.2–1.8 kg. (Results of 1. Okumura, Tasaki and Saito, 1977; 2. Randall, 1978; 3. Zimmerman and Snetsinger, 1975; 4. Van Kampen, unpublished; 5. Vo, Boone and Johnston, 1978; 6. Wilson, McNally and Ota, 1957 and 7. Lillie et al., 1976)

If drinking is initiated by a deficit in the body water pool, then the sharp increase in water intake at high ambient temperatures may be caused by the increased water loss required for thermoregulation. Hill, Powell and Charles (1979) have stated that food intake is one of the most important factors which affect water intake. If it were the only factor then one should expect a high water intake at low ambient temperatures and a low water intake and a constant water to food ratio at high ambient temperatures. However, *Figure 8.3* indicates the reverse.

Sykes (1977) suggested that following a sudden increase in ambient temperature there may be a temporary water deficit, but M. van Kampen (unpublished) found that when the temperature was raised from 20 to 35 °C in the morning, after 8 hours White Leghorn hens had a higher body weight increase and a lower food intake than the controls at 20 °C. *Figure 8.4* shows that on the first day after a temperature increase the water intake is higher than on subsequent days, whereas the body weight has not changed. It is not likely, therefore, that there will be a water deficit, unless the deficiency is even more severe on the other days. But if that is the case a reduction in urine production would be expected, though van Kampen (1981) found a highly significant correlation between water intake and urine production over a temperature range from 5 to 35 °C. However, a sudden decrease in ambient temperature from 35 to 5 °C did cause a temporary water deficit, because (as shown in *Figure 8.4*) on the first day the water

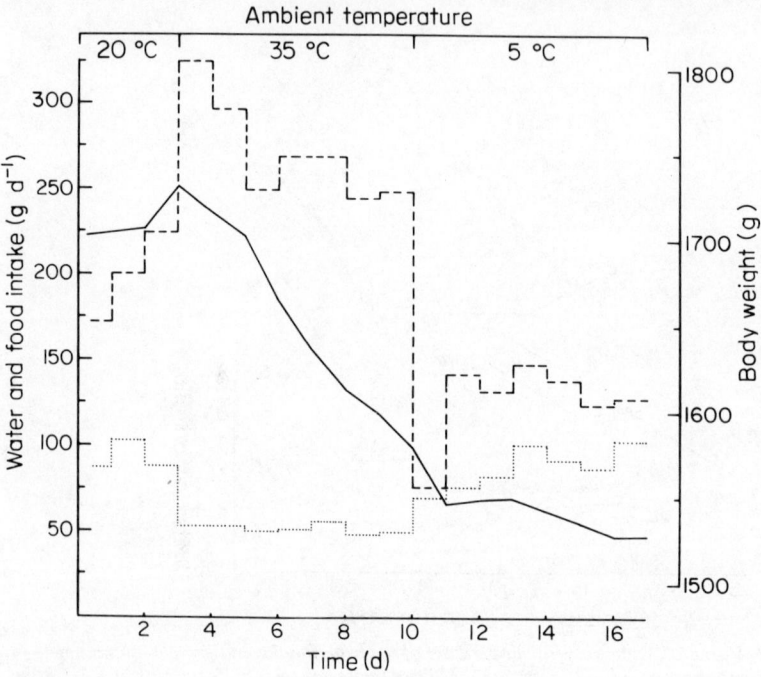

Figure 8.4 Acclimation of White Leghorn hens to sudden temperature changes. (———) Body weight; (. . . . .) food intake; (- - - - -) water intake

intake was so reduced and the food intake increased that the water to food ratio was only 1.07. Consequently, the body weight decreased even more than on the previous days at 35 °C.

DRINKING-WATER TEMPERATURE

In general, water temperature follows ambient temperature. A reduction in water intake at a low ambient temperature may also be caused by the water temperature *per se*. Gold *et al.* (1977) reported a 25 per cent reduction in rats' water intake when the water was cool, while Gentle (1979) has observed a similar response in the fowl. White Leghorn hens at an ambient temperature of 35 °C drank 10 per cent less during the first 8 hours when the temperature of the water was 0 °C than when it was 35 °C (M. van Kampen, unpublished). If the reduced uptake is temporary, as in rats, and if the amount of water intake is positively correlated with the food intake, then in order to maintain the production level at high ambient temperature, one should supply cool water not only on an exceptionally hot day but, if possible, one or two days before the hot weather occurs. At high ambient temperatures heat is dissipated principally by evaporation. By supplying cold water, the respiratory rate of the birds may decrease because there will be less heat left to dissipate after the water has been warmed to body temperature. Concurrent with a reduction in respiratory movements will be a reduction in energy cost and perhaps more energy available for production. Wilson and Edwards (1952) also found a significantly lower body temperature in controls than in hens under thermal stress allowed cool water, but no significant increase in water and food consumption. A recent seven-month study with White Leghorn hens under heat stress given access to cool water, has shown that food intake was improved and the drop in laying performance and in body weight were reduced (Musharaf and Janssen, personal communication). Conversely, warm drinking water at low ambient temperature may improve the efficiency of food use, as less energy is needed to warm water to body temperature.

However, when the water temperature is too high the water intake will be reduced and restricting the water intake may lead to a reduced food intake (Savory, 1978) and reduced efficiency of food use.

WATER TEMPERATURE PREFERENCE

When they have been denied access to water for 16 hours overnight, White Leghorn hens exhibit no significant preference between drinking water at 0 °C, 20 °C or 35 °C, whatever the ambient temperature (*Table 8.5a*). However, with free access at regular intervals of 15 minutes per hour for 8 hours, the intake of water at 35 °C was significantly the lowest at all the ambient temperatures (5, 20 and 35 °C). Only at an ambient temperature of 35 °C did the hens prefer the water at 0 °C to that at 20 °C (*Table 8.5b*).

Table 8.5 INTAKE OF DRINKING WATER OF DIFFERENT TEMPERATURES, EXPRESSED AS A PERCENTAGE OF TOTAL WATER INTAKE, AT VARIOUS AMBIENT TEMPERATURES (T_a)

T_a (°C)	*Water intake (%) for a water temperature of:*		
	0 °C	*20 °C*	*35 °C*
(a) 5	41.2[a]	44.0[a]	14.8[a]
20	41.0[a]	36.4[a]	22.6[a]
35	33.2[a]	33.3[a]	33.5[a]
(b) 5	50.0[a]	40.0[a]	10.0[b]
20	40.4[a]	49.9[a]	9.7[b]
35	55.7[a]	37.7[b]	6.6[c]

White Leghorn hens were allowed access to water for 15 minutes after a deprivation period of 16 hours (a) and at 1 h intervals for 15 min during 8 hours (b). M. van Kampen (unpublished) The letters in a horizontal column indicate significant differences (e.g. ab) and lack of significance (aa); $n = 12$, $p < 0.05$

WATER VAPOUR PRODUCTION

With increasing ambient temperature, up to body temperature, sensible heat loss declines and evaporative heat loss (hence evaporative water loss) becomes important (van Kampen, 1974; Richards, 1976), particularly above 25 °C as indicated in *Figure 8.1*. In this range feather cover is disadvantageous and the internal heat transport (convection via blood) will be directed more to the head appendages (van Kampen, 1974; van Handel-Hruska, Wegner and Nordstrom, 1977) and respiratory system (Wolfenson *et al.*, 1978).

Respiratory evaporation provides most latent heat loss in the fowl above 25 °C. It depends on the water vapour pressures of the inspired and expired air and on the respiratory minute volume, and hence the metabolic rate, which is affected by the plane of nutrition and the state of acclimation. Evaporative water loss decreases with rising ambient water vapour pressure, by 0.7 mg of water for each gram of live-weight per hour per kilopascal increase in water vapour pressure (Richards, 1976). Representative values are given in *Table 8.6* and *Figure 8.1*.

Table 8.6 WATER VAPOUR PRODUCTION PER HEN (g d^{-1}) AT DIFFERENT AMBIENT TEMPERATURES

T_a (°C)	van Es et al. *(1973)*	van Kampen *(1980)*	Richards *(1976)*
5	120–220	42	37
20	120–220	62	59
35	220–280*	133	204

*Measured at 30 °C

When humidity is raised at temperatures above 25 °C a new balance between heat loss and heat production can be achieved by lowering the insulation of the feathers, by reductions in heat production and hence food intake, by raising body temperature or by increased ventilation of the respiratory tract. The last three are most effective but depress production. Increasing humidity therefore has a similar effect to raising ambient temperature.

When water vapour lost from droppings and the watering system is included the production per bird housed is much higher, as shown by the values in *Table 8.6* (van Es *et al.*, 1973). The author has found that the water content of fresh excreta produced at ambient temperatures of 5 and 35 °C was 76% and 86%, respectively, and that the water lost by evaporation from excreta was high and greatly influenced by the relative humidity. There was no reduction in excreta output at high ambient temperatures (*Figure 8.3*) if the evaporation of water from excreta was prevented, although food intake was reduced. The measured excreta output will be less if excreta are collected on open trays (Lillie *et al.*, 1976), owing to water loss by evaporation.

Air velocity

Forced convection reduces the boundary layer around the body. If the air temperature is lower than the skin temperature heat loss will be increased. But body temperature may be raised if the air temperature exceeds that of the skin (Drury and Siegel, 1966), when the only way the animal can lose heat is by evaporation. Increasing air velocity when the ambient temperature is above the critical temperature but below body temperature will reduce heat stress. However, below the critical temperature such an increase in heat loss results in an increase of heat production and therefore of food intake and, if the ambient temperature is very low, loss in body weight.

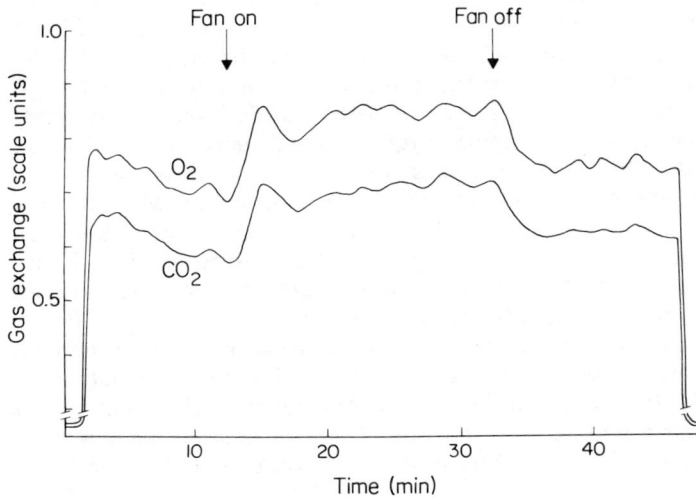

Figure 8.5 Influence of forced convection on the CO_2 production and O_2 consumption of a sitting hen

A 1.4 m s^{-1} air flow incident on either the head or the body of a sitting White Leghorn hen at 20 °C causes in either case an immediate increase of 15 per cent in total heat production as compared with still air (inferred from gas exchanges, *Figure 8.5*), while increasing air velocity, even from 0.1 to 0.3 m s^{-1} at a fluctuating temperature of 26–35 °C, results in a 9 per cent increase in food intake, a 5 per cent increase in egg weight and reduced body weight loss (Zimmerman and Snetsinger, 1975). Chappell (1980) found a linear relationship between heat loss and the square root of air velocity for Arctic-breeding shore birds. McDonald (1978) has suggested that ME intake in domestic fowl is also linearly related to the square root of air velocity, and that a decrease in ambient temperature from 35 to 5 °C at a constant wind speed would increase the extra ME intake requirement 8.5-fold. However, most analyses of convective heat loss from intact animals or of heat transfer through excised coat samples have assumed the square root relation expected from engineering studies of heat transfer from smooth uninsulated bodies. The response is certainly an inverse curve, but Campbell, McArthur and Monteith (1980) have recently re-examined the literature of forced convection from animals (and man) and found that, for all but naked animals, the published measurements are a statistically better fit to an increase in conductance that is linearly proportional to wind speed.

Whatever the precise relation, if air temperature is lower than body temperature an increase in air movement is equivalent to a decrease in ambient temperature.

Activity

Physical activity may also cause differences in metabolic rate and food efficiency, even within a strain. Movement may cause forced convection, which disturbs the boundary layer of air and also the feather cover. This results not only in an increase in heat loss or heat gain but also in heat production due to the activity. In fowl, standing up increases heat loss by 20–40 per cent (DeShazer, Jordan and Suggs, 1970) and this contributes to the heat produced by activity *per se* (Dawson and Hudson, 1970). Birds forced to move at very low temperatures expend more energy than do motionless birds, while at high temperatures the extra heat loss will be lower than the extra heat produced, and hence forced movement will lower the upper lethal temperature. Hens in a climatic chamber huddle at low temperatures and are restless at high temperatures. Food-restricted chicks are more active than chicks fed *ad libitum,* but these in turn are more active than starving chicks (author's unpublished observation).

No direct measurements of the activity of layers at different ambient temperatures are known to the present author. However, calculations of additional activity above that shown by starving fowl (for hens fed at maintenance for a temperature range of 15–33 °C) suggest that activity is lowest in the range 27–33 °C (Balnave, 1974). Measurements of the variation

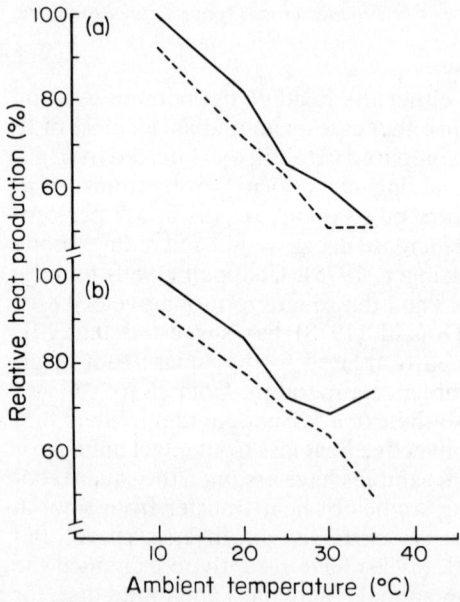

Figure 8.6 Relative heat production of (a) five single chickens and (b) a group of five chickens during light (———) and dark (- - - -) at different ambient temperatures

of heat production by starving White Leghorn hens with ambient temperature suggested that activity was almost constant over a temperature range from 10 to 30 °C, though minimal for the single birds and for the group at 25 °C.

The measurements shown in *Figure 8.6* were made during the day in both light and dark environments, to which the birds were subjected for short periods. Three periods of 15 minutes were used at each temperature, namely an equilibration period, and a dark and a light period in an alternate sequence. The analysis assumes that a difference in heat production between the dark and light period is due to activity. The higher difference for the group at 35 °C, in comparison with the single bird, suggests restlessness; but between 30 and 35 °C heat production increases in general, and a temperature–time effect may have influenced the results. The linear decrease in heat production with increasing ambient temperature is unlikely to be related to changes in activity. However, both indirect measurements suggest a minimal activity in the temperature range 25–32 °C, where body temperature starts to rise and metabolic rate is becoming minimal (*Figure 8.1*).

Conclusions

The responses shown in *Figure 8.1* depend on such factors as air velocity, state of acclimation and relative humidity, as well as ambient temperature. There is not a parallel relationship between ambient temperature and heat production, respectively, and ME intake. In other words, the net energy for egg production is not constant without a concurrent change in body mass. In the temperature range 10 to 30 °C the energy available for egg production and the egg production *per se* are almost constant, while ME intake decreases with increasing temperature. Food costs per egg are minimal at 30 °C but not the costs per kilogram of egg production, because of the inverse relationship between egg size and ambient temperature. Above 30 °C there is a tendency towards lower egg production and a progressive decline in egg size. In contrast, below 10 °C egg production is depressed but not egg size. Above 30 °C and below 10 °C both food and body energy are used for egg production.

Water intake, water to food ratio, water content of excreta and evaporative water loss increase with increasing ambient temperature.

Controlling the quantity and temperature of drinking water results in water saving, a lower production of excreta and a higher efficiency of food use.

References

AL-SOUDI, K. A. and AL-JEBOURI, M. A. (1979). *Wld's Poult. Sci. J.* **35**, 227–235

ARIELI, A., MELTZER, A. and BERMAN, A. (1979). *Br. Poult. Sci.* **20**, 505–513

BALNAVE, D. (1974). In *Energy Requirements of Poultry*, pp. 25–46. Ed. by T. R. Morris and B. M. Freeman. British Poultry Science, Edinburgh

BALNAVE, D., FARRELL, D. J. and CUMMING, R. B. (1978). *Wld's Poult. Sci. J.* **34**, 149–154

BORDAS, A., OBEIDAH, A. and MÉRAT, P. (1978). *Annls Génét.* **10**, 233–250

BRAY, D. S. and GESELL, J. A. (1961). *Poult. Sci.* **40**, 1328–1335

BRUCKNER, J. H. (1936). *Poult. Sci.* **15**, 417–418

BYERLY, T. C. (1979). In *Food Intake Regulation in Poultry*, pp. 327–363. Ed. by K. N. Boorman and B. M. Freeman. British Poultry Science, Edinburgh

CAMPBELL, G. S., McARTHUR, A. J. and MONTEITH, J. L. (1980). *Boundary-Layer Met.* **18**, 485–493

CAMPOS, A. C., WILCOX, F. H. and SHAFFNER, C. S. (1962). *Poult. Sci.* **41**, 856–865

CHAPPELL, M. A. (1980). *Comp. Biochem. Physiol.* **65A**, 311–317

DAVIS, R. H., HASSAN, O. E. M. and SYKES, A. H. (1972). *J. agric. Sci., Camb.* **79**, 363–369

DAVIS, R. H., HASSAN, O. E. M. and SYKES, A. H. (1973). *J. agric. Sci., Camb.* **81**, 173–177

DAWSON, W. R. and HUDSON, J. W. (1970). In *Comparative Physiology of Thermoregulation*, vol. 1, pp. 224–310. Ed. by G. C. Whittow. Academic Press, London

DeSHAZER, J. A., JORDAN, K. A. and SUGGS, C. W. (1970). *Trans. ASAE* **13**, 82–84

DRURY, L. N. and SIEGEL, H. S. (1966). *Trans. ASAE* **9**, 583–585

EMMANS, G. C. (1974). In *Energy Requirements of Poultry*, pp. 79–90. Ed. by K. N. Boorman and B. M. Freeman. British Poultry Science, Edinburgh

GENTLE, M. J. (1979). *Br. Poult. Sci.* **20**, 533–539

GOLD, R. M., LAFORGE, R. C., SAWCHENKO, P. E., FRASER, J. C. and PYTKO, D. (1977). *Physiol. Behav.* **18**, 1047–1053

HARRISON, P. C. and BIELLIER, H. V. (1969). *Poult. Sci.* **48**, 1034–1045

HILL, J. A., POWELL, A. J. and CHARLES, D. R. (1979). In *Food Intake Regulation in Poultry*, pp. 231–257. Ed. by K. N. Boorman and B. M. Freeman. British Poultry Science, Edinburgh

HILL, A. T. and RICHARDS, J. F. (1975). *Poult. Sci.* **54**, 1704–1706

HILLERMAN, J. P. and WILSON, W. O. (1955). *Am. J. Physiol.* **180**, 591–595

LAMOREUX, W. F. (1943). *Endocrinology* **32**, 497

LEESON, S. and MORRISON, W. D. (1978). *Poult. Sci.* **57**, 1094–1096

LEESON, S., SUMMERS, J. D. and MORAN, E. T., Jr (1976). *Wld's Poult. Sci. J.* **32**, 185–195

LILLIE, R. J., OTA, H., WHITEHEAD, J. A. and FROBISH, L. T. (1976). *Poult. Sci.* **55**, 1238–1246

McDONALD, M. W. (1978). *Wld's Poult. Sci. J.* **34**, 209–221

MacLEOD, M. G., TULLETT, S. G. and JEWITT, T. R. (1979). *Br. Poult. Sci.* **20**, 521–531

MAXWELL, B. F. and LYLE, J. B. (1957). *Poult. Sci.* **36**, 921–922

MOWBRAY, R. M. and SYKES, A. H. (1971). *Br. Poult. Sci.* **12**, 25–29

MUELLER, W. J. (1961). *Poult. Sci.* **40**, 1562–1571

NICHELMANN, M., THOMAS, E. and LYHS, L. (1974). *Mh. VetMed.* **29**, 656

NORDSTROM, J. O. (1973). *Poult. Sci.* **52**, 1687–1690

OKUMURA, J., TASAKI, I. and SAITO, K. (1977). *Jap. Poult. Sci.* **14**, 217–222

O'NEILL, S. J. B., BALNAVE, D. and JACKSON, N. (1971). *J. agric. Sci., Camb.* **77**, 293–305

PAYNE, C. G. (1966a). *Wld's Poult. Sci. J.* **22**, 126–139
PAYNE, C. G. (1966b). In *Proc. 13th World's Poultry Congress, Kiev*
RANDALL, M. C. (1978). *Poult. Notes*, 7–12
RICHARDS, S. A. (1976). *J. agric. Sci., Camb.* **87**, 527–532
RICHARDS, S. A. (1977). *J. agric. Sci., Camb.* **89**, 393–398
ROMIJN, C. and LOKHORST, W. (1961). *Tijdschr. Diergeneesk.* **86**, 153–172
SAVORY, C. J. (1978). *Br. Poult. Sci.* **19**, 631–641
SHANNON, D. W. F. and BROWN, W. O. (1969). *Br. Poult. Sci.* **10**, 13–18
SIMON, J. (1973). *4th European Poultry Conference, London*, pp. 203–210
SMITH, A. J. and OLIVER, J. (1972). *Rhod. J. agric. Res.* **10**, 3–21
SPILLER, R. J., DORMINEY, R. W. and ARSCOTT, G. H. (1976). *Poult. Sci.* **55**, 1871–1881
SYKES, A. H. (1972). In *Egg Formation and Production*, pp. 187–196. Ed. by B. M. Freeman and P. E. Lake. British Poultry Science, Edinburgh
SYKES, A. H. (1977). In *Nutrition and the Climatic Environment*, pp. 17–29. Ed. by W. Haresign, H. Swan and D. Lewis. Butterworths, London
VAN ES, A. J. H., VAN AGGELEN, D., NIJKAMP, H. J., VOGT, J. E. and SCHEELE, C. W. (1973). *Z. Tierphysiol. Tierernähr. Futtermittelk.* **32**, 121–129
VAN HANDEL-HRUSKA, J. M., WEGNER, T. N. and NORDSTROM, J. O. (1977). *Fedn Proc. Fedn Am. Socs exp. Biol.* **36**, 524
VAN KAMPEN, M. (1974). In *Energy Requirements of Poultry*, pp. 47–59. Ed. by T. R. Morris and B. M. Freeman. British Poultry Science, Edinburgh
VAN KAMPEN, M. (1976). In *Progress in Biometeorology*, pp. 158–166. Ed. by S. W. Tromp. Swets & Zeitlinger, Amsterdam
VAN KAMPEN, M. (1977). *Tijdschr. Diergeneesk.* **102**, 504–514
VAN KAMPEN, M. (1980). *Arch. Geflügelk.* **44**, 124–128
VAN KAMPEN, M. (1981). *Br. Poult. Sci.* **22**, 17–23
VO, K. V., BOONE, M. A. and JOHNSTON, W. E. (1978). *Poult. Sci.* **57**, 789–803
WILSON, W. O. and EDWARDS, W. H. (1952). *Am. J. Physiol.* **169**, 102–107
WILSON, W. O., ITOH, S. and SIOPES, T. D. (1972). *Poult. Sci.* **51**, 1014–1023
WILSON, W. O., McNALLY, E. H. and OTA, H. (1957). *Poult. Sci.* **36**, 1254–1261
WOLFENSON, D., BERMAN, A., FREI, Y. F. and SNAPIR, N. (1978). *Comp. Biochem. Physiol.* **61A**, 549–554
WOLFENSON, D., FREI, Y. F., SNAPIR, N. and BERMAN, A. (1979). *Br. Poult. Sci.* **20**, 167–174
ZIMMERMAN, R. A. and SNETSINGER, D. C. (1975). *Ralston Purina Company Report*. St Louis, Missouri

9

THE CLIMATIC REQUIREMENTS OF THE PIG

W.H. CLOSE
*Agricultural Research Council's Institute of Animal Physiology,
Babraham, Cambridge*

Introduction

The efficiency of pig production is dependent upon the extent to which
nutrients in the feed are used for the maintenance and production of tissue
within the animal. Any factors which divert nutrients from these processes
lead to a reduction in the efficiency of feed utilization and hence in the
growth of the animal. As the environment is an important influence on the
animal's energy expenditure and as this energy is derived from nutrients in
the feed, it follows that the thermal effects of the environment will have
direct consequences for the partition of the animal's energy intake: between
that retained as growth and that dissipated as heat, and therefore lost to
production.

The animal's environment can be most precisely controlled within buildings, and recent years have seen an increasing number of intensive husbandry systems where larger numbers of animals are kept at greater stocking
densities than in the past. Within these systems the animal's freedom of
choice of living conditions is restricted and it is therefore important that the
effects of the environment on the animal's health, welfare and productivity
should be properly understood. To maintain optimum production it is
necessary to maintain optimum climatic conditions consistent with the
requirements of the animal. The questions which arise are: what constitutes
an optimum environment? what factors influence it? and what are the
consequences for the animal's growth and development should conditions
fall below those required for optimum production? These criteria will be
discussed in the light of recent developments in the understanding of the
climatic requirements of the pig.

The primary influence of the environment on the productivity of animals
is by way of their heat exchange. This is regulated so that heat produced
within the body is equal to the heat lost from the body, enabling body
temperature to be maintained within relatively narrow limits. The regulation
of heat exchange and the pathways through which it occurs have been
discussed at a previous Easter School (Monteith and Mount, 1974) and will
not be specifically discussed in this chapter. Further comprehensive reviews
have been written by Mount (1968, 1979), Curtis (1970) and Ingram (1974,
1976).

Thermal neutrality and critical temperature

In terms of the efficient use that an animal makes of its feed, interest lies in determining the range of environmental conditions over which energy expenditure is minimal. The influence of the environment on the animal's energy exchange can be illustrated most precisely in relation to environmental temperature (*Figure 9.1*). Under cold conditions heat loss or heat

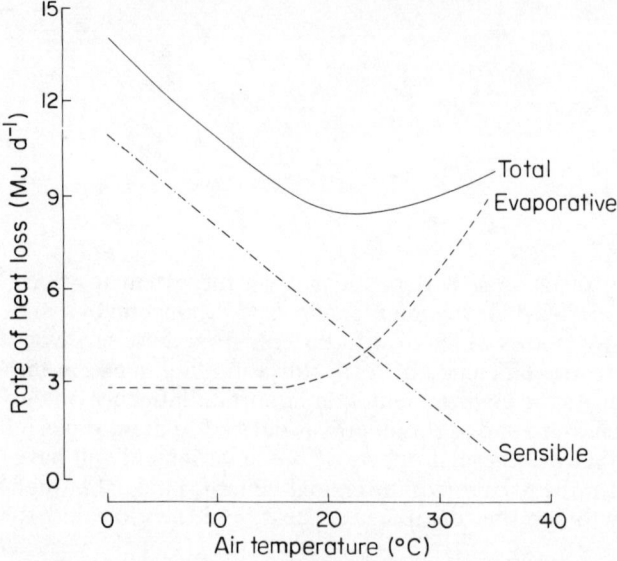

Figure 9.1 The relation between air temperature (in °C) and the components of heat loss (MJ d^{-1}) in the 35 kg pig. The feed intake is twice maintenance, $2F_m$ (where $F_m = 440$ kJ $kg^{-0.75}$ d^{-1} of ME). (Values calculated from the equations provided by Close and Mount, 1978)

production is high, and if the animal is on a fixed energy intake the energy available for growth is minimal. As the environmental temperature is increased the animal's heat loss decreases until a range of temperatures is reached where it is at a minimum. This range is called the *thermally neutral* (or thermoneutral) range, with the lower and upper limits called the lower and upper critical temperatures, respectively (Mount, 1974). Within the thermal neutral zone the energy available for growth is optimal. Above the thermal neutral zone and as environmental temperature increases the animal's body temperature rises, with a concomitant increase in heat loss. If this situation persists and the animal is not cooled it will eventually die.

In pig production, the interest lies in the determination of the lowest temperature at which heat loss is minimal; that is, the lower critical temperature. It is often referred to as being a precisely defined temperature below which the animal's heat loss or heat production begins to rise, at a uniform rate, from a constant level within the zone of thermal neutrality. In practice, no such well-defined critical temperature is obtained as the behavioural, postural and activity patterns of the animal can influence the rate and level of heat loss. Thus the relation between heat loss and environmental temperature is curvilinear, so that it becomes necessary to make an estimate of 'effective critical temperature'. This phenomenon can be seen in *Figure 9.1*, which also shows the partition of heat loss into its non-evaporative and evaporative components. The role of the evaporative heat loss in the thermoregulatory characteristics of the animal, particularly in relation to the animal's lower critical temperature, has been discussed by Mount (1974). It has been suggested that an environment cannot be considered as thermally neutral where the rate of evaporation has increased through panting or sweating, although heat loss may be minimal. However, for the purposes of animal production the zone of thermal neutrality, defined as the zone of minimum heat loss or heat production, represents the range of temperature over which the efficiency of energy utilization is at a maximum and the energy available for growth is optimal. This definition is implicit in the ensuing discussions, in which the lower critical temperature, T_{cl}, is referred to as the 'critical temperature'.

The position of the thermally neutral zone, and hence critical temperature, on the ambient temperature scale is influenced by animal, nutritional and environmental factors. It is useful to know the extent to which temperatures below the critical level cause an increase in the energy required for maintenance and thermoregulation, because this indicates what reduction in growth is to be expected. The significance of these factors must be realized before a quantitative analysis can be made in respect of the relative effects of the environment.

Factors influencing the critical temperature

ANIMAL FACTORS

Animal factors influencing the critical temperature include body weight, thermal insulation, group size and the condition of the animal. The critical temperature decreases as the body weight and thermal insulation increase. Thus, the individual newborn pig, which has little subcutaneous fat, has a critical temperature in the region of 34 °C (Mount, 1968). However, as the animal increases in size, the critical temperature decreases from approximately 21 °C at a body weight of 20 kg to 20 and 18 °C at 60 and 100 kg, respectively (Holmes and Close, 1977). These values are specific to singly housed animals at an energy intake of twice maintenance ($2F_m$, where $F_m = 420$ kJ kg$^{-0.75}$ d^{-1} of metabolizable energy (ME)), when maintained at low levels of air movement and when mean air and mean radiant temperature are equal. This decrease in critical temperature is associated with a greater thermal insulation in the larger animal, due to its increased subcutaneous fat content and to a change in the radius of curvature of the animal's body with increase in size. The relation between critical temperature and thermal insulation, collected from a number of sources, can be seen in *Figure 9.2*. Critical temperature (T_{cl},°C) was related to thermal insulation, I (m^2 K W^{-1}), according to the equation

$$T_{cl} = 41.5 - 88.3\,I \qquad (9.1)$$

Each 0.1 m^2 K W^{-1} increase in insulation was therefore associated with an 8.8 K decrease in critical temperature. (The standard error of the first

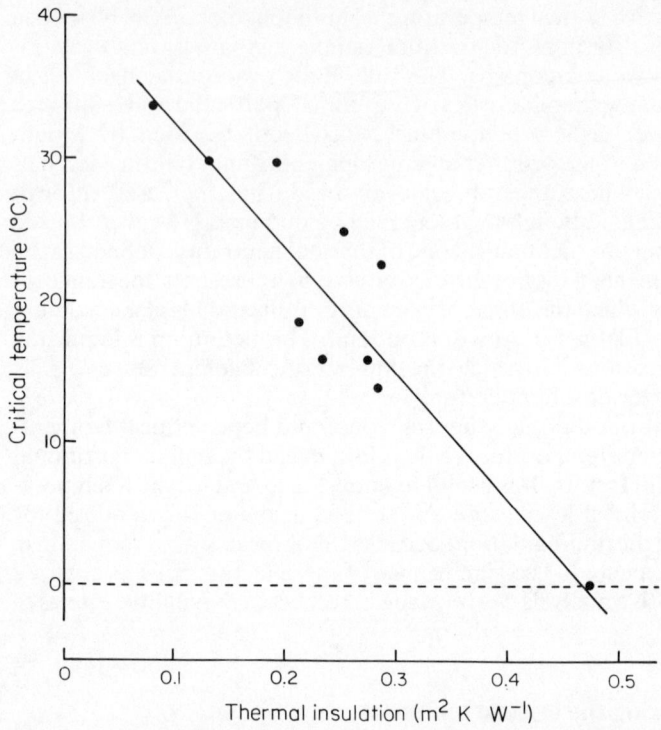

Figure 9.2 The relation between critical temperature (°C) and thermal insulation (m² K W⁻¹) in the pig (equivalent body weight range 1–200 kg). (Values taken from the following sources: Bond, Kelly and Heitman, 1952; Irving, Peyton and Monson, 1956; Ingram, 1964; Mount, 1964; Fuller and Boyne, 1972; Stombaugh et al., 1973; Holmes and McLean, 1977; Hovell and MacPherson, 1977 and Stombaugh and Roller, 1977)

constant in equation 9.1 is ± 3.2, and that of the second is ± 12.5, with r = 0.93.) However, it may be that not all the change in critical temperature and thermal insulation is associated with variations in body weight. For example, over comparable weight ranges, Holmes and McLean (1974) calculated the critical temperatures of thin and fat sows to be 19 and 14 °C respectively with corresponding thermal insulation values of 0.21 and 0.28 m² K W⁻¹. This demonstrates that differences in critical temperature may be attributable in part to the body composition of the animal.

The grouping of animals, and the postural and behavioural effects associated with it, can markedly affect the rate of heat production and critical temperature. For example, increasing the group size from 1 to 4 or 9 animals, at a feed intake of 45 g d⁻¹ per kilogram of body weight lowered the critical temperature from 19 to 16 and 14 °C, respectively (Mount, 1975). In addition, the rate at which heat production increased per 1 K below the critical temperature was also reduced. Close and Mount (1978) calculated that each additional increase in group size between 1 and 9 animals was associated with a 7 per cent reduction in the rate at which heat production increased below the critical temperature. However, the extent to which this occurs for group sizes greater than 9 animals is not known. Very large groups of animals may create unfavourable social environments and this may increase their heat production. For this reason Sainsbury (1972) has recommended group sizes of 12–20 pigs per pen.

NUTRITIONAL FACTORS

The plane of nutrition influences the critical temperature, so that the higher the level of feeding, the lower the critical temperature. Within and above the zone of thermal neutrality there is no extra thermoregulatory heat demanded of the animal and the additional heat associated with the productive processes has to be dissipated. Below the zone of thermal neutrality part of this increased heat production is used to compensate for some of the extra thermoregulatory heat demanded of the animal. The higher the level of feeding, the greater is the degree to which all the thermoregulatory heat requirement can be met by such heat production, and this has the effect of lowering the animal's critical temperature.

The relation between feeding level and critical temperature in the pig has been reported by Verstegen (1971), Close, Mount and Start (1971), Verstegen *et al.* (1973) and Close and Mount (1978). For an individually maintained 35 kg animal, Close and Mount (1978) have calculated that at intakes between the fasting and ad-libitum levels, critical temperature falls by 1 K for each 201 kJ $kg^{-0.75}$ d^{-1} increase in ME intake. This value will vary according to the animal's body weight, its group size and degree of thermal insulation. In practice, however, it is necessary to have an indication of the relation between ME intake and critical temperature so as to determine how this changes with variations in the body weight of the animal. Estimates of heat production or heat loss in relation to environmental temperature and ME intake have therefore been collected from the literature and combined on a body weight basis both for individual pigs and those in groups. The ensuing values of thermoneutral heat production, ranges of effective critical temperature and increase in heat production per 1 K below the effective critical temperature are given in *Table 9.1*. The relation between ME intake and critical temperature is further illustrated in *Figure 9.3* for groups of animals at several body weights. These values show the very wide range of critical temperature appropriate to the different levels of ME intake throughout the life of a pig. There is little information available for sows, although the results of Holmes and McLean (1974), Hovell and MacPherson (1977) and Holmes and Close (1977) indicate critical temperatures within the range 15–20 °C.

The rate at which ME intake should be increased in order to effect a 1 K decrease in critical temperature can be calculated from the values presented in *Table 9.1*. Each 1 K decrease in the critical temperature of individual animals was associated with a 300, 167 and 144 kJ $kg^{-0.75}$ d^{-1} increase in ME intake at body weights of 5, 35 (mean of 20–50) and 75 (mean of 50 100) kg, respectively. For groups of animals the corresponding increases in ME at the two higher body weights were 100 and 75 kJ $kg^{-0.75}$ d^{-1}, respectively. For thin sows the mean increase in ME intake required to reduce critical temperature by 1 K was 86 kJ $kg^{-0.75}$ d^{-1} whereas for fat sows it was 72 (Holmes and Close, 1977).

Table 9.1 THERMONEUTRAL HEAT PRODUCTION RATES, RANGE OF EFFECTIVE CRITICAL TEMPERATURES AND MEAN INCREASE IN HEAT PRODUCTION BELOW THE EFFECTIVE CRITICAL TEMPERATURE IN RELATION TO BODY WEIGHT AND METABOLIZABLE ENERGY (ME) INTAKE

Body weight (kg)	ME intake ($kJ\ kg^{-0.75}\ d^{-1}$):						Range of effective critical temperatures (°C)	Mean increase in heat loss per 1 K below the effective critical temperature ($kJ\ kg^{-0.75}\ d^{-1}\ K^{-1}$)
	Fasting	Maintenance	900	1100	1300	1500		
		Heat production ($kJ\ kg^{-0.75}\ d^{-1}$)						
5	426	531	609	632	666	700	30–35	–
15	423	584	678	737	791	844	28–21	25
20–50 (individual)	360*	440	615	650	730	765	26–17	20
20–50 (group)	360	440	620	660	715	780	26–9	11
50–100 (individual)	320*	420	625	640	780	–	23–14	15
50–100 (group)	320	420	630	660	760	–	23–8	7

*Fasting heat production assumed equal for both individual and group of animals

The following sources have been used in compiling the above information: Tangl (1912); Capstick and Wood (1922); Deighton (1923, 1929); Breirem (1936, 1939); Brody and Hall (1936); Bond, Kelly and Heitman (1952); Mount (1959, 1960); Holmes and Mount (1967); Jenkinson, Young and Ashton (1967); Bowland et al. (1970); Jordan and Brown (1970); Kielanowski and Kotarbinska (1970); Mount and Stephens (1970); Close, Mount and Start (1971); Sharma, Young and Smith (1971); Verstegen (1971); Fuller and Boyne (1972); McCracken and Gray (1972); Burlacu et al. (1973); Holmes (1973, 1974); Holmes and Breirem (1974); Jordan (1974); Close and Mount (1975, 1978)

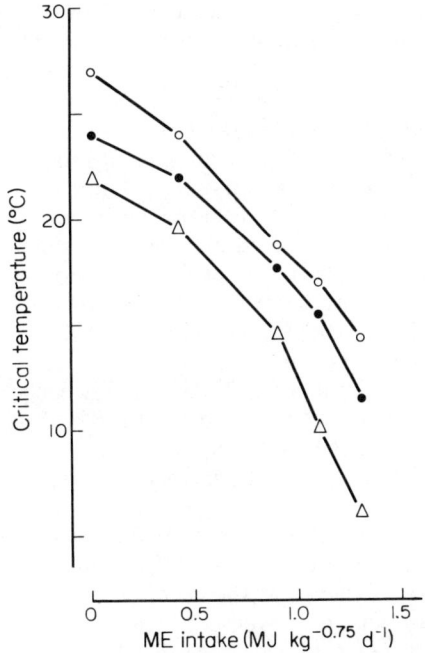

Figure 9.3 The relation between critical temperature (°C) and metabolizable energy (ME) intake in MJ kg⁻⁰·⁷⁵ d⁻¹ for pigs housed in groups. The lines are for different body weights. (○) 20 kg; (●) 60 kg; (△) 100 kg. (From Close, 1980 courtesy of Cambridge University Press)

The influence of these animal and nutritional factors allows precise calculations of critical temperature to be made under standardized environmental conditions. These relate to environmental conditions where the air temperature is constant, mean radiant and air temperatures are equal, still air conditions (air movement less than 0.20 m s⁻¹) and the animals kept on a dry insulated floor but without bedding. Under practical conditions, however, it is necessary to predict the likely changes that occur should there be departures from the standardized conditions, so that the extent of the changes can be measured in terms of equivalent changes in air or critical temperature.

THERMAL EFFECTS OF VARIOUS ENVIRONMENTAL COMPONENTS

Air movement

An increase in air movement disrupts the thermal insulation provided by the boundary layer of air around the animal, causing an increase in convective heat loss. This has been demonstrated by the technique of schlieren photography by Stephens and Start (1972). Mount (1979) has calculated that the insulation provided at the boundary layer under still air conditions is 0.11 m² K W⁻¹ but on exposure to rapid air movements this is reduced to almost zero. From the relation between thermal insulation and critical temperature (equation 9.1) it may be calculated that the effects associated with the removal of the boundary layer insulation would be equivalent to a 10 K increase in critical temperature, and this is in agreement with recent observations of Close, Heavens and Brown (1981).

The extent to which air movement causes an increase in heat loss is dependent upon an animal's body weight, the temperature of exposure and whether it is kept individually or in a group. Younger animals are more susceptible to changes in air movement than older animals (Mount and Ingram, 1965; Holmes and Mount, 1967). Similarly, low air velocity changes are proportionately more effective in increasing heat loss than similar changes at high wind speed (Mount, 1966). The effect of air movement is also temperature-dependent, with the higher losses occurring at lower environmental temperatures (Mount and Ingram, 1965). In these investigations the rate at which heat loss increased in response to an increase in air movement has been calculated. For individual 25 kg animals each 0.05 m s^{-1} increase in air movement above the wind speed at which forced convection predominates (0.20 m s^{-1}; Mount, 1977) was equivalent to a 1 K decrease in air temperature; for 60 kg pigs the effect was less, each 0.10 m s^{-1} increase being equivalent to a 1 K decrease. The effect will, however, diminish as air movement rate decreases (Mount, 1975). Values for groups of animals are greater than those for individual animals. Verstegen and van der Hel (1977) have shown that the increase in heat production associated with an increase in air movement from 0.15 to 0.45 m s^{-1} is equivalent to a 1.4 K decrease in temperature. These values relate to draughts of air provided at the same temperature as that in the pen. In practice, however, they may often be at a lower temperature, so that the heat loss is increased, resulting in a further lowering of the equivalent air temperature.

Air and structure temperature

The difference between air and structure temperature determines the rate at which heat is lost by radiation and convection. Mount (1964) investigated the extent to which the radiant environment influenced the heat loss of 1-week-old pigs, by varying air and wall temperature independently of one another. When wall temperature was reduced below that of the air, total heat loss increased because of the increase in radiant heat loss. He concluded that a 1–2 K change in the mean temperature of the surroundings was equivalent in its thermal effect to a 1 K change in air temperature. This compares well with a figure of 2 K determined by Holmes and McLean (1977). The significance of the change in radiant heat loss for the overall heat exchange of piglets has been studied by Stephens and Start (1970). The provision of a radiant heat source resulted in 47 and 18 per cent reductions in heat production when measured on a concrete floor at environmental temperatures of 10 and 20 °C, respectively. When straw bedding was supplied the corresponding values were 35 and 21 per cent.

In order to predict the effect of the radiant environment on the animal's heat loss and critical temperature it is necessary to know the air, wall and external temperatures. Holmes and Close (1977) calculated the effective environmental temperature as the weighted mean of air and wall temperature, whereas Mount (1975) calculated it as a function of internal and external air temperature. There is good agreement between the two methods over a wide range of internal and external climatic conditions. The direction of the effect is dependent on whether external or surface temperature is higher or lower than internal air temperature. The effect will be reduced in a well-insulated house, where the radiant temperature approaches the air temperature.

Bedding and floor type

The nature of the floor determines the extent of conductive heat loss, and as approximately 20 per cent of the animal's body can be in contact with the floor, there may be considerable heat loss through it (Bond, Kelly and Heitman, 1952). On a cold, uninsulated floor, the conductive heat loss may represent 20–25 per cent of the animal's heat loss. Under hot conditions, however, it may be a valuable avenue of heat dissipation. Comparisons between different floor types show the significant changes in both heat loss and critical temperature associated with bedding. For example, heat loss to a wooden floor at 20 °C was similar to that to a concrete floor at 30 °C (Mount, 1967). Stephens (1971) showed that moving piglets onto a straw floor at 10 °C had an effect equivalent to that of concrete at 18 °C. At higher air temperatures the effect was reduced. The benefit derived from keeping piglets on floors with high insulative properties is therefore lessened as the environmental temperature is increased.

Verstegen and van der Hel (1974) measured the heat production rates of 40 kg pigs kept in groups of nine on floors of asphalt, straw and concrete slats, constructed within a large group calorimeter. They calculated effective critical temperatures of 11–13 °C on straw, 14–15 °C on asphalt and 19–20 °C on concrete slats. Below these levels the heat production on the concrete was 7 per cent higher than that on either straw or asphalt. On the concrete slats energy retention was reduced, resulting in a 10 per cent decrease in body weight gain.

The degree of wetness of the floor will also have a considerable effect upon the animal's heat loss. If the pigs have to lie on a wet floor, conduction to the floor and evaporation from the animal's surface will increase. This will have a beneficial effect in hot, humid conditions. However, under cold conditions the effect would be detrimental to the animal. In the cold, groups of animals would be able to compensate to some degree, by altering their behavioural and postural characteristics to minimize heat loss. For young individual animals, Mount (1975) has calculated that a cold, wet floor may have an effect equivalent to a 7–10 K decrease in air temperature, although for groups the effect would be less.

Relative humidity

The humidity of the air can have an important influence on the wellbeing of the animal at high environmental temperatures. Morrison, Bond and Heitman (1967) have shown that as relative humidity increased from 30 to 90%, at an environmental temperature of 30 °C, the animal became more dependent on cutaneous water loss, even though respiration rate had almost doubled. The importance of cutaneous water loss, in particular that associated with wallowing, has been discussed by Ingram (1965a, b). Heitman and Hughes (1949) found that at 35 °C the animal's respiration rate and body temperature were reduced and rate of gain increased on a wetted floor as compared with a dry floor. At this temperature similar effects were noted on increasing the air velocity.

Other factors, which may indirectly influence the rate of heat loss through effects associated with changes in the behaviour, health and welfare of the animals, have been discussed by Done and Wijeratne (1972), Ewbank (1972) and Sainsbury (1972).

PRACTICAL IMPORTANCE OF 'DEVIATIONS' IN THE ENVIRONMENT

Knowledge of the various animal, nutritional and environmental factors allows calculations to be made of the equivalence between deviations in the environment and changes in the standardized environment or critical temperature. Equivalent air temperatures can then be calculated as those at which minimal levels of heat loss can be maintained under different environmental and housing conditions. The example given in *Figure 9.4* is for 60 kg

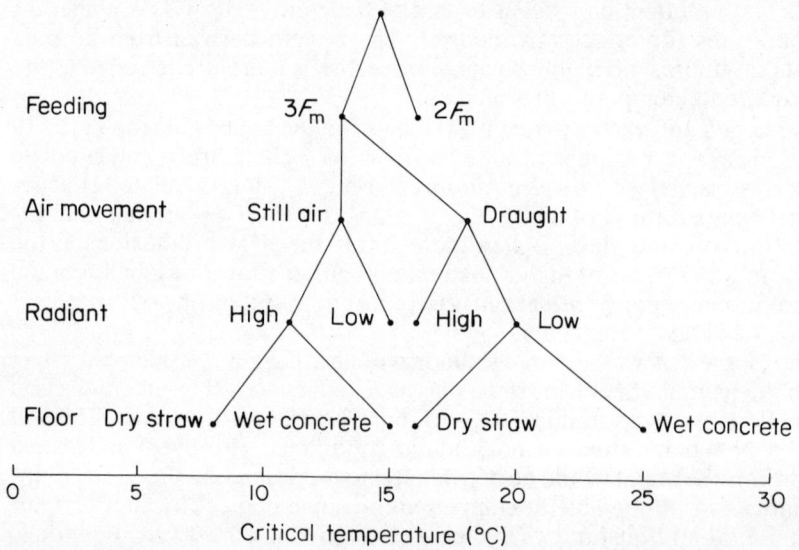

Figure 9.4 A diagrammatic representation of the effects of nutritional and environmental factors influencing the critical temperature of a group of nine 60 kg pigs. Feed intakes, 2 F_m and 3 F_m (where F_m = 440 kJ kg$^{-0.75}$ d^{-1} of ME); air movement, still air (0.1 m s^{-1}) and draughts (0.8 m s^{-1}); radiant temperature, high (summer) and low (winter); floor, dry straw and wet concrete

pigs at a feeding level of 3F_m (F_m = 440 kJ kg$^{-0.75}$ d^{-1} of ME), kept in a group of nine and exposed to different combinations of air movement, radiant environment and floor type. The extent to which the different environmental components influence the critical temperature is shown: on the left-hand side are optimum environmental factors which exert minimal thermoregulatory demand, while on the right-hand side are unfavourable environmental conditions where the thermoregulatory demand upon the animal is high. The limits over which minimum heat losses occur are very wide, and in this

Table 9.2 THE CALCULATED RANGES OF EFFECTIVE CRITICAL
TEMPERATURE FOR PIGS OF DIFFERENT BODY WEIGHTS UNDER DIFFERENT
COMBINATIONS OF NUTRITION AND HOUSING

Body weight (kg)	Group size	Range of effective critical temperature* (°C)
20	1	14–33
	9	10–31
60	1	11–31
	9	8–28
100	1	7–28
	9	2–25

*Low values refer to high feed intake at optimal environmental conditions, while high values refer to low feed intake in adverse environmental conditions

Range of variables

Feed intake ($F_m = 440\,kJ\,kg^{-0.75}\,d^{-1}$ of ME)	$2F_m–3F_m$
Air movement (m s^{-1})	0.1–0.8
Radiant environment (summer or winter; °C)	–2 to +2
Floor type	ranging from clean, dry straw to wet, cold concrete

example, range from 8 to 25 °C. Corresponding values have been calculated for animals of different body weights and group size and these are presented in *Table 9.2*. These ranges of values are of similar magnitude to those calculated from the model recently developed by Bruce and Clark (1979). If the air temperature for each particular housing condition is less than those given in *Table 9.2*, energy from the feed will be used to compensate for the temperature deficit. This leads to a reduction in growth rate and a deterioration in feed conversion efficiency.

Effects of the environment on production characteristics

BODY-WEIGHT GAIN

As the level of feed intake is the first limiting factor influencing the animal's growth and development, it follows that the extent to which the environment influences growth rate should be assessed at different levels of feeding. The rate at which the animal's growth changes in response to different environmental/plane of nutrition combinations has therefore been predicted using a multiple-regression analysis technique from data compiled from a study of the literature. The ensuing predictions are presented in *Figure 9.5*. The range of body weights was 20–105 kg, with a mean of 55 kg. The temperature at which growth rate appeared optimal varied with level of feed intake, ranging between 13 and 30 °C. Above and below these optimal temperatures growth rate was reduced. The sensitivity of the reduction in body-weight gain to temperatures below the optimum was also dependent on the level of feeding: between 13 and 5 °C the rate of gain decreased by 1.1, 0.9, 0.8 and 0.6 g kg$^{-0.75}$ d^{-1} per 1 K at feed intakes of 40, 80, 120 and 160 g kg$^{-0.75}$ d^{-1}, respectively.

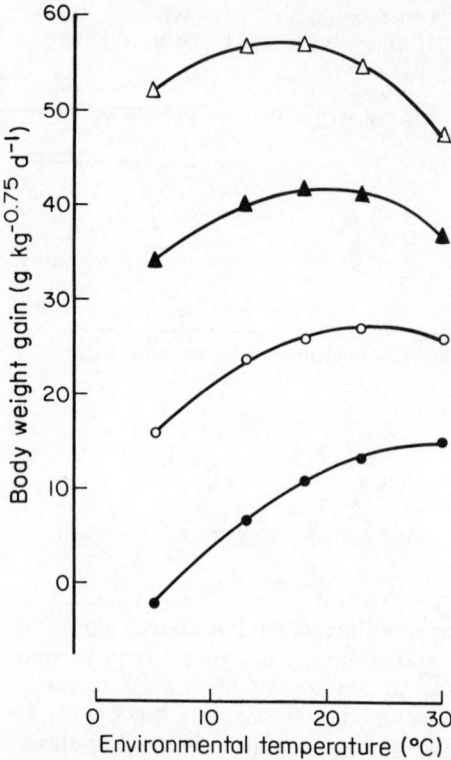

Figure 9.5 Body-weight gain of pigs, 20–105 kg body weight, in relation to environmental temperature (°C) and level of feed intake. Both body-weight gain and feed intake are expressed relative to 'metabolic body size', that is, g kg$^{-0.75}$ d^{-1}. In these limits the lines correspond to: (●) 40; (○) 80; (▲) 120; (△) 160 g feed kg$^{-0.75}$ d^{-1}. (The following sources of information were used: Heitman and Hughes, 1949; Hicks and Webster, 1968; MacGrath et al., 1968; Jensen et al., 1969; Sugahara et al., 1970; Fuller and Boyne, 1971; Morrison and Mount, 1971; Verstegen et al., 1973; Close, Mount and Brown, 1978; Verstegen, Brascamp and van der Hel, 1978; Phillips et al., 1979 and Stahly, Cromwell and Aviotti, 1979). The general equation relating body weight gain (ΔW) to environmental temperature (T) and feed intake (I) was: $\Delta W = 26.26(\pm 0.32) + 1.48(\pm 0.17)T - 0.015(\pm 0.005)T^2 + 0.45(\pm 0.01)I - 0.00020(\pm 0.00003)I \times T^2$ (r = 0.88) (The SEs of the coefficients are given in parentheses)

As there was a wide variation in the body weights of the animals within individual experiments considered in the above analysis, it was not possible to calculate the rate of change of body-weight gain at specific body weight ranges. An alternative method was therefore used: the procedure reported by Holmes and Close (1977), in which the increase in heat loss calculated from *Table 9.1*, and hence the decrease in energy retention per 1 K below the critical temperature, is divided by the energy content of the body-weight gain. The values derived by this method are presented in *Table 9.3* for an intake equivalent to $2.5F_m$. The reduction in growth rate is greater in smaller animals and this may be associated with differences in the nature of the tissues deposited by the animals. It is interesting to note that the values for the 55 (mean of range 20–105 kg) and 60 kg pig derived by the two methods are similar at a mean feed intake of 100 g kg$^{-0.75}$ d^{-1}; that is, approximately $2.5F_m$.

Table 9.3 THE REDUCTION IN GROWTH RATE AND THE ADDITIONAL FEED REQUIREMENTS OF THE PIG PER °C DECREASE IN TEMPERATURE BELOW THE CRITICAL TEMPERATURE

	Body weight (kg):		
	20	60	100
Reduction in growth rate*			
($g\,d^{-1}$)	14	12	8
($g\,d^{-1}$ per kg body weight)	0.70	0.20	0.08
Additional feed requirements†			
($g\,d^{-1}$)	14	20	20
($g\,d^{-1}$ per kg body weight)	0.70	0.33	0.20

*Calculated on the basis that the energy content of the body-weight gain is 8, 16 and 24 $kJ\,g^{-1}$ at 20, 60 and 100 kg body weight respectively
†Calculated on the basis that the partial efficiency of energy utilization below thermal neutrality is 0.80 and that the energy content of the feed is 12 $kJ\,g^{-1}$ fresh weight

PROTEIN AND FAT GAIN

The depression in growth rate associated with a decrease in environmental temperature results from a reduction in the rates at which both protein and fat are deposited. Of the two components, protein appears to be less seriously affected. A number of estimates have been made of the extent to which protein retention decreases with temperature. These range from –0.7 to +2.8 $kJ\,kg^{-0.75}\,d^{-1}$ per 1 K change in temperature (Fuller and Boyne, 1971; Verstegen *et al.*, 1973; Gray and McCracken, 1974; Close, Mount and Brown, 1978; Phillips *et al.*, 1979). These differences in protein retention may be attributed to several factors, notably the breed of the animal, its body weight, the level of feeding and the concentration, digestibility and availability of the protein in the feed. Fat deposition, on the other hand, is more severely influenced by environmental temperature, decreasing by between 6.7 and 17.7 $kJ\,kg^{-0.75}\,d^{-1}$ per 1 K decrease in environmental temperature (Verstegen *et al.*, 1973; Gray and McCracken, 1974; Close, Mount and Brown, 1978; Phillips *et al.*, 1979). The rates of protein and fat deposition also depend upon the level of feed intake. Close, Mount and Brown (1978) have calculated the temperature equivalence of these tissues in terms of feed intake. The reductions in protein and fat deposition as a result of a 1 K decrease in environmental temperature below 15 °C, were equivalent to 4 and 28 g reductions in daily feed intake, respectively. The corresponding value for body-weight gain was 20 g. The temperature-dependent change in fat deposition was thus greater than those for protein and for body-weight gain as a whole.

CARCASS CHARACTERISTICS

The variations in the protein and fat gains of the animals reflect changes in carcass composition. On the basis that the rate of fat deposition is more severely affected than that of protein, it may be concluded that exposure to cold environments results in leaner carcasses. Chemical analysis of whole-body tissue and the measurement of backfat thickness as indicators of carcass fatness support this view (Holme and Coey, 1967; MacGrath *et al.*, 1968; Weaver and Ingram, 1969; Jensen *et al.*, 1969; Sugahara *et al.*, 1970; Holmes, 1971; Hacker, Stefanovic and Batra, 1973; Verstegen, Brascamp and van der Hel, 1978). The results of Sørensen (1962) and Hale, Johnson and Warren (1968), on the other hand, suggest an increase in carcass fatness. However, it is likely that the latter effects were compounded by level of feeding, as under ad-libitum conditions the effect of the cold environment was to increase feed intake relative to that at the higher temperatures. Exposure to high environmental temperatures also influences carcass composition, through its effect on the animal's feed intake.

Many of the environmental effects on carcass composition act through their effect on feed intake. But temperature *per se* has been shown to influence the morphological characteristics of the animal. Pigs kept under cold conditions have shorter ears, denser coats, shorter limbs and fewer blood vessels in the skin than those under warm conditions (Fuller, 1965; Ingram and Weaver, 1969). Carcass length may also be reduced (Holme and Coey, 1967). Prolonged exposure to cold environments produces a high degree of unsaturated fatty acids in the fat depots of the animals, and this has been shown to influence the melting point and physical characteristics of the fat within the carcass (MacGrath *et al.*, 1968; Fuller, Duncan and Boyne, 1974). The water-binding capacity of the muscle appears to increase with increase in environmental temperature (Comberg, Stephan and Späth, 1971, 1972). Muscle quality may also be influenced by environmental temperature in some breeds of pigs, because of their inability to respond to sudden changes in high temperatures. This causes stress in the animal, a condition known as malignant hyperthermia syndrome, and affects the muscle quality after slaughter, producing a pale, soft and exudative condition known as PSE.

Practical implications

The general requirement for farm practice is to determine the range of environments and nutrition that allows maximum efficiency in the utilization of feed. In this respect, the prediction of equivalent air or critical temperatures allows different environmental situations to be assessed. Should the air temperature fall below the range required for optimum utilization of feed, then for any given environmental–nutritional combination action can be taken to improve the environment. The possibilities include structural alterations to the building, the provision of supplementary heating and increasing the animal's feed allocation. Estimates of the increased feed allocation can be calculated if the increase in heat loss, the partial efficiency of energy retention below the critical temperature and the energy content of the feed are known (*Table 9.3*). The value of 0.33 g of feed per day for each kilogram of body weight per °C calculated as a mean throughout the fattening period, is in agreement with that determined for groups of pigs by Mount (1975). The values for individually maintained animals are greater than this and vary from 0.7 to 0.4 $g\,kg^{-1}\,d^{-1}$ per °C

(Mount, 1975; Close and Mount, 1978; Close, 1980). Increasing the feed allocation may also promote a more desirable carcass as the combination of colder temperatures plus higher feed intake increases protein deposition at a greater rate than that of fat, resulting in an enhanced lean : fat ratio (Close, 1980).

When the air temperatures are above those given in *Table 9.2* for any given combination of nutrition and environment, the requirement of the animal is to limit its heat production in order to avoid hyperthermia. Any phenomenon which increases the animal's energy expenditure therefore militates against this process. Thus feed intake is reduced and the animal eats little and often, so that the heat increment associated with feeding is reduced. It will also avoid any unnecessary activity. As the animal relies on its evaporative heat exchange for maintenance of homeothermy it follows that methods which enhance water vapour removal help adaptation and increase production. The use of wallows, hoses or automatic sprinklers and the control of relative humidity have been shown to improve production under tropical conditions (Hsia, Fuller and Koh, 1974; Morrison, Bond and Heitman, 1968; Morrison *et al.*, 1972). The use of shades to reduce the level of solar radiation and the provision of draughts of air around the animal may also be effective in alleviating the heat burden of the pigs.

Conclusions

The foregoing account described the progress that has been made in defining the climatic requirements of the pig under different combinations of nutrition and environment. The use of equivalent air or critical temperatures enables comparison of different housing systems and conditions, although conclusions based on one situation do not necessarily apply to another. Information obtained from these studies, particularly on the exchange of energy between the pig and its environment, permits a rational approach to be used in the design of buildings and in deciding the type of housing, management and husbandry practices most suited to the different types of production. Climatic studies should be used to increase further the understanding of the nutritional–environmental interactions, not only in relation to housing requirements but also in relation to other factors which may impose limitations on production, namely, adaptation, welfare and disease. What is still therefore required is information which can be applied in farm practice and in this respect climatic studies will make a valuable contribution to research in pig production.

References

BOND, T.E., KELLY, C.F. and HEITMAN, H., Jr. (1952). *Trans. ASAE* **8**, 167–169
BOWLAND, J.P., BICKEL, H., PFIRTER, H.P., WENK, C.P. and SCHÜRCH, A. (1970). *J. Anim. Sci.* **31**, 494–501
BREIREM, K. (1936). *Tierernährung* **8**, 436–498
BREIREM, K. (1939). *Tierernährung* **11**, 487–528

BRODY, S. and HALL, W.C. (1936). *Bull. Univ. Missouri agric. exp. Sta.*, No. 166, pp. 45–65

BRUCE, J.M. and CLARK, J.J. (1979). *Anim. Prod.* **28**, 353–370

BURLACU, G., BAIA, G., DUMITRA, I., DOINA, M., TASCENCO, V., VISAN, I. and STOICA, I. (1973). *J. agric. Sci., Camb.* **81**, 295–302

CAPSTICK, J.W. and WOOD, T.B. (1922). *J. agric. Sci., Camb.* **12**, 257–268

CLOSE, W.H. (1980). *Proc. Nutr. Soc.* **39**, 169–175

CLOSE, W.H., HEAVENS, R.P. and BROWN, D. (1981). *Anim. Prod.* **32**, 75–84

CLOSE, W.H. and MOUNT, L.E. (1975). *Br. J. Nutr.* **34**, 279–290

CLOSE, W.H. and MOUNT, L.E. (1978). *Br. J. Nutr.* **40**, 413–421

CLOSE, W.H., MOUNT, L.E. and BROWN, D. (1978). *Br. J. Nutr.* **40**, 423–431

CLOSE, W.H., MOUNT, L.E. and START, I.B. (1971) *Anim. Prod.* **13**, 285–294

COMBERG, G., STEPHAN, E. and SPÄTH, H. (1971). *Züchtungskunde* **43**, 225–267

COMBERG, G., STEPHAN, E. and SPÄTH, H. (1972). *Züchtungskunde* **44**, 402–415

CURTIS, S.E. (1970). *J. Anim. Sci.* **34**, 576–587

DEIGHTON, T. (1923). *Proc. R. Soc., Ser. B* **95**, 340–355

DEIGHTON, T. (1929). *J. agric. Sci., Camb.* **19**, 140–184

DONE, J.T. and WIJERATNE, W.V.S. (1972). In *Pig Production*, pp. 53–70. Ed. by D.J.A. Cole. Butterworths, London

EWBANK, R. (1972). In *Pig Production*, pp. 129–142. Ed. by D.J.A. Cole. Butterworths, London

FULLER, M.F. (1965). *Br. J. Nutr.* **19**, 531–546

FULLER, M.F. and BOYNE, A.W. (1971). *Br. J. Nutr.* **25**, 259–272

FULLER, M.F. and BOYNE, A.W. (1972). *Br. J. Nutr.* **28**, 373–384

FULLER, M.F., DUNCAN, W.R.H. and BOYNE, A.W. (1974). *J. Sci. Fd Agric.* **25**, 205–210

GRAY, R. and McCRACKEN, K.J. (1974). *Publs Eur. Ass. Anim. Prod.*, No. 14, pp. 161–164

HACKER, R.R., STEFANOVIC, M.P. and BATRA, T.R. (1973). *J. Anim. Sci.* **37**, 739–744

HALE, O.M., JOHNSON, J.C., Jr. and WARREN, E.P. (1968). *J. Anim. Sci.* **27**, 1577–1582

HEITMAN, H., Jr. and HUGHES, E.H. (1949). *J. Anim. Sci.* **8**, 171–181

HICKS, A.M. and WEBSTER, A.J.F. (1968). *University of Alberta Feeders' Day*, pp. 16–17

HOLME, D.W. and COEY, W. (1967). *Anim. Prod.* **9**, 209–218

HOLMES, C.W. (1971). *Anim. Prod.* **13**, 521–528

HOLMES, C.W. (1973). *Anim. Prod.* **16**, 117–124

HOLMES, C.W. (1974). *Anim. Prod.* **19**, 211–220

HOLMES, C.W. and BREIREM, K. (1974). *Anim. Prod.* **18**, 313–316

HOLMES, C.W. and CLOSE, W.H. (1977). In *Nutrition and the Climatic Environment*, pp. 51–74. Ed. by W. Haresign, H. Swan and D. Lewis. Butterworths, London

HOLMES, C.W. and McLEAN, N.R. (1974). *Anim. Prod.* **19**, 1–12

HOLMES, C.W. and McLEAN, N.R. (1977). *Trans. ASAE* **20**, 527–528

HOLMES, C.W. and MOUNT, L.E. (1967). *Anim. Prod.* **9**, 435–452

HOVELL, F.D.DeB. and MacPHERSON, R.M. (1977). *J. agric. Sci., Camb.* **89**, 513–522

HSIA, L.C., FULLER, M.F. and KOH, F.K. (1974). *Trop. Anim. Hlth Prod.* **6**, 183–187

INGRAM, D.L. (1964). *Res. vet. Sci.* **5**, 357–364

INGRAM, D.L. (1965a). *Res. vet. Sci.* **6**, 9–17

INGRAM, D.L. (1965b). *Nature, Lond.* **207**, 415–416

INGRAM, D.L. (1974). In *Heat Loss from Animals and Man*, pp. 233–254. Ed. by J.L. Monteith and L.E. Mount. Butterworths, London

INGRAM, D.L. (1976). In *Progress in Biometeorology*, pp. 148–157. Ed. by S.W. Tromp. Swets & Zeitlinger, Amsterdam

INGRAM, D.L. and WEAVER, M.E. (1969). *Anat. Rec.* **163**, 517–524

IRVING, I., PEYTON, L.J. and MONSON, M. (1965). *J. appl. Physiol.* **9**, 421–426

JENKINSON, G.M., YOUNG, L.G. and ASHTON, G.C. (1967). *Can. J. Anim. Sci.* **47**, 217–226

JENSEN, A.H., KUHLMAN, D.E., BECKER, D.E. and HARMAN, B.G. (1969). *J. Anim. Sci.* **29**, 451–456

JORDAN, J.W. (1974). *Publs Eur. Ass. Anim. Prod.*, No. 14, pp 189–192

JORDAN, J.W. and BROWN, W.O. (1970). *Publs Eur. Ass. Anim. Prod.*, No. 13, 161–164

KIELANOWSKI, J. and KOTARBINSKA, M. (1970). *Publs Eur. Ass. Anim. Prod.*, No. 13, 145–148

McCRACKEN, K.J. and GRAY, R. (1972). *Proc. Br. Soc. Anim. Prod.* **1**, 139

MacGRATH, W.S., Jr., VAN DER NOOT, G.W., GILBREATH, R.L. and FISHER, H. (1968). *J. Nutr.* **96**, 461–466

MONTEITH, J.L. and MOUNT, L.E. (Eds) (1974). *Heat Loss from Animals and Man*. Butterworths, London

MORRISON, S.R. and MOUNT, L.E. (1971). *Anim. Prod.* **13**, 51–57

MORRISON, S.R., BOND, T.E. and HEITMAN, H., Jr. (1967). *Trans. ASAE* **10**, 691–692

MORRISON, S.R., BOND, T.E. and HEITMAN, H., Jr. (1968). *Trans. ASAE* **11**, 526–528

MORRISON, S.R., HEITMAN, H., Jr., GIVENS, R.L. and BOND, T.E. (1972). *Trop. Agric.* **49**, 31–35

MOUNT, L.E. (1959). *J. Physiol., Lond.* **147**, 333–345

MOUNT, L.E. (1960). *J. agric. Sci., Camb.* **55**, 101–105

MOUNT, L.E. (1964). *J. Physiol., Lond.* **170**, 286–295

MOUNT, L.E. (1966). *Q. Jl exp. Physiol.* **51**, 18–26

MOUNT, L.E. (1967). *Res. vet. Sci.* **8**, 175–186

MOUNT, L.E. (1968). *The Climatic Physiology of the Pig*. Arnold, London

MOUNT, L.E. (1974). In *Heat Loss from Animals and Man*, pp. 425–439. Ed. by J.L. Monteith and L.E. Mount. Butterworths, London

MOUNT, L.E. (1975). *Livest. Prod. Sci.* **2**, 381–392

MOUNT, L.E. (1977). *Anim. Prod.* **25**, 271–279

MOUNT, L.E. (1979). *Adaptation to Thermal Environment*. Arnold, London

MOUNT, L.E. and INGRAM, D.L. (1965). *Res. vet. Sci.* **6**, 84–91

MOUNT, L.E. and STEPHENS, D.B. (1970). *J. Physiol., Lond.* **207**, 417–427

PHILLIPS, P.A., YOUNG, B.A., McQUITTY, J.B. and HARDIN, R.T. (1979). Paper No. 79-4002 presented at the 1979 Summer Meeting of ASAE and CSAE (June 1979)

SAINSBURY, D.B. (1972). In *Pig Production*, pp. 91–105. Ed. by D.J.A. Cole. Butterworths, London.

SHARMA, V.D., YOUNG, L.G. and SMITH, G.C. (1971). *Can. J. Anim. Sci.* **51**, 761–770

SØRENSEN, P.H. (1962). In *Nutrition of Pigs and Poultry*, pp. 88–103. Ed. by J.T. Morgan and D. Lewis. Butterworths, London

STAHLY, T.S., CROMWELL, G.L. and AVIOTTI, M.P. (1979). *J. Anim. Sci.* **49**, 1242–1251

STEPHENS, D.B. (1971). *Anim. Prod.* **13**, 303–313

STEPHENS, D.B. and START, I.B. (1970). *Int. J. Biomet.* **14**, 275–283

STEPHENS, D.B. and START, I.B. (1972). *Cornell Vet.* **62**, 20–26

STOMBAUGH, D.P., ROLLER, W.L., ADAMS, T. and TEAGUE, H.S.C. (1973). *Am. J. Physiol.* **225**, 1192–1198

STOMBAUGH, D.P. and ROLLER, W.L. (1977). *Trans. ASAE* **20**, 1110–1118

SUGAHARA, M., BAKER, D.H., HARMON, B.G. and JENSEN, A.H. (1970). *J. Anim. Sci.* **31**, 59–62

TANGL, F. (1912). *Biochem. Z.* **44**, 252–278

VERSTEGEN, M.W.A. (1971). Thesis. *Meded. LandbHogesch. Wageningen*

VERSTEGEN, M.W.A., BRASCAMP, E.W. and VAN DER HEL, W. (1978). *Can. J. Anim. Sci.* **58**, 1–13

VERSTEGEN, M.W.A., CLOSE, W.H., START, I.B. and MOUNT, L.E. (1973). *Br. J. Nutr.* **30**, 21–35

VERSTEGEN, M.W.A. and VAN DER HEL, W. (1974). *Anim. Prod.* **18**, 1–11

VERSTEGEN, M.W.A. and VAN DER HEL, W. (1977). *Publs Eur. Ass. Anim. Prod.* No. 15, pp. 347–350

WEAVER, M.E. and INGRAM, D.L. (1969). *Ecology* **50**, 710–713

10

THERMAL INFLUENCES ON RUMINANTS

B.A. YOUNG
A.A. DEGEN
Department of Animal Science, University of Alberta, Edmonton, Alberta, Canada

Introduction

Animal housing or other management practices are usually based on compromises between gains in animal productivity and the capital, fuel and other inputs required to modify and maintain the desired system. Rational selection among management alternatives necessitates an understanding of the functional relationships between animal productivity and environmental variables. Within the complexity of interactions between the animals and their surroundings, thermal factors have received substantial research attention.

This chapter is restricted to the influences of the thermal environment, particularly cold, on energy metabolism and productivity of ruminants. Minimal attention will be given to the processes of physical heat exchange and the immediate acute responses of animals to sudden changes in their thermal environment as these processes and their consequences for ruminants have been the subject of several reviews at this and past Easter Schools (Monteith and Mount, 1974; Webster, 1976; Blaxter, 1977). The intention of this chapter is to focus on some of the longer-term chronic effects of the thermal environment and the practical consequences of physiological adaptation to ruminant productivity and nutritional efficiency.

Partition of dietary energy

Figure 10.1 is a slightly modified version of the classical breakdown of the dietary energy intake of animals via digestible and metabolizable energy to its potential use for maintenance and production. This schematic representation is a useful basis for identifying the main influences of the thermal environment on animals. A brief explanation and interpretation of *Figure 10.1* is warranted before considering influences of the environment.

The dietary energy intake (*intake energy*) of an animal is a function of the nature of the food (combustible energy density) and amount ingested. Ingested food is not completely digested, absorbed or available to the animal. Because of variations among feeds and digestive systems among animal species it has been customary to evaluate feeds in terms of their ability to provide useful energy. For example, feed energy values may be

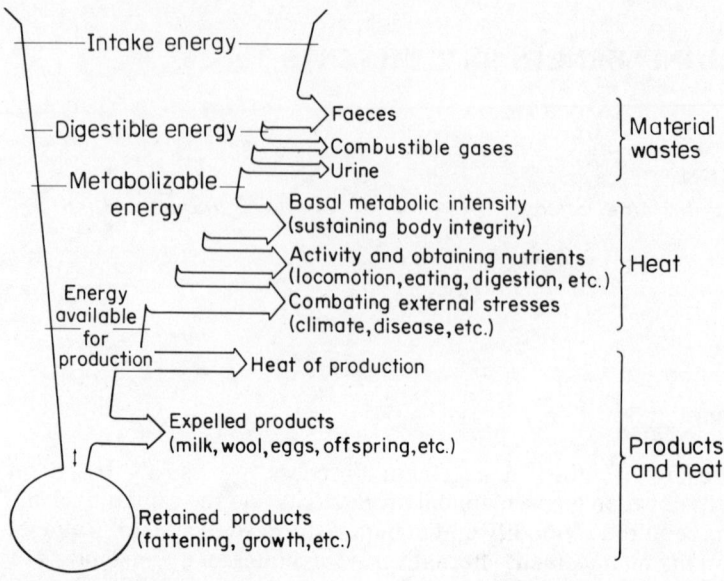

Figure 10.1 Funnel model of the partition of dietary energy in animals. (After Young, 1975a)

expressed in terms of their digestible or metabolizable energy content. By definition the *metabolizable energy* intake of an animal represents the energy available to that animal for maintenance and productive functions.

Maintenance functions include:

(1) *basal metabolic intensity*, which is represented by the energy utilized in sustaining body integrity by vital processes;
(2) *voluntary activity and obtaining nutrients*, including the activities of seeking, digesting and converting food into metabolizable forms;
(3) *combating external stresses* imposed on the animal.

In virtually any environment, animals are consistently faced with stresses of varying types and magnitudes. Although the physiological consequences of many stresses are still poorly understood, as a consequence of stress there is often an increase in catabolism and a reduction in metabolizable energy (ME) available for production. The ME oxidized for maintenance processes is released from the animal as heat (*maintenance heat*) and is ultimately disposed of to the environment through thermal exchange mechanisms. The amount of ME utilized for the maintenance processes of a producing animal may be different from that of a non-producing animal of the same type and size. The very fact that the animal is producing probably increases its vital functions and requirements, necessitating greater utilization of energy for maintenance. Debate arises as to whether this increase should be charged to maintenance *per se* or to production. For the purpose of the present discussion it is considered to be a part of the maintenance requirement.

Production utilizes ME, but energy is available for production only after the maintenance needs of the animal are met. Furthermore, the energy available for production is not entirely incorporated into animal products, be the products retained (growth, fattening, etc.) or expelled (milk, pelage, eggs, offspring, etc.), because of the inefficiencies of productive processes (*heat of production*).

Typically, an animal retains energy in glycogen or lipid when ME intake exceeds the animal's immediate needs. Conversely, retained energy is mobilized when the animal's demand is greater than the energy available from food. Dairymen, for example, often allow their cows to accumulate body energy reserves for mobilization and utilization during peak lactation, when maximum intake is insufficient to meet the immediate requirements for maintenance and milk production.

In summary, *Figure 10.1* represents the intake of dietary energy and its partition through the major routes of energy usage. In a cold environment the heat from maintenance and productive processes may be of considerable value to the animal in maintaining body temperature, saving the animal the need to produce extra body heat by shivering or other cold-induced thermogenic processes. On the other hand, in moderate and hot environments thermoregulatory mechanisms are activated to dissipate excess heat from the body to maintain homeothermy and prevent a rise in body temperature. Thus, heat which under one set of conditions is beneficial may under other circumstances be waste or, in the case of a hot environment, a burden to the animal. The thermal environment, by influencing energy intake and its partition within the animal, affects the amount of ME available for production and has a broader effect than implied in the single component of 'combating external stresses' indicated in *Figure 10.1*.

Heat exchange

Many reviews and papers on the effects of the environment on animal productivity have been based on an examination of the immediate calorimetric responses of animals when exposed to various ambient temperatures (*see* Monteith and Mount, 1974; Robertshaw, 1974; Johnson, 1976; Haresign, Swan and Lewis, 1977). Often overlooked, but of major consequence when interpreting the practical meaning of results from measurements of physical heat exchange, is the short-term nature of calorimetric studies upon which these measurements are based. Animal responses are usually measured during exposure to test conditions lasting only a few hours, during which time there is little opportunity for the animal to develop physiological adaptations. However, an understanding of these short-term measurements of physical heat exchange is fundamental to the understanding of longer-term consequences of the thermal environment for ruminants.

The basic concept of heat exchange between an animal and the thermal environment is illustrated in *Figure 1.2* in Chapter 1 (by Robertshaw) and relies on the premise of a *zone of thermoneutrality* where, by definition, an animal's metabolic heat production is constant and independent of the ambient temperature. In this scheme there are zones above and below thermoneutrality where the animal's heat production is dependent upon the environmental temperature. The lower border of the zone of thermoneutrality is called the *lower critical temperature* and is defined as the temperature below which an animal must increase its rate of metabolic heat production (by *cold thermogenesis*) to maintain homeothermy. Below the lower critical temperature metabolic heat production becomes increasingly dependent upon the ambient temperature. However, a point of maximal heat production (*summit metabolism*) is reached and continued exposure to lower temperatures results in hypothermia, a reduced capacity to produce heat and, if not reversed, the death of the animal.

Of considerable importance in determining the lower critical temperature is the metabolic heat production of an animal in its thermoneutral environment and the thermal insulation of the hair coat and superficial tissue of the animal. The reviews edited by Monteith and Mount (1974) discuss the factors influencing the lower critical temperatures of various animals. Estimates of the lower critical temperature are usually stated in terms of dry, still-air temperatures, i.e. conditions which can be generated within climatic chambers. However, in practice, the environment experienced by the animal is rarely well specified by the measured air temperature, because of variables such as wind, moisture and radiation. Wind, for example, lessens the insulative value of the hair coat and may substantially increase the rate of loss of body heat. Likewise, a coat wet from rain or melting snow can have its insulative value reduced and, through the latent heat of melting or evaporation, markedly increase the thermal demand on an animal.

Lower critical temperature estimates for young lambs and calves are usually about 10 °C and decrease as the animals grow. Typical values for dry, pregnant beef cows during winter are between −10 and −20 °C, while for high-producing dairy cows and feedlot beef cattle, measurements suggest values which are usually between −20 and −40 °C. Estimates of lower critical temperature for full-fleeced sheep are also particularly low (−20 to −30 °C), whereas in recently shorn sheep the lower critical tempratures are raised to +20 °C or more. These lower critical temperatures should be considered only as indicators of cold susceptibility for, in reality, actual values may vary somewhat depending upon the housing conditions, breed, nutrition, thermal adaptation and animal behaviour.

Physiological adaptation

A physiological adaptation is an environmentally induced change in an animal which favours survival. In presenting this definition Prosser (1958) drew a distinction from the term 'response', which refers to a direct reaction to an environmental stimulus that may or may not be adaptive. Physiological adaptation refers to changes in one or more functional systems within the individual that persist beyond the immediate relief of the disturbance. Long-term exposure to a specific environment allows an animal to adapt. There then may be compensation and restoration of biological displacements (responses) that occurred at the beginning of an exposure, but of greater significance is the increased ability of the animal to withstand exposures to more severe conditions than prior to adaptation.

Physiological adaptation to cold is well documented for small mammals (*see* Dill, 1964). The more obvious changes that directly improve an animal's ability for thermal regulation and survival include increases in thermal insulation, in appetite and in metabolic intensity. Increased metabolic intensity is reflected as an increase in thermoneutral metabolic rate as well as an increase in summit metabolism when challenged by severe cold. During the past few years reports have indicated that farm animals may also adapt physiologically to the thermal environment to which they are exposed. Rather than attempting to give an account of the interactions and synchronized array of neural, endocrine and other changes occurring with adaptation, we shall focus attention in the following sections on those aspects of adaptation which may have practical consequences to ruminant productivity.

In natural environments, adaptation is likely to be continually changing, for as the animal adjusts the thermal environment continues to change. However, animals often show adaptive changes in anticipation of a potential stress. For example, the shortening daily photoperiod acts as a stimulus for animals to develop their winter hair coat prior to the coldest part of the winter. The involvement of photoperiod in other cold-adaptive processes is largely unknown, although there is evidence from studies on small mammals that photoperiod has an influence on metabolic heat production (Dill, 1964).

By reference to *Figure 10.1*, several components of dietary energy within animals can be identified as being influenced by the processes of physiological adaptation to cool or cold environments. Basically, the adaptive changes are centred about the animal's defence against a threat to homeothermy.

INSULATION

The growth and retention of a long, thick winter hair coat is seen in many species and contributes to their increased thermal insulation and ease of maintaining homeothermy in cold conditions. In cattle, seasonal changes in hair cover are influenced by daily photoperiod as well as by ambient temperature (Yeates and Southcott, 1958; Webster, Chlumecky and Young, 1970). From data of Webster, Chlumecky and Young (1970), the rate of growth of new hair appears to be inversely related to day length, and not affected by cold *per se*. However, the rate of shedding of hair does seem to be associated with the thermal status of the animal. Animals that are cold, whether because of ambient conditions or because of a low plane of nutrition and reduced metabolic heat production, tend to retain their hair coat in situations where animals at a higher feeding level and heat production shed theirs. In domestic sheep, shedding of wool is minimal and the amount of external insulation is usually a consequence of a management decision. The rate of wool growth is, however, influenced by photoperiod (Hutchinson, 1965).

In addition to the obvious advantages of additional external insulation during winter, there are suggestions that tissue insulation also increases as a consequence of prolonged exposure and adaptation to cold (Webster, Chlumecky and Young, 1970; Webster, 1976). Various mechanisms have been suggested for this enhanced tissue insulation, including alterations in peripheral vasomotor control, structural changes in the skin and subcutaneous fat, and habituation resulting in reduced perception of cold

(Goldsmith, 1974; Robertshaw, 1974). However, the nature of the change in the superficial tissues of ruminants with cold adaptation remains uncertain.

Substantial insulative adaptive changes are seen in cattle wintered in prairie Canada, where air temperatures remain below freezing for 100 days or more during the winter and where during January and February air temperatures of −30 to −40 °C are not uncommon. Data from the University of Alberta (Webster, Chlumecky and Young, 1970) indicate that, relative to housed control cattle kept at 20 °C, similar young growing cattle kept outside substantially increased their tissue and external insulation. As a consequence there was a fall of more than 15 K in their estimated lower critical temperature during mid-winter. This adaptation for increased cold-hardiness was accompanied by a shift of the whole range of thermoneutrality to lower temperatures, for when in mid-winter the outside group of cattle were temporarily moved indoors to the 20 °C environment they were heat-stressed, markedly increasing their respiratory rate, heart rate and rectal temperature. The adaptation developed to conserve body heat during the winter could not immediately be modified to accommodate the warmer conditions in which the housed cattle were thermally comfortable.

METABOLIC INTENSITY

One of the characteristic phenomena observed in cold-adapted animals is an increased thermoneutral (resting) or basal metabolic rate. Such an increase in metabolic intensity is clearly evident from many studies on small mammals and birds (Dill, 1964; Jansky, 1971; Smith *et al.*, 1972) but reports on larger mammals including man have not been as consistent (Jansky, 1973; Goldsmith, 1974). In human studies, conflict in interpretation has arisen because of cold-induced behavioural changes and the general reluctance of subjects to be exposed to test situations for long durations. However, studies have clearly established a seasonal increase in basal metabolism during winter and a decrease during summer than cannot be explained by dietary or behavioural changes (Hong, 1973).

There are a few studies available on changes in metabolic intensity in domestic ruminants as a consequence of prolonged exposure to warm or cold environments. Kibler *et al.* (1965) reported that the metabolic rate and heart rate of lactating cows were reduced about 15 per cent by prolonged exposure (9 weeks) to chamber temperatures of 29 °C when compared with exposure to 18 °C. Slee and Sykes (1967) and Sykes and Slee (1969) deduced on the basis of changes in heart rate that the metabolic intensity in sheep was increased as a consequence of prolonged prior cold exposure. Similarly, results based on calorimetric measurements on sheep (Webster, Hicks and Hays, 1969) and young cattle (Webster, Chlumecky and Young, 1970) indicated an increase in resting metabolic rate following prolonged cold exposure. Unfortunately, in the above calorimetric studies of Kibler *et al.* (1965) and Webster and co-workers there was possible confounding with the stimulation of appetite and increased feed intake. In two more recent studies at our laboratory we examined the influence of prior thermal exposure on the metabolic intensity (resting metabolic rate in thermoneutral conditions) of beef cows on constant rations (Young, 1975b, c). Data from these studies

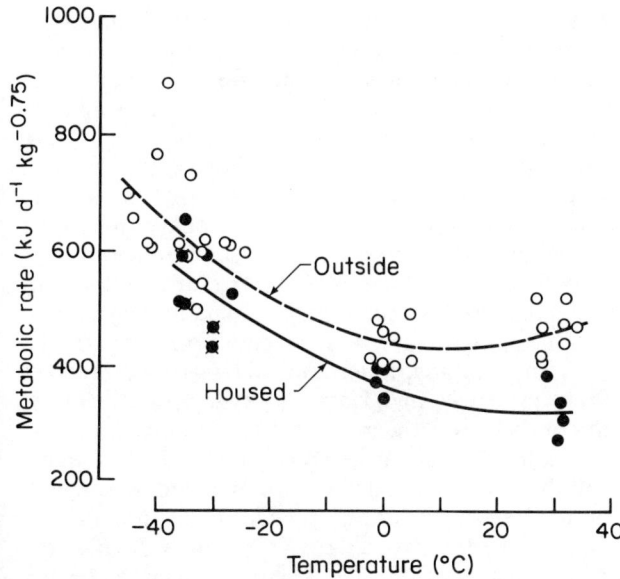

Figure 10.2 Relationship between the metabolic rate of beef cows, 22 hours post-feeding, and ambient test temperature. Measurements were made during a 5 hour period at the test temperature. At other times the housed cows (●) were kept at 18 ± 2 °C while the outside cows (O) were exposed to the natural winter condition of –10 to –48 °C. Some housed cows (⊗) failed to maintain homeothermy during tests at temperatures less than –25 °C. (From Young, 1975b)

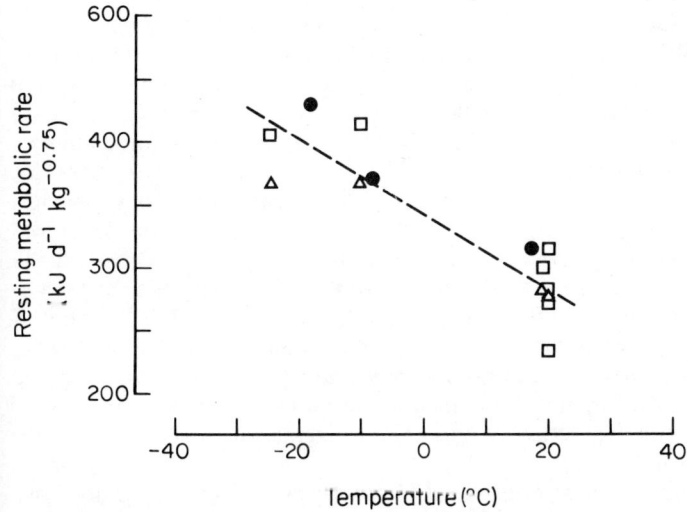

Figure 10.3 Effect of cold adaptation on the thermoneutral, resting metabolic rate of beef cows, 22 h post-feeding. The data were derived from regressions of ambient test temperature on the metabolic rate of housed and outside groups of cows (●) (Young, 1975b; see also Figure 10.2) and from individual cows exposed for prolonged periods in controlled temperature chambers (△,□) (Young, 1975c). The line represents an increase in resting metabolic rate of 2.9 kJ d^{-1} $kg^{-0.75}$ for each degree Celsius decrease in temperature

are summarized in *Figures 10.2* and *10.3*, and clearly indicate an increase in the metabolic intensity in cattle after prolonged exposure to cold in either climatic chambers or natural outdoor winter conditions. The percentage increase in thermoneutral resting metabolic rate in cows was similar to increases reported in cold-exposed small birds and mammals.

A retained response in metabolic intensity, and not simply an acute response, is commensurate with metabolic adaptation to the thermal environment. An increased metabolic intensity under resting non-stressful conditions is believed to be indicative of an increased summit metabolism, such as would be demanded during extreme cold exposure. In fact, there are numerous studies showing that cold-adapted small mammals have a higher summit metabolism and are able to survive at much colder temperatures than similar animals without prior adaptation to cold (Jansky, 1971, 1973; Smith *et al.*, 1972). Webster, Hicks and Hays (1969) reported increased summit metabolism in cold-adapted sheep. Summit metabolism is, however, difficult to reach in cold-adapted cattle. During collection of the data shown in *Figure 10.2*, all the cows that had been cold-adapted were able to maintain normal rectal temperatures during exposure to temperatures of –25 °C and below, while several of the housed cows were unable to maintain homeothermy at these low temperatures. The ability of sheep (Sykes and Slee, 1969; Berman, Chauca and Bligh, 1976) and cattle (Young, 1975b; Gonyou, Christopherson and Young, 1979) to withstand increasingly low temperatures without shivering as winter progresses, or during prolonged cold exposure in a climatic chamber, is further evidence of adaptive change. However, the relative contributions of increased insulation and increased metabolic intensity to the cold-adapted animal's reduced dependence on shivering thermogenesis are uncertain.

Disastrous death losses in livestock occasionally occur following sudden storms and cold weather (Hutchinson, 1968; Blaxter, 1977). Presumably, cold-adapted animals would be in a better physiological state to withstand acute exposure to extreme conditions than would be animals without cold adaptation. Thus, there is a distinct advantage to the cold-adapted animals in terms of survival in cold climates. The advantage, however, does not come without cost, for with an increase in metabolic intensity there is also an increase in maintenance energy requirement, even during relatively mild periods of winter when there is no direct challenge or cold stress on the animal. The maintenance energy requirement during winter has been estimated to be increased by approximately $2.9 \text{ kJ kg}^{-0.75} \text{ d}^{-1}$ for each 1 K decrease in the ambient temperature to which cattle are adapted (Young and Christopherson, 1974; Young, 1975a; *see also Figure 10.3*). The increase in maintenance energy requirement may be looked upon as an insurance premium paid for protection of the animal against the risk of calamitous weather situations.

The mechanisms of increased metabolic intensity with cold adaptation are currently under research investigation in several laboratories. From this research there appears to be a distinct difference between the processes in small and larger species. Brown fat and the calorimetric response to catecholamines seem to be an integral part of the cold-adaptation mechanism in small mammals and possibly also in the newborn of larger mammalian species (Jansky, 1973; Smith *et al.*, 1972; *see also* Chapter 7). However, these mechanisms appear to be lacking in adult ruminants, although there is evidence of increased catecholamine activity during cold exposure (Thompson, 1973; Johnson, 1976). Among the endocrine responses, increased thyroid activity is reported for most species, but as yet the exact role of thyroid hormones in cold adaptation and associated processes is not well understood. Adaptations to particular thermal conditions include changes in appetite, body fluids and digestive functions. These changes are interrelated with energy metabolism and also have direct consequences for ruminant production, particularly in terms of utilization of dietary energy.

APPETITE AND DIGESTIVE FUNCTION

The influences of the thermal environment on the appetite of animals are well recognized (Baile and Forbes, 1974). Appetite is usually stimulated by cold and is progressively depressed by increased exposure to warm and hot conditions. Appetite drive has been assumed to reflect the energy metabolism demands of the animal in the various thermal environments. What, however, are the changes in the animal that allow for the expression of appetite? Recent research from our laboratory and elsewhere has shed some light on the digestive and associated changes in ruminants which may explain, in part, the increased food intake during cold exposure. When ruminants are cold-exposed there is an increase in rumination activity (Gonyou, Christopherson and Young, 1979) and reticulo-rumen motility (Westra and Christopherson, 1976; Gonyou, Christopherson and Young, 1979) and a decrease in retention time of digesta (Westra and Christopherson, 1976; Kennedy, Christopherson and Milligan, 1976; Kennedy, Young and Christopherson, 1977). There is also a reduction in the animals' extracellular fluid volume, particularly in the reticulo-rumen (Kennedy, Christopherson and Milligan, 1976; Degen and Young, 1980). Furthermore, the mechanisms by which the thermal environment influences digestive function appear to involve thyroid hormones (Levin, 1969; Miller *et al.*, 1974; Westra and Christopherson, 1976; Kennedy, Young and Christopherson, 1977).

An interesting aside, with practical economic consequences to livestock producers, arises from the above-mentioned fluid volume changes occurring with the onset of a cold stress. Degen and Young (1980) showed that the reticulo-rumen fluid volume decreased by 25–30 per cent during cold exposure, which was reflected as a 3–5 per cent reduction in live-weight of the animal. This loss in live-weight occurs during initial exposure to cold and is primarily associated with a reduced water intake. Upon re-exposing the animals to warmer conditions, the fluids lost during cold exposure are rapidly regained, mainly through a marked increase in water consumption. While the changes in live-weight largely reflect losses and gains of gut fluid, the consequences to those buying or selling animals on a live-weight basis can be economically important. Furthermore, the changes in gut volume result in a higher dressing percentage of carcasses from animals exposed to cold prior to slaughter.

With the marked reduction in volume of gut contents when animals are initially exposed to cold there is an increase in faecal output. Observations

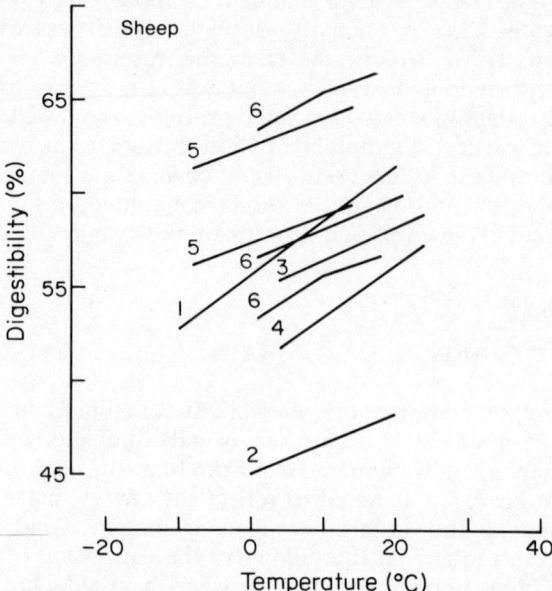

Figure 10.4 Ration dry-matter digestibility, in relation to the ambient temperature to which sheep were exposed when consuming the ration. (Examples from different roughage feeds and trials: 1. Christopherson, 1976; 2. Kennedy, Christopherson and Milligan, 1976; 3. Kennedy, Young and Christopherson, 1977; 4. Kennedy and Milligan, 1978; 5. Nicholson, McQueen and Burgess, 1980; 6. Westra and Christopherson, 1976)

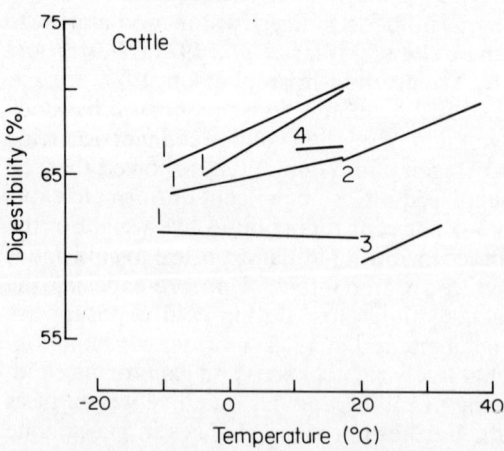

Figure 10.5 Ration dry-matter digestibility, in relation to the ambient temperature to which cattle were exposed when consuming the ration. (Examples from different roughage feeds and trials: 1. Christopherson, 1976; 2. Colditz and Kellaway, 1972; 3. McDowell et al., 1969; 4. Wöhlbier and Schneider, 1965)

by Fuller and Cadenhead (1969) of an apparent reduction in ration digestibility as a consequence of cold exposure may have been associated with the transient change in gut fill. However, several other studies on both cattle and sheep exposed to cold conditions for prolonged periods, either outdoors or in temperature-controlled chambers, also indicate a depression in digestibility of roughage feeds (0.1–0.2 digestibility units per °C; *Figures 10.4* and *10.5*). The reduced digestibility of feeds is apparently independent of the level of feed intake and is most probably associated with the increased rumination, gut motility and rate of passage through the gastro-intestinal tract observed during cold exposure. The more rapid passage of digesta results in a decrease in total organic matter degradation in the reticulo-rumen but, at the same time, an increase in efficiency of microbial synthesis (Kennedy, Christopherson and Milligan, 1976; Kennedy and Milligan, 1978). Kennedy and Milligan deduced from the partition of methane production that, for sheep on a pelleted brome grass ration, the fermentation in the reticulo-rumen was decreased by about one-third during prolonged cold exposure, but hind-gut fermentation was unaltered.

From the data contained in *Figures 10.4* and *10.5* there remains open the question of the importance of ration type on the influence of ambient temperature on ratio digestibility. All data in *Figures 10.4* and *10.5* were for roughage-based rations. In two recent experiments in our laboratory involving concentrate rations, ambient temperature had no significant influence on ration digestibility. In both experiments the rations were pelleted and composed of equal portions of barley grain and dehydrated lucerne meal (protein 15.5%; acid detergent fibre 16.8%). In the first experiment 16 sheep were exposed in climate chambers to ambient temperatures of 0, 10, 20 or 30 °C and offered daily rations of 500, 750, 1000 or 1250 g of the concentrate feed. After adjustment to the chamber temperatures, ration dry matters were determined in each sheep by the total faecal collection method. Regression analyses showed no significant effect of temperature or feeding level on ration digestibility (*Table 10.1*). In the second experiment 8 sheep were offered 1000 g of the concentrate ration daily and exposed to 0 and 30 °C to assess the effect of ambient temperature on ration digestibility and rate of passage of liquid (^{51}Cr-EDTA) and solid (^{103}Ru-phenanthroline) digesta from the reticulo-rumen; no significant effect was found (*Table 10.2*).

The adaptive significance of the cold-induced changes in digestive function is revealed when calculations are made taking into account both the

Table 10.1 REGRESSIONS OF AMBIENT TEMPERATURE (T; °C) AND INTAKE OF A BARLEY GRAIN–LUCERNE MEAL RATION (F; g d^{-1}) ON DRY-MATTER DIGESTIBILITY (D; %). REGRESSION COEFFICIENTS WERE NOT SIGNIFICANTLY DIFFERENT FROM ZERO

$D = 71.9 + 0.031\ T$ ± 0.042	$r^2 = 0.04$
$D = 76.5 - 0.004\ F$ ± 0.001	$r^2 = 0.52$
$D = 76.1 + 0.017\ T - 0.004\ F$ $\pm 0.031 \quad \pm 0.001$	$r^2 = 0.52$

Table 10.2   EFFECT OF AMBIENT TEMPERATURE ON THE DIGESTIBILITY OF A CONCENTRATE RATION (1 kg d⁻¹) AND ON THE RATE OF FLOW OF LIQUID (^{51}Cr-EDTA) AND SOLID (^{103}Ru-PHENANTHROLINE) DIGESTA FROM THE RETICULO-RUMEN

Exposure temp. (°C)	Dry-matter digestibility (%)	Half-time of marker (h)	
		^{51}Cr-EDTA	^{103}Ru-Phenanthroline
0	72.0	23.5	27.3
30	72.7	26.1	31.1
(SEM	±0.34	±2.6	±3.1)

Temperature treatment not significant

drop in digestibility of the roughage ingested and the increase in appetite during cold exposure (Christopherson, 1976; Kennedy and Milligan, 1978). For sheep exposed to 0 °C the result is an overall increase in digestible organic matter intake of about 4 per cent when intake is not restricted, but the food is digested at a lower efficiency than by similar sheep exposed to 20 °C. The actual ME intake could be increased by more than 4 per cent, since methane production is decreased slightly by exposure to cold.

Conclusions

The thermal environment has immediate as well as long-term adaptive influences on animals. High-producing ruminants are cold-hardy because of their well-developed insulation and relatively high rates of heat production. They are not often exposed to temperatures below their lower critical temperature and rarely require an immediate increase in their energy requirements because of cold weather. There are, of course, situations with stock which are young or ill and with non-producing animals without access to adequate shelter where there is a direct challenge to homeothermy necessitating an increase in shivering thermogenesis.

In contrast, the longer-term adaptive changes that occur in ruminants as well as in other animals during chronic exposure to even relatively mild cold conditions are of substantial importance to the livestock producer. The adaptive changes are probably responsible for the marked seasonal changes in efficiency seen in many commercial systems where high-producing animals are exposed to the natural climatic conditions. Elam (1971) reported a 14–20 per cent lower feed-to-gain ratio by cattle during winter than during summer in large commercial feedlots in southern California and in the midwestern USA. Similarly, data compiled by Knox and Handley (1973) from northern Colorado feedlots, and Milligan and Christison (1974) from the University of Saskatchewan feedlot also show marked seasonal fluctuations in cattle performance. A summary of data from research feedlots located on the great plains of North America extending from Texas to Saskatchewan indicated that while satisfactory daily weight gains can be achieved by cattle at all locations, there were marked differences between summer and winter performances, especially in feed conversion efficiency in the northern regions (Young, 1980). Regression equations relating thermal-stress indices to performance of cattle in commercial and semi-commercial feedlots indicate that 50–70 per cent of the variation in performance can be accounted for by climatic variables (Petritz, 1972; Knox and Handley, 1973; Milligan and Christison, 1974; Ames, Brink and Schalles, 1975).

Adaptation to cold involves a complex of neural, endocrine and functional changes in ruminants that increase their ability to withstand cold. The practical changes from an animal production viewpoint are the increases in metabolic intensity (maintenance energy requirement) and an increase in appetite but, at the same time, a possible decrease in digestive efficiency. Overall there may not be a decrease in animal productivity during cold exposure if the increase in feed intake can compensate for the increase in energy requirement. With prolonged exposure to cold and development of cold adaptation the risk to the animal of a calamity from a sudden cold spell is lessened by the insurance paid as extra feed by the farmer. While this extra feed cost for ruminants exposed to cold may be avoided by the provision of elaborate housing, the capital and maintenance costs, particularly the cost of heating fuel, may be uneconomic relative to the cost of the extra feed required by the unhoused animal.

References

AMES, D.R., BRINK, D.R. and SCHALLES, R.R. (1975). *J. Anim. Sci.* **41**, 262

BAILE, C.A. and FORBES, J.M. (1974). *Physiol. Rev.* **54**, 160–214

BERMAN, A., CHAUCA, D. and BLIGH, J. (1976). *Israel J. med. Sci.* **12**, 892–895

BLAXTER, K.L. (1977). In *Nutrition and the Climatic Environment*, pp. 1–16. Ed. by W. Haresign, H. Swan and D. Lewis. Butterworths, London

CHRISTOPHERSON, R.J. (1976). *Can. J. Anim. Sci.* **57**, 201–212

COLDITZ, P.J. and KELLAWAY, R.C. (1972). *Aust. J. agric. Res.* **23**, 717–725

DEGEN, A.A. and YOUNG, B.A. (1980). *Can. J. Anim. Sci.* **60**, 33–41

DILL, D.B. (1964). *Adaptation to the Environment.* American Physiological Society, Washington DC

ELAM, C.J. (1971). *J. Anim. Sci.* **32**, 554–559

FULLER, M.F. and CADENHEAD, A. (1969). In *Energy Metabolism of Farm Animals*, pp. 455–460. Ed. by K.L. Blaxter, J. Kielanowski and G. Thorbek. Oriel Press, Newcastle upon Tyne

GOLDSMITH, R. (1974). In *Heat Loss from Animals and Man*, pp. 311–319. Ed. by J.L. Monteith and L.E. Mount. Butterworths, London

GONYOU, H.W., CHRISTOPHERSON, R.J. and YOUNG, B.A. (1979). *Appl. Anim. Ethol.* **5**, 113–124

HARESIGN, W., SWAN, H. and LEWIS, D. (1977). *Nutrition and the Climatic Environment.* Butterworths, London

HONG, S.K. (1973). *Fedn Proc. Fedn Am. Socs exp. Biol.* **32**, 1614–1622

HUTCHINSON, J.C.D. (1965). In *Biology of the Skin and Hair Growth*, pp. 565–574. Ed. by A.G. Lyne and B.F. Short. Angus & Robertson, Sydney

HUTCHINSON, J.C.D. (1968). *Aust. J. exp. Agric. Anim. Husb.* **8**, 393–400

JANSKY, L. (1971). *Non-shivering Thermogenesis.* Academia, Prague

JANSKY, L. (1973). *Biol. Rev.* **48**, 85–132

JOHNSON, H.D. (1976). *Progress in Biometeorology*, Vol. 1, Part 1, Div. B. Swets & Zeitlinger, Amsterdam

KENNEDY, P.M., CHRISTOPHERSON, R.J. and MILLIGAN, L.P. (1976). *Br. J. Nutr.* **36**, 231–242

KENNEDY, P.M. and MILLIGAN, L.P. (1978). *Br. J. Nutr.* **39**, 105–117

KENNEDY, P.M., YOUNG, B.A. and CHRISTOPHERSON, R.J. (1977). *J. Anim. Sci.* **45**, 1084–1090

KIBLER, H.H., JOHNSON, H.D., SHANKLIN, M.D. and HAHN, H. (1965). *Research Bulletin of the University of Missouri Agricultural Experiment Station*, No. 893

KNOX, K.L. and HANDLEY, T.M. (1973). *J. Anim. Sci.* **37**, 190–199

LEVIN, R.J. (1969). *J. Endocr.* **45**, 315–348

McDOWELL, R.E., MOODY, E.G., VAN SOEST, P.J., LEHMANN, R.P. and FORD, G.L. (1969). *J. Dairy Sci.* **52**, 188–194

MILLER, J.K., SWANSON, E.W., LYKE, W.A., MOSS, B.R. and BYRNE, W.F. (1974). *J. Dairy Sci.* **57**, 193–197

MILLIGAN, J.D. and CHRISTISON, F.I. (1974). *Can. J. Anim. Sci.* **54**, 605–610

MONTEITH, J.L. and MOUNT, L.E. (Eds) (1974). *Heat Loss from Animals and Man.* Butterworths, London

NICHOLSON, J.W.G., McQUEEN, R.E. and BURGESS, P.L. (1981). *Can. J. Anim. Sci.*, in press

PETRITZ, D. (1972). *PhD Thesis*, University of Illinois, Urbana

PROSSER, C.L. (1958). *Physiological Adaptation.* American Physiological Society, Washington DC

ROBERTSHAW, D. (1974). *Environmental Physiology.* Physiology Series 1, vol. 7. University Park Press, Baltimore

SLEE, J. and SYKES, A.R. (1967). *Anim. Prod.* **9**, 333–347

SMITH, R.E., HANNON, J.P., SHIELDS, J.L. and HORWITZ, B.A. (1972). *Int. Symp. Environmental Physiology: Bioenergetics.* Federation of American Societies for Experimental Biology, Washington DC

SYKES, A.R. and SLEE, J. (1969). *Anim. Prod.* **11**, 65–75

THOMPSON, G.E. (1973). *J. Dairy Res.* **40**, 441–473

WEBSTER, A.J.F. (1976). In *Principles of Cattle Production*, pp. 103–120. Ed. by H. Swan and W.H. Broster. Butterworths, London

WEBSTER, A.J.F., CHLUMECKY, J. and YOUNG, B.A. (1970). *Can. J. Anim. Sci.* **50**, 89–100

WEBSTER, A.J.F., HICKS, A.M. and HAYS, F.L. (1969). *Can. J. Physiol. Pharmac.* **47**, 553–562

WESTRA, R. and CHRISTOPHERSON, R.J. (1976). *Can. J. Anim. Sci.* **56**, 699–708

WÖHLBIER, W. and SCHNEIDER, W. (1965). In *Energy Metabolism*, pp. 405–418. Ed. by K.L. Blaxter. Academic Press, London

YEATES, N.T.M. and SOUTHCOTT, W.H. (1958). *Proc. Aust. Soc. Anim. Prod.* **2**, 102–103

YOUNG, B.A. (1975a). 'Some physiological costs of cold climates'. *Special Report of the University of Missouri Agricultural Experiment Station*, No. 175

YOUNG, B.A. (1975b). *Can. J. Anim. Sci.* **55**, 619–625

YOUNG, B.A. (1975c). *Can. J. Physiol. Pharmac.* **53**, 947–953

YOUNG, B.A. (1980). *J. Anim. Sci.* **51**, 811–815

YOUNG, B.A. and CHRISTOPHERSON, R.J. (1974). In *Livestock Environment*, pp. 75–80. American Society of Agricultural Engineering, St Joseph, Michigan

IV

OPTIMAL HOUSING ENVIRONMENTS FOR TEMPERATE AND COOL CLIMATES

11

PRACTICAL VENTILATION AND TEMPERATURE CONTROL FOR POULTRY

D. R. CHARLES
Agricultural Development and Advisory Service, Shardlow, Derby

Introduction

In the poultry industry, food accounts for about 70 per cent of the overall cost of production. Thus the margin of egg or meat sales less food cost is a useful index of efficiency, and the economic optimum temperature is usually defined as that which maximizes the margin. Experimental data are now available to quantify the effects on performance of deviations from the optimum temperature, and meteorological data can be used to estimate the frequency and magnitude of such deviations, so that the appropriate levels of expenditure for environmental control can be assessed. On this basis, the UK laying and broiler industries require a high degree of control, and rather precisely controlled ventilation systems have been developed. Less elaborate systems may be adequate in areas which have different combinations of climate, feed price and product prices.

Requirements

LAYING HENS

Payne (1967), Mowbray and Sykes (1971), Smith and Oliver (1972a, b), Emmans and Charles (1977) and A. Marsden and T.R. Morris (personal communication) have all found a reduction of about 1.5 $g\,d^{-1}\,K^{-1}$ per bird in ad-libitum feed intake with increases in temperature over the range 15–25 °C. The depression in feed intake per degree increases at higher temperatures. Up to about 30 °C this important feed saving is not associated with a depression of production if the intake of essential nutrients is maintained; indeed, annual production is probably increased by about one egg per hen housed per degree. *Figure 11.1* (from Emmans and Charles, 1977) shows some effects of temperature on food intake.

There is no clearcut temperature beyond which production is depressed. The upper limit depends on the strain of bird, feathering, acclimatization, and on bird numbers per cage. It can be considered to be exceeded if prolonged and severe panting occurs. Egg weight is also reduced by high temperatures, by about 0.3 $g\,K^{-1}$ per egg. Taking all these factors into account has led to a recommendation, for many UK market circumstances,

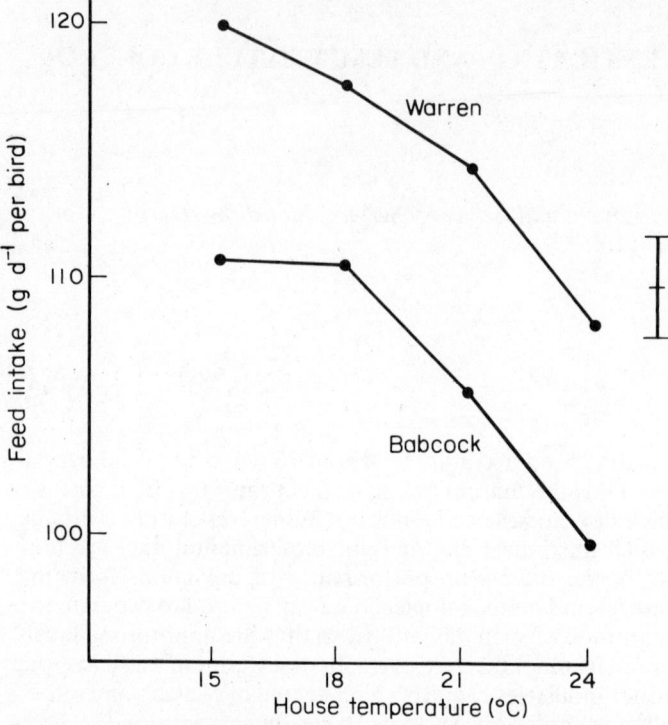

Figure 11.1 Effect of temperature on the feed intake of layers

of a minimum house temperature of 21 °C. For the past few years ventilation system design has attempted to achieve this by the conservation of bird heat and without fossil fuel heating.

BROILERS—BROODING

Experimental evidence on brooding temperature requirement is rare, partly because the behaviour of the birds is normally a satisfactory guide. For convectional heating systems Felton (1974) suggested 27 °C at 1 day old, gradually being reduced to the finishing regime at the rate of 1 K per day. For radiant systems unpublished work by P. G. Spencer at Gleadthorpe showed that 20 °C is probably high enough for the house dry-bulb air temperature, if radiant heat is also present from brooders. Ventilation system designs attempt to achieve these regimes for least fuel cost. Some of the designs below are likely to be associated with efficient fuel use.

BROILERS—POST-BROODING

There has been much more experimental work on the post-brooding period, because the bulk of the food consumption occurs at that stage. Experiments up to 1975 (reviewed by Charles, 1979) suggested an optimum of 20–25 °C. In recent large experiments, Charles, Groom and Bray (1981) and Wathes *et al.* (1981) found that body weight and food consumption (to 49 days) decreased linearly with increasing temperature, within the range 19–26 °C. The rates were 16.2 and 52.9 g K^{-1} per bird respectively. At prices typical at the time of writing the economic optimum temperature was 21 °C, but Wathes *et al.* developed a multi-dimensional model of the response data, also taking into account killing age and feeding regime, allowing recalculation of margins for any set of prices. Some of the responses of broilers to temperature were described by the following relationships. They predict the body weight (W) in kilograms per bird and the cumulative food intake (F) also in kilograms per bird, and apply to broilers fed a finisher diet of 12.9 MJ kg^{-1} and 197 g protein per kilogram. Other models were developed for other diets.

$$W = 0.041T - 1.373 - 0.0016(T - 21.75)^2 + 0.074a - 0.0013Ta \quad \text{(females)} \tag{11.1a}$$

$$W = 0.041T - 1.499 - 0.0016(T - 21.75)^2 + 0.085a - 0.0013Ta \quad \text{(males)}$$

$$F = 0.111T - 6.739 + 0.211a + 0.0014(a - 49)^2 - 0.0034Ta \quad \text{(females)} \tag{11.1b}$$

$$F = 0.111T - 7.202 + 0.229a + 0.0014(a - 49)^2 - 0.0034Ta \quad \text{(males)}$$

where a is the age in days and T is the mean dry-bulb air temperature, in °C.

TURKEYS

Albuquerque *et al.* (1978) found that the food intake of turkeys also decreased with increasing temperature at the rate of 3 g d^{-1} K^{-1} per bird.

Table 11.1 RECOMMENDED DRY-BULB AIR TEMPERATURES FOR POULTRY

Class of stock	Recommended temperature, T_* (°C)	Factors affecting T_*
Laying hens	Minimum normally 21 °C, but could be in range 19–24 °C. Maximum 30 °C	Prices of feed and eggs. Price difference between egg weight grades. Price difference between food protein levels
Broilers (a) brooding	27–28 °C at 1 day old reduced by 0.5–1 K per day, or with radiant brooders, 20–21 °C outside brooder area	The behaviour of the birds should be used as the ultimate guide, not the thermometer reading.
(b) finishing	21–23 °C	Killing age, prices of food and meat. Females are likely to require higher temperature than males
Broiler breeders	20 °C approx.	
Turkeys (a) brooding	29 °C at 1 day old, reduced by 3 K per week	The behaviour of the birds should be used as the guide
(b) finishing	17–20 °C	Prices of meat and feed

This requires 300 g more food per °C per bird between the ages of 8 and 24 weeks. Total weight gain decreased by approximately 30 g K^{-1} per bird, above 18 °C, but was unaffected by temperatures between 10 and 18 °C up to 18 weeks of age. The weight gain of older birds (after 18 weeks of age) was depressed with increasing temperature throughout the range used (10–35 °C). Current unpublished work from the Gleadthorpe Poultry Unit shows a depression of cumulative food intake of 150 g K^{-1} per bird to 14 weeks of age, and a reduction of 75 g K^{-1} in bird body weight at the same age.

Table 11.1 summarizes temperature recommendations for poultry, based on the literature reviewed above.

Achieving the requirements

In temperate climates the economic optimum temperatures, described above, are achieved through thermostatic modulation of the ventilation rate. The balance between the heat input from the birds' metabolism and the heat losses through the ventilation air and the structure is adjusted, by increasing ventilation rate as temperature rises and decreasing it as temperature falls. The limits on this procedure are imposed by the maximum and the minimum ventilation requirements of the birds. Heaters should be used only when ventilation is at its minimum, and in the UK they are unnecessary for laying hens in thermally efficient houses.

Minimum ventilation rates are defined by the minimum amounts of air needed to provide oxygen, remove carbon dioxide and ammonia, and to support maximum levels of production. The choice of the correct minimum is critical in cold weather, because the lower the value used, the warmer the inside temperature obtainable without the use of heaters. For brooding hens, the lower the minimum ventilation rate, the lower the brooding cost. Minimum air requirements for layers and broilers have now been determined in experiments using large numbers of birds, where performance was measured as well as air composition (Emmans and Charles, 1977; Charles, Scragg and Binstead, 1981). The rates are system-dependent, since the limiting factor tends to be ammonia and this is influenced by such factors as manure handling and litter condition. Note that the rates are better expressed per unit of food consumed rather than per unit of live-weight, since food intake is directly related to metabolism. On this basis, a minimum value of 2 m^3 s^{-1} per tonne of food used per day is suitable for all weights and probably all species of poultry. Alternatively, but perhaps less conveniently, minimum ventilation rates can be expressed relative to live-weight as 1.5×10^{-4} m^3 kg$^{-0.75}$ s^{-1}. The minimum should be exceeded if the ammonia concentration reaches 25 ppm. Considerable care with inlet design and effective wind exclusion are needed in order to achieve these rates reliably.

For temperate climates the maximum ventilation rate is defined approximately by the amount of air required to maintain the inside equilibrium temperature within 3 K of outside temperature. For hot climates lower rates are sometimes appropriate, when used with evaporative cooling of very hot incoming air. The required maximum for temperate climates has traditionally been calculated using heat balance equations of the type reviewed by

Charles (1970) and brought up to date by Saville, Clark and Charles (1978). Such equations also predict the feasible temperature lift (ΔT) at minimum ventilation rate.

$$\Delta T = \frac{Q_s}{1200\dot{V} + UA} \tag{11.2}$$

where ΔT is the temperature lift above the outside temperature, in kelvins; Q_s is the bird's sensible heat output in watts per bird (e.g. about 8.5 for a well-feathered 1.8 kg layer); $\dot{V}$ the ventilation rate in $m^3 s^{-1}$ per bird; U the average thermal transmittance of walls and roof ($W\,m^{-2}\,K^{-1}$); and A is the exposed area of walls and roof per bird in square metres, normally greater than the floor area per bird by a factor of 1.2–1.7, depending on the shape of the building.

Equation 11.2 suggests that a laying house can be held at the target minimum temperature of 21 °C at outside temperatures ranging from just below 0 up to about 18 °C, assuming a stocking rate of about 20 birds per square metre (i.e. three-tier cages) and a U value for the walls of 0.6 $W\,m^{-2}\,K^{-1}$. At lower outside temperatures house temperature cannot be maintained without added heat, and at high outside temperature inside temperature will rise unless cooling is used. However, this oversimplified equation takes no account of heat storage in the mass of the building and birds, or of evaporative heat loss from the droppings, the birds and water spillage. Minor omissions from the equation are floor heat losses and light-bulb heat production, but these are relatively trivial and usually of opposite sign. For example, in a battery house stocked at 20 birds per square metre of floor there would be a metabolic heat input of between 80 and 160 $W\,m^{-2}$, compared with a heat input of 2.5–5 $W\,m^{-2}$ from the electric lamps. Solar heat gain is also usually ignored for buildings with good insulation in temperate climates, though it may be significant in clear summer weather. Of greater importance is the uncertainty in the value of Q_s, which fluctuates diurnally (Lundy, MacLeod and Jewitt, 1978) and varies with the feather cover of the birds (Emmans and Charles, 1977). Despite these imprecisions, practical experience in a wide range of commercial houses suggests that 21 °C can be held throughout much of the year in the UK in buildings where the assumptions implicit in equation 11.2 hold. *Table 11.2* summarizes the literature on metabolic heat production by the birds, and also their production of evaporated moisture, which is needed for the calculation of house psychrometrics.

Figure 11.2 shows that the temperature lift is very sensitive to changes in ventilation rate close to the minimum, therefore wind exclusion is important in achieving precise temperature control. *Figure 11.3* shows the sensitivity of the temperature lift to stocking density at low ventilation rates. At high ventilation rates temperature lift is not much affected by changes in either stocking density or ventilation rate. Obviously, if ventilation rates must be increased in order to remove a contaminant such as ammonia, this may limit the temperature lift attainable.

Table 11.3 gives approximate performances of agricultural propeller fans against representative working pressures, and *Table 11.4* lists some suggested maximum and minimum ventilation requirements for various classes of poultry.

Table 11.2a PUBLISHED ESTIMATES OF SENSIBLE HEAT OUTPUT OF POULTRY

Stock	Body wt. (kg per bird)	Experiment temp. (°C)	Sensible heat output, Q_s (watts per bird)	Source*	Notes	Suggested value of Q_s for use in calculations (watts per bird)
Layers	1.6 (white birds)	23	8.7 / 7.4 / 8.7	A	1 per cage / 2 per cage / 3 per cage	At a corridor temp of 20°C for multi-bird cages; white birds 7.0, brown birds 8.5. Increase by 25 per cent for poorly feathered birds
	2.3 (brown birds)	15 / 20 / 25	9.9 / 8.8 / 5.0	B	1 per cage. Well feathered	
	2.0 (brown)	15 / 20 / 25	14.9 / 13.4 / 9.5	B	1 per cage. Poorly feathered	
	1.5 (white)	20	Max. (day) 8.0 / Mean 6.5 / Min. (night) 5.0	C	1 per cage	
	2.3 (brown)	20	Max. 9.7 / Mean 7.9 / Min. 5.8			
Broiler breeders	3	20	12	D		10.5, increased by 10 per cent for poorly feathered birds
Broilers	0.04	29	0.7	E		At typical recommended temperatures
	0.36	25	3.4			Body wt. (kg per bird) / Q_s (watts per bird)
	0.95	19	6.7			0.05 / 0.7
	1.36	19	8.6			0.1 / 1.2
	2.04	19	9.6			0.5 / 3.4
	1.6	16	6.7	F, G		
	1.5	29	4.7			

Stock	Body wt. (kg per bird)	Experiment temp. (°C)	Sensible heat output, Q_s (watts per bird)	Source*	Notes	Suggested value of Q_s for use in calculations (watts per bird) Body wt.	Q
Broilers (males)	0.05	25	0.7	H	These values allow for clustering behaviour by the birds	1	4.5
	0.12	24	1.3			1.5	5.2
	0.40	23	3.4			2	5.8
	2.04	16	5.6				
(females)	0.05	24	0.7				
	0.11	24	1.2				
	0.35	22	2.6				
	1.73	17	5.5				
Turkeys	0.11	35	0.5	I			0.5
	0.24	32	1.4				1.4
	0.42	29	2.0				2.0
	0.63	27	3.7				3.7
	0.96	24	5.6				5.6
	2	20	9.7	D			9.7
	4	20	16.3				16.3
Ducks	2	7	24.6	J			24.6
		11	23.4				23.4
		17	13.0				13.0
		22	12.0				12.0
		26	10.6				10.6

*See source listings at foot of *Table 11.2b*

Table 11.2b PUBLISHED ESTIMATES OF EVAPORATIVE MOISTURE OUTPUT OF POULTRY

Stock	Body wt. (kg per bird)	Experiment temp. (°C)	Partial pressure of moisture (kPa)	Moisture output (g d⁻¹ per bird)	Source*	Suggested values for output (g d⁻¹ per bird) Temp. (°C)	Output
Layers		10	0.6	17	K		
		20	0.6	20		20	40
		30	0.6	34		25	75
		35	0.8	90		30	100
		40	1.2	164		35	175
		15		53	L		
		20	0.5–1.1	58			
		25	0.9–1.7	106			
		30	1.2–2.0	154			
		35	1.5–2.4	205			
Broilers	2.04	19		74	E, F, G	(For 2 kg bird at 20 °C)	
	1.6	16		21			75
	1.5	29		340			
Turkeys	0.42	29		24	Calc. from I		24
	0.96	24		21			21
	2	20		36			36
	4	20		61			61

*Sources: A—Olson, De Shazer and Mather (1974); B—Richards (1977); C—Lundy, MacLeod and Jewitt (1978); D—calculated on the basis of metabolic body size; E—Longhouse et al. (1968); F—Reece, Deaton and Bouchillon (1969); G—Deaton, Reece and Bouchillon (1969); H—Wathes (1978); I—De Shazer, Olson and Mather (1974); J—Pugh (1978); K—van Kampen (1974); L—Richards (1976)

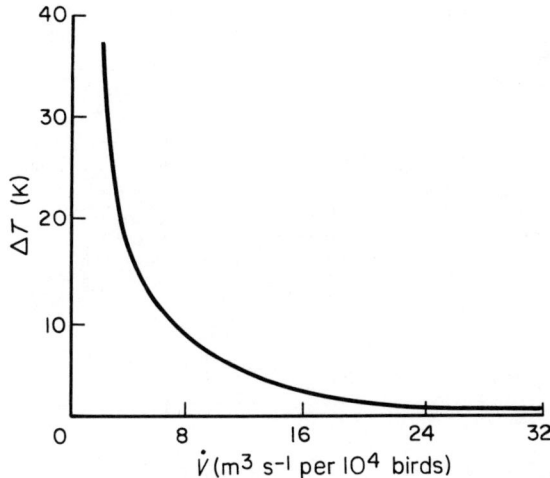

Figure 11.2 Effect of ventilation rate on temperature lift above outside. The lift is calculated for a stocking density of 20 birds per square metre, building heat transmittance U of 0.5 W m⁻² K⁻¹, and sensible heat loss H$_s$ of 9 W per bird

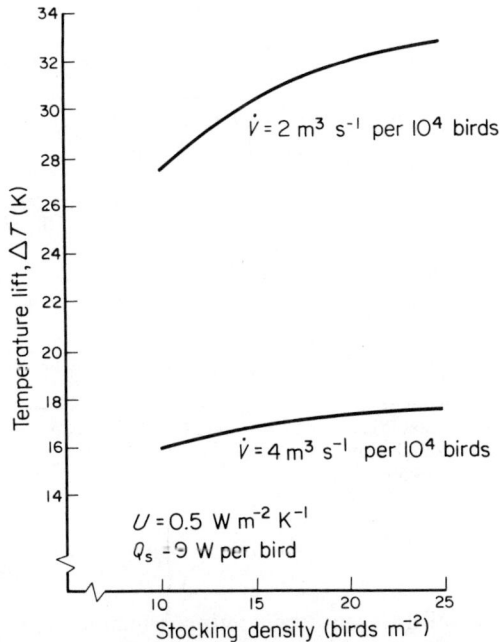

Figure 11.3 Effect of stocking density and ventilation rate on temperature lift. Ventilation rate $\dot{V}$ is taken as 2 and 4 m³ s⁻¹ per 10⁴ birds, building heat transmittance U as 0.5 W m⁻² K⁻¹, sensible heat loss Q$_s$ as 9 W per bird

Table 11.3 FAN PERFORMANCE: THE EFFECTS OF REPRESENTATIVE
WORKING PRESSURES ON THE EXPECTED PERFORMANCE OF
AGRICULTURAL PROPELLER FANS

Diameter (mm)	rev min⁻¹	Airflow ($m^3\ s^{-1}$ per fan)			
		Working pressure 50 Pa	Working pressure 75 Pa	Working pressure 100 Pa	Working pressure 125 Pa
380	1400	0.9	0.8	0.6	0.5
460	1400	1.7	1.6	1.4	1.2
610	700	1.6	1.1	0.0	0.0
610	900	2.6	2.3	2.0	1.6
760	700	3.7	3.1	2.5	0.0
760	900	5.2	5.0	4.7	4.2

Table 11.4 VENTILATION REQUIREMENTS OF POULTRY OF VARIOUS
WEIGHTS

Stock	Weight or age	Maximum requirement		Minimum requirement	
		$m^3\ s^{-1}$ per 10 000 birds	Fans per 10 000 birds*	$m^3\ s^{-1}$ per 10 000 birds	Fans per 10 000 birds*
Pullets and	1.8 kg	24	9	2.1	0.8
hens	2	26	10	2.3	0.9
including	2.5	32	12	2.5	1
breeders	3	36	14	2.7	1.1
	3.5	38	15	3.3	1.3
Broilers	1 day old			0.13	0.05
	1 week			0.3	0.1
	2			0.5	0.25
	3			1.3	0.5
	4			1.6	0.6
	5			2	0.8
	6			2.3	0.9
	7			2.6	1
	8	24	9	2.9	1.1
Turkeys	0.5 kg	17	6.5	1.3	0.5
	2	33	12	2.2	0.8
	5	42	16	2.8	1.1
	11	75	28	5	1.9

The maxima are the values associated with acceptable levels of temperature lift above outside in
summer; the minima are those required for gas exchange and air hygiene
*Refers to 610 mm fans at 900 rev min⁻¹, at a working pressure of 50 Pa

Practical ventilation systems

Ventilation systems based on several different engineering principles are
used in the poultry industries of temperate countries, but whatever
mechanical arrangement is chosen, certain design constraints should be met,
in view of the requirements reviewed above. These are listed in the following
numbered paragraphs.

(1) If it is intended to use lighting programmes involving day lengths shorter than natural day length then the light intensity should be below 0.4 lx during the dark part of the light/dark programme (Morris and Owen, 1966). Spencer and Charles (1968) found that even in houses with powered ventilation, failure of light exclusion was then widespread; and therefore that failure properly to exploit lighting programmes was common. More recent ventilation systems are probably better in this respect, because attention to detail on wind and temperature control is also conducive to effective light exclusion.

(2) Ventilation systems should be capable of providing appropriate temperature control through accurate regulation of ventilation rate. This needs not only accurate thermostatic controls for the fans, but also wind-proofed inlets and outlets.

(3) Since temperature control is important, the ventilation system should provide uniform temperature within the building by virtue of good air distribution.

Systems that meet these criteria are available based both on fan extraction and on pressurized fan input, since the criteria are difficult to achieve by natural ventilation. Design details of these systems are described by other authors in this volume. There is now considerable experience in the British

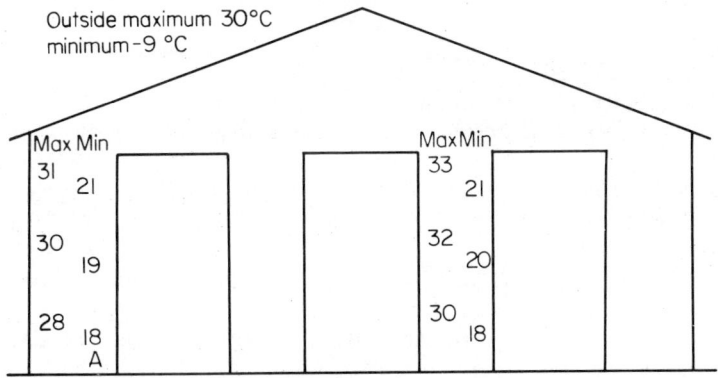

Figure 11.4 Temperature distribution in a building ventilated by the high-speed inlet system in hot and cold weather. The effect of outside daily minimum temperature (T_o, °C) on inside daily minimum (T_i, °C) at position A (one of the coldest places in the house) was as follows for 28 consecutive cold nights:

$$T_i = 0.09T_o + 18.98; r = 0.63, SD = \pm 0.56$$

poultry industry of operating both extraction and pressurized systems. Successful examples of pressurized systems include the glass fibre ceiling system and polyethylene duct systems. A thermally efficient light-proof extraction system is the high-speed inlet jet. *Figure 11.4* shows some internal dry-bulb air temperatures (measured in the gangways between the cages) in a laying house without artificial heating, ventilated by this system.

Efficacy of control

Despite the numerous imperfections in the simplification of the house heat balance equation (equation 11.1), its predictions of ΔT are not unreasonable. Experience of a wide range of laying houses suggests that 21 °C can be maintained without artificial heat throughout most of a British winter in well-designed buildings, and 30 °C is rarely exceeded in summer in laying houses in the UK. Similarly, broilers can be held at 21 °C during finishing in good buildings. During brooding a fuel heat input is inevitably necessary, because the amount of metabolic heat available is very small. The amount of fuel heat required (Q_a, in watts per bird) is given by the relationship

$$Q_a = 1200\dot{V}\,\Delta T + UA - Q_s \tag{11.3}$$

where all other terms are as for equation 11.2. When equation 11.3 is being used to calculate brooder sizes it is wise to add an allowance of about 20–25 per cent to cover extreme weather and imperfections of ventilation control, which are very hard to avoid. Summer cooling is not justified in the UK, and probably not in most temperate climates.

References

ALBUQUERQUE, K. DE, LEIGHTON, A. T., Jr, MASON, J. P., Jr and POTTER, L. M. (1978). *Poult. Sci.* **57**, 353–362

CHARLES, D. R. (1970). *Wld's Poult. Sci. J.* **26**, 422–434

CHARLES, D. R. (1979). *25th Poultry Convention, Massey University, New Zealand*

CHARLES, D. R., GROOM, C. M. and BRAY, T. S. (1981). *Br. Poult. Sci.*, in press

CHARLES, D. R., SCRAGG, R. H. and BINSTEAD, J. A. (1981). *Br. Poult. Sci.*, in press

DEATON, J. W., REECE, F. N. and BOUCHILLON, C. W. (1969). *Poult. Sci.* **48**, 1529–1582

DE SHAZER, J. A., OLSON, L. L. and MATHER, F. B. (1974). *Poult. Sci.* **53**, 2047–2054

EMMANS, G. C. and CHARLES, D. R. (1977). In *Nutrition and the Climatic Environment*, pp. 31–50. Ed. by W. Haresign, H. Swan and D. Lewis. Butterworths, London

FELTON, K. E. (1974). *Scientific Article No. A1982*, Maryland Agricultural Experimental Station

LONGHOUSE, A. D., OTA, H., EMERSON, R. E. and HEISHAM, J. O. (1968). *Trans. ASAE* **11**, 694–700

LUNDY, H., MacLEOD, M. G. and JEWITT, T. R. (1978). *Br. Poult. Sci.* **19**, 173–186

MORRIS, T. R. and OWEN, V. M. (1966). *Proc. 13th World's Poultry Congress*, pp. 458–461

MOWBRAY, R. M. and SYKES, A. H. (1971). *Br. Poult. Sci.* **12**, 25–29

OLSON, L. L., DE SHAZER, J. A. and MATHER, F. B. (1974). *Trans. ASAE* **17**, 960–967

PAYNE, C. G. (1967). In *The Physiology of the Domestic Fowl,* pp. 235–241. Ed. by C. Horton-Smith and E. C. Amoroso. Oliver & Boyd, Edinburgh

PUGH, M. (1978). *BSc Thesis,* University of Nottingham

REECE, F. N., DEATON, J. W. and BOUCHILLON, C. W. (1969). *Poult. Sci.* **48**, 1297–1393

RICHARDS, S. A. (1976). *J. agric. Sci., Camb.* **87**, 527–532

RICHARDS, S. A. (1977). *J. agric. Sci., Camb.* **89**, 393–398

SAVILLE, C. A., CLARK, J. A. and CHARLES, D. R. (1978). Paper presented to WPSA (UK Branch) spring meeting

SMITH, A. J. and OLIVER, J. (1972a). *Rhod. J. agric. Res.* **10**, 3–21

SMITH, A. J. and OLIVER, J. (1972b). *Rhod. J. agric. Res.* **10**, 43–60

SPENCER, P. G. and CHARLES, D. R. (1968). *Wld's Poult. Sci. J.* **24**, 318

VAN KAMPEN, M. (1974). In *Energy Requirements of Poultry,* pp. 47–59. Ed. by T. R. Morris and B. M. Freeman. British Poultry Science, Edinburgh

WATHES, C. M. (1978). *PhD Thesis,* University of Nottingham

WATHES, C. M., GILL, B. D.. CHARLES, D. R. and BACK, H. L. (1981). *Br. Poult. Sci.,* in press

12

VENTILATION AND TEMPERATURE CONTROL CRITERIA FOR PIGS

JAMES M. BRUCE
Scottish Farm Buildings Investigation Unit, Craibstone, Bucksburn, Aberdeen

Introduction

The main reason why we provide ourselves and animals with buildings is to give shelter from the weather. It follows from this that if we wish to produce effective and efficient buildings, we must understand:

(1) how to describe and quantify the climate;
(2) how to describe and quantify the animals' responses to climatic and thermal stimuli;
(3) how to describe and quantify the internal climatic environment as produced by the interaction of climate, buildings and animals.

The pig's responses to climate and its thermal environment are complex, varied and can be acute. The lack of a significant hair coat makes the pig very susceptible to cold and its inability to sweat in response to a thermal stimulus (Ingram, 1967) means that it is readily heat-stressed in the absence of a wallow. The energy balance of a pig is strongly affected by temperature and so the temperature can be regarded as a criterion of fundamental biological and agricultural importance. The control of temperature in a pig building is commonly achieved by varying the ventilation rate. This indicates that the temperature criteria ought to be considered first and from these the ventilation criteria can be derived.

Temperature criteria for pigs

A pig produces waste heat as an unavoidable consequence of its metabolic processes. The amount of heat generated will depend primarily on the size of the pig, the metabolizable-energy intake and the efficiency with which the pig uses the energy in its feed. The pig is a homeotherm and must maintain its deep-body temperature within close limits, at about 39 °C. The waste heat is available to keep the pig in thermal comfort. However, if the demand of the thermal environment is greater than the normal heat production then the pig must increase its heat production or its resistance to heat loss, or both. Ultimately, when the physiological and behavioural responses of the pig have maximized the resistance to heat loss, the pig must produce more heat. This may mean very little to the pig as a biologically programmed survival machine but within an agricultural context it is highly undesirable.

On the other hand, the pig may be producing heat at a rate which is greater than that at which it can be lost. In these hot conditions the pig must minimize the resistance to heat flow and if necessary it must reduce its heat production by decreasing feed (energy) intake. Generally, this is also economically undesirable.

Within limits a pig can adjust its resistance to heat flow. It can, for example, vasoconstrict the peripheral tissues of its body and huddle with other pigs when cold. When hot it can vasodilate, remove itself from the proximity of other pigs and pant, although the pig is by no means an efficient panter like the dog (Hales, 1974). Failure by the pig to regulate the deep-body temperature means death within hours rather than days. Failure on the part of the stockman to control the energy demand of the thermal environment can lead to the death of baby pigs (Sainsbury, 1972) and growing pigs, which are stress-susceptible (Lucke, Hall and Lister, 1979), and will at the very least lead to decreased productivity.

Air temperature lends itself as a criterion of thermal comfort since it quantifies the potential for heat loss and is readily and cheaply measured, but the heat transfer characteristics of the pig and its surroundings are also very important in determining the heat flow. It has therefore become clear that it is necessary to relate temperature criteria to other environmental variables (Mount, 1960, 1967; Stephens, 1971; Sainsbury, 1972; Verstegen and van der Hel, 1974; Holmes and Close, 1977, Bruce and Clark, 1979). Also, it has been shown (SBI, 1967) that a diurnally fluctuating temperature is equivalent to a steady temperature of the mean value.

The rest of this section is devoted to explaining a method of calculating the temperature criteria for pigs and tabulating some representative values. Only a selection of values can be tabulated because there is literally an infinite set of temperature criteria. This is also why it is so important to have a method or model for calculating temperature criteria for any given set of circumstances.

THERMONEUTRAL HEAT PRODUCTION MODEL

Idealized graphical relationships based on observation are in common use to describe a pig's thermal response to temperature (Mount, 1968; Holmes and Close, 1977). The thermoneutral zone is defined as that within which the heat production of a pig is independent of the temperature and it may be identified with the zone of best productivity. Within this zone the heat produced by a pig is determined by its live-weight and feed intake. The zone is bounded by a lower critical temperature and an upper critical temperature. A model based on the commonly used concepts of maintenance and efficiency of utilization of metabolizable energy has been developed (Bruce and Clark, 1979). In this model the thermoneutral heat production in watts is given by

$$Q_{tn} = 11.57 \left[F_m + (1-k)(F - F_m) \right] \qquad (12.1)$$

where k is the efficiency of utilization of the feed increment above mainten-ance. For pigs of weight $W \geqslant 20$ kg, the maintenance energy requirement F_m (MJ d^{-1}) is $0.44W^{0.75}$ and for $W < 20$ kg, $F_m = 0.64W^{0.66}$. The efficiency of utilization of metabolizable energy, k in equation 12.1, is dimensionless with a value of less than unity. For $W < 20$ kg and $W > 100$ kg, $k = 0.75$, while for $20 \leqslant W \leqslant 100$ kg, k is given by $(0.625 + 0.00142W)$. The meta-bolizable-energy intake, F in equation 12.1, is in MJ d^{-1}.

This model has been validated only for pigs of live-weight between 20 and 90 kg fed on barley-based diets. The root-mean-square error of prediction for 62 measurements of heat production was 5.7 W, or 3.2% (Bruce and Clark, 1979).

HEAT LOSS MODEL

A model was also developed which allows the heat loss from a pig to be calculated as a thermal demand by the environment (Bruce and Clark, 1979). The model variables are air temperature, air velocity, floor type, live-weight and the number of pigs in the group, and the model can accommodate huddling and postural behaviour. The total heat loss from the pig, Q (in watts), is given by

$$Q = A\left\{\left[1 + \frac{A_f}{A}\left(\frac{I_a - I_f}{I_b + I_f}\right) - \frac{A_c}{A}\right](T_b - T_a) + EI_a\right\}/(I_a + I_b) \qquad (12.2)$$

The other symbols in equation 12.2 are defined as follows: For pigs the surface area A (in m^2) is given by $0.09W^{0.67}$ and A_f and A_c are the areas of a pig's surface in contact with the floor and with other pigs respectively, also in m^2. The insulation of the air interface I_a (in m^2 K W^{-1}) is given by $(5.3 + 15.7 u^{0.6} W^{-0.13})^{-1}$, where u is the air velocity in m s^{-1}. The body tissue insulation I_b (in m^2 K W^{-1}) is given by $0.02W^{0.33}$. The effective floor insulation I_f in the same units is given by $I_{f45}(W/45)^{0.33}(5 A_f/A)N^{0.5}$, where N is the number of pigs in a group. The measurement of I_{f45} is described elsewhere (Bruce, 1977). In calculations, the deep-body temperature, T_b, is taken as 39 °C and T_a is the air temperature in degrees Celsius. The latent heat flux, E, is in W m^{-2}. A_f/A is taken as 0.2 for $I_f \geqslant I_a$ or 0.1 for $I_f < I_a$. A_c/A is given by $0.15 (N - 1)/N$.

For heat loss below the lower critical temperature this model has been validated for pigs of between 20 and 80 kg live-weight in groups of from 1 to 9. The floors had a range of I_{f45} from 0.07 to 0.5 m^2 K W^{-1}. The root-mean-square error of prediction for 78 measurements of heat loss was 8.9 W, or 5.6% (Bruce and Clark, 1979).

MODEL FOR CRITICAL TEMPERATURES

Equation 12.2 may be rewritten to bring the air temperature, T_a, to the left-hand side. Then, using the thermoneutral heat production from equation 12.1 for Q_{tn} allows calculation of the critical temperature of a pig $(T_a = T_c)$.

$$T_c = T_b - \cfrac{Q_{tn}(I_a + I_b) - EAI_a}{A\left[1 + \cfrac{A_f}{A}\left(\cfrac{I_a - I_f}{I_b + I_f}\right) - \left(\cfrac{A_c}{A}\right)\right]} \qquad (12.3)$$

This expression can be used as a basis for calculating the lower critical temperature, T_{cl}, the upper critical temperature in dry conditions, T_{cd}, and the upper critical temperature in wet conditions, T_{cw}. To be able to use equation 12.3 it is necessary to establish a suitable choice for the evaporative heat loss, E. It has been shown previously (Bruce and Clark, 1979) that the data for minimum evaporative heat loss from pigs at low temperatures (Mount, 1968; Close, 1971; Hovell, Gordon and MacPherson, 1977) can be represented by $E = 8.0 + 0.07W$ for live-weight 1–170 kg. When E is substituted in equation 12.3 the lower critical temperature may be calculated. *Table 12.1* shows the highest measured evaporative heat losses,

Table 12.1   HIGHEST MEASURED EVAPORATIVE HEAT LOSS FROM PIGS

Live-weight (kg)	Temperature (°C)	Evaporative heat loss (W m⁻²)	Source
25–85	23	32–37	Fuller and Boyne (1972)
91	20	31	Hovell, Gordon and MacPherson (1977)
90	29	37	Morrison, Bond and Heitman (1967)
(1–7 days)	36	21	Mount (1968) p. 134
20	30	52	p. 137
60	30	42	p. 137
6–14	40	42	Stombaugh and Roller (1977)

from a variety of sources, required for calculation of the upper critical temperature in dry conditions. There is no clear indication of a correlation with live-weight, so the mean value of 37 W m⁻² is adopted. Even if the 20 °C and 23 °C data were omitted the mean value would not alter by much. It is therefore assumed that the maximum evaporative heat loss from any pig is 37 W m⁻² in dry conditions. This heat will be lost mainly from the respiratory tract but is expressed per unit area of the external surface. It is also assumed that a pig under heat stress will adopt a fully recumbent posture, so that $A_f/A = 0.2$, and that the tissue insulation is reduced by vasodilation to a negligible amount ($I_b \simeq 0$).

In hot, wet conditions some animals sweat, some pant and some spread saliva on their bodies (Hales, 1974). Pigs do none of these effectively. They do, however, wallow in moisture, mud, urine or faeces. Very high rates of heat loss are achieved in this way (Ingram, 1965a) provided that the humidity of the ambient air is not very high. In the UK an infrequently used system of keeping pigs is the 'sweat-box' or 'Turkish bath' house. In such a building pigs are exposed to high temperatures and very high humidities; generally the floor is covered in urine and faeces. In some hot countries water sprinkling is practised to help pigs remain free of heat stress. Even in buildings where no wallowing is intended, if pigs feel too warm they will wallow in the dunging and drinking area or indeed defaecate indiscriminately in the pen. Oddly, this is regarded as misbehaviour when, in fact, the pig is merely striving to carry out a perfectly normal type of behaviour for which it is adapted and to achieve evaporative cooling from its skin. To obtain estimates of the upper critical temperature in wet conditions using equation 12.3, it is again assumed that $I_b = 0$, $A_f/A = 0.2$ and that evaporation from the respiratory tract is at a rate of 37 W m^{-2} (referred to the external surface), as for hot, dry conditions. The critical temperature is not very sensitive to the last assumption. Heat will be transferred from the skin in both sensible and latent form, so an estimate of the evaporation rate from a wet pig is needed. From the analogy between heat and mass transfer (Monteith, 1973), we can write the mass transfer rate ε for water vapour from a pig's skin, in kg m^{-2} s^{-1}, as

$$\varepsilon = 0.23 \, Re^{0.6} \, D \, (\chi_{sk} - \chi_a)/d \tag{12.4}$$

where Re is the Reynolds number, χ_{sk} and χ_a are the absolute concentrations of water vapour at the skin surface and in the air, respectively, D is the diffusion coefficient for water vapour in air, and d is the diameter of the pig.

Bruce and Clark (1979) gave $d = 0.052W^{0.33}$, so equation 12.4 reduces to

$$\varepsilon = 15.6 \times 10^{-3} \frac{u^{0.6}}{W^{0.13}} (\chi_{sk} - \chi_a) \tag{12.5}$$

where the density of saturated water vapour at a temperature T can be approximated by $\chi = 2 \times 10^{-3} (T - 15)$ for $30\,°C \leqslant T \leqslant 40\,°C$. Substituting χ_{sk} into equation 12.5 and multiplying by the latent heat of evaporation (2.42 MJ kg^{-1}) gives the latent heat loss from the skin, E (in W m^{-2}), as

$$E = 75.6 \frac{u^{0.6}}{W^{0.13}} (T_{sk} - T_a) \tag{12.6}$$

The latent heat transfer coefficient ($75.6u^{0.6}/W^{0.13}$) can then be added to the denominator of I_a for high-temperature wet conditions:

$$I_a = \frac{1}{5.3 + 91.3\left(\frac{u^{0.6}}{W^{0.13}}\right)} \tag{12.7}$$

The upper critical temperature in wet conditions, T_{cw}, can be calculated by substituting the above expression for I_a into equation 12.3. This applies for temperatures between 30 and 40 °C and assumes 100% relative humidity in the ambient air. If the relative humidity is less than 100%, the latent heat loss can be considerably enhanced.

RECOMMENDATIONS FOR CRITICAL TEMPERATURES

The previous sections have outlined the basis for estimating the lower and upper critical temperatures under wet and dry conditions. These are given in

Table 12.2 CRITICAL TEMPERATURES FOR PIGS ON A SOLID CONCRETE FLOOR

Live-weight (kg)	Number in group	Lower critical temp. (°C) Feed level (multiple of maintenance):				Upper critical temp. (dry) (°C) Feed level (multiple of maintenance):				Upper critical temp. (wet) (°C) Feed level (multiple of maintenance):			
		1	2	3	4	1	2	3	4	1	2	3	4
1	1	32	31	29	27	37	36	36	35	38	38	37	37
	10	31	29	26	24	35	34	32	30	37	37	36	35
5	1	30	28	26	23	37	35	34	33	38	37	36	36
	10	28	25	22	20	34	32	30	28	37	36	35	34
10	1	29	27	24	21	36	35	34	32	38	37	36	36
	10	27	24	21	17	34	32	29	27	37	36	35	34
20	1	29	26	22	18	36	35	33	31	38	37	36	35
	15	26	21	16	11	34	31	28	24	37	36	34	33
40	1	27	23	19	15	36	34	32	30	37	36	35	34
	15	24	18	13	7	33	30	26	23	37	35	34	32
60	1	26	22	18	14	35	33	31	29	37	36	35	34
	15	22	16	11	5	33	29	26	22	36	35	34	32
80	1	25	21	17	13	35	33	31	29	37	36	35	34
	15	21	15	10	4	32	29	26	23	36	35	34	32
100	1	24	21	17	13	35	33	31	29	37	36	35	34
	15	19	14	9	4	32	29	26	23	36	35	34	32
140	1	23	19	14	10	34	32	30	28	37	36	35	34
	5	20	12	9	3	32	30	26	23	36	34	33	32
180	1	22	18	13	8	34	32	29	27	36	35	34	33
	5	19	10	7	1	32	27	25	22	36	34	33	32

Table 12.3   CRITICAL TEMPERATURES FOR PIGS ON A PERFORATED METAL FLOOR

Live-weight (kg)	Number in group	Lower critical temp. (°C) Feed level (multiple of maintenance):				Upper critical temp. (dry) (°C) Feed level (multiple of maintenance):				Upper critical temp. (wet) (°C) Feed level (multiple of maintenance):			
		1	2	3	4	1	2	3	4	1	2	3	4
1	1	30	28	25	23	35	33	31	29	37	37	36	35
	10	29	26	23	20	34	32	29	27	37	36	36	35
5	1	28	26	23	20	34	32	30	28	37	36	36	35
	10	27	24	20	17	34	31	29	26	37	36	35	34
10	1	28	25	21	18	34	32	30	27	37	36	35	35
	10	26	22	19	15	33	31	28	26	37	36	35	34
20	1	28	24	20	15	35	32	29	26	37	36	35	34
	15	26	21	17	12	34	31	28	24	37	36	34	33
40	1	26	22	17	13	34	32	29	26	37	36	35	34
	15	24	19	14	8	34	30	27	24	37	35	34	32
60	1	25	20	16	11	34	31	28	26	37	36	35	33
	15	23	17	12	7	33	30	27	24	36	35	34	32
80	1	24	20	15	11	34	31	29	26	37	36	35	33
	15	22	16	11	6	33	30	27	24	36	35	34	32
100	1	23	19	15	11	33	31	29	26	37	36	35	34
	15	21	16	11	6	33	30	27	24	36	35	34	33
140	1	22	17	13	8	33	30	28	25	36	35	34	33
	5	20	14	9	3	32	29	26	23	36	35	33	32
180	1	21	16	11	6	33	30	27	25	36	35	34	33
	5	19	13	7	1	32	29	26	23	36	34	33	32

Table 12.4 CRITICAL TEMPERATURES FOR PIGS ON A STRAW-BEDDED FLOOR

Live-weight (kg)	Number in group	Lower critical temp. (°C) Feed level (multiple of maintenance):				Upper critical temp. (dry) (°C) Feed level (multiple of maintenance):				Upper critical temp. (wet) (°C) Feed level (multiple of maintenance):			
		1	2	3	4	1	2	3	4	1	2	3	4
1	1	30	27	25	22	34	32	30	28	37	37	36	35
	10	27	23	20	17	33	30	27	24	37	36	35	34
5	1	27	24	21	18	34	31	29	26	37	36	35	35
	10	24	20	16	12	32	29	26	23	37	36	35	34
10	1	26	23	19	16	33	31	28	26	37	36	35	34
	10	23	18	14	9	32	29	26	22	37	35	34	33
20	1	26	22	17	12	34	31	28	24	37	36	35	34
	15	23	17	11	4	33	29	25	20	37	35	34	32
40	1	24	20	14	9	34	30	27	23	37	36	34	33
	15	20	13	7	0	32	28	24	19	36	35	33	32
60	1	23	17	12	7	33	30	27	23	37	36	34	33
	15	18	12	5	-2	32	27	23	19	36	35	33	32
80	1	22	16	11	6	33	30	27	23	37	35	34	33
	15	17	10	4	-2	31	27	23	20	36	35	33	32
100	1	21	16	11	6	32	30	27	24	36	35	34	33
	15	16	10	4	-2	31	27	24	20	36	35	33	32
140	1	19	14	8	2	32	29	26	22	36	35	34	32
	5	15	8	2	-5	31	27	23	19	36	34	33	31
180	1	18	12	6	0	32	28	25	22	36	35	33	32
	5	14	5	-1	-8	30	26	22	18	35	34	32	31

Tables 12.2–12.4. Not all the values tabulated are necessarily of practical interest, but they are provided for completeness. The data show very good agreement with those of Holmes and Close (1977) even though the methods of derivation are dissimilar.

Typical piglets and growing pigs would be fed 2.5–3.5 times maintenance, the heavier pigs being given the relatively lower feed intake. Pregnant sows need be fed only a little more than maintenance, while lactating sows ought to be fed about 4 times maintenance.

The value of pigs' huddling in groups for resisting heat loss is very great, and this behaviour can be used to good advantage in designing for winter conditions, as it reduces their lower critical temperature. However, in hot conditions pigs will wish to separate, to promote cooling. This is sometimes possible only if they spread into dunging areas. On fully perforated floors, when heavily stocked, such isolation of individuals becomes impossible. This results in decreased upper critical temperatures and, consequently, an increased incidence of heat stress. Management must therefore be prepared to adjust stocking density according to season.

SOME PRACTICAL IMPLICATIONS

Growing pigs

Suppose a pig producer kept his pigs in groups of 15 and fed at 3.3 times maintenance from 20 to 60 kg and then held the feed level constant through to 100 kg live-weight. This top level of feeding would be 2.5 kg d⁻¹ at 12.5 MJ of metabolizable energy per kilogram. *Figure 12.1* shows the lower

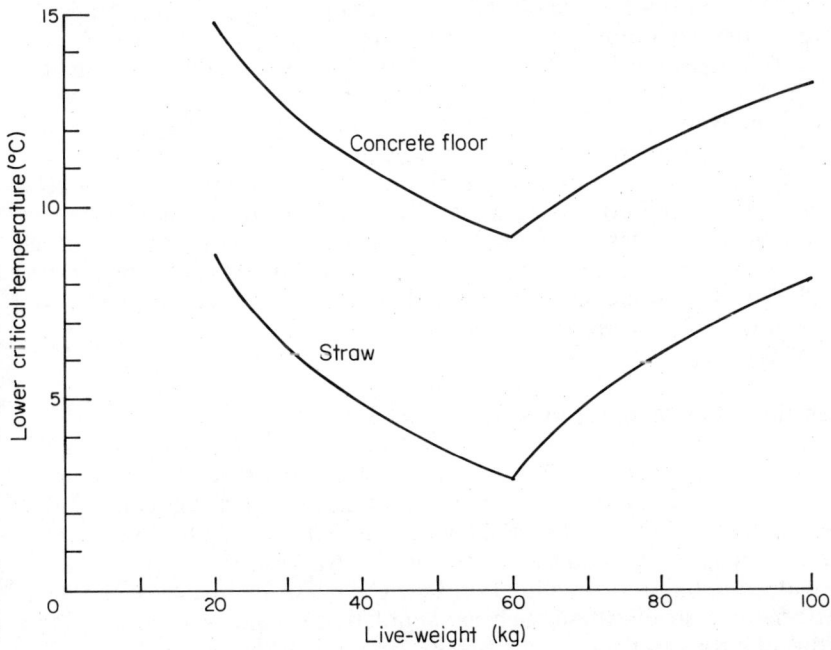

Figure 12.1 Lower critical temperatures for a group of 15 growing pigs. Energy intake is 3.3 times maintenance up to 60 kg live-weight, then restricted to 31 MJ d⁻¹. Air velocity = 0.15 m s⁻¹

critical temperature. On the restricted intake the heavier pigs have the highest critical temperature. Therefore, if, as is common, the fattening house took 40 kg pigs at entry through to 90 kg at slaughter, then the control temperature should be the lower critical temperature of the heaviest pigs. On a concrete floor the critical temperature would be 12.5 °C for 90 kg pigs and 11 °C for 40 kg pigs. It would be quite wrong to use a control temperature based on the *average* weight: this would be 10 °C.

Pregnant sows

A very common method of housing pregnant sows is to tether or crate them individually on concrete floors. In a good environment little more than a maintenance ration needs to be fed to a pregnant sow in order to maintain her condition from one parity to the next. If a 140 kg sow is fed 1.2 times maintenance (1.7 kg of meal) then her lower critical temperature is estimated to be 22 °C. This is considerably higher than farmers generally keep, or are advised to keep, their sows, but is in good agreement with results of Holmes and Close (1977) and Hovell, Gordon and MacPherson (1977). If the sow is kept at 15 °C, which is quite typical, then she will require an additional 0.56 kg of feed per day or 2.26 kg in total. It is usual for farmers to feed normal sows 2.25–2.5 kg per day: but do they realize that they are feeding to compensate for cold stress?

Lactating sows and litters

From *Tables 12.2–12.4* we can see that a 140 kg sow, fed 4 times maintenance in dry conditions, has an *upper* critical temperature in the range 22–28 °C, depending on what type of floor she has to lie on. A young litter would have a *lower* critical temperature in the range 25–31 °C. This means that if the whole building were heated to satisfy the piglets the sow would probably be heat-stressed. This is avoided by creating a warmer microclimate for the litter, so that the temperature is not too high for the sow. This is easily accomplished. For example, the effect of a radiant lamp on the heat production of piglets was found to be equivalent to an increase of about 16 K in air temperature (Stephens and Start, 1970). Sows on a high ration will be prone to some heat stress at high temperatures, and this is observed from time to time in practice.

Ventilation criteria for pigs

Sällvik (1979) has pointed out that, although ventilation in livestock buildings is used primarily to control temperature and humidity, other environmental factors within the building are affected. Of high importance are the air velocity and gas and dust concentrations. It is very difficult to provide a definition of ventilation but it is possible to state a high-level performance specification such as: ventilation should provide a suitable aerial environment for the livestock, for the stockman, for the building materials and for the equipment within the building. By and large the modern materials and equipment used in pig buildings are able to withstand the aggressive environment. I am not at all sure that stockmen are. How-ever, the scope of this discussion will be restricted to the pigs and their needs.

The air velocity in the vicinity of pigs affects their energy balance and is affected by the ventilation rate and the position, shape and size of inlets. Considerable effort has gone into defining how these variables interact (Randall, 1975; Boon, 1978). *Figure 12.2* illustrates the effect of air velocity on the lower critical temperature according to equation 12.6. High air velocities are detrimental in cool conditions but can be advantageous in hot conditions.

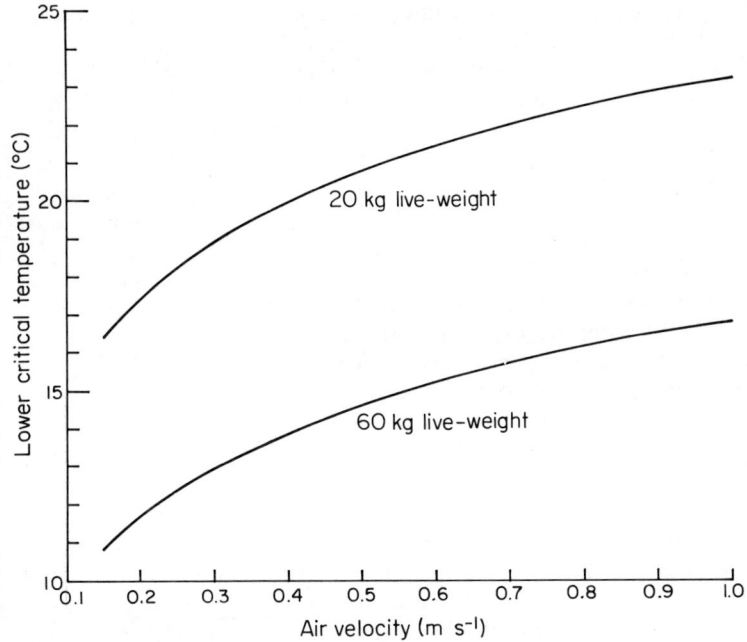

Figure 12.2 Effect of air velocity on the lower critical temperature for pigs of live-weight 20 kg on metal mesh and live-weight 60 kg on concrete. The energy intake is 3 times maintenance and the pigs are in groups of 15

The pig's responses to its environment are complex and it is difficult to assess the importance and effect of individual components of the environ-ment (Baxter, 1969). The greatest body of current knowledge is related to the thermal consequences of the aerial environment and the emphasis in practice is still in this area. However, work has been proceeding on the measurement and effects of dust and gases such as hydrogen sulphide and ammonia.

DUST

Honey and McQuitty (1976) have provided an excellent review of dust in the animal environment. These authors discussed the nature of dust and methods of sampling it, the significance of dust with regard to hygiene and its effect on the respiratory system together with the importance of particle size. Pig buildings are shown to have the highest dust concentrations of livestock buildings, though the factors that determine the amount of dust produced are not well defined. Factors implicated include animal activity, temperature, relative humidity, ventilation rate, stocking density and air space per animal, feeding method, type of feed and type of bedding. Honey and McQuitty (1976) emphasized the need for more detailed studies. Their review suggested that, while no effect of dust on productivity could be demonstrated, dust has an important influence on the incidence of respiratory disease. Micro-organisms have been shown to be associated with dust. These authors considered dust a significant contaminant in livestock buildings but were unable to quantify the limits of acceptable concentrations.

Small changes in relative humidity were found to have a significant effect on settled dust (Honey and McQuitty, 1979); lower humidity resulted in more settled dust. Self-feeding by pigs produced higher aerial dust concentrations than floor-feeding, but a greater amount of settled dust was associated with the latter. Doubling air-flow rate had no significant effect, but associated with the lower air-flow rate was a greater proportion of dust particles of a size able to penetrate to the lungs. The ventilation rate was the least important factor affecting dust concentration. Activity, temperature, relative humidity, amount of feed given, method of feeding and live-weight of pigs were all more important, in the order given.

A theory for dust aerodynamics was put forward by Baturin (1972), who also suggested methods of removing dust from industrial premises.

GASES

The composition of normal outdoor air is approximately 78% nitrogen, 21% oxygen, 1% argon and 0.03% carbon dioxide, plus traces of other inert gases. The more important gaseous contaminants produced in pig houses are ammonia, NH_3; hydrogen sulphide, H_2S; and carbon dioxide, CO_2.

Industrially, carbon dioxide is classified as a simple asphyxiant. Generally, at 1–2% concentration there is little effect on humans; severe headache and nausea will occur at 3–5% and 10% causes unconsciousness. The concentration of carbon dioxide in exhaled air is about 5%. A normal concentration for carbon dioxide in a pig respiration chamber is between 0.5% and 1% (Verstegen *et al.*, 1976). A threshold limit of 0.5% is given by Baxter (1969) and Nordstrom and McQuitty (1976) but, on the arbitrary basis that we should design so as to provide conditions an order of magnitude removed from those that are known to have detrimental effects, I suggest that 0.3% should be the normal design limit, with intermittent excursions to 0.5%.

Ammonia is classed as an irritant and even a short exposure can have serious effects, especially at high concentrations (Nordstrom and McQuitty, 1976). The gas is detectable by smell at 5–50 ppm (0.0005–0.005%), irritating to mucous surfaces at 100–500 ppm, causes severe eye irritation, coughing and frothing at the mouth with possible fatalities at 2000–3000 ppm and at 10 000 ppm (1%) is rapidly fatal. Ammonia and its effects on animals have been reviewed by Baxter (1969) and Nordstrom and McQuitty (1976). In pigs ammonia has been associated with reduced appetite, convulsion and irregular breathing (Stombaugh, Teague and Roller, 1969). Curtis *et al.* (1975) found that concentrations up to 50 ppm did not affect productivity nor the structure of the respiratory tract. Tail-biting was not induced by ammonia concentrations of up to 60 ppm while activity was reduced (Verstegen *et al.*, 1976). Nordstrom and McQuitty (1976) quoted 25 ppm as the threshold limit concentration for workers exposed for 8 hours per day during a 5 day week. As an interim measure we could perhaps consider 20 ppm as a limit for long-term exposure of pigs. However, it does not seem possible at present to predict the rate at which ammonia will be produced in a given environment (Robertson, 1971). It is therefore not practicable to base ventilation rate on ammonia levels.

Anaerobic protein degradation and sulphate reduction in livestock manure produce hydrogen sulphide. The production rate is dependent on temperature and dilution (Barber and McQuitty, 1974). Hydrogen sulphide is an irritant at sublethal concentrations but its action in acute poisoning overshadows the irritant action. It has an extremely offensive odour but continuous exposure fatigues the olfactory sense (Nordstrom and McQuitty, 1976).

The least odour detectable occurs at 0.01–0.7 ppm and it is offensive at 3–5 ppm. Irritation of the eyes and respiratory tract will occur on exposure to 50–100 ppm for 1 hour. Exposure from 8 to 48 hours at a concentration of 150 ppm can be fatal, while rapid death ensues at 700–2000 ppm. Robertson (1971) reported that pigs experience loss of appetite and photophobia at 20 ppm and vomiting and diarrhoea at 50–200 ppm. As with ammonia, it is not possible to estimate in advance the evolution of hydrogen sulphide and therefore it cannot be used as a basis for ventilation calculations. A level of 5 ppm may be appropriate for long-term exposure of pigs. Lethal concentrations of hydrogen sulphide have occurred in practice, and lethal levels have been measured in association with slurry mixing (Ober, 1970; Nordstrom and McQuitty, 1976).

QUANTITATIVE ROLE OF VENTILATION

There appear to be only three useful quantitative criteria that can currently be applied to calculate the ventilation rate of pig buildings. These are:

(1) the concentration of carbon dioxide;
(2) the relative humidity and occurrence of condensation;
(3) the temperature, which should lie between the lower and upper critical temperatures for pigs.

For a given set of conditions one of these criteria may overrule the others. For example, if in winter the lower critical temperature cannot be obtained without exceeding the carbon dioxide limit then supplementary heat may be necessary. On the other hand, temporary excursions above the carbon dioxide limit may be acceptable.

Ventilation and carbon dioxide

An approximate relationship between metabolism and carbon dioxide production can be given, although it depends on which particular chemical processes are going on in the body. About 45×10^{-9} m^3 s^{-1} of carbon dioxide will be generated for each watt of energy produced by a pig. If Q is the total metabolic heat production in the building, measured in watts, the approximate ventilation rate (in m^3 s^{-1}) to maintain a given CO_2 concentration is given by

$$\dot{V} = 45 \times 10^{-9} \, Q \left(\frac{100}{c - 0.034} \right) \tag{12.8}$$

where c is the required percentage concentration in the building as a volume fraction and 0.034 is the percentage concentration in the air introduced to the building. For $c = 0.3\%$ the ventilation rate is $16.7 \times 10^{-6} Q$ while for $c = 0.5\%$ the ventilation rate is $9.6 \times 10^{-6} Q$. The heat production in thermoneutral conditions is given by equation 12.1 and is a function of live-weight and metabolizable-energy intake only (it is tabulated in *Table 12.5*). In the absence of feed intake information, 3 times maintenance may be assumed for pigs weighing less than 100 kg. The ventilation rate to give 0.3% CO_2 can then be calculated from equation 12.8 as 0.19×10^{-3} m^3 s^{-1} kg$^{-2/3}$ or 0.67 m^3 h^{-1} kg$^{-2/3}$. For pregnant sows these values may be multiplied by a factor of $^2/_3$.

Ventilation and moisture

Water vapour production in a livestock building is an extremely difficult thing to estimate, because water is transferred to the air not only from the respiratory tract and the skin but also from urine, faeces and other diverse sources. *Table 12.5* shows the estimated latent heat production from the pig itself at the lower and upper dry critical temperatures. It is suggested that linear interpolation between the two critical temperatures should be used to estimate the latent heat production from the pig at intermediate temperatures. In addition, 20 per cent of the sensible heat loss of the pig (i.e. $0.2 \, Q_s$) may be assumed to be converted to latent heat within the building, on the basis of the author's unpublished estimates. The total latent heat produced in the building, Q_{eb}, will therefore be given (in watts) by

$$\begin{aligned} Q_{eb} &= N_b(0.2Q_s + Q_e) \\ &= N_b(0.2Q + 0.8Q_e) \end{aligned} \tag{12.9}$$

where Q_e, the latent heat loss per pig (in watts), is given by EA and N_b is the total number of pigs. Q_s is the sensible heat loss, in watts. The latent heat production, Q_e (in watts) is converted to water vapour (in kg s^{-1}) by multiplying by 0.41×10^{-6}.

The relative humidity is not normally a biologically important feature of the climatic environment for pigs, mainly because the pig has no sweating response and is also not a good panter. The dry-bulb temperature is therefore far more important than the wet-bulb temperature in dry conditions (Ingram, 1965b; Roller and Goldman, 1969). In hot, wet conditions the

Table 12.5 HEAT PRODUCTION FROM PIGS

Live-weight (kg)	Thermoneutral heat production (sensible + latent) (W)*				Latent heat production (W)†	
	Feed level (multiple of maintenance):				At the lower critical temp.	At the upper dry critical temp.
	1	2	3	4		
1	7.4	9.3	11.1	13.0	0.7	3.3
5	21	27	32	38	2.2	9.8
10	34	42	51	59	3.7	15.6
20	48	65	82	98	6.3	25
40	71	107	133	158	11.5	39
60	110	142	173	195	17.1	52
80	136	172	207	243	23	63
100	161	198	236	274	30	73
140	207	259	311	363	44	91
180	250	313	375	438	60	108

*Multiply by 45×10^{-9} to give amount of CO_2 produced ($m^3\,s^{-1}$)
†Multiply by 0.41×10^{-6} to give amount of H_2O produced ($kg\,s^{-1}$)

wet-bulb temperature would, of course, be very important. In dry pig buildings there is no apparent biological reason why humidity should be controlled other than to prevent chronic condensation and fogging. Fogging will cool a pig. The small droplets of water in the air deposit on the pig's skin and will require latent heat to evaporate them. On the other hand high humidity—or 'sweat-box' conditions—has been claimed to be beneficial in purging the air of bacteria-carrying particles and producing a lowered incidence of respiratory infection (Gordon, 1963). Condensation and relative humidity are very sensitive to the internal temperature of buildings and with the trend towards high insulation levels in pig buildings it will become easier to cope with the moisture balance problems that have beset pig housing in the past. There is therefore no reason why a relative humidity of 90–95% cannot be used as the criterion for ventilation design in extreme winter conditions, given a vapour barrier and a high level of insulation.

Ventilation and temperature

Ventilation influences and controls the temperature in a pig building by adjusting the energy balance. The sensible heat lost from a building has two pathways: the dominant one is the ventilating air and the minor pathway is conduction through the building structure. Models are available for estimating the internal climate of livestock buildings both for fixed conditions (Bruce, 1974) and for dynamic conditions (Albright and Scott, 1974; Barlott and McQuitty, 1976; Feddes, Campbell and McQuitty, 1978).

For constant conditions, and ignoring radiation effects, the sensible heat balance of a building, Q_{sb}, is written as

$$Q_{sb} = (\varrho c_p \dot{V} + \Sigma U_i A_i) \Delta T \qquad (12.10)$$

where ϱ is the density of air (in $kg\,m^{-3}$), c_p is the specific heat of air (in $J\,kg^{-1}\,K^{-1}$), $\dot{V}$ is the ventilation rate (in $m^3\,s^{-1}$), U_i is the heat transfer coefficient of the ith building element (in $W\,m^{-2}\,K^{-1}$), A_i is the area of the ith building element (in m^2) and ΔT (in K) is the temperature difference between the inside and outside air.

To illustrate the results of the application of equation 12.10 together with the model for heat production and critical temperatures, three cases are considered:

Case 1:   1000 pigs at 20 kg live-weight kept on metal mesh floors in groups of 15 and fed 3 times maintenance.

Case 2:   500 pigs at 60 kg live-weight kept on concrete floors in groups of 15 and fed 3 times maintenance.

Case 3:   100 pregnant sows at 140 kg live-weight kept individually on concrete floors and fed 1.2 times maintenance.

All the buildings were assumed to have an overall heat transfer coefficient (U value) of $0.4\,W\,m^{-2}\,K^{-1}$, which in the UK is a good standard of insulation. *Figures 12.3–12.5* show graphically the relationship between inside temperature, outside temperature and ventilation rate. The upper (dry) and lower critical temperatures are shown, as these provide limiting criteria for the inside temperature. An arbitrary design upper limit for outside temperature of 22 °C has been shown. Within the UK, only in the Greater London area is

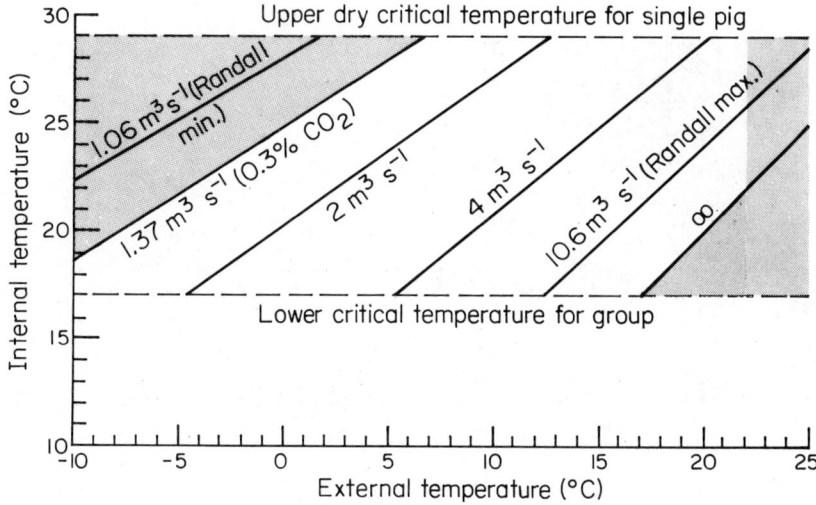

Figure 12.3 Temperature and ventilation relationships for pigs of live-weight 20 kg: case 1

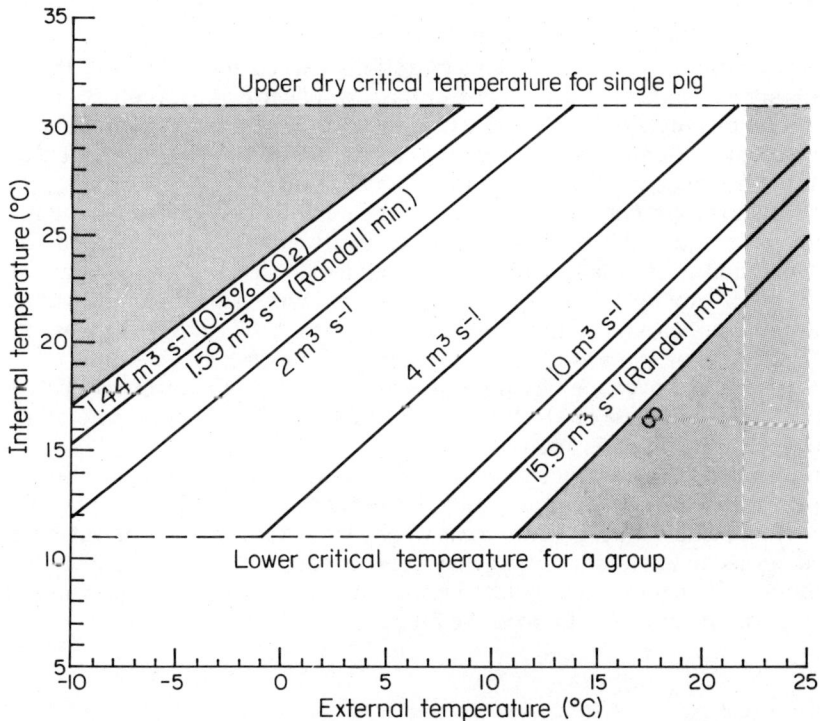

Figure 12.4 Temperature and ventilation relationships for pigs of live-weight 60 kg: case 2

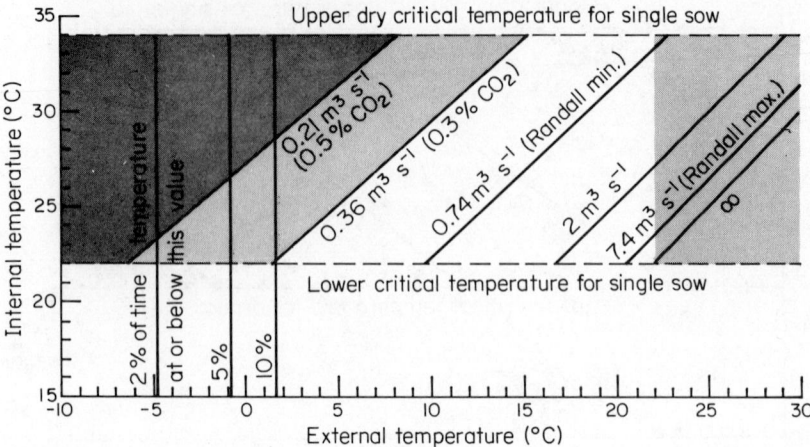

Figure 12.5 Temperature and ventilation relationships for a sow of live-weight 140 kg: case 3

the average maximum daily temperature in July higher than this figure (Lacy, 1977). Also shown in *Figure 12.5* are the percentages of a year for which the outside temperature at Edinburgh Airport (Turnhouse) is equal to or less than the indicated temperature (Lacy, 1977). It is possible to create overlays of temperature probabilities for diagrams of the type shown in *Figures 12.3–12.5* for any site, so that decisions on ventilation rates can be made with some understanding of the risks.

The conditions when CO_2 concentrations exceed the 0.3% and 0.5% levels are also indicated. In the case of the sows (*Figure 12.5*), ventilation for thermoneutrality would result in CO_2 concentrations above 0.3% for about 10 per cent of the time, with excursions above 0.5%. The alternatives are to heat the building or feed more to the animals.

Also shown on *Figures 12.3–12.5* are the minimum and maximum ventilation rates quoted from other sources by Randall (1977) (marked for convenience as 'Randall min.' and 'Randall max.'). These rates correspond to 53×10^{-6} $m^3 kg^{-1} s^{-1}$ and ten times that amount. This minimum ventilation rate appears to be unnecessarily low for the 20 kg and 60 kg pigs but far too high for the sows. The minimum and maximum ventilation rates should not, however, be based on live-weight or any other arbitrary basis; they should be based on diagrams such as *Figure 12.3–12.5* or some other appropriate bioclimatic analysis.

Internal relative humidity curves or a condensation limit can also be plotted or overlaid on diagrams such as *Figures 12.3–12.5*. These can also be used to decide limiting ventilation rates. However, we cannot yet base predictions on ammonia, hydrogen sulphide, odour, dust or disease hazard contours. These are nonetheless important and we must move towards a better understanding of them in the future.

References

ALBRIGHT, L. D. and SCOTT, N. R. (1974). *Trans. ASAE* **17**, 88–98
BARBER, E. M. and McQUITTY, J. B. (1974). *Hydrogen Sulphide Evolution*

from Anaerobic Swine Manure. Department of Agricultural Engineering, University of Alberta

BARLOTT, P. J. and McQUITTY, J. B. (1976). *Jnl Can. agric. Engng* **18**, 31–35

BATURIN, V. V. (1972). *Fundamentals of Industrial Ventilation*, 3rd edn. Pergamon Press, Oxford

BAXTER, S. H. (1969). *The Environmental Complex in Livestock Housing*. Scottish Farm Buildings Investigation Unit, Aberdeen

BOON, C. R. (1978). *J. agric. Engng Res.* **23**, 129–139

BRUCE, J. M. (1974). *Farm Bldg Prog.* **36**, 15–18

BRUCE, J. M. (1977). *Farm Bldg R & D Stud.* **8**, 9–15

BRUCE, J. M. and CLARK, J. J. (1979). *Anim. Prod.* **28**, 353–369

CLOSE, W. H. (1971). *Anim. Prod.* **13**, 295–302

CURTIS, S. E., ANDERSON, C. R., SIMON, J., JENSEN, A. H., DAY, D. L. and KELLY, K. W. (1975). *J. Anim. Sci.* **41**, 735–739

FEDDES, J. J. R., CAMPBELL, W. D. and McQUITTY, J. B. (1978). *Fortran IV Program to Predict Heating and Ventilating Requirements of Totally Enclosed Farm Buildings*. Department of Agricultural Engineering, University of Alberta

FULLER, M. F. and BOYNE, A. W. (1972). *Br. J. Nutr.* **28**, 373–384

GORDON, W. A. M. (1963). *Br. vet. J.* **119**, 263–273

HALES, J. R. S. (1974). In *Environmental Physiology*, pp. 107–163. Ed. by D. Robertshaw. Butterworths, London

HOLMES, C. W. and CLOSE, W. H. (1977). In *Nutrition and the Climatic Environment*, pp. 51–74. Ed. by W. Haresign, H. Swan and D. Lewis. Butterworths, London

HONEY, H. F. and McQUITTY, J. B. (1976). *Dust in the Animal Environment*. Department of Agricultural Engineering, University of Alberta

HONEY, H. F. and McQUITTY, J. B. (1979). *Jnl Can. agric. Engng* **21**, 9–14

HOVELL, F. D. DeB., GORDON, J. G. and MacPHERSON, R. M. (1977). *J. agric. Sci., Camb.* **89**, 523–533

INGRAM, D. L. (1965a). *Nature, Lond.* **207**, 415–416

INGRAM, D. L. (1965b). *Res. vet. Sci.* **6**, 9–17

INGRAM, D. L. (1967). *J. comp. Path. Ther.* **77**, 93–98

LACY, R. E. (1977). *Climate and Building in Britain*. HMSO, London

LUCKE, J. N., HALL, G. M. and LISTER, D. (1979). *Ann. NY Acad. Sci.* **317**, 326–337

MONTEITH, J. L. (1973). *Principles of Environmental Physics*. Arnold, London

MORRISON, S. R., BOND, T. E. and HEITMAN, H. (1967). *Trans ASAE* **10**, 691–692, 696

MOUNT, L. E. (1960). *J. agric. Sci., Camb.* **55**, 101–105

MOUNT, L. E. (1967). *Res. vet. Sci.* **8**, 174–186

MOUNT, L. E. (1968). *The Climatic Physiology of the Pig*. Arnold, London

NORDSTROM, G. A. and McQUITTY, J. B. (1976). *Manure Gases in the Animal Environment*. Department of Agricultural Engineering, University of Alberta

OBER, J. (1970). *Bauen Lande* **21**, 32–35

RANDALL, J. M. (1975). *J. agric. Engng Res.* **20**, 199–215

RANDALL, J. M. (1977). *A Handbook on the Design of a Ventilation System for Livestock Buildings using Step Control and Automatic Vents*. National Institute of Agricultural Engineering, Silsoe, Bedford

ROBERTSON, A. M. (1971). *Farm Bldg R & D Stud.* **1,** 17–28

ROLLER, W. L. and GOLDMAN, R. F. (1969). *Trans. ASAE* **12,** 164–169

SAINSBURY, D. W. B. (1972). *Pig Production,* pp. 91–106. Ed. by D. J. A. Cole. Butterworths, London

SÄLLVIK, K. (1979). *Principles for Mechanical Exhaust Ventilation Systems in Animal Houses.* Department of Farm Buildings, Swedish University of Agricultural Sciences, Uppsala

SBI (1967). 'Experiments on piggery climatic conditions 1957–61.' *SBI–Landbrugsbyggeri 25,* Copenhagen

STEPHENS, D. B. (1971). *Anim. Prod.* **13,** 303–313

STEPHENS, D. B. and START, I. B. (1970). *Int. J. Biomet.* **14,** 275–284

STOMBAUGH, D. P. and ROLLER, W. L. (1977). *Trans. ASAE* **20,** 1110–1118

STOMBAUGH, D. P., TEAGUE, H. S. and ROLLER, W. L. (1969). *J. Anim. Sci.* **28,** 844–847

VERSTEGEN, M. W. A. and VAN DER HEL, W. (1974). *Anim. Prod.* **18,** 1–11

VERSTEGEN, M. W. A., VAN DER HEL, W., JONGEBREUR, A. A. and ENNEMAN, G. (1976). *Z. Tierphysiol. Tierernähr. Futtermittelk.* **37,** 255–263

13

OPTIMAL HOUSING CRITERIA FOR RUMINANTS

A.J.F. WEBSTER
*Department of Animal Husbandry, University of Bristol,
School of Veterinary Science, Langford, Bristol*

Introduction

The design of any housing system to create an environment for optimal animal production involves much more than just the regulation of energy and moisture exchange by the control of temperature and ventilation. There are, moreover, certain fundamental differences between ruminant animals on the one hand and pigs and poultry on the other, which makes the approach to environmental control quite different for the two groups of livestock. Pigs and poultry raised out of doors in temperate and cool climates maintain homeothermy in most circumstances by regulating heat production, Q, in order to keep *warm (Figure 13.1a)*. The consequent increase in Q can usually be achieved with little metabolic distress but at some cost in

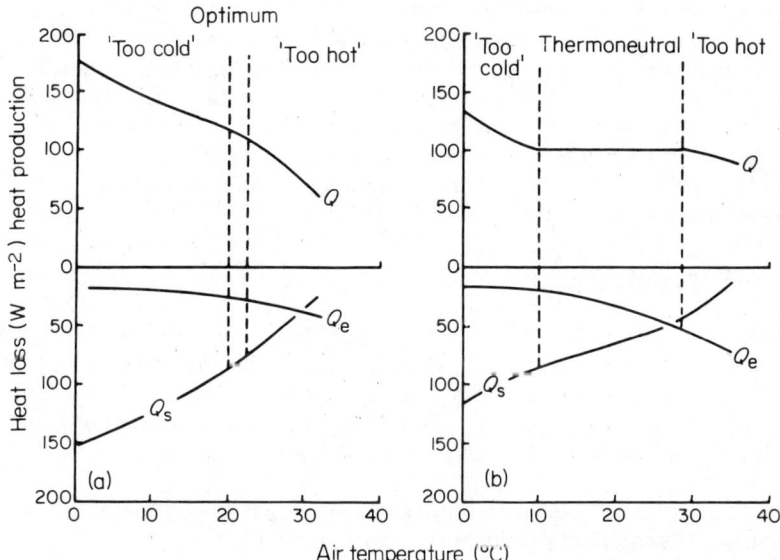

Figure 13.1 Patterns of heat exchange in farm animals. (a) Group I, animals that maintain homeothermy primarily by regulating heat production, Q, e.g. pigs and chickens. The vertical lines indicate the narrow zone of air temperature wherein food conversion efficiency is optimal. (b) Group II, animals that maintain homeothermy primarily by regulating evaporative heat loss, Q_e, e.g. horses and ruminants. Other symbol: Q_s, sensible heat loss. (From Webster, 1980)

terms of food energy. Intensive poultry and pig houses have therefore been designed to maximize feed conversion efficiency by regulating temperature and ventilation so as to keep the animals in an environment that is just on the point of being too hot for optimal efficiency of production. It should not be assumed that a slightly cooler environment would be uncomfortable to a pig or chicken with adequate food (Webster, 1980).

In most cool, temperate environments, ruminants maintain homeothermy by regulating evaporative heat loss, Q_e, at little metabolic cost in order to keep *cool* (*Figure 13.1b*). Moreover, ruminants have a very marked ability to alter their zone of thermal neutrality in response to previous thermal history. Yearling cattle in Alberta, Canada can shift both the upper and lower limits of their thermoneutral zone by about 20 K during adaptation to severe winter conditions (Webster, Chlumecky and Young, 1970; *see also* Chapter 10). In other words, there are no absolute criteria for the thermal requirements of any particular class of ruminant livestock (except perhaps the newborn animal); they depend to a very large extent on what the animal has grown accustomed to.

The fundamental difference between the environmental requirements of ruminants and those of pigs and poultry concerns the prevention of disease. Modern intensive pig and poultry producers are able to operate closed herds or purchase their stock from certified specific-pathogen-free units. Vaccines against respiratory diseases have been particularly effective in poultry and if disease enters a unit (or even if it doesn't) the animals can be medicated *en masse* with antibiotics. However, ruminants such as young calves, for example, arrive on a rearing unit from a variety of sources, which are often unknown, and they undoubtedly carry infection onto the premises. The development of vaccines for respiratory diseases in ruminants has been conspicuously unsuccessful, probably mainly because of the multifactorial nature of the diseases (Morzaria *et al.*, 1978), and regular oral antibiotic cover is incompatible with normal rumen development.

As a general rule, therefore, environmental control for ruminants should be directed primarily towards ensuring optimal health; heat flow and energy conservation assume relatively minor importance.

Definition of requirements

The proposed revisions to the UK Codes of Welfare for Farm Livestock (discussed by Moss in Chapter 26) state that the prime requirement of any system of housing and husbandry is that it 'should meet the health and behavioural needs of the animals'. To be more specific, the environmental needs for the individual animal may be defined as follows:

(1) *Thermal comfort*. The environment must not be so hot nor so cold as significantly to affect production or cause discomfort.
(2) *Physical comfort*. The space available to the animal and the floors and surfaces with which it makes contact should be such as to prevent acute injury or chronic discomfort.
(3) *Disease control*. The building should be designed in such a way as to minimize disease, either by reducing the spread of infection or by avoiding stresses liable to decrease resistance to infection, or both.

(4) *Behavioural satisfaction.* The animal should not seriously be impeded from performing most socially acceptable patterns of spontaneous behaviour that it would perform in an unrestricted environment. It should also be reasonably free from fear.

These four definitions are, of course, little more than honest expressions of good intent. They are not necessarily the same as the requirements for optimal production and they are not necessarily compatible since, for example, it is not, on balance, in the interests of the animal to let it run around and play if in doing so it contracts a fatal case of enteritis. In order to convert these needs into scientifically based recommendations for housing design, it is necessary to examine the evidence. In the case of thermal comfort the evidence is quite extensive; in all other cases it is more or less inadequate.

Thermal requirements

Table 13.1 lists some examples of the calculated lower critical temperatures (T_{cl}) for ruminants standing up in a dry enclosure, i.e. in circumstances where heat loss can be described simply in terms of air temperature (T_a, °C) and air movement (u, m s^{-1}), uncomplicated by complex problems of conduction, radiation and evaporative exchange. The derivation of these values is given by Webster (1974). However, *Table 13.1* differs in some degree from Table 2 in that previous publication, having been updated and modified to include more recent information on thermal insulation in young cattle (*see* Webster, Gordon and McGregor, 1978), and values for sheep (Joyce and Blaxter, 1964; Alexander, 1974) and red deer (Simpson *et al.*, 1978).

In an unheated building at low air movement (u = 0.2 m s^{-1}), the only cattle likely to experience cold sufficient to elevate Q are newborn calves, or young calves whose metabolic rate is low by virtue of starvation, diarrhoea or emaciation. Such animals can undoubtedly be stressed by cold in unheated buildings and may require special attention. Increasing air movement to 2.0 m s^{-1}, which is not uncommon in draughty buildings, increases T_{cl} in the newborn calf from +9 to +17 °C. This illustrates the importance of ensuring that adequate ventilation for calf houses is achieved without allowing excess air movement to impinge on the calves; a draught is simply ventilation in the wrong place.

By 1 month of age the healthy calf with a good appetite has a T_{cl} close to 0 °C and is not likely to be stressed by cold while indoors. Well-grown veal calves, by virtue of their very high energy intake and heat production, are particularly tolerant to cold and, by the same criteria, sensitive to heat. The traditional belief that veal calves should be kept in a warm environment because they are not ruminants and are therefore somehow more sensitive to cold is a myth.

No other class of cattle is likely to experience a systemic stress of cold when standing up in a dry enclosure unless air movement is exceptionally high. For the dairy cow, however, cold stress should not be considered as a systemic but as a local problem. Q in the high-yielding cow is, again, very high and so

Table 13.1 LOWER CRITICAL TEMPERATURE OF RUMINANTS HOUSED IN CONDITIONS OF VERY LOW AIR MOVEMENT (0.2 m s⁻¹) AND IN A DRAUGHT (2 m s⁻¹)

	Body weight (kg)	Coat depth (cm)	Heat production ($W\,m^{-2}$)	Thermal insulation ($m^2\,K\,W^{-1}$)		Lower critical temperature (°C)	
				I_t	I_e†	$u = 0.2\,m\,s^{-1}$	$u = 2.0\,m\,s^{-1}$
Cattle							
newborn calf	35	1.2	100	0.09	0.25	+9	+17
1 month old	50	1.4	120	0.10	0.26	0	+9
veal	100	1.2	154	0.25	0.25	−14	−1
Store cattle, maintenance	250	2.0	157	0.17	0.32	−32	−20
Beef cow, maintenance	450	2.9	107	0.22	0.36	−17	−9
Dairy cow, 22 kg milk per day	500	1.2	154	0.20	0.25	−26	−13
Sheep							
ewe, maintenance	50	6.0	75	0.15	0.68	−11	−4
ewe, shorn	50	1.0	75	0.15	0.19	+17	+20
newborn lamb, coat dry	4	0.8	80	0.10	0.19	+19	+24
growing lamb, 0.2 kg gain per day	35	4.0	100	0.13	0.48	−13	−3
Red deer							
calf, 0.2 kg gain per day	45	1.6	135	0.10	0.28	−7	+3

† I_e measured in still air (0.2 m s⁻¹)

I_t, tissue insulation; I_e, external insulation; u, wind speed

T_{cl} is low. Milk synthesis, however, depends on blood flow to the mammary gland, which is reduced by local cooling (Thompson and Thompson, 1977). The production of dairy cows has been shown to fall at T_a below about 0 °C (MacDonald and Bell, 1958), although it is obvious that direct chilling of the udder depends as much on the thermal properties of the floor as on T_a.

Sheep have a relatively lower heat production per unit area than cattle because their fleece provides excellent thermal insulation. The housed adult sheep in full fleece is seldom, if ever, cold. There has, from time to time, been a fashion for shearing ewes at housing with the intention of increasing the size and viability of lambs, presumably by stimulating appetite in late pregnancy (Kneale and Bastiman, 1977). Undeniably, there are cases where ewes intensively housed over winter when in full fleece are subjected to mild heat stress to which they might respond by reducing food intake and this is obviously undesirable in late pregnancy. Moreover, in dry, draught-free conditions inside a sheep house a shorn ewe, adequately fed, will not be intolerably stressed by low air temperatures alone. Shearing the ewe increases T_{cl} in still air initially to about +17 °C. This undoubtedly increases appetite (Davey and Holmes, 1977), but it is likely from the first principles of energy exchange that a ewe shorn in January would have a metabolizable-energy requirement for maintenance from January to March about 25 per cent greater than that of one in full fleece. This seems to be an unacceptable increase in feed costs relative to the small potential gain in lamb crop. It would be better to ensure adequate natural ventilation in the building to avoid any loss of appetite due to heat stress.

One of the main reasons for housing ewes is to ensure individual care at lambing time. Prevention of hypothermia in the newborn lamb is, of course, a part of this, but only a small part. It is common practice, unless the weather is extremely severe, to turn ewes and their lambs out of doors as soon as possible to minimize the risk of (especially) enteric disease in the lambs. Values in *Table 13.1* indicate that a lamb born in late winter or early spring in the UK would be almost continuously below T_{cl} for at least the first month of life. In this case it is considered good husbandry to permit a significant degree of cold stress in order to reduce the risk of disease. The same logic has rarely been applied to calf husbandry.

Table 13.1 also compares T_{cl} for growing lambs and red-deer calves at about halfway to maturity, a condition they might expect to reach at the onset of their first winter. The red deer, which has less insulation but a higher thermoneutral heat production than the lamb, has a calculated T_{cl} in still air of −7 °C compared with −13 °C for the lamb. However, the fat reserves of the deer calf entering its first winter are only about 20 per cent of those of the lamb, so the deer is far more prone to energy exhaustion resulting from prolonged cold and undernutrition.

To summarize this section: there is no class of healthy ruminant for which the direct effects of low air temperature *per se* are likely to cause intolerable stress in the temperate and cool zones of the world. Moreover, the effects of T_a on food conversion efficiency below T_{cl} are likely to affect only the smallest animals and at a time when their daily intake is very small relative to lifetime requirements. Thus there are no sound economic grounds for providing any more environmental control for the healthy ruminant animal than shelter from excessive air movement and precipitation.

Physical comfort and floor type

The design of the fixtures and fittings within a cattle building so as to minimize the risk of injury or chronic discomfort is largely a matter of common sense, although much can be learnt about cubicle design, for example, from a study of the 'space envelope' occupied by a cow when standing, lying or, more importantly, changing position (Rogerson, 1972).

Floor design is critical to thermal comfort, physical comfort, health and security. Again, before examining the scientific evidence it pays to apply a little common sense. The floor is the surface on which the animal stands, walks, lies down and passes excreta. It must, therefore, according to the needs of the moment, be unyielding, non-slippery and well-drained or comfortably soft, warm and dry and easy to clean by machinery. No single material from concrete to a grass meadow meets all these specifications, and it is probably fruitless to contemplate one.

Table 13.2   EFFECT OF THE FLOOR TYPE ON THE LOWER CRITICAL TEMPERATURE FOR CONVENTIONALLY REARED CALVES AT ABOUT 2 WEEKS OF AGE AND FOR VEAL CALVES AT ABOUT 10 WEEKS OF AGE

	Critical temperature (°C)	
	2-week-old calves	*10-week-old veal calves*
Standing	10	−3
Lying		
dry concrete	17	+6
dry straw	7	−8
damp straw		
wooden slats		
rubber mat	10	−3
asphalt		

From Mitchell (1976)

The thermal properties of floors for livestock have been studied by Bruce (1979) using a heat loss simulator. *Table 13.2* (from Mitchell, 1976) provides examples of the extent to which heat loss by conduction from individual calves to different types of floor affects cold tolerance, as indicated by T_{cl}. Wooden slats, asphalt, rubber mats or damp straw can be said to be neutral types of floor because total sensible heat loss from a calf (or an adult dairy cow) would not differ significantly according to whether the animal was standing up or lying down. Concrete, whether dry or wet, has a very high thermal conductivity (24 W m^{-2} K^{-1} dry; 32 wet) which is not improved by underfloor insulation since most of the conductive heat loss from the individual animal is transmitted laterally. A calf compelled to lie on concrete is cold when T_a falls below about +17 °C. Bruce (1979) has calculated that for a calf of about 50 kg a bed of dry straw decreases heat loss by an amount equivalent to a decrease of only 3 K in T_{cl} compared with that of a calf lying on wooden slats. While this is undoubtedly a true measure of decreased conduction of heat from the under-surface of a heat source to the ground, it almost certainly underestimates the cold-ameliorating effects of deep, dry straw for a young calf, for three reasons. First, a calf lying in deep straw will have a greater contact area than the heat loss simulator on a flat surface.

Secondly, a deep straw bed will generate its own warmth by fermentation, and finally, air movement at ground height, and thus convective heat losses, will be less for a calf lying in straw than on wooden slats. Moreover, calves tend to spend a longer time lying down on straw than on other surfaces and this too conserves energy.

It is reasonable to suppose that direct contact of the udders of dairy cows with cold floors might have a deleterious effect on production by reducing blood flow or precipitating mastitis. The evidence is, however, scanty and confusing, not least because of the complex aetiology of mastitis. There is no consensus of opinion as to whether it is better for the cow to have a straw bed that is warm, comfortable, but unhygienic or an unbedded cubicle that is hard but clean (*see* Ekesbo, 1968; Marr, 1978). In the latter case asphalt, or rubber mats, are undoubtedly less cold than concrete but it is again reasonable to assume that for a heavy, large-jointed animal like a cow the physical discomfort of lying on an unyielding surface far surpasses any thermal discomfort.

The evidence that concrete and similar floors predispose to injuries, especially to feet and teats, is unequivocal. Injuries to teats occur when cows are changing position. A cow lying on a hard, cold floor is not only more likely to injure her teats each time she changes position, she also changes position more often, presumably because she is less comfortable. For other details of effects of floor type on injuries to dairy cows, *see* Ekesbo (1968). Murphy (1978) has made a very thorough survey of the incidence of injury and infection to the feet of beef cattle fattened on concrete slats or in straw

Table 13.3 INCIDENCE (%) OF INJURY AND INFECTION TO THE FEET OF BEEF CATTLE KEPT ON CONCRETE SLATS OR ON STRAW

	Incidence (%)	
	Concrete slats	*Straw*
1. Conditions observed after slaughter		
Flattening of sole	90	20
Overgrowth of heel horn	22–35	1–6
Abrasion of abaxial wall	57	46
Necrotic pododermatitis	77	64
Traumatic pododermatitis	8–12	3–4
Septic traumatic pododermatitis	20	6
2. Clinical lameness	4.8	2.4

From Murphy (1978)

yards, assessed by inspection after slaughter. *Table 13.3* shows that all conditions were more prevalent on concrete slats. Some conditions, such as flattening of the sole and overgrowth of heel horn, cannot be said to be true injuries but they do predispose to more severe conditions such as bruising, penetration of the sole and sepsis. However, the actual incidence of observed clinical lameness was only 4.8% on slats and 2.4% on straw. It has been thought by some that regular abrasion is good for the feet of cattle and sheep. Murphy has shown, however, that while abrasion stimulates hoof growth, the hardness of the hoof is inversely proportional to its growth rate. Concrete can undoubtedly be an excellent, durable, non-slip surface for cattle to walk about on from time to time; it can never make a satisfactory bed.

Environment and health

All animals coexist with a mass of potentially pathogenic micro-organisms. Indeed, infection may be said to be the natural state and disease simply the result of loss of equilibrium between the host, the parasites and the environment (Top, 1964). The extent to which this concept holds true depends of course on the nature of the disease. Certain conditions, like smallpox in man, have a high morbidity which appears to be little affected by environment, but can be traced to a single variety of organism and effectively treated by vaccination. However, the major enteric and respiratory diseases of housed cattle appear to be associated with a multiplicity of organisms (Omar, 1966) and it is extremely difficult to reproduce clinical disease by experimental infections, even when heroic doses of mixed organisms are used. Moreover, the use of multivalent vaccines containing five viral antigens has not been shown to have a significant effect on the incidence of respiratory disease in calves (Morzaria et al., 1978). The incidence of most important enteric and respiratory diseases in young cattle is undoubtedly dominated by non-specific factors of climate, housing and husbandry (Martin, Schwabe and Franti, 1975a, b). This chapter will consider only aspects of housing. Loosemore (1964) included in his comprehensive review of calf diseases other important husbandry factors such as ensuring adequate colostrum, proper feeding, regular cleansing and disinfection, minimizing transport stresses, purchasing calves from as few sources as possible and application of appropriate preventive medicine on arrival.

The literature relating environment and housing to disease in cattle is quite copious but not very coherent. In the UK, the death rate in calves to 12 weeks of age is about 5%, with, of course, enormous variations (Withers, 1952; Kilkenny and Rutter, 1975). This inevitably makes the experimental study of predisposing factors very difficult and most of the information that exists has come from clinical reports in which the information has inevitably been incomplete. In order to extract what one can from field observations of this nature, it is necessary first to consider ways and means by which environmental variables can influence the pattern of infection and disease. These are as follows.

FACTORS AFFECTING THE SURVIVAL AND SPREAD OF MICRO-ORGANISMS

Pathogenic micro-organisms may be transferred from one animal to another

(1) by direct contact: mouth to mouth, faeces to mouth, mouth to teat to mouth;
(2) by short-range direct transmission of organisms in droplet form (sneezing and coughing);
(3) by airborne diffusion of organisms, mostly in aerosol form.

A number of factors in the environment may control one or other of the routes of transmission.

Clearance

If one could attribute infection entirely to diffuse airborne transmission, the magnitude of the challenge to any animal would be determined simply by the concentration of pathogenic organisms in the environment. This can, in theory, be calculated quite simply.

At equilibrium, the rate of production of pathogens in a confined space (q, s^{-1}) must equal the sum of death rate within that space (R_m, s^{-1}) plus clearance by ventilation ($C_o\dot{V}$). Thus

$$q = R_m + C_o\dot{V} \tag{13.1}$$

where C_o is the concentration of organisms expressed as number per unit volume and $\dot{V}$ is the ventilation rate (m^3 s^{-1}). Also,

$$q = bkN \tag{13.2}$$

where b is the rate of emission of organisms from an infected animal (s^{-1}), k is the proportion of infected animals and N the number of animals in the building. If the volume of the building is V (in m^3) and t is time elapsed (in seconds), the concentration of pathogenic organisms in the building at any one time can be written

$$C_o = \frac{bk(N/V)}{R_m + (\dot{V}/V)} [1 - \exp\{-(R_m + \dot{V}/V)t\}] \tag{13.3}$$

For this version of the classical clearance equation, I am indebted to Dr J.M. Bruce of the Scottish Farm Buildings Investigation Unit. The term within the square brackets is governed mainly by $\dot{V}/V$ but, for practical purposes, it resolves to unity in times that are very short relative to the residence of animals in buildings. At equilibrium C_o is a function of stocking density (N/V) and ventilation rate expressed as number of air changes per second or per hour ($\dot{V}/V$).

There is some dispute over the extent to which respiratory infections, for example, in calves are transmitted by the airborne route. To a first approximation, droplets of diameter over 100 μm shed by an animal during coughing or sneezing probably travel less than 1 m before dropping to the floor, while droplets of less than 100 μm can be borne as aerosols in air currents for as long as 24 hours (Wells, 1955). It is reasonable to suppose therefore that most exhaled droplets, which are in the second category, have the ability to diffuse evenly throughout the air and so will obey equation 13.3.

When calves are housed in groups, or in individual pens from which they can contact one another, they are obviously able to transmit organisms by direct contact. This mode of transmission is undoubtedly of importance in the transmission of coliforms and salmonellae, which infect the digestive tract (Linton *et al.*, 1974), although an environment that predisposes to infection is not necessarily the same as one that predisposes to disease. On balance, it would seem that penning calves individually confers no consistent advantage in terms of reducing the incidence of respiratory disease (Miller *et al.*, 1978). Moreover, there is clear clinical evidence that inadequate ventilation and overstocking markedly increase the incidence of calf pneumonias. Taken together, these observations support the conclusion that clinical respiratory disease in young cattle is transmitted primarily by the airborne route. It follows therefore that steps taken to reduce stocking density (N/V) and increase clearance rate ($\dot{V}/V$) are likely to be of major benefit in reducing the challenge from respiratory pathogens, whether calves are housed singly or in groups.

DEATH RATE

The effects of temperature and relative humidity (RH) on the survival of some common pathogens are illustrated in *Table 13.4*. Mycoplasma species and *Escherichia coli* appear to survive and multiply best at about 15 °C (Wray, 1975) while survival of viruses outside the host is largely unaffected by temperature. Death rates for both *E. coli* and mycoplasmas are more

Table 13.4　SOME EFFECTS OF TEMPERATURE AND RELATIVE HUMIDITY ON THE DEATH RATES OF PATHOGENS

	Death rate of:			
	E. coli	*Mycoplasma*	*Rhinovirus*	*PI3, IBR*
Temperature effect	Least at 15 °C	Greater at 10 °C than at 15 °C	Unaffected by temp.	Unaffected by temp.
Relative humidity effect:				
RH 50%	Slow	Slow	Decreasing	Increasing
RH 70%	Rapid	Rapid	↓	↓
RH 80%	Slow	Slow		

rapid at 70% RH than at higher or lower values (Strange and Cox, 1976). Rhinoviruses appear to survive better at high humidities, IBR and PI3 viruses at low humidities (Buckland and Tyrrel, 1962). In these last examples, these differences in survival time reflect the association between seasonal patterns of RH and disease incidence. There is the further possibility that at high RH there is accelerated sedimentation of airborne pathogens in large aerosols (Beer, Mealhorn and Arnold, 1975). However, Jericho, Langford and Pantekoek (1977) were unable to observe any consistent effects of temperature and RH on the recovery of *Pseudomonas haemolytica* from aerosols.

Other factors that may influence the death rate of organisms include ionization and ultraviolet radiation. Both have been used to control infection in animals, the former in recent years (Jensen and Curtis, 1976), the latter since the beginning of agriculture.

BEHAVIOUR

If calves are penned individually they are obviously restricted from widely disseminating any infectious organisms by contact. Osborne (1975) showed, for example, that the spread of *Salmonella* infection in calves was almost 100% when they were group-housed but almost zero when they were penned individually. However, although infection was universal, very few calves contracted clinical disease. My own clinical observations of veal calves have shown that clinical salmonellosis and other enteric infections can move from pen to pen. This may be due, in part, to the abnormal behaviour patterns of veal calves housed in individual wooden crates. Claire Saville (unpublished results) has shown, for example, that veal calves at 10–14 weeks of age restricted in crates spend 44 per cent of daylight hours licking, chewing or sucking themselves, their crates or any accessible bit of their neighbours. Veal calves group-housed on straw spent only 11 per cent of the time in such activities.

The spread of respiratory pathogens may also be affected by behaviour. For example, cattle exposed to cold are more likely to congregate in groups and this would tend to increase the spread of infection by short-range direct transmission of organisms.

LOCAL RESISTANCE TO INFECTION

In order to cause disease, invading organisms must first colonize the sensitive tissues of the host. The ability of calves to clear organisms from the gut or respiratory tract is obviously a major factor in local resistance to infection. Veit, Farrell and Troutt (1978) have studied the pulmonary clearance by calves of *Serratia marcescens*, a non-pathogenic organism to which they had not previously been exposed. Mean clearance rate was about 50% per hour for calves at 2 or 7 weeks of age. Regional values for mean per cent retention (MPR) showed that clearance was slower from the ventral and cranial regions of the lungs than from dorsal and caudal areas, suggesting an inverse relationship between O_2 tension and MPR. Hypoxia has been shown to reduce tracheal mucociliary rates and to reduce alveolar macrophage phagocytic activity (Thomson and Gilka, 1974). The relatively poor clearance from the cranioventral areas of the bovine lung may relate to the high incidence of pneumonic lesions in this area (Smith, Jones and Hunt, 1972). It is almost certain that environmental factors affect pulmonary clearance rate, though it is difficult to prove experimentally. Jericho and Darcel (1978) were unable to observe any consistent effects of temperature and humidity on the response of the bovine respiratory tract to aerosols containing herpes virus, although Jericho and Magwood (1977) did observe a reduction in goblet cells and polymorphs in the respiratory epithelium of calves kept in a hot (30 °C), dry (RH 35%) environment.

Cattle exposed to cold show a degree of pulmonary hypoxia because pulmonary ventilation rate does not keep pace with increased tissue metabolism. It is possible therefore that prolonged mild-cold exposure might contribute to respiratory disease by reducing pulmonary clearance of organisms (Will *et al.*, 1978).

It is again reasonable to suppose that pollutants such as dust, ammonia and hydrogen sulphide may predispose to respiratory infection either by damaging the mucous membranes or specifically by inhibiting clearance mechanisms. Positive ions in the atmosphere have been shown to reduce tracheal ciliary activity and mucous secretion (Krueger and Smith, 1958).

This process can be reversed by negative ionization. Studies of air pollution in cattle units (Nordstrom and McQuitty, 1975) or of negative ionization in pig units (Jensen and Curtis, 1976) have failed to show consistent effects in terms of production or health, although Pritchard *et al.* (1980) have recently observed that filtration of the air entering an intensive veal unit appeared to reduce slightly not only the incidence but also the severity of pneumonia. It may be that filtration was removing pollutants that interfered with pulmonary clearance. It is unlikely that the answers to questions of this nature will come from attempts to induce disease in small experimental units but rather from a study, on the one hand, of the local physiological and immunological responses of animals to pollutants and infective agents, complemented by field epidemiological studies of large numbers of animals in comparable environmental conditions. Such experiments will never establish direct cause and effect, just as the direct causal relationship of smoking and bronchocarcinoma has not yet been proven, but ultimately the evidence should be sufficiently strong to suggest suitable remedial action.

In young calves, local resistance to enteric disease depends in large part on keeping pathogens from entering the small intestine, either by overflow from the abomasum or by migration upwards from the large intestine. When calves feed naturally in a 'little and often' fashion, from their mothers or from an automatic milk dispenser, the abomasum is not overloaded and natural acid secretions ensure that pH is rapidly restored to a value below about 4.5 that is sufficient to kill orally ingested pathogens. Bucket feeding once or twice a day is more likely to allow pathogens to enter the ileum or undigested material to enter the hindgut and allow potentially pathogenic bacteria therein to multiply. It is probably for this reason that ad-libitum teat feeders for calves, whether using warm milk from dispensing machines or cold, acidified milk from buckets, have become accepted so widely in recent years. A recent unpublished survey of over 10 000 calves made by British Denkavit revealed a death rate to 12 weeks of 2.5% for calves raised on teat feeders as against the national average of about 5.5% (Withers, 1952; Kilkenny and Rutter, 1975).

Cold·stress has been shown to increase gut motility and decrease digestibility and rumen retention time in sheep and in cattle (Young, Chapter 10). It is therefore possible that cold stress might contribute to enteric disease in calves by accelerating the passage of undigested food and undegraded bacteria into the ileum.

SYSTEMIC RESISTANCE TO DISEASE

It is undoubtedly the case that a severe acute stress such as heat, cold or any other trigger to the alarm phase of the general adaptation syndrome of Selye (1950) can seriously reduce the ability of an animal to resist a concurrent challenge by pathogenic organisms. Marcus and Miya (1956) showed that Coxsackie B virus killed 20 per cent of mice at 20 °C but 100 per cent of mice exposed to 4 °C on the day of the challenge. However, after 10 days of adaptation to 4 °C the resistance of the mice to challenge was comparable to that at 20 °C. This suggests that systemic resistance to infection is seriously reduced in the alarm, and presumably in the exhaustion phases of the general adaptation syndrome, but that after adaptation resistance may be normal (*see also* Webster, 1970). The stress imposed by Marcus and Miya was severe. There is, however, good evidence to show that less severe systemic cold stresses do not significantly affect the incidence of infection of man with the viruses of the common cold (Andrewes, 1967).

Comparable evidence for cattle is scanty. While Stott *et al.* (1976) showed that acute heat stress in calves reduced the production of immunoglobulins (IgG) and increased mortality, the evidence presented earlier in this chapter suggests that for initially healthy and well-fed calves the severity of cold stress experienced in the cool, temperate zones is likely to be slight and not sufficient to be a serious predisposing factor to respiratory disease. Clinical evidence supports this conclusion. In many areas of the mid-west of the USA and Canada, where the winters can be very cold, young dairy calves are raised out of doors in individual hutches where they can seek shelter from wind and precipitation but experience ambient temperature. Although these units are very labour-intensive, death rates from respiratory or enteric disease are negligible.

It is sometimes thought that the stress of fear imposed on calves transported in the first days of life might reduce their resistance to infection. However, Stephens and Toner (1975) showed that heart rate and cortisol secretion in calves were little affected by transport, which suggests that they did not find it a particularly frightening experience. Transport does, however, ensure that nearly all calves are infected (if not sick) on arrival at a rearing unit and if the period of transportation is prolonged many individuals may be in a state of exhaustion and thus more susceptible to disease.

IMPLICATIONS FOR HOUSING DESIGN

In the cool, temperate zone, air temperature is one of the least important aspects of the environment for housed cattle. It has been clearly shown that provision of suitable bedding and a draught-free environment can reduce the direct stress of cold on even the young calf to negligible proportions in terms of animal discomfort or feed costs. In terms of ventilation therefore there need be no compromise (as is necessary with poultry and pigs) between the need to provide sufficient fresh air and the need to ensure an adequate temperature lift within the building. Since air temperature within the building can be kept close to ambient, ventilation at all times can be kept at a rate sufficient to minimize the incidence of respiratory disease in the building. Unfortunately, one does not know for sure what this rate is. *Table 13.5* provides some empirical space and ventilation recommendations for calf houses taken in part from Mitchell (1976) and in part from his and my clinical experience. The minimum ventilation rate of $0.01 \text{ m}^3 \text{ s}^{-1}$ per calf ($0.6 \text{ m}^3 \text{ min}^{-1}$ per calf) is designed to provide six air changes per hour when air space per calf is 6 m^3. This can certainly be achieved within calf rearing units by natural or fan ventilation without causing draughts, defined as air movement $>0.5 \text{ m s}^{-1}$ at calf height. Excess air movement or cold downdraughts from air inlets can be avoided by, for example, building kennels for individual calves.

Table 13.5   SUGGESTED HEATING, AIR SPACE AND VENTILATION
REQUIREMENTS FOR CALVES TO 12 WEEKS OF AGE

---

1. Air temperature: unregulated

2. Minimum air space per calf: 6 m³

3. Ventilation rate: 0.01–0.03 m³ s⁻¹ per calf

4. Air changes per hour: 6–18

5. Air movement at calf height: < 0.5 m s⁻¹

These minimum space and ventilation standards are no guarantee of
freedom from serious epidemics of respiratory disease. Moreover, despite
the implications of equation 13.3, stocking densities allowing less than 6 m³
per calf appear to be associated with a very high incidence of respiratory
disease even if ventilation rate is greatly in excess of the minimum suggested
(Miller *et al.*, 1978). It is possible that in these circumstances short-range
transmission of infected droplets assumes a major importance.

The factors that determine air movement and mixing in a building are
considered elsewhere in this book. In practice, the effectiveness of ventilation systems for livestock buildings is usually assessed on the basis of criteria
such as CO_2 and ammonia, but these are rather poor criteria since they do
not reach harmful proportions in calf houses until the ventilation rate is well
below the minima recommended in *Table 13.5*. Moreover, they are inadequate markers of the transmission of respiratory disease for they reflect
diffusion of pathogens only in aerosol form. They provide no indication of
contagion or short-range transmission and they may or may not reflect the
extent to which the 'quality of fresh air' may affect pulmonary clearance of
pathogens. Such effects of the environment on the aetiology of respiratory
disease in young cattle have always have been recognized in scientific and
clinical investigations but they have usually been considered in a subjective
fashion or excluded from the scientific discussion as being too complex for
evaluation. My intention in this chapter has been to recognize the complexity
of the problem. However, in circumstances where the environment, in its
broadest sense, can be controlled or subjected to controlled experimental
variation (as for example in a calf rearing unit), I do believe that this
problem is capable of resolution by conventional scientific methods.

References

ALEXANDER, G. (1974). In *Heat Loss from Animals and Man*, pp. 173–203.
    Ed. by J.L. Monteith and L.E. Mount. Butterworths, London
ANDREWES, C.H. (1967). *Biometeorology*, **2**, 56–62
BEER, K., MEALHORN, G. and ARNOLD, H. (1975). *Mh. VetMed.* **30**, 406–409
BRUCE, J.M. (1979), *Farm Bldgs Prog.* **55**, 1–4
BUCKLAND, F.E. and TYRRELL, D.A.J. (1962). *Nature, Lond.* **195**, 1063–1064
DAVEY, A.W.F. and HOLMES, C.W. (1977). *Anim. Prod.* **24**, 355–362
EKESBO, I. (1968). *Acta Agric. scand.*, Suppl. 15

JENSEN, A.H. and CURTIS, S.E. (1976). *J. Anim. Sci.* **42**, 8–11

JERICHO, K.W.F. and DARCEL, C. Le Q. (1978). *Can. J. comp. Med.* **42**, 156–157

JERICHO, K., LANGFORD, E.V. and PANTEKOEK, J. (1977). *Can. J. comp. Med.* **41**, 211–214

JERICHO, K. and MAGWOOD, S.E. (1977). *Can. J. comp. Med.* **41**, 369–379

JOYCE, J.P. and BLAXTER, K.L. (1964). *Br. J. Nutr.* **18**, 5–27

KILKENNY, J.B. and RUTTER, J.M. (1975). *General assessment of the situation in the United Kingdom—'Perinatal ill-health in calves'*, pp. 30–38. Ed. by J.M. Rutter. Commission of the European Communities, Compton, Berkshire

KNEALE, W.A. and BASTIMAN, B. (1977). *Expl Husb.* **32**, 70–74

KRUEGER, A.P. and SMITH, R.F. (1958). *J. gen. Physiol.*, **42**, 69–82

LINTON, A.H., HOWE, K., PETHIYAGODA, S. and OSBORNE, A.D. (1974). *Vet. Rec.* **94**, 581–585

LOOSEMORE, R.M. (1964). *Vet. Rec.* **76**, 1335–1363

MacDONALD, M.A. and BELL, J.M. (1958). *Can. J. Anim. Sci.* **38**, 160–170

MARCUS, S. and MIYA, F. (1964). 'Effect of different routes of challenge of Coxsackie B virus on cold-stressed mice'. *Rep. AAL TDR-642*, Arctic Aeromedical Laboratory. H. Wainwright, Alaska

MARR, A. (1978). *Vet. Rec.* **102**, 132–134

MARTIN, S.W., SCHWABE, C.W. and FRANTI, C.E. (1975a). *Am. J. vet. Res.* **36**, 1105–1109

MARTIN, S.W., SCHWABE, C.W. and FRANTI, C.E. (1975b). *Am. J. vet. Res.* **36**, 1111–1114

MILLER, W.B., HARKNESS, J.W., RICHARDS, M.S. and PRITCHARD, D.G. (1978). In *Current Topics in Veterinary Medicine*, vol. 3, pp. 181–194. Ed. by W.B. Martin. Martinus Nijhoff, The Hague

MITCHELL, C.D. (1976). *Calf Housing Handbook*, 2nd edn. Scottish Farm Buildings Investigation Unit. Scottapress, Aberdeen

MORZARIA, S.P., MAUND, B.A., RICHARDS, M.S. and HARKNESS, J.W. (1978). In *Current Topics in Veterinary Medicine*, vol. 3, pp. 497–508. Ed. by W.B. Martin. Martinus Nijhoff, The Hague

MURPHY, P. (1978). Paper presented at a Symposium, *Injuries to Animals due to Floor Surfaces, November 1978, Fulmer Grange, Slough*

NORDSTROM, G.A. and McQUITTY, J.B. (1975). *Proc. Can. Soc. agric. Engng* 75–212

OSBORNE, A.D. (1975). In *Perinatal Ill-health in Calves*, pp. 60–63. Commission of the European Communities, Compton, Berkshire

OMAR, A.R. (1966). *Vet. Bull.*, *Weybridge* **36**, 259–273

PRITCHARD, D.G., CARPENTER, G.A., MORZARIA, S.P., HARKNESS, J., RICHARDS, M.S. and BREWER, J.I. (1980). *Res. vet. Sci.*, in press

ROGERSON, P.D. (1972). *Farm Bldg R & D Stud.* No. 3, 3–18. Scottish Farm Buildings Investigation Unit

SELYE, H. (1950). *Stress*. Acta, Montreal

SIMPSON, A.M., WEBSTER, A.J.F., SMITH, J.S. and SIMPSON, C.A. (1978). *Comp. Biochem. Physiol.* **60**, 251–256

SMITH, H.A., JONES, H.C. and HUNT, R.D. (1972). In *Veterinary Pathology*, 4th edn, p. 1104. Lea & Febiger, Philadelphia

STEPHENS, D.B. and TONER, J.N. (1975). *Appl. Anim. Ethol.* **1**, 233–243

STOTT, G.H., WIERSMA, F., MENEFEE, B.E. and RADWANSKI, F.R. (1976). *J. Dairy Sci.* **59**, 1306–1311

STRANGE, R.E. and COX, C.S. (1976). In *The Survival of Vegetative Microbes*, pp. 111–154. Ed. by T.R.G. Gray and J.R. Postgate. 26th Symp. Soc. Gen. Microbiol., Cambridge University Press, Cambridge

THOMPSON, G.E. and THOMPSON, E.M. (1977). *J. Physiol., Lond.* **272**, 187–196

THOMSON, R.G. and GILKA, F. (1974). *Can. vet. J.* **15**, 99–107

TOP, F.H. (1964). *Archs envir. Hlth* **9**, 699–723

VEIT, H.P., FARRELL, R.L. and TROUTT, H.F. (1978). *Am. J. vet. Res.* **39**, 1646–1650

WEBSTER, A.J.F. (1970). In *Resistance to Infectious Disease*, pp. 61–80. Ed. by R.H. Dunlop and H.W. Moon. Saskatoon Modern Press, Saskatoon

WEBSTER, A.J.F. (1974). In *Heat Loss from Animals and Man*, pp. 205–231. Ed. by J.L. Monteith and L.E. Mount. Butterworths, London

WEBSTER, A.J.F. (1980). In *Nutritional Physiology of Farm Animals*. Ed. by J.A.F. Rook. Longmans, London

WEBSTER, A.J.F., CHLUMECKY, J. and YOUNG, B.A. (1970). *Can. J. Anim. Sci.* **50**, 89–100

WEBSTER, A.J.F., GORDON, J.G. and McGREGOR, R. (1978). *Anim. Prod.* **26**, 85–92

WELLS, W.F. (1955). *Airborne Contagion and Air Hygiene.* Harvard University Press, Cambridge, Massachusetts

WILL, D.H., McMURTRY, I.F., REEVES, J.T. and GROVER, R.F. (1978). *J. appl. Physiol.* **45**(3), 469–473

WITHERS, F.W. (1952). *Br. vet. J.* **108**, 382–405

WRAY, C. (1975). *Vet. Bull.* **45**, 543–550

V

HOUSING ENVIRONMENTS FOR HOT CLIMATES

14

POULTRY HOUSING PROBLEMS IN THE TROPICS AND SUBTROPICS

W.K. SMITH
*Poultry Research Centre, Massey University, Palmerston North,
New Zealand**

Introduction

Over the past decade there has been a rapid expansion of poultry production
in some tropical countries. The problems of housing the various types of
poultry for intensive production in a variety of tropical climates are different
from those in cool and temperate climates. When we strive to achieve in the
tropics the potential production from each type of poultry, by the provision
of, among other things, an equable climatic environment, we have to manage the converse of the now refined procedures of heat conservation practised in cool climates—the protection from and dissipation of excess heat. In
certain tropical climates it may be possible to maintain indoor temperatures
cooler than outdoor temperatures without the use of mechanical refrigeration. Within the tropics, high dry-bulb temperatures need to be combated
with both high and low humidities, and a greater diversity of methods must
be used to achieve control of the indoor climate.

The problems of housing and managing poultry in the various tropical
climates have been considered by Castello (1964), Ben-Adam (1964), Bond
(1967), Ferguson (1970), Robinson (1978a–c), El Boushy and van Marle
(1978), and Costa and Hunton (1979). In addition, Williamson and Payne
(1978) and Oluyemi and Roberts (1979) have dealt with some aspects of
poultry housing in wider reviews of animal husbandry and poultry production in the tropics, respectively. Ferguson (1970) has considered the relevance to poultry housing of some of the methods used to diminish the effects
of heat in human dwellings in the tropics. Clearly, the design of buildings for
these climates is important, but although there is a wealth of theoretical and
practical knowledge of housing, this knowledge is applied irregularly.
Designers and builders make avoidable mistakes even today, owing to
inadequate appreciation of local conditions and the needs of the birds.
Again, the majority of tropical countries have areas that may be climatically
suited to poultry production, but which are geographically isolated from
towns or conurbations. Poultry units tend to be sited near the population
centres, which may be climatically less suitable.

In this chapter I shall first attempt to differentiate between different types
of tropical climate, so that special problems in particular climates can be

* Present address: West of Scotland Agricultural College, Ayr.

easily stated. Attention will then be focused on the problems of housing poultry in the tropics and the effects of high ambient temperatures on production. Lastly, I shall consider the means by which the climates in the tropics can be modified in order to allow a higher level of production from poultry.

Tropical climates

It is frequently convenient to classify tropical climates into two groups—hot wet climates and hot dry climates (Ferguson, 1970). This is a simplistic view, and the two types of climate may be experienced in the same locality at different times of the year. It is neither sensible nor practically rewarding to plan housing for one climate type alone.

In planning to cope with the heat from the radiation of the sun, designers may overlook other aspects of climate such as lighting, wind and rain. Although these latter difficulties are not confined to the tropics, the engineering, construction and management solutions to them must be compatible with the solution to the problems of heat. Conversely, it is universally recommended that houses should be aligned east–west so as to minimize the heating effect of short-wave radiation on walls, but this may involve missing out on the cooling effect of a local wind, something that could on balance be of greater benefit than the ideal house alignment. The problem of intense short-wave radiation on walls may often be diminished by shading with a roof overhang or trees and shrubs, a solution that may be applied with benefit on the coast of Tanzania, for example.

Even if a designer is equipped with a list of criteria to help him choose one or a combination of solutions for a building in a particular locality, in many instances the cost of a given solution may exclude it from consideration. This applies, however, to every building in its planning stage, and for a given cost the efficacy of the finished building depends entirely on the skill of the designer.

A more fundamental approach in the context of this chapter is to examine in more detail the classification of tropical climates. For each climate type the best solution to housing poultry may then be applied, within reason, wherever that climate type occurs.

CLIMATE CLASSIFICATION

Latitude zones

The tropics may be broadly defined by latitude zones. However, the classical concept that the 'tropics' lie between the tropics of Cancer and Capricorn has little relevance in the study of climates. Three broad zones may be defined (Strahler, 1965):

(1) Equatorial: a zone extending 10° north and south of the equator. Insolation is intense throughout the year, and day and night are roughly of equal duration.

(2) Tropical zones north and south: zones astride the tropics of Capricorn and Cancer in the latitudes 10° to 25°. In these zones the sun's path lies close to the zenith at one solstice but is appreciably lower at the other solstice. There are marked seasonal cycles but the annual total insolation is large.
(3) Subtropics: these are transitional areas between the tropics and the cooler mid-latitudes, lying between 25° and 35°; they may extend a few degrees poleward or equatorward of these parallels.

About 57 per cent of the earth's land area lies within the zones from 35°S to 35°N.

First-level classification

The Koeppen–Geiger system of climate classification, although it has its shortcomings and has undergone substantial modifications, seems to have achieved wide acceptance in atlases and geography books (e.g. Anon., 1968; Anon., 1976; Strahler, 1965; Critchfield, 1974). The climates that are of broad importance when we consider the tropics are those designated below as A, B and C. A and C are defined by average temperatures, while B is defined by precipitation : evaporation ratios.

A—Tropical climates with each month having an average temperature above 18 °C; there is a large annual rainfall, which exceeds the annual evaporation.
B—Dry climates with mean annual temperatures <18 °C (k subclass) or >18 °C (h subclass) and in which the potential evaporation exceeds rainfall; there is no water surplus and no permanent streams.
C—Warm mesothermal climates in which the mean temperature of the coolest month is between –3 °C and +18 °C.

Second-level classification

The second level of classification is based on the amounts and distribution of rainfall.
   Type A climates have subdivisions f, m and w.

Af —climates with no dry season;
Am—climates with a short dry season, with monsoon-type rainfall in the wet season;
Aw —climates in which rainfall of the driest month of the cold season is one-tenth or less of the rainfall of the wettest month of the hot season.

Type B climates have subdivisions W and S based on precipitation : evaporation ratios, but generally the division is as follows:

BW—climates with a rainfall of <250 mm year^{-1} (arid);
BS —climates with a rainfall of 375–750 mm year^{-1} (steppe).

Type C climates have subdivisions s, w and f.

Cs —climates with a dry summer in which the rainfall in the warmest month
      is less than one-third of the rainfall in the wettest month of the cold
      season, and is less than 40 mm;
Cf —as for Af;
Cw —as for Aw.

Third-level classification

The third level of classification is based on temperature, and subdivisions
are recognized in the B and C climates.
    The type B climates have two subclasses h and k, the temperature char-
acteristics of which are given above.
    The type C climates have two subclasses a and b:

a—climates whose warmest summer month has an average temperature of
    >22 °C
b—climates whose warmest summer month has an average temperature of
    <22 °C.

    In order to provide a 'boundary' between the subtropics and the cooler
climates, in the context of this chapter, I have used arbitrarily the boundary
between h and k subclasses in type B climates and the boundary between a
and b subclasses in type C climates.

CHARACTERISTICS OF TROPICAL CLIMATES

The tropical and subtropical climates defined above can be described very
briefly (after Strahler, 1965).

Wet equatorial—Af, Am.   These are generally in the region of ±5° of the
equator, where temperatures average close to 27 °C the year round, with an
annual range of 2–4 K only, which is less than the diurnal range of 8–11 K.
The rainfall exceeds 2000 mm year^{-1}.
    In Asia between 10°N and 25°N or on the west coasts of India and
south-east Asia the climate is warm and wet with a large annual rainfall but
with a dry month having less than 60 mm of rain. The dry season is in the
low-sun period when north-east monsoon winds prevail.
    Along the east coasts of Central and South America, Madagascar, south-
east Asia, the Philippines and north-east Australia, in the latitudes 10°–25°,
the easterly trade winds produce heavy rainfall when the moist air rises up
coastal slopes. During the high-sun solstice period, tropical cyclones are
frequent in these areas. In these higher latitudes, the annual average temperatures may be 5–7 K cooler than in the zone within 5° of the equator.

Tropical desert and steppe—BW, BS.   These very dry climates dominate
much of the land areas in the latitude belts from 15° to 35°. These regions
contain truly arid zones with a rainfall of <250 mm year^{-1} (BWh) and the
steppes, or semi-arid zones, with a rainfall of 375–750 mm year^{-1} (BSh). In
the extremely arid regions the relative humidity falls to 15–20% at 1300
hours in the hottest months. Rainfall is sporadic and variable from year to
year. Temperatures in the high-sun months will have a daily mean around

37.5 °C, with a mean daily maximum of 45 °C. The annual range of monthly average temperatures is 17–22 K, and the average daily range is similar.

The wetter steppe climates border the tropical deserts to the north, south and east (and locally with altitude). Equatorward the steppe climates merge with a tropical wet–dry climate (Aw; *see below*). Poleward the steppes graduate into a Mediterranean climate (Cs) in many places. The steppes equatorward have a higher annual average temperature than desert climates, and consequently a smaller annual range. At the border of Bs and Aw climates, the hot, wet season may be potentially the most difficult to manage stock in.

Tropical wet–dry—Aw. Between the wet Af and Am and dry BW and BS climates there are intermediate zones having a tropical wet–dry climate with a wet season during the high-sun period and a dry season at the low-sun period. In Central and South America, Africa and Australia, Aw climates occur in the latitude belts of 5° to 25° while in Asia they are pushed to 10°N–30°N. Except in Asia, the higher dry-bulb temperatures occur in the period approaching high sun or before the wet season, and the average daily temperature in this season is 27–30 °C. In the low-sun seasons of wet–dry climates temperatures drop by 4–6 K. The Asian Aw climate is controlled by the monsoon and there is an extreme contrast between wet and dry seasons.

The poleward extension of the Aw climate is the Cw climate, which has a greater annual temperature range and cooler low-sun temperatures. The Cwa climates are included in the subtropical groups as having warmest monthly average temperatures of 22 °C or above.

Subtropical, humid summer—Cf. In the south-east of the USA, eastern China, Japan, south-east Brazil, Uruguay, eastern Argentina and eastern Australia, rainfall is adequate all year but is highest in the warm months. Temperatures average between 25 and 30 °C in summer, and at times the climate resembles a wet equatorial one. The annual range of temperature is 10–20 K so the winters are mild to cool and humid. In south-east Asia the climate is modified by intensive monsoon development, the winters are drier and the summer is wetter.

Subtropical, dry summer—Cs. In latitudes 30°–45° on western coasts are zones subject to a wet winter and a dry summer, in contrast to the Cf climate. Such zones are located between a dry west-coast tropical desert on the equatorial side and a wet west coast on the poleward side. The warmest months' average temperatures vary from 22 to 27 °C. The climate is typified in the Mediterranean lands and is commonly called by that name. The annual range of temperature varies from 13 to 16 K.

The problems of poultry production in the tropics

At any particular geographical location the combination of total daily solar radiation, solar altitude and azimuth, cloud cover, rainfall distribution and intensity, wind speed and direction, ground cover and shade, and the thermal characteristics of the housing, will determine the 'effective temperature' for the birds in an enclosed space. There are some features of tropical climates that cause grave problems to a producer, but these are by no means unique to the areas within the arbitrarily chosen boundaries considered here.

A high rainfall, for instance, is a feature of cool, wet climates on the west coasts of South America, Canada and southern New Zealand. Scarcity of water is not confined to tropical deserts: the west coast of South America, for instance, has areas of zero rainfall as well. Destructive ants or beetles are also found in temperate and cool regions. The winds in the turbulent 'mid-forties' can be as destructive as the tropical cyclones.

The main problem in the tropics is one of heat, manifested as a high wet-and/or dry-bulb temperature. The magnitude of the problem in a particular climate can be gauged from climate records. The temperature records may be summarized to provide outdoor minimum or maximum temperatures for a specific month or season which may be utilized in designing a building, and which are termed *design temperatures*. As the design temperatures are based on the frequency distribution of hourly temperatures, the design temperature is expressed as that exceeded during a certain percentage of hours during winter or summer. Those which are commonly used are the 1%, 2½% or 5% design temperatures. For summer they refer to the percentage of hours during which the temperature will be greater than the design temperature, while for winter they refer to the percentage of hours during which the temperature will be less than the design temperature. The design dry-bulb temperatures are useful for summer and winter, while the design wet-bulb

Table 14.1   THE 2½% DESIGN TEMPERATURES FOR THE SUMMER AND WINTER MONTHS, AND THE DAILY RANGE OF DRY-BULB TEMPERATURES AT SELECTED LOCATIONS WITH TROPICAL AND SUBTROPICAL CLIMATES

Location		Climate type	2½% Design temperature (°C)			
			Winter dry-bulb	Summer dry-bulb	Summer wet-bulb	Dry-bulb daily range
Kisangani, Zaire	0 °N	Af	20.0	32.8	26.7	10.6
Singapore	1 °N	Af	22.2	32.8	27.2	7.8
Belem, Brazil	1 °S	Am	21.7	31.7	26.1	10.6
Columbo, Ceylon	7 °N	Am	21.1	31.7	26.7	8.3
Manila, Philippines	15 °N	Aw	23.3	33.3	27.2	11.1
Recife, Brazil	8 °S	Aw	21.1	30.6	25.5	5.6
Baghdad, Iraq	33 °N	BWh	1.7	43.9	22.2	18.9
Karachi, Pakistan	25 °N	BWh	10.6	36.7	27.8	7.8
New Delhi, India	29 °N	BSh	5.0	41.7	27.8	14.4
Accra, Ghana	6 °N	BSh	20.6	32.2	26.1	7.2
Brisbane, Queensland	29 °S	Cfa	8.3	31.1	24.4	10.0
Buenos Aires, Argentina	35 °S	Cfa	1.1	31.7	24.4	12.2
Perth, W. Australia	32 °S	Csa	5.6	35.6	23.3	12.2
Athens, Greece	38 °N	Csa	2.2	33.9	21.7	10.0

Latitudes are given to the nearest degree. Climate type according to the Koeppen–Geiger classification system
Reprinted with permission from the 1977 Fundamentals Volume, *ASHRAE Handbook and Product Directory*

temperatures are of particular value for summer conditions. The summer and winter design temperatures at 14 locations representing seven distinct climate types are given in *Table 14.1*. These data are the least that are needed before making an assessment about the suitability of a locality for a poultry unit.

It is possible that in Af or Am climates, where the diurnal variation of dry-bulb temperature may be greater than or equal to the annual range, poultry will become acclimatized to the constantly high and uniform temperatures. The egg production of modern strains of laying hens in Af climates can be comparable with that of the same strains in cool and temperate climates (Sugandi, Bird and Atmadilaga, 1975). The egg production of the Shaver 288 strain of laying hen achieved on a commercial egg-laying unit in an Am climate in Brazil (S.M. Costa, personal communication) is compared in *Figure 14.1* with that achieved at the West of

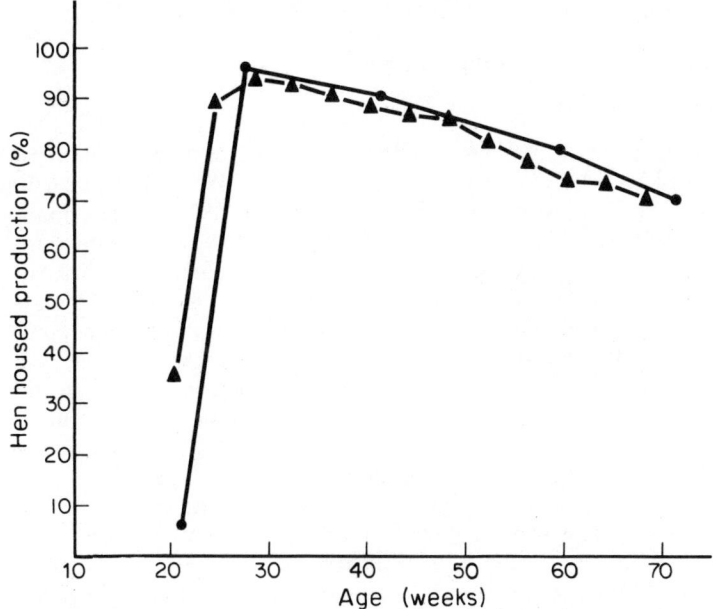

Figure 14.1 The performance of the same strain of laying hens in a tropical and a cool climate, recorded during 1978/79. Location (●) Belem, Brazil, (▲) Auchincruive, Scotland. (Courtesy of S. M. Costa, Shaver Breeding Farms Ltd, Ontario, Canada; and P. Dun, Poultry Department, West of Scotland Agricultural College, Ayr, Scotland)

Scotland Agricultural College (P. Dun, personal communication) during 1978/79. The performance of the hens was similar at the two locations. The flocks produced 294 and 290 eggs in 52 weeks from the onset of production with an average egg weight of 59.5 g and 60.9 g for the Brazilian and Scottish flocks respectively.

Where the seasonal rainfall differences become marked, as in some Am and in most Aw climates, the wet season may depress productivity. In the Hawaiian Aw climate, egg production of modern strains is reportedly less than that in a drier climate with the same dry-bulb temperature range (Francis, Reid and Wilson, 1972). Similarly, according to Howes (1963), the onset of the wet season in the Am climate of Trinidad causes reduced egg production, the depression depending on the age of the bird at the onset of the wet season. Although the onset of the wet season may bring a slightly lower dry-bulb temperature in Am and Aw climates, there is a marked fall only in regions with a high monsoon rainfall; the increase in the wet-bulb temperature accompanying the wet season may be the cause of the decrease in production. Rao, Kumarasamy and Rathnasabapathy (1966) found that egg production in an Aw climate was negatively correlated (–0.898) with dry-season temperatures, which ranged from 30.5 to 41 °C, but that humidity was without significant additional effect. However, egg production reached a minimum in the wet south-west monsoon season. The effect of the wet monsoon season was highlighted by Quarishi, Thatte and Desai (1973), who found that flocks of laying hens commencing lay during the wet monsoon season had a lower egg production.

The effects of high ambient temperatures

RATE OF EGG PRODUCTION

Under constant temperatures, egg production will decrease when the temperatures exceed 26–30 °C (Wilson, McNally and Ota, 1957; Howes, Grub and Rollo, 1961; Francis, Reid and Wilson, 1972; Smith, 1973; Zimmerman, Snetsinger and Greene, 1975; Lillie *et al.*, 1976). However, when temperatures are variable, as they are under all tropical climates, it is difficult to establish the daily maximum above which egg production is affected and, if egg production is reduced, what magnitude of diurnal variation is needed to alleviate or negate the reduction. One of the difficulties in assessing the effects of high ambient temperatures on any production trait is to reconcile the results of studies using constant and diurnally fluctuating temperatures. Following extensive field studies in Guatemala, Squibb (1959) suggested that the failure of high temperatures and humidities to affect egg production adversely must lie in the magnitude of the observed diurnal ranges. He concluded that heat stress was greater when the highest temperatures were at the same time in association with a narrow diurnal variation of temperature. There have been a number of studies of the effect of cycling temperature regimes. Wilson, Itoh and Siopes (1972) found that a diurnally cycling regime between 10 and 30 °C depressed egg production as much as did a constant 34 °C. By contrast, Mowbray and Sykes (1971) have shown that the egg production from hens in a diurnal cycle ranging from 13 to 35 °C was similar to that from hens in a regime varying between 10 and 15 °C. Mueller (1961, 1967) found that egg production of hens in a 13–32 °C regime was similar to that of hens held at a constant 13 °C, while Francis, Reid and Wilson (1972) reported that a regime of 21–38 °C was detrimental to production. It seems apparent that as temperatures increase above a threshold point somewhere between 26 and 29 °C, a diurnally cycling temperature is needed to maintain 'normal' egg production. As the daily maximum increases, the magnitude of the diurnal variation must increase to avoid a depression of production. It is also apparent that when the daily maximum rises above an upper threshold, production is depressed regardless of the magnitude of the diurnal variation.

A wide diurnal variation does have benefits for the heat-stressed hen, as Squibb (1959) suggested. Howes, Grub and Rollo (1965) have shown this experimentally with hens housed at 26.7 °C and exposed twice a week for a duration of 2 hours to 32.5 °C or 37.5 °C over a six-month period. After the high-temperature exposure the hens were given either 26.7 °C again, or an immediate 2 hour period of cooling to either 18 °C or 10 °C and followed by a return to 26.7 °C. The results are shown in *Figure 14.2*. Production

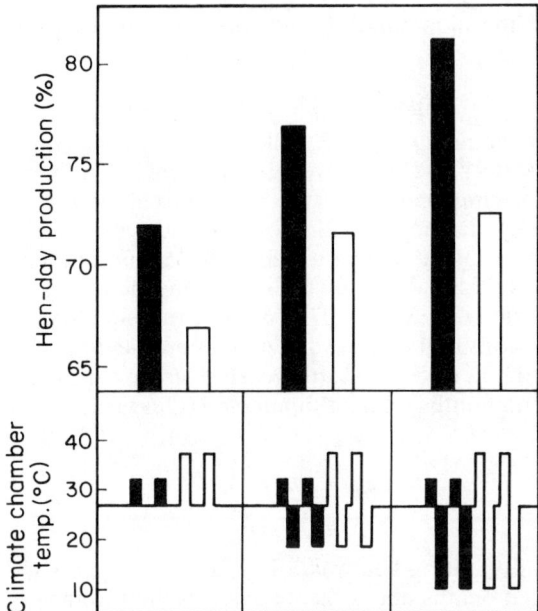

Figure 14.2 The egg production of White Leghorn pullets over a six-month period housed in climate chambers held at 26.7 °C and given a two-hour heating exposure twice a week to 32.5 °C (■) or 37.5 °C (□) with or without an immediate two-hour period of cooling to 18 °C or 10 °C dry-bulb temperature, as indicated by the histograms. (Data from Howes, Grub and Rollo, 1965)

responded to the increasing cooling at both levels of heat exposure. The extra heat stress of 37.5 °C depressed production substantially below that observed with the more moderate heat exposure.

At increasingly higher constant temperatures the depression of production is progressively more severe. Howes, Grub and Rollo (1965) found that depression of production, as measured by the number of days taken for egg production to decline to 50 per cent, is accelerated as temperatures increase. The higher temperatures act as though the bird were undergoing an increasing rate of senescence. Within a strain of hens, those with a smaller body weight are able to produce more than heavier birds under continued exposure to high temperatures (Horst and Petersen, 1975).

EGG WEIGHT

The effect of high environmental temperatures on egg size has been fully reviewed by Smith (1974). He concluded that above 26 °C 80 per cent of the decrease in egg weight was due to heat stress and only 20 per cent to inadequate energy in the diet. Heat stress was the main cause for the loss of shell weight, the reduction of which is due to alkalosis caused by excessive loss of CO_2. In a study to determine the cause and magnitude of the weight loss, Smith (1969) concluded that it depended on:

(1) the energy balance set by the interaction between genotype, environment and diet; and
(2) the energy expense of thermoregulation and the General Adaption Syndrome.

At temperatures above 10 °C, egg weight shows a gradual decrease (Wilson, Itoh and Siopes, 1972; Emmans and Dun, 1973; Lillie *et al.*, 1976; Haughen, 1977). From 26 to 35 °C, the decrease in egg weight accelerates (Smith and Oliver, 1972). When the maximum temperature in a diurnal cycle rises above the 26–29 °C threshold, leading to a depression in the rate of egg production, egg weight is maintained at that expected for a constant temperature equal to the average of the diurnally cycling regime (Hutchinson, 1953; Mueller, 1967; Mowbray and Sykes, 1971; Wilson, Itoh and Siopes, 1972). Under tropical conditions that cause a decrease in production, as in the south-west monsoon and hot, dry season, the yearly average egg weight of pullets hatched in different months remains unaffected (Quarishi, Thatte and Desai, 1973).

MORTALITY

In a study of poultry mortality during heatwaves in the USA, Squibb and Wogan (1960) concluded that deaths due to heat stress resulted from very sharp increases in daily maximum temperatures following periods of more moderate environment. Other variables such as relative humidity and wind speed were not correlated with reported mortality. Earlier, Squibb (1959) had noted that adverse effects were not observed where very high temperatures were recorded because of wide diurnal temperature ranges. More recently, Mannel (1971) reported that in Iraq, long periods with maximum temperatures of 37–38 °C caused higher mortality than short peaks of 40–41 °C, regardless of house design. In a similar climate in Saudi Arabia, Appleman and Bonhof (1971) found that mortality resulted from sudden rises in relative humidity. The climate chamber work of Wilson, Itoh and Siopes (1972) suggests that pullets may be more susceptible to heat death at the onset of or just after attaining sexual maturity.

FERTILITY AND HATCHABILITY

Marked seasonal differences have been reported in the fertility and hatchability of eggs laid in Aw climates (Kothandaraman, 1973; Kumaraswamy and Rathnasabapathy, 1973). The male may be less affected than females by temperatures up to 30 °C (Long and Godfrey, 1952; Perek and Snapir, 1963), but both semen production and the fertilizing capacity of that semen may be reduced when temperatures reach 38 °C (Boone and Huston, 1963).

In a diurnally fluctuating environment of 21–38 °C, Clark and Sarakoon (1967) found that the high temperatures did not affect semen volume or sperm concentration as compared with a control of 21 °C. The fertility of females was reduced in the warm environment, even if they were inseminated from males held at 21 °C. Conversely, the fertility of females held at 21 °C was unaffected by insemination from males held in the warm environment. The lower sensitivity of the male component of fertility to high environmental temperatures is in agreement with the earlier work of Huston and Carmon (1958). The insensitivity of the fowl testes to high temperatures is in direct contrast to the sensitivity of most mammalian testes, which need to be kept cooler than the deep-body temperature to maintain normal production of semen (*see* Chapter 16).

GROWTH

Meat chickens

After the brooding is completed at 3–4 weeks, maximum growth rate of meat chickens is achieved with constant temperatures in the range 15–20 °C (Ota and McNally, 1965; Yoshida *et al.*, 1968; Ota, Whitehead and Lillie, 1973; Deaton, Reece and McNaughton, 1978). For each increment above 20 °C and decrement below 15 °C the change in growth rate is similar but, over a wide range in temperature, feed efficiency improves as temperature increases. When the average daily temperature is kept as near as possible to 'optimum' for growth from hatching to marketing age, growth and feed efficiency will be affected if the diurnal variation in temperature exceeds 11 K (Siegel and Drury, 1970). However, if only the post-brooding stage is considered the diurnal variation may be widened to about 16 K without seriously affecting growth rate or feed conversion efficiency (Griffin and Vardaman, 1970).

Above an average temperature of 20 °C, growth is depressed even if the daily minimum is in the zone of temperatures for maximum growth rate. Maximum dry-bulb temperatures of up to 40 °C can be tolerated without increased mortality provided the diurnal variation is at least 16 K (Harris, Dodgen and Nelson, 1974). For daily maximum temperatures above 30 °C, food intake and growth increase as the diurnal variation widens (Griffin and Vardaman, 1971a; Harris, Dodgen and Nelson, 1974).

In tropical climates the roofs of poultry houses are often inadequately insulated. Although dry-bulb temperatures may not be excessive, the long-wave radiant heat load from the roof can increase the black-bulb temperature sufficiently to cause growth depression and substantial mortality, with the latter more severe in males (Griffin and Vardaman, 1971b).

At temperatures above 29 °C, a relative humidity that is constantly above 80% or below 40% will permanently affect growth. Abrupt increases in temperature combined with low (40%) or high (80%) relative humidity cause a marked depression of growth and feed efficiency, the hot and wet conditions having a greater depressing effect (Godfrey and Winn, 1966).

Laying chickens

In much of the early work on the effect of high environmental temperatures on growth, light breeds of chickens, such as White Leghorns, were compared with heavy breeds, such as White Plymouth Rocks and New Hampshires (Kempster and Parker, 1936; Kempster, 1941). During summer, when average temperatures increased above 26.7 °C, growth rate fell below normal (Kempster and Parker, 1936) and the depression of growth was more pronounced with older, and therefore heavier, birds (Kempster, 1941). More recently, Wilson (1967) reported a depression of the growth of White Leghorns in constant temperatures above 21 °C. Stockland and Blaylock (1974) demonstrated that pullets reared at 29.4 °C had a smaller body weight at 20 weeks, later onset of sexual maturity and subsequently produced smaller and fewer eggs than pullets reared at 18.3 °C. In the Cwa climate of the south-eastern USA some concern has been expressed in the industry that spring-hatched pullets are too small at sexual maturity to sustain maximum egg production (McNaughton *et al.*, 1977).

Turkeys

The growth rate of turkeys is best within the temperature range 10–20 °C (Hellickson *et al.*, 1967; Albuquerque *et al.*, 1978); a range twice as wide as that suited to meat chickens. The depression of growth by high temperatures (up to 35 °C; Albuquerque *et al.*, 1978) is as marked with turkeys as it is with meat chickens. Up to a certain stage of growth, feed efficiency improves as temperature increases, and in the later stages of growth, feed efficiency reaches a maximum somewhere between 21 and 27 °C (Hellickson *et al.*, 1967; Albuquerque *et al.*, 1978).

Housing in tropical climates

In all tropical climates the adverse effects of excessive heat on production can be reduced by appropriate housing and by cooling the enclosed spaces. The nature of the building and the means of cooling employed will depend largely on the finances available. At the outset the choice of site, situation of the building on the site, and landscaping can substantially diminish the problem of excess heat. A careful choice of site might take advantage of lower daily maximum temperatures and a prevailing local wind. A site suited to growing a green soil cover and shading trees would be especially advantageous in a B or Cs climate.

A variety of methods and materials can be used in building to reduce or minimize the effect of the hottest part of year and day. There are also many ways a building and stock can be cooled. In Af, Am, Aw, BWh and BSh climates cooling may be a daily routine. For poultry housed in C climates cooling may be needed in only the hottest months or brought into use only on the hottest days.

THE BUILDING AND BUILDING SITE

The exchange of radiation between a building and the surroundings is shown diagrammatically in *Figure 14.3*. It can be shown by calculation alone that in the tropics a vertical surface receives the least energy and a horizontal

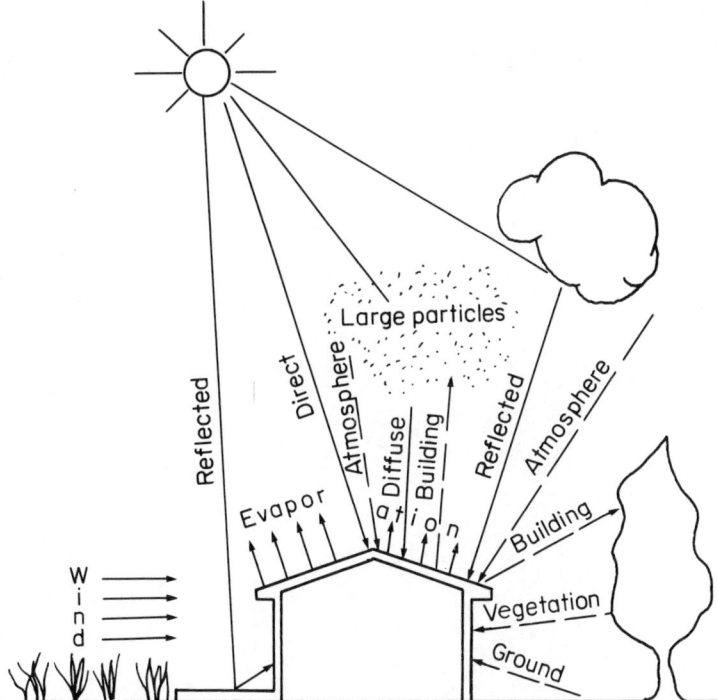

Figure 14.3 Channels of heat transfer for a building and the environment. (———) Short-wave radiation; (– – –) long-wave radiation

surface the most. Furthermore, a vertical surface facing the equator receives less direct solar radiation than one facing east or west. *Table 14.2* shows the mean daily direct insolation on vertical surfaces facing equatorward, east or west respectively, and the total insolation on a horizontal surface computed for various latitudes and clear skies (Anon., 1975). For most sites a building would receive less direct radiation if orientated east–west rather than north–south (Bond, 1967; Costa and Hunton, 1979). Diffuse radiation contributes one-fifth to one-sixth of the total irradiance and is available even when the sky is overcast. Sometimes a small amount of cloud in the sky increases the total irradiance (Monteith, 1973), but usually clouds reduce the total solar irradiance. At the same time the loss of heat from the earth is reduced, thereby reducing the diurnal range. The sky temperature is less than ambient by about 20 K over the range 0–20 °C (Monteith, 1973); at 38 °C it is about 16 K cooler and at 43 °C about 12 K cooler (Bond, Kelly and Heitman, 1958). Nevertheless, for a bird in shade, exposure to a clear sky will allow greater long-wave radiation losses than a shaded yet warmer wall. With complete cloud cover the cloud surface is frequently only about 2 K cooler than the ambient temperature. Ambient temperatures may exceed 40 °C on a clear day, and if cloud cover becomes complete in the late afternoon, the low long-wave radiation losses will result in only a small nocturnal fall in temperature. Under such conditions it would be wise to operate cooling devices all night.

248

Table 14.2 THE DIRECT SOLAR INTENSITIES (in W m^{-2}) ON VERTICAL SURFACES AND TOTAL INTENSITY ON A HORIZONTAL SURFACE AT THE EQUINOXES

Orientation	Latitude (deg)	Sun time (h):										
		7	8	9	10	11	12	13	14	15	16	17
South (N. Hemisphere) North (S. Hemisphere)	0	0	0	0	0	0	0	0	0	0	0	0
	10	25	65	105	135	155	160	155	135	105	65	25
	20	45	125	200	260	300	315	300	260	200	125	45
	30	60	175	285	370	425	445	425	370	285	175	60
	40	70	210	350	460	530	550	530	460	350	210	70
East	0	500	645	595	445	240	0	0	0	0	0	0
	10	495	640	590	440	240	0	0	0	0	0	0
	20	480	630	585	435	235	0	0	0	0	0	0
	30	450	605	570	425	230	0	0	0	0	0	0
	40	415	570	545	410	220	0	0	0	0	0	0
West	0	0	0	0	0	0	0	240	445	595	645	500
	10	0	0	0	0	0	0	240	440	590	640	495
	20	0	0	0	0	0	0	235	435	585	630	480
	30	0	0	0	0	0	0	230	425	570	605	450
	40	0	0	0	0	0	0	220	410	545	570	415
Horizontal	0	185	450	685	870	1000	1015	1000	870	685	450	185
	10	180	440	670	855	985	1020	985	855	670	440	180
	20	170	415	635	810	925	965	925	810	635	415	170
	30	150	375	575	735	835	870	835	735	575	375	150
	40	130	315	495	630	725	750	725	630	495	315	130

Reproduced with permission from Section A6 of the *CIBS Guide* published by the Chartered Institution of Building Services

The atmosphere gains most of its energy indirectly, via convection and thermal radiation from the earth's surface. In the tropics the surface of concrete can reach temperatures well above 60 °C. The temperature gradient over the first 2–3 m, however, is likely to be of the order of 30 K. With enclosed, fan-ventilated buildings an air inlet well above ground level is desirable in these circumstances. However, the temperature of air close to turf is essentially the same as ambient (Bond, 1967) and large areas of grass, while green, can provide cooler air downwind (Ferguson, 1970).

The direct irradiance of a wall can be reduced substantially by shades on the building. For most of the day a roof overhang will shade a vertical wall from direct radiation, but reflected short-wave radiation and long-wave radiation emitted from the ground and surroundings can still represent a substantial heat load. The amount of overhang needed to shade a wall, especially one facing east or west, can be readily calculated from solar altitude angles for a specific time, date and latitude. For instance, in the latitudes 20° to 35° for the 90 days of the highest noon sun (±45 days of the summer solstice), the sun's zenith angle is above 45° from 0900 to 1500 hours.

A shade with an angle of 45° allows the use of the minimum of materials to shade the whole wall. With the sun above 45° zenith angle, the shade length needed is 70 per cent of the wall height. Before and after 0900 to 1500 up to half of the wall would be unshaded. From 0700 to 0900 hours and 1500 to 1700 hours, when the radiation intensity on east- and west-facing walls is greatest (Anon., 1975), only the bottom quarter of the wall would be exposed.

SOURCES OF RADIANT HEAT WITHIN A BUILDING

The sources of radiant heat load on a chicken outside and within a shelter have been investigated by Bond (1967). By intercepting the sun with a shade the radiant heat load may be reduced by 30 per cent. Within the boundaries of the shade the radiant heat load on a chicken comes from four sources: the atmosphere, the roof, the shaded ground and sunny ground. The amount of radiation received by a bird under a shade depends on the fraction of total energy from the various sources intercepted by the bird, which in turn depends on the bird's height above the ground. The higher it is, the more the total radiation load increases. Long-wave radiation represents about 80 per cent of the total radiation heat load. The sunny-ground component is at a maximum when the ground surface is bare; a cover of turf will reduce this load by about a quarter.

The heat load on poultry will be at a minimum under shade when the distance from the ceiling is at a maximum. Consequently, birds on the floor will be relatively cooler than those in elevated cages. Under certain production situations it will be advantageous to keep birds on litter; this is a preferred practice in the hotter Aw climate regions of India, for instance (A. Mitra, personal communication).

The radiant heat load may be further reduced by shading the building with trees and shrubs. Trees capable of shading the walls and roof would be most advantageous for buildings with a north–south orientation, so placed to take advantage of local prevailing winds. In addition to the shade offered by trees and shrubs, the microclimate they produce may be undervalued. As soon as the ground surface is covered with vegetation, the heat-absorbing surface, previously the soil, is transferred to the top of the plants. Solar radiation absorbed by the foliage is largely carried away by forced convection, with only a fraction being reradiated to cooler bodies (Deering, 1952).

The high foliage density of some shrubs may prevent air moving through them. This is generally a disadvantage, though such plants could provide a valuable windbreak in certain locations (Ferguson, 1970). In contrast, loosely foliated shrubs can shade the walls of a building while allowing the passage of air. The layer of air beneath trees and moving through shrubs receives the water vapour of transpiration, which has a substantial cooling effect, so that if such shading trees and shrubs are present, shade temperatures are less than on open turf (Deering, 1952).

ENCLOSING A SHELTER

As mentioned above, a substantial part of the radiation incident on a chicken beneath a shelter comes from the atmosphere and sunny ground. Where it is impractical to use shrubs to shade the perimeters of a shelter, the radiation may be reduced by intercepting the reflected short-wave and emitted long-wave radiation by walls. The ideal placement of the solid parts of the walls will depend on the location of the stock: on the ground or in cages. Depending on their separation, wooden slats will slightly reduce long-wave radiation, but substantially reduce short-wave radiation (Bond, 1967). The complete enclosure of a shelter with walls could reduce the radiant heat load by at least 10 per cent if the results of Hahn, Bond and Kelly (1963) are capable of extrapolation from three to four walls. Further reduction of the radiant heat loads on animals in a completely enclosed shelter may be obtained by the use of different materials in construction, reflective and low-emissivity paints with which to coat interior and exterior building surfaces, and the use of water to cool the building surfaces.

MATERIALS

The materials that can be chosen fall into two categories: those that affect the flow of heat from the exterior to the interior; and those which, by the nature of their surface properties, reduce absorption and emission of radiation. The flow of heat through a building material is governed by the thermal conductivity, k (in $W\,m^{-1}\,K^{-1}$), the density, ϱ (in $kg\,m^{-3}$), and the specific heat, c (in $J\,kg^{-1}\,K^{-1}$), of the material. The most significant property of a material with respect to heat flow is its thermal diffusivity, $k/\varrho c$, which has units of $m^2\,s^{-1}$. Walls made of materials with a low thermal diffusivity produce large time lags between the outdoor and indoor maximum temperatures, and reduce the amplitude of diurnal variation. Poultry buildings constructed with such materials would, without complicating factors, have a reduced diurnal variation compared with the outdoor climate. This could be advantageous in A and B climates, which have maximum temperatures likely to produce severe and possibly fatal heat stress. Human dwellings constructed along these lines have a reduced indoor variation in temperature; though Ferguson (1970) has pointed out that the dampening of the outdoor fluctuations in temperature depends on relatively low ventilation rates. Thus this type of construction may be applied only to poultry houses with very small flocks. With large flocks the much higher ventilation rates needed, according to Ferguson (1970), would reduce the effectiveness of the low-thermal-diffusivity walls, especially those of a high heat capacity, such as masonry. Under these circumstances the indoor dry-bulb temperature will be similar to that outside.

However, when the solar radiation is intense, the combination of the high outdoor dry-bulb temperature and the radiation governs the heat flow. The calculated dry-bulb temperature that would produce the same heat flow, were the radiation exchanges absent, is termed the 'sol-air' temperature. It is the difference between 'sol-air' temperature and the indoor temperature, together with the materials used in construction, that govern the flow of heat. It would always be preferable to use materials with a low rather than a high thermal diffusivity. For common construction materials the thermal diffusivities are in the order wood < asbestos cement < dense concrete < steel < aluminium (Hassall, 1973).

SURFACES AND SURFACE COATINGS

There is substantial scope for reducing the radiation, both short- and long-wave, to which the enclosed stock are exposed. All forms of radiation when striking a material are either reflected from, transmitted through, or absorbed by that material. All materials reflect, transmit and absorb incident radiation differently, but the sum of the proportions that are reflected, absorbed and transmitted is always unity. The differences depend on the absolute temperature of the material, its chemical and physical characteristics, and the wavelengths of the incident radiation. The absorbed energy is transformed into heat and is conducted away to cooler parts of the material or emitted as long-wave radiation. Each material has a different capacity to emit long-wave radiation, its *emissivity,* ε, being the ratio of its radiating power to the radiating power of a black body, for which $\varepsilon = 1.0$ (Anderson, 1977).

Ferguson (1970) has suggested that it would be disadvantageous to use highly reflective materials on ceilings because of the reflection of short- and long-wave radiation back onto housed stock. Ferguson (1970) and Williamson and Payne (1978) have also suggested that ceilings should be dark in order to reduce reflection. However, the dark ceiling might, on balance, reradiate more long-wave radiation than a highly reflective material. Robinson (1978a) has determined the radiant heat flow from various poultry house ceiling materials, during summer. Radiant heat incident from the highly reflective surface of aluminium foil was much less than from other surfaces, including dark ones. This was due in part to the low emittance value (0.05) of the aluminium foil surface.

Materials for and coatings on the outside of buildings would absorb less heat if they had a low absorptivity for short-wave radiation and also had a high emissivity. On the other hand, materials and coatings on the inside would reduce the radiation load on the housed stock if they had a high short-wave absorptance and a low emissivity. Anderson (1977) has arranged materials according to the absorptivity : emissivity ratio. Materials with a low ratio would be suitable for use on the outside of buildings. For instance, white paint on aluminium and whitewash on galvanized iron have ratios of 0.22 and 0.24 respectively. Materials with a high ratio would be suitable for use on the inside of buildings. For instance, new and old galvanized iron, and aluminium foil have ratios of 5.0, 2.9 and 3.0 respectively. There are some special materials that are particularly suited as solar-energy collectors because they combine the attributes of a high absorptance of solar radiation and low emittance of infrared radiation. For instance, black nickel oxide on aluminium has a ratio of about 15.

Were the high-ratio special materials competitively priced, then they could be used to advantage on the inside of animal houses in the tropics. Otherwise, galvanized iron is to be commended because of its universal availability, although where available, aluminium foil bonded onto plywood is a better alternative. A material with reinforced bituminous paper sandwiched between two aluminium foil layers, similar in construction to the roofing material tested by Robinson (1978a), has a special attraction because of its value in decreasing the summer *U* value of a wall or roof with one or more air spaces (Hassall, 1973). The radiation load on a bird in a house may therefore be reduced by use of appropriate materials on the inside and outside surfaces of a building.

Methods of cooling

In any particular building, the comfort and performance of a bird can be improved by cooling the air or building by water evaporation and by increasing air movement.

EVAPORATIVE COOLING

Roofs may be cooled substantially by the use of roof-top sprinklers (Sutton, 1952; Robinson, 1978a). In climates where water is in short supply, sprinkling is an extravagance (Ben-Adam, 1964) since other means of cooling are as effective, but can be a very useful emergency measure where a primary cooling system has failed or no other system exists.

Where buildings are not fully enclosed, a fogging system of cooling may be used, and this is even more effective than roof-top sprinkling (Hart and Wilson, 1954). Wilson, Hart and Woodard (1957) reported that foggers eliminated mortality due to heat prostration up to an outdoor dry-bulb temperature of 41 °C. Robinson, however, found that although birds showed relief when foggers were used (Robinson, 1978b), the lowered dry-bulb temperature (and presumably the relief of the birds) did not persist when they were switched off (Robinson, 1978c). With frequent use, moreover, both feed and droppings can get very wet, though foggers can be made that use a small amount of water and can be controlled by a thermostat and time clock. Costa and Hunton (1979) have suggested that foggers should be operated for 15 minutes each hour when temperatures are above 32 °C.

When buildings are completely enclosed it is possible to incorporate various methods of cooling into the ventilation system. Evaporative cooling is effective in improving environmental conditions and the performance of the stock (Heywang, 1947; Drury, Brown and Driggers, 1964) whenever the outdoor temperature exceeds 30 °C (Robinson, 1978b). Hobgood (1961) compared evaporative cooling by means of coolers or foggers, evaporative pad and fan, and refrigerated air cooling, concluding that the pad and fan were the most effective, and Mentzer and Dale (1960) also favoured pad and fan evaporative cooling as a means of improving the environment.

While evaporative cooling is very effective in B climates (Ferguson, 1970) and in certain A and C climates in the dry season, it would seem of little benefit in the humid tropics (Oluyemi and Roberts, 1979). El Boushy and van Marle (1978) have suggested that as a rule of thumb, relative humidity should not exceed 65% or 75% for stock on litter or in cages, respectively. Reece, Deaton and Lott (1970) simulated evaporative cooling by starting from a dry-bulb temperature of 35 °C with 18 °C or 21 °C dewpoint. The performance of broilers was observed at a constant wet-bulb temperature with successively higher constant relative humidities yet lower dry-bulb temperatures. Growth was stimulated and feed efficiency improved as dry-bulb temperatures decreased and relative humidity increased up to 80%.

AIR MOVEMENT

In all climates, an increase in air velocity will increase heat losses. Siegel and Drury (1968) demonstrated that velocities up to 2.5 m s^{-1} will diminish the physiological response to high dry-bulb temperatures up to a level of 40 °C. Above 40 °C, where the temperature is equal to or greater than body temperature, increasing the air velocity exacerbates the physiological responses. Drury (1966) showed that the benefit to weight gains of broilers at air velocities up to 2.5 m s^{-1} at 18 °C dewpoint, over the range of dry-bulb temperatures from 21 to 35 °C, was nearly proportional to the square root of the velocity.

In the humid tropics the use of a high air velocity to cool the stock may be the only possible way of improving performance, if climate modifications by the various housing and landscaping techniques have been exhausted. Where it is possible to use both evaporative cooling and an increased air velocity to cool the stock (and a production unit in an Aw climate might benefit from a dual system), evaporative cooling could be used until the indoor relative humidity reached 70–75%. Following this, an increased air velocity would be the last means available for cooling on a large scale, other than air conditioning.

Monteith (1973) has shown that radiation heat losses per unit of surface area are independent of the size of the animal while convective heat losses are determined in part by geometry and scale. The heat losses by radiation and convection of 'birds' of various sizes can be calculated using the equations and data given by Monteith (1973). *Figure 14.4* compares the variation of heat losses by radiation and convection with body size for three representative air velocities. The body size range represents a 1-day-old chick to

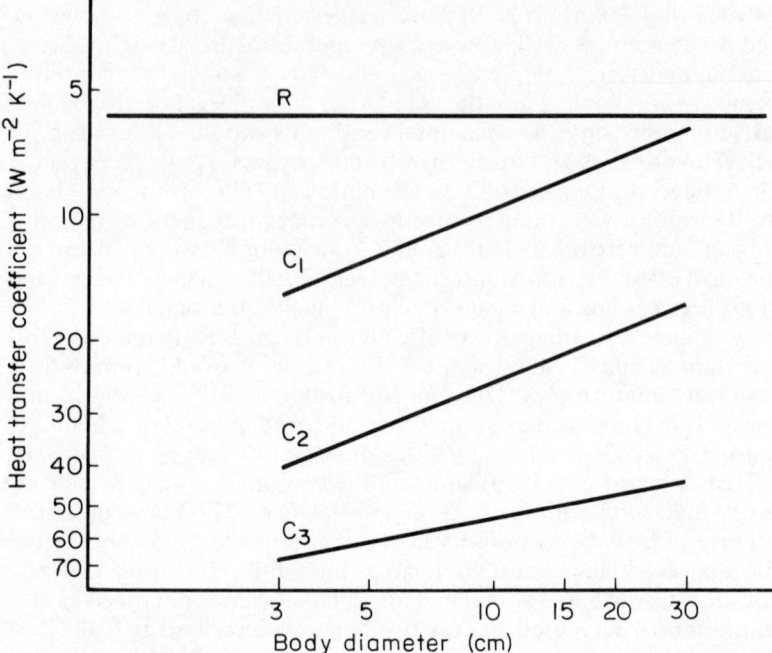

Figure 14.4 The effect of body size and air velocity, u, on the transfer coefficients for radiation and convective heat losses. Line R, radiation; line C_1, $u = 1$ m s^{-1}; line C_2, $u = 5$ m s^{-1}; line C_3, u = 9 m s^{-1}

a small turkey. For a young bird, increasing the air velocity above 1 m s^{-1} would be a more effective way to increase heat loss than decreasing the wall temperatures. However, for a large turkey it would be equally effective to decrease wall temperatures and increase air speed to increase heat losses.

In Am or Af climates the growth rate and feed conversion of young broilers, for instance, could be improved by increasing the rate of air movement alone, rather than also employing evaporative cooling, up to the practical limit of high relative humidity.

Conclusions

The partition of tropical climates into three classes A, B, and C, which are divided further into subclasses on the basis of rainfall and temperature enables areas of land to be mapped according to these climate subclasses. While the recognition that an area has a particular type of climate will not help a farmer or company produce eggs or meat, it may be of assistance in planning for the siting of a particular enterprise in the future. In certain tropical climates the level of egg production is similar to that which can be achieved in temperate climates, but in much of the tropics not only is the level of output lowered (egg production and meat production) but reproduction is also adversely affected and mortality due to heat stress is common.

In the tropics the unique problem for the designer of housing for poultry is one of heat, manifested as a high wet- and/or dry-bulb temperature. In designing a poultry house to reduce the effect of heat it is useful to proceed from the area surrounding the house to the house structure and then to the enclosed volume of the house.

The radiation incident on a building may be reduced by orientating it in an east–west direction and by shading the walls with a roof overhang or by growing trees and shrubs at appropriate distances from the building. The temperature of the ground surface and that of the air above it will be less with a covering of turf rather than with bare soil, concrete or asphalt. Evapotranspiration from the trees, shrubs and turf gives an additional cooling effect to the air ventilating the buildings.

The absorption of radiation by roofing materials may be reduced to a minimum by using materials or coatings that have a high reflectivity of short-wave radiation and a high emissivity of long-wave radiation. At any time during the day a roof may be cooled by water sprinklers; however, a good deal of water is used and therefore as a means of cooling, roof sprinkling would find limited application, except in an emergency. Enclosure of the building by walls made of wooden slats reduces the amount of short-wave radiation on stock, but apart from that it cannot be regarded as a serious means of reducing the heat load. But in an open-sided shelter or in a shelter enclosed by wooden slats, water foggers controlled by time-clock and thermostat are an effective means of cooling, although care must be taken to avoid excessive use, which leads to wet feed and manure. The materials lining the ceiling and walls of a building can also influence the radiation load on stock. Highly reflective materials incorporated in the roof structure increase the level of insulation when heat flow is downwards. Ceiling linings with a low emissivity, such as aluminium foil attached to plywood, are effective in reducing the heat load.

Stock housed on the floor are further from a ceiling than those in cages and in summer those on the floor would receive less radiation from the ceiling. Stock on the floor and in cages would benefit from increased air movement through the use of fans although only at air temperatures less than the body temperature of the bird.

The most expensive means of cooling stock, without considering mechanical refrigeration, is the complete enclosure of a building with insulated walls and the installation of an evaporative cooling system in addition to the mechanical ventilation system. If all the above means of reducing the radiant heat load on stock and buildings were incorporated into such a building, the productivity could be improved in even the hottest climates, and the survival of the housed stock ensured.

References

ALBUQUERQUE, K. DE, LEIGHTON, A.T., Jr, MASON, J.P., Jr and POTTER, L.M. (1978). *Poult. Sci.* **57**, 353–362

ANDERSON, B. (1977). *Solar Energy. Fundamentals in Building Design.* McGraw-Hill, London

ANON. (1968). *Man's Domain. A Thematic Atlas of the World.* McGraw-Hill, London

ANON. (1975). *CIBS Guide A6 Solar Data*. The Chartered Institution of Building Services, London

ANON. (1976). *The New Zealand University Atlas*. Hicks Smith, with George Philip, London

APPLEMAN, H. and BONHOF, B.J. (1971). *Nether. J. agric. Sci.* **19**, 204–210

BEN-ADAM, Z. (1964). *2nd European Poultry Conference, World's Poultry Science Association, Bologna*, pp. 73–79. Accademia Nazionale di Agricoltura, Bologna

BOND, T.E. (1967). *Environment Control in Poultry Production*, pp. 200–211. Ed. by T.C. Carter. Oliver & Boyd, Edinburgh

BOND, T.E., KELLY, C.F. and HEITMAN, H. (1958). *J. Hered.* **49**, 75–79

BOONE, M.A. and HUSTON, T.M. (1963). *Poult. Sci.* **42**, 670–676

CASTELLO, J.A. (1964). *2nd European Poultry Conference, World's Poultry Science Association, Bologna*, pp. 19–72. Accademia Nazionale di Agricoltura, Bologna

CLARK, C.E. and SARAKOON, K. (1967). *Poult. Sci.* **46**, 1093–1098

COSTA, M.S. and HUNTON, P. (1979). *6th Latin America Poultry Congress, Lima, Peru.* In press

CRITCHFIELD, H.J. (1974). *General Climatology*, 3rd edn. Prentice-Hall, Englewood Cliffs, New Jersey

DEATON, J.W., REECE, F.N. and McNAUGHTON, J.L. (1978). *Poult. Sci.* **57**, 1070–1074

DEERING, R.B. (1952). *Proc. Symposium on Housing and Buildings in Hot-humid and Hot-dry Climates.* National Research Council, National Academy of Sciences, Washington, DC

DRURY, L.N. (1966). *Trans. Am. Soc. agric. Engrs* **9**, 329–332

DRURY, L.N., BROWN, R.H. and DRIGGERS, J.C. (1964). *Ga Agric. Exp. Stn Bull. NS 115*

EL BOUSHY, A.R. and VAN MARLE, A.L. (1978). *Wld's Poult. Sci. J.* **34**, 155–171

EMMANS, G.C. and DUN, P. (1973). *Gleadthorpe Experimental Husbandry Farm Poultry Booklet No. 1.* Ministry of Agriculture, Fisheries and Food, London

FERGUSON, W. (1970). *Trop. Anim. Hlth Prod.* **2**, 44–58

FRANCIS, D.W., REID, B.L. and WILSON, W.O. (1972). *New Mexico agric. exp. Stn Bull. 601*

GODFREY, E.F. and WINN, P.N., Jr (1966). *Feedstuffs, Minneap.* **38** (52), 34

GRIFFIN, J.G. and VARDAMAN, T.H. (1970). *Poult. Sci.* **49**, 387–392

GRIFFIN, J.G. and VARDAMAN, T.H. (1971a). *Poult. Sci.* **50**, 463–466

GRIFFIN, J.G. and VARDAMAN, T.H. (1971b). *Poult. Sci.* **50**, 459–463

HAHN, L., BOND, T.E. and KELLY, C.F. (1963). *Calif. Agric.* **17**, 10–11

HARRIS, G.C., Jr, DODGEN, W.H. and NELSON, G.S. (1974). *Poult. Sci.* **53**, 2204–2208

HART, S.A. and WILSON, W.O. (1954). *Rep. Wld's Poult. Congr., Edinburgh* **10**, 330–332

HASSALL, D.N.H. (1973). *Reflective Insulation and the Control of Thermal Environments.* St Regis–ACI, Sydney

HAUGHEN, A.E. (1977). *Aktuelt fra Landbrukdepartmentets Opplysningst-jeneste nr 2*

HELLICKSON, M.H., BUTCHBAKER, A.F., WITZ, R.L. and BRYANT, R.L. (1967). *Trans. Am. Soc. agric. Engrs* **10**, 793–795

HEYWANG, B. (1947). *Poult. Sci.* **26**, 20–24

HOPGOOD, P. (1961). *Disease, Environmental and Management Factors Related to Poultry Health.* Publication Agricultural Research Service USDA 45–2, Washington, DC

HORST, P. and PETERSEN, J. (1975). *Arch. Geflügelk.* **39**, 225–231

HOWES, J.R. (1963). *Wld's Poult. Sci. J.* **20**, 8–15

HOWES, J.R., GRUB, W. and ROLLO, C.A. (1961). *Poult. Sci.* **40**, 1416 (abstr.)

HOWES, J.R., GRUB, W. and ROLLO, C.A. (1965). *Highlts agric. Res.* **12** (2), 10

HUSTON, T.E. and CARMON, J.R. (1958). *Physiol. Zoöl.* **31**, 232–235

HUTCHINSON, J.D.C. (1953). *Poult. Sci.* **32**, 692–696

KEMPSTER, H.L. (1941). *Univ. Mo. agric. exp. Stn Res. Bull. 423*

KEMPSTER, H.L. and PARKER, J.E. (1936). *Univ. Mo. agric. exp. Stn Res. Bull. 247*

KOTHANDARAMAN, K. (1973). *Cheiron* **2**, 167

KUMARASWAMY, K. and RATHNASABAPATHY, V. (1973). *Cheiron* **2**, 83–89

LILLIE, R.J., OTA, H., WHITEHEAD, J.A. and FROBISH, L.T. (1976). *Poult. Sci.* **55**, 1238–1246

LONG, E. and GODFREY, G. (1952). *Poult. Sci.* **31**, 665–673

McNAUGHTON, J.L., KUBENA, L.F., DEATON, J.W. and REECE, F.N. (1977). *Poult. Sci.* **56**, 1391–1398

MANNEL, K.H. (1971). *Technical Report Animal Husbandry Research Training Project, Baghdad,* No. 39. FAO, Rome

MENTZER, J.R. and DALE, A.C. (1960). *Agric. Engng, St Joseph, Mich.* **41**, 816–819

MONTEITH, J.L. (1973). *Principles of Environmental Physics.* Arnold, London

MOWBRAY, R.M. and SYKES, A.H. (1971). *Br. Poult. Sci.* **12**, 25–29

MUELLER, W.J. (1961). *Poult. Sci.* **40**, 1562–1571

MUELLER, W.J. (1967). *Poult. Sci.* **46**, 82–88

OLUYEMI, J.A. and ROBERTS, F.A. (1979). *Poultry Production in Warm Wet Climates.* Macmillan, London

OTA, H. and McNALLY, E.H. (1965). Paper No. 65–411, American Society of Agricultural Engineers Summer Meeting

OTA, H., WHITEHEAD, J.A. and LILLIE, R.J. (1973). Paper No. 73–4557, American Society of Agricultural Engineers Winter Meeting

PEREK, M. and SNAPIR, N. (1963). *Br. Poult. Sci.* **4**, 19–26

QUARISHI, S.R., THATTE, V.R. and DESAI, R.T. (1973). *Indian Poult. Gaz.* **57**, 6–11

RAO, F.A., KUMARASAMY, K. and RATHNASABAPATHY, V. (1966). *Indian vet. J.* **43**, 636–644

REECE, F.N., DEATON, J.W. and LOTT, B.D. (1970). Paper 70–412, American Society of Agricultural Engineers Annual Meeting

ROBINSON, D. (1978a). *Poult. Fmr, Sydney* **46** (July 22), 8–10

ROBINSON, D. (1978b). *Poult. Fmr, Sydney* **46** (Aug. 26), 12–13

ROBINSON, D. (1978c). *Poult. Fmr, Sydney* **46** (Aug. 5), 13–14

SIEGEL, H.S. and DRURY, L.N. (1968). *Poult. Sci.* **47**, 1230–1235

SIEGEL, H.S. and DRURY, L.N. (1970). *Poult. Sci.* **49**, 238–244

SMITH, A.J. (1973). *Trop. Anim. Hlth Prod.* **5**, 259–271

SMITH, A.J. (1974). *Trop. Anim. Hlth Prod.* **6**, 237–244

SMITH, A.J. and OLIVER, J. (1972). *Rhod. J. agric. Res.* **10**, 43–60

SMITH, W.K. (1969). *PhD Thesis,* University of Sydney

SQUIBB, R.L. (1959). *J. agric. Sci., Camb.* **52**, 217–222

SQUIBB, R.J. and WOGAN, G.N. (1960). *Wld's Poult. Sci. J.* **16**, 126–137

STOCKLAND, W.L. and BLAYLOCK, L.G. (1974). *Poult. Sci.* **53**, 1174–1187

STRAHLER, A.N. (1965). *Introduction to Physical Geography.* John Wiley, London

SUGANDI, D., BIRD, H.R. and ATMADILAGA, D. (1975). *Poult. Sci.* **54**, 1107–1114

SUTTON, G.E. (1952). *Proc. Symposium on Housing and Buildings in Hot-humid and Hot-dry Climates.* National Research Council, National Academy of Sciences, Washington, DC

WILLIAMSON, G. and PAYNE, W.J.A. (1978). *An Introduction to Animal Husbandry in the Tropics*, 3rd edn. Longmans, London

WILSON, W.O. (1967). *Ground Level Climatology.* American Association for the Advancement of Science, Washington, DC

WILSON, W.O., HART, S.A. and WOODARD, A.E. (1957). *Poult. Sci.* **36**, 606–613

WILSON, W.O., ITOH, S. and SIOPES, T.D. (1972). *Poult. Sci.* **51**, 1014–1023

WILSON, W.O., McNALLY, E. and OTA, H. (1957). *Poult. Sci.* **36**, 1254–1261

YOSHIDA, M., HOSHII, H., KOSAKA, K. and MORIMOTO, H. (1968). *Jap. Poult. Sci.* **5**, 81–90

ZIMMERMAN, R.A., SNETSINGER, D.G. and GREENE, D.E. (1975). *Poult. Sci.* **54**, 1831 (abstr.)

15

THE HOUSING OF LARGE MAMMALS IN
HOT ENVIRONMENTS

W.V. MACFARLANE
Waite Agricultural Research Institute, University of Adelaide, Australia

Introduction

The approach to sheltering livestock, particularly the larger forms (cattle, horses, sheep, pigs, camels and elephants), in hot environments, is different from that appropriate in cool or cold regions. Reduction of heat storage, and the use of reflective surfaces, protection from radiant energy, increased air movement, high ceilings and space between animals, are desirable physical aspects of designs for the tropics. An environment is hot when animals gain heat and either seek to avoid it, or employ cooling mechanisms: retention of heat is a stress leading to physiological strain. The three major categories of hot environment are the hot wet, hot dry and the hot intermittently wet and dry regions. But as Köppen and others (Koeppe and de Long, 1958) have pointed out, classification is not simple and there are intermediate climates. Each type presents different problems. Shelter is provided for many reasons ranging from protection against predators to provision of shade; the housing of animals for intensive production of proteins, or for research; the holding of camels and horses for the army or for races; ceremonial elephants and the display of animals in zoos.

Benefits from commercial housing or cooling of large animals that have been found or suggested include:

(1) increased growth rate and more efficient conversion of food;
(2) greater fecundity, less fetal loss, longer oestrus, less male sterility;
(3) greater milk production and longer milking periods;
(4) cleaner wool, with greater yield; less parasitism.

There is not necessarily a gain in productivity, however, and there is always a cost both in construction and in energy for bringing supplies to the animals. There are also the metabolic, infective and social diseases of intensive production on the other side of the balance.

In spite of this diversity, an attempt will be made to bring out the physical and biological principles of housing animals in the tropics and to consider the energy costs. At the same time, as Marcus Vitruvius (30 BC) recommended, efforts should be made to ensure the strength, utility and grace of the houses constructed.

Origins, ecotypes and domestication of livestock

Virtually all domesticated livestock evolved in climates that were at least intermittently hot. Primordial cattle and pigs belonged to the wet tropics of south Asia (*Table 15.1*). Elephants radiated from an equatorial hearth, while sheep and goats originated in the dry Caucasian uplands (Zeuner, 1963) where summers are hot. Horses and camels came to Eurasia (three million years ago) from hot, dry areas in mid-western North America. Although the hearth of evolution determines the essential physiological settings for turnover of water, energy and protein in each genus, as well as of salt tolerance (Macfarlane, 1978), most genera migrated from their primordial hearth, so that modified ecotypes developed in cooler regions with different humidities. Among livestock the essential rank order of metabolic functions is illustrated by the rate of water turnover (*Table 15.1*), which is high in jungle types and low in desert forms. Oxygen consumption rates are dependent upon the turnover of both water and protein. So the energetics of housing different ecotypes vary with their evolutionary inheritance (Ucko and Dimbleby, 1969; Herre and Röhrs, 1977; Macfarlane, Howard and Good, 1974; Macfarlane, 1976, 1978).

Table 15.1 EVOLUTIONARY HEARTH AND ECOPHYSIOLOGICAL CHARACTERISTICS OF THE MAJOR LARGE DOMESTICATED MAMMALS

Genus	Hearth	Latitudes	Photoperiodic response		Water turnover
			Coat shedding	Breeding season	
Bos (cattle)	South Asia	0–15	Spring	All year	++++
Elephas (elephant)	Africa	0–10	–	All year†	++++
Sus (pig)	South Asia	0–10	–	All year	++++
Equus (horse)	NW America	25–30	Spring	All year	+++
Ovis (sheep)	Caucasia	35–40	Spring (lost)*	Autumn	++
Capra (goat)	Caucasia	35–40	Spring	Autumn	++
Camelus (camel)	NW America	20–30	Spring	All year†	+

*Coat shedding persists in Wiltshire and Navajo sheep, but is lost in other domestic breeds.
†Oestrus occurs when there has been a good supply of food for at least two months but it is non-seasonal if there is no failure of food supplies.

Domestication of mammals began about 12 000 years ago when goats and sheep were brought into human groups in the Euphrates–Zagros region. Cattle, horses, elephants and camels were domesticated later, from 9000 to 6000 years BP. Wild animals sought their own shade or shelter; but in human social structures they became property. With that territorial component, the first animal housing was probably built against predators. Then shelter from the sun had to be provided if the animals were held enclosed.

Mammals that evolved near the equator are not influenced by photo-period, rather the coat is shed throughout the year and reproduction is continuous if there is food. The pelage is short and shiny to reflect solar radiation (as in tropical antelope and deer). Pigs and elephants have little pelage, but they wallow for cooling and they breed throughout the year, unless food supplies fail. Animals that evolved more than 10° from the equator show some degree of photoperiodism, affecting coat type or repro-duction. The camel sheds its coat in spring, but breeds at any time of the year when its pituitary gland is roused by adequate food. Cattle also shed their coats with increasing day length, but they breed through the whole year like horses. Sheep and goats originally had spring coat-shedding but this was lost in domestic selection. At the equator they breed continuously, while the reproductive period becomes later in the autumn in high latitudes, so that the young are born in spring when feed occurs. Animal housing in hot environments, therefore, must either be suitable for animals carrying uni-form insulation or for others with seasonal changes of coat. Young will be born and lactation ensue in tropical livestock at any season, except for some sheep and goats.

Shelter in hot environments

Housing for livestock in the heat benefits both man and beast in a variety of ways. First, it helps the prevention of predation by carnivores or by human raiders, which is common in equatorial Africa. Some Somalis, Masai and Samburu make formidable *bomas* of acacia to hold sheep and cattle over-night (*Figure 15.1a*). Sheep folds of stone serve the same purpose in more arid areas.

Secondly, it provides shelter from solar radiation, the heat of the air, rain and wind. Such shelter is provided for sheep and cattle throughout South Asia, India and Central America. Generally, in regions with hot, dry sum-mers such as North Africa and the Middle East, shade is important in relieving stress from insolation (*Figures 15.1b–15.1d*).

Thirdly, housing may be used because of limited territory and the need to exclude stock from crops, as in the more densely populated areas of India or Indonesia. In such cases food is gathered and brought to the animals. For example, the difficulties of feeding cattle in Bombay, as well as of dealing with faecal output from city cows and of preventing interference with traffic, led to the setting up of the Aarey Milk Colony near Bombay thirty years ago. Some twenty thousand cattle and buffalo were housed hygenically (*Figure 15.2*) in roofed structures with open walls, and selective breeding was practised. The animals were stall-fed. By these means, limited amounts of land maintained large herds.

Intensive production is unusual in the tropics, but enterprises for the production of beef, wool, milk and pig meats have grown up to cater for the luxury market. Control of disease, nutrition and reproductive processes is undertaken as well as control of the environment. Superfine Merino sheep staples-grown indoors without breaks or dirt are in some demand to provide a replacement for cashmere. Housing sheep in hot regions is economic when wool prices are above $20 kg^{-1}.

262

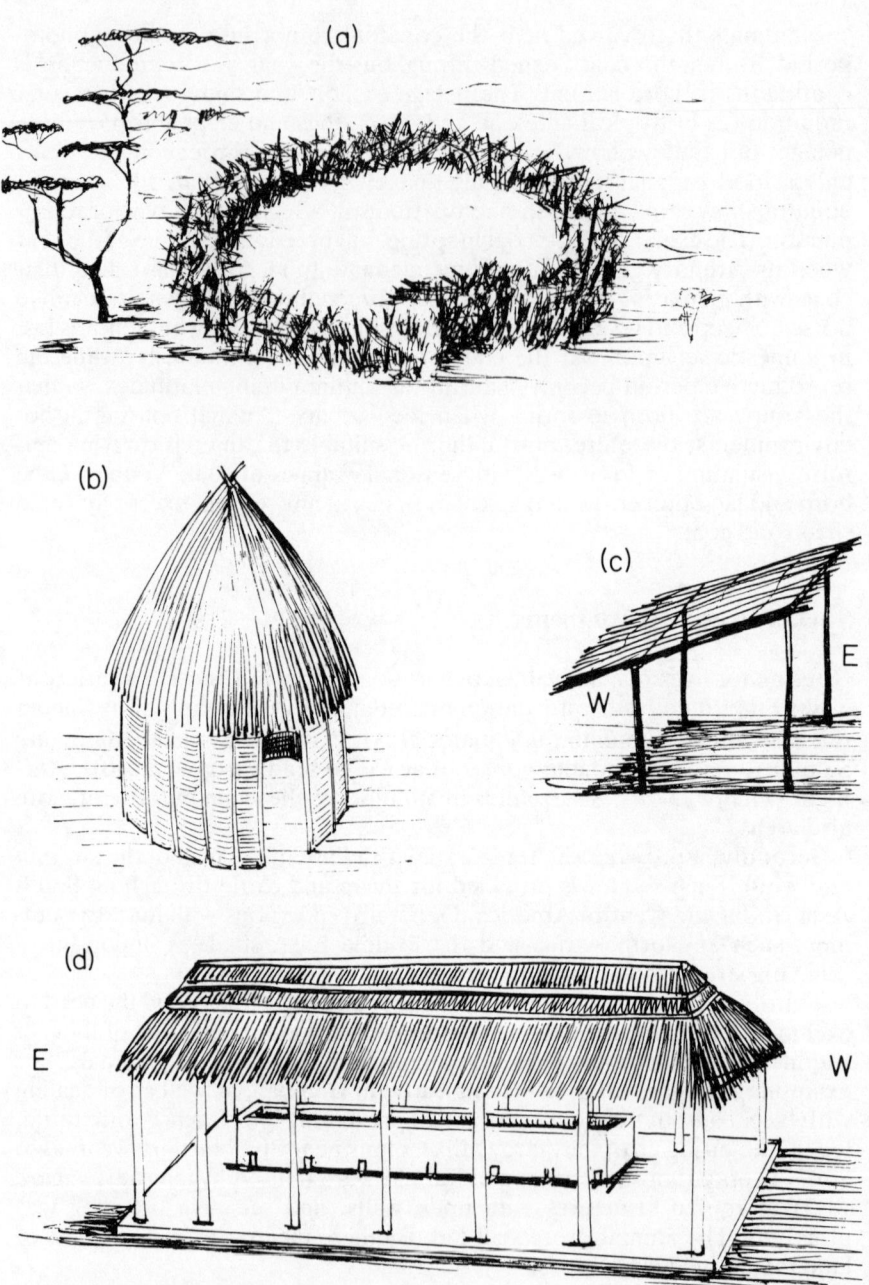

Figure 15.1 (a) Shelter from predators (carnivores and man) was probably the first accommodation made for domestic animals. The boma is still important in pastoral Africa, and the thorny bush deters both man and beast from taking livestock. Stone versions were made in the Middle East and Europe. (b) Central African wattle and daub hut with thatch. Good rejection of radiation but poor air flow. (c) Basic shade, with thatch to absorb heat and with lower side of roof to the west. (d) Recommended Indian Standard (ISI, 1978) cattle shelter. Some protection by eaves and good ventilation—but exposed to east and western solar input. The eaves should be wider and lower

Figure 15.2 Aarey Dairy Colony near Bombay (lat 19 °N). Bos indicus *breeds and* Bubalus *were moved from town and housed in stalls, under asbestos roofs ventilated at the ridge, with open sides. The long axis of buildings is N–S. Good yields of milk were obtained*

Housing may also be used for specialized purposes. Controlled quarantine holding and laboratory areas are being developed for moving livestock between different countries. Luxury quarters for raising polo, military or ceremonial animals have been adapted for hot climates. In the oil-rich arid zone, quarters for horses and camels are often opulent, while sheep and goats are sheltered prior to their conversion to meat. Convective and evaporative cooling are used.

Species ranking in reaction to hot environments

The four major large animal groups likely to be housed in the tropics are pigs, cattle, sheep and goats. These animals may be placed in rank order of tolerance to the major components of hot environments. Air temperature, humidity, long-wave and short-wave radiation and air velocity affect each type of animal differently according to its metabolic rate, cooling mechanisms, insulation and behaviour.

AIR TEMPERATURE

When adult pigs, cattle, sheep and goats are exposed to 40 °C at low humidity, the pig shows strain first because of its high dependence on respiratory cooling. It pants rapidly, seeks to wallow and lies down on any cool substrate. White pigs with and without shelter showed differences in skin temperature when ambient temperature was above 22 °C. At 27 °C the sheltered pigs were 2–5 K cooler on the back than those in the sun, but at

33 °C the difference was only 1.5 K, at 36 °C it was 1 K, and there was no difference at an ambient air temperature of 41 °C. At 40 °C ambient in the sun, pigs need a shower or wallow to cool the skin (B. Stone, personal communication).

Cattle have a larger mass (which takes longer to heat up), a relatively small surface (*Table 15.2*) and use both sweating and panting for cooling, so in spite of a relatively high metabolic rate, they are less affected than pigs. Sheep, although small, have low metabolic rates, effective panting and some

Table 15.2   SURFACE TO MASS RATIOS OF MAMMALS OF DIFFERENT SIZES

Animal	Weight (kg)	Surface area (m²)	(kg m⁻²)	(cm² kg⁻¹)
Kid	5	0.22	22.7	440
Lamb	10	0.34	29.4	340
Sheep	50	1.04	48.1	208
Pig	100	1.59	62.8	159
Horse	500	4.66	107.2	93
Ox	1000	7.46	134.0	75

Animals such as camels with long neck and legs would have greater areas for a given mass than these approximations. The ratio of surface to mass influences the gain or loss of radiation, the convective exchange of heat, the ratio of heat storage to heat exchange and the metabolic rate, and the rate of water turnover

sweating, so they are less affected by heat than cattle. Arid-zone goats, with even lower metabolic rates and less insulation than sheep, cool by panting and sweating. They are the most tolerant of heat among the four genera.

HUMIDITY

Wet-bulb temperatures over 27 °C and relative humidities over 70% (vapour pressure above 2.4 kPa) increase the importance of convective and conductive cooling, especially for the pig, which normally compensates for heat stress by wallowing behaviour. Desert sheep and goats find high humidities difficult and a sheep with wool insulation is less able to cope than goats with short coats.

RADIANT ENERGY

In shelters, most radiant exchange is by long-wave radiation ($\lambda > 3\,\mu m$), which is absorbed by all coats regardless of colour, as though they were black bodies. Since the surface area collecting radiation determines responses, heating is greatest in small and least in large animals.

WIND

Sparsely haired pigs and buffalo lose heat most easily by convection. Buffalo sweat, but pigs do not, so in hot environments buffalo benefit most from air movement. Shearing of sheep or goats aids convective cooling indoors, and their relatively large surface to mass ratio aids heat loss. Although cattle and buffalo sweat, relative to sheep they have only about one-third the surface area per kilogram (*Table 15.2*). At temperatures below 15–20 °C wind affects species in the opposite order, so that pigs are most easily cold-stressed while cattle are little affected.

Environmental physiology of larger mammals

The major functional variables to be considered in designing shelters for the tropics are the body size, physiological ecotype, age, insulation, thermo-regulatory mechanisms, lactation and the nutritional status of the animals. These contribute either to the input or to the dissipation of heat and water from the animals. The main thermal (Mount, 1979) and ecophysiological aspects (Macfarlane, 1976, 1978) of livestock have been reviewed recently.

EFFECTS OF SIZE ON ENERGY AND WATER TURNOVER

Amongst homeotherms, size in itself is a major determinant of the rate of energy turnover (Brody, 1945; Adolph, 1949; Schmidt-Nielsen, 1975). Although the average rate of *heat production* per unit mass declines with increasing size (as mass$^{0.75}$), at any given body weight ecotypes differ at least twofold in basal metabolism (kJ kg^{-1}) (*Table 15.3*). Desert goats and Merino sheep have metabolisms about 15 per cent less than the overall average of

Table 15.3   APPROXIMATE OUTPUTS OF HEAT AND WATER OF ADULT LIVESTOCK IN WARM ENVIRONMENTS (*c.* 30 °C)

Species	Weight (kg)	Daily heat output			Daily water turnover*		
		(kJ)	(kJ kg^{-1})	(kJ kg$^{-0.75}$)	(ml kg^{-1})	(l)	(ml kg$^{-0.82}$)
Elephant	3672	206 000	56	436	68	250	300
Horse	703	49 900	71	360	78	61	280
B. taurus	600	50 600	84	418	145	87	460
B. indicus	530	41 000	77	372	113	60	350
Camel	500	23 000	46	219	36	18	110
Donkey	300	20 000	67	278	70	21	200
Pig	120	10 000	83	277	108	13	250
Sheep (wet)	50	6 200	124	326	88	4.4	180
Sheep (arid)	45	4 850	108	280	64	2.9	130
Goat (milk)	45	5 210	115	300	91	4.1	180
Goat (arid)	30	3 360	112	262	47	1.4	90

*Metabolic, food and drinking water combined, i.e. the amount released daily to the environment

Measurements have been made on different breeds of animal, in different thermal conditions on diverse foods and by different methods so that it is difficult to produce a unified table. There is a 20–30 per cent range on either side of the values given, because of individual and breed variation and differences in nutrition. The estimates given (based on Brody, 1945; Macfarlane and Howard, 1972; Macfarlane, 1976) lie at the upper range of water turnovers—for evaporative cooling—and the lower range of metabolism, for hot climates. Metabolic rates are standard basal values

295 kJ kg^{-1} d^{-1} (Brody, 1945) while wet-country animals such as deer or Romney sheep are 30 per cent or more above the average. A 500 kg camel is 30 per cent below the average but cattle are above it by 50 per cent or more. Hereford or Shorthorn breeds from cool, wet environments have some 30 per cent greater metabolic turnover than *Bos indicus* types. The eland is in the same class as *Bos indicus* cattle (Macfarlane and Howard, 1972) while the oryx rates with the camel (King, 1979).

It is still not clear *a priori* by what mechanism metabolic intensity falls with increase in body mass. It is not a matter only of surface area, nor of potential diffusion rates across lung or vascular beds. McMahon (1973) argued that the 0.75 exponent can be derived from mechanical structure. This, however, would not explain the two- to threefold difference between animals of the same size, from different evolutionary hearths.

Under natural selection, animals with inappropriate rates of use of water, energy, oxygen and protein for their size in any resource system, would be eliminated. In the wet tropical areas food and water are always abundant, allowing high rates of use of both energy and water. Buffalo, cattle, pigs and deer evolved there. Their thyroxine turnover is proportional to food intake and weight (Good, Howard and Macfarlane, 1974). Linked with these is a high rate of use of protein, and an intolerance of salt. In the wet tropics, salt is leached from soil, so vegetation contains about one-tenth the sodium concentration found in similar plants of the temperate zone. Jungle mammals do not pump salt out effectively, so an excess is toxic. Low salt tolerance and high water turnover evolved also with a limited ability to concentrate the urine (Macfarlane, 1971), as in pigs and cattle.

The exponent for water turnover among the main types of livestock is mass$^{0.82}$. This differs from that for metabolic rate, probably because water has functions other than those directly involved in metabolism, such as evaporative cooling and vascular regulation of temperature (Macfarlane, 1976, 1978). Water leaves the body through sweat glands and kidney, but it diffuses also through the skin and the respiratory tract. Skin permeability is greater in animals with high rates of water turnover than amongst arid-zone animals with low rates (Haines *et al.*, 1974). Reabsorption of water from the renal tract is paralleled by the concentration of faecal material in the colon. Jungle and wet-tropical animals have dilute urine and high water content in the faeces, while desert camels or oryx have more concentrated urine and they stipate faeces down to 40% water. The rate of water use may also be linked with the relative lengths of the small and large intestines (Ledger, 1968). Water-dependent cattle and water-buck have a small gut about 3.7 times longer than the colon and they produce fluid faeces. Desert oryx (1.4×) and kob (1.8×) have relatively less ileum and more colon for water reabsorption and have lower rates of water turnover (Macfarlane, 1968; King, 1979).

Size (together with ecotype) determines the release of heat, water vapour and urinary and faecal water into the environment of housed animals (*Table 15.3*). Size affects the *surface* area of the interface with the environment and thus the heat exchange rates. The larger the animal (*Table 15.2*), the less the surface per unit of mass, the greater the heat capacity, and the slower the rates of cooling or heating. As a result of the fall of metabolic intensity with increased size, similar amounts of heat are discharged into an enclosure by a tonne of beef cattle or a tonne of sheep. On the other hand, camels, which are about the same size as cattle, produce about half as much heat per tonne. The heat production of pigs is high, since they originated in the wet tropics, and at the time when they are usually sold (before reaching full adult size) they have two or three times as much water and heat output per kilogram as adult forms. The heat and water input to an enclosure is, therefore, relatively great in pig houses.

Size and heat tolerance are closely related. As Priestley (1971) pointed out, in still air in the sun, the body temperature of a sheep without evaporative cooling would rise by 20 K in a day, while a mouse would reach 150 °C. Both would be dead. Radiant loss from animals is linearly proportional to surface area, but convective removal of heat is approximately proportional to length$^{0.7}$, the exponent depending upon air speed and body size, so small animals are heavily dependent on convection for cooling, and less dependent than large animals on radiation.

Size enters into the pattern of heat exchange in several ways. Among pigs and cattle, large size goes with relatively lower metabolic rates and often with more insulating fat. Convective and radiation losses of heat are reduced relatively with size, and conduction to the floor also decreases; at night, piglets need bedding or an insulated floor even in warm environments. Larger pigs or cattle lying down in the tropics can lose excess heat to a concrete or wet floor. Much of the heat dissipation from a prone pig through a concrete floor takes place laterally (Bruce, 1979). Triform slats reduce heat transfer but wet concrete has less thermal resistance (3.1 m^2 K W^{-1} as against 7.3 m^2 K W^{-1}). These estimates of heat loss are described by Bruce, in Chapter 12.

For hot climates, floors of low thermal resistance are desirable, to cool animals, except for those that are small or newborn. But wet floors encourage skin infection, fungi and pathogens. A 45 kg pig was found by Bruce (1979) to have a thermal resistance of 1.2 m^2 K W^{-1} so that a 'thermally neutral' floor should have a similar value. Materials with floor resistances below 1.2 m^2 K W^{-1} would be cool. These include concrete, wood, concrete slats and mesh floors. Size affects the neutral value of the substrate by almost an order of magnitude, since a 1 kg pig of thermal resistance 3.5 m^2 K W^{-1} and a 1000 kg bullock (0.5 m^2 K W^{-1}) have very different thermal characteristics. Thus straw and concrete respectively would provide neutrality for the 1 kg piglet and for the 1000 kg animal.

The use of concrete floors allows easy cleaning, which can be mechanized (Gribble, 1973), but it is hard for both standing and sitting upon. These floors act as thermal mass, helping cooling if wet (Starr, Neubauer and Melzer, 1978) and slowing temperature changes. Deep litter for cattle works well if the climate is reasonably dry. In a hot, wet area, however, the chance of fungal and bacterial growth is much greater. The spread of tuberculosis and brucellosis can be troublesome in such circumstances, unless there is good ventilation.

For pigs, triangular-section reinforced-concrete floor slats seem to be very effective. With a 15 mm gap between the components and a bar width of 5–7 cm, a useful outflow of urine and faeces is obtained with very little danger of feet being caught. The substrate is cool and conducts heat from the pigs. Since the gap between the floor members slopes away on either side, there is little retention of excreta. Effluent can be arranged to flow gravitationally to pits, where methane, bacterial protein or compost is produced.

INGESTION AND EGESTION

The ecological origin of an animal determines its food and water preferences. Cattle and pigs, with high water turnover rates, prefer food which is wet. They will eat dry food, of course, but when offered a choice, wetter forms of food are chosen (Yang, 1979). Cattle and pigs have wet muzzles with glands of salivary type secreting a high-potassium fluid under the influence of acetylcholine. The skin of the muzzle is protected by mucous fluid from damage by wet pasture or mud (Brewer and Macfarlane, 1967). Sheep, horses and camels grazing drier food have muzzles covered with fine hair and lack external salivary glands.

When pigs were offered dry food pellets (13% water) or the same pellets mixed with water, the pigs chose food with 40–70% water. This is like porridge (Yang, 1979). A gruel with 85% water spilt from the mouth and was rejected. Dry feeding of pigs is being displaced by wet-feed systems, because of easier distribution, and because most by-products used for pig feeding (whey, fruit pulps) are watery (Braude and Rowell, 1967). There is less dust, pigs suffer fewer gastro-oesophageal ulcers, and achieve about 3 per cent greater efficiency of conversion when using wet food. They also like the texture of porridge, which is more like their forest food. Pig housing should therefore be designed for the pumping of wet food, and for disposal of copious urine and wet faecal output.

Camels, on the other hand, will ignore succulent vegetation and choose dry or salty foodstuffs, while sheep and goats accept a wide range of water content in grasses, from 10% to 85% water, with little discrimination.

The humidity of animal house atmospheres and the amount and type of egesta are affected by water turnover rates (Macfarlane, 1978). Water vapour leaves the animal via the respiratory tract, by diffusion through the skin (Haines *et al.*, 1974) and as sweat. In addition, water in urine and faeces vaporizes in hot environments, adding to the humidity in housing. Partition of water to respiratory, skin, urine and faecal outlets differs with ecotype and temperature. Cattle have high rates of water turnover (*Table 15.3*), and at high temperatures there are both respiratory and sweating vapour losses, as well as large amounts of urine and faecal water: their faeces are 70–80% water. Sheep, however, sweat less and have more respiratory water loss than cattle. In the heat, urine and faecal water is rapidly reduced, while sheep normally produce stipated faeces down to 55% water. The camel sweats, but does not pant, and has a lower urine output than the other animal types. It also produces faecal stipes with as little as 45% water. In contrast, pigs respond to hot environments by respiratory cooling, which is not very effective (Ingram, 1965), while urine and wet faeces continue to be produced even though the animal is hot. Pigs have sweat glands (Jenkinson, 1972), but these are not activated by heat even though adrenalin does produce secretion from them.

The amount of urine and the type of faeces therefore influence the design of housing. Pigs readily choose dunging areas, and these can be arranged so that drainage of their large volumes of urine and relatively watery faeces occurs readily through grids. The dry faecal stipes of sheep and goats will roll and can easily be washed across concrete floors, but grids need to be of appropriate size for each type of animal if the stipes are to pass through.

Responses to heat

INSULATION

Exposure to heat quickly induces the growth of shorter pelage on all animals. Animals in the field are exposed to greater extremes of temperature than those in shelters, and grow more fur and fat in cold seasons. This insulation may be unsuited both to hot seasons and to shelter. The insulating qualities of hair coats in cattle, horses, camels and goats are primarily modulated by day length. Short, shiny summer coats are brought on by increasing day length, but temperature has a separate influence on the growth and disposition of hairs. In hot environments the hair is shorter and lies flatter than in a cold environment (Macfarlane, 1968). The hairs on animals from hot countries are shiny, to reflect solar energy, which can enter tropical shelters in surprising amounts. Coats of all colours have similar properties for wavelengths above 1 µm since they act almost as black bodies; but in the visible range, colour and structure of coat are important. Energy in the 0.4–1.0 µm range is reflected by shiny hair, and there is greater absorption by matt-finished curly hair. In pigs, the reflectance of the skin between 0.4 and 1.0 µm is 51% for white, and 7% for black pigs, while they both reflect only about 5 per cent of the thermal radiation energy (Bond, 1967). In animal shelters, the main radiation is in the long-wavelength (thermal) range, coming from sources such as the roof or the ground, which may reach 50–70 °C on hot days (Bond, 1967). Coat colour then makes little difference to heat absorption, but the insulating properties of the coat will affect the heat load.

For indoor living in hot environments, it is preferable that animals with short coats should be selected. The type of pelage is genetically determined, and selection of tropical types of coat may increase the efficiency of animals for hot countries.

The insulation of the vascular and fatty tissues is also modifiable. Most tropical animals have little fat under the skin. According to Ledger (1968), antelope and goats in Kenya have between 1% and 7% of fat in the carcass while zebu cattle may reach over 35% of fat. The breeding of less fat animals is becoming commercially important, as it is realized that much of the Western Hemisphere pattern of mortality derives from eating an excess of animal fats. On the same pasture, Merino and Dorset sheep differ in growth, with the Merino producing one part of fat to two of protein, and the Dorset, two of fat to one of protein. Goats normally carry less than 10% of fat (like antelope), but pigs have been selected for the production of fat. Wild pigs, however, are lean, and domestic pigs have little hair so their insulation derives largely from the shell of the body. Fat as an insulator has an overall thermal conductivity about half that of muscular tissue (Burton and Edholm, 1955).

Vascular insulation arises from vasomotor control. Extremities such as the nose, ears and feet of livestock have strong vasoconstrictor regulation of blood flow. A sheep standing in snow maintains a foot temperature near 1 °C, but the temperature of the skin on its back is around 34 °C. If shorn, the sheep shivers, but the back remains at 32–33 °C since this area is normally protected by wool and lacks vasoconstrictor control. In a hot environment the feet, ears and nose heat up to near the core temperature, while skin temperature of the back rises. The vasodilation of all these areas results in heated blood moving to the surface, whence sweating extracts heat for transmission to the atmosphere.

DISTANCE BETWEEN ANIMALS

Radiant energy is exchanged between animals, and between animals and their surroundings, in the thermal infrared, which peaks at about 10 μm for surface temperatures of 30–37 °C. Thus space between animals is important, as it affects radiant heat losses as well as convective removal of heat and water. Huddling is therefore disadvantageous for the dissipation of heat in the tropics. For sweating to be effective for animals housed in hot wet or dry climates a free flow of unsaturated air is a primary need. For this reason shelters designed for cloudless hot regions are without walls and the roof is high. Animals can then radiate heat to the cool sink of deep space. Height also allows hot, humid air to rise above the animals, and reduces the radiant load from the roof (Bond, 1967). When there is cloud, high roofs without walls are not an advantage, since clouds reflect solar energy into the shelters.

EVAPORATIVE COOLING

Size also enters into the patterns of evaporative cooling in animals (Bligh, 1972). Few animals weighing less than 3 kg use panting or sweating in temperature regulation. Sweating for cooling is greater the larger the animal, so that cattle and camels depend more on sweating than sheep and goats which, like pigs, use respiratory cooling more (Macfarlane, 1968).

In evolution, large animals were probably at a selective disadvantage when running if the respiratory tract had to supply oxygen, remove carbon dioxide and cool the body at the same time, so sweat dominates the evaporative cooling of mammals of 50 kg and above. In the heat, sweat gland cells are driven by adrenalin to secrete continuously (as in cattle, donkeys and camels), but intermittent myoepithelial contractions in sheep and goats produce phasic sweating patterns that decline after one to two hours (Jenkinson, 1972). In man, sweat rates reach over 2 litres $m^{-2} h^{-1}$, while cattle are the next greatest secretors of sweat (600 ml $m^{-2} h^{-1}$), then come camels (which sweat but do not pant), sheep and goats, with maximum rates in the order of 250 ml $m^{-2} h^{-1}$.

Bos indicus cattle produce more sweat than *Bos taurus* and they have larger sweat glands of great density. A cow sweating at 600 ml $m^{-2} h^{-1}$ and panting would produce vapour at rates of up to 6 kg h^{-1}. During the six hottest hours of the day, over 30 litres of water could be vaporized; and at least as much water again would pass as urine and faeces, some of which would vaporize. At 35 °C, each animal could add about 7 litres of water per hour to the air. This is an essential parameter to be considered in the levels of convection needed in animal housing. The range of respiratory, sweat and excretory water likely to be added to the air by goats, pigs and camels, as well as cattle, is indicated in *Table 15.3*.

In summer, a residual winter coat in camels, or long wool in sheep, makes sweating less effective. But in the sun, shorn sheep are overheated as radiation penetrates to the skin, though in the shade they have the advantage of easier heat dissipation. A housed sheep is better shorn in hot environments since radiant gain is low and evaporation is more effective.

ENDOCRINE ADJUSTMENTS TO HEAT

Extracellular fluid

Mammals react to heat by reducing most endocrine functional levels (Macfarlane, 1963; Pennycuik, 1964). The exceptions are the renin–angiotensin–aldosterone system and vasopressin, which increase in plasma concentration to adjust the extracellular volume (ECV) during evaporative cooling. As water is lost by transpiration through the skin or respiratory tract, vasopressin rises to retain water (Macfarlane *et al.*, 1958), especially in panting animals such as sheep and goats. As the ECV decreases, aldosterone acts to retain sodium, which holds water in the ECV. Vasopressin increases potassium secretion by the kidney in ruminants, and so reduces the body content of this ion, which is present to excess in ruminant food (Macfarlane, 1976).

Animals increase ECV by 20–30 per cent within a day or two of exposure to heat, through retention of sodium and thus also of water (Macfarlane, 1964). Renal monitoring of plasma sodium combined with volume receptors in the atrium, result in the enzyme renin's being released from juxtaglomerular cells of the kidney. This converts plasma protein to angiotensin I, a peptide, which is further split in the lungs to angiotensin II, which releases aldosterone from the adrenal cortex. Sodium is then retained by the kidney and ECV expands. When cool conditions return there is a lag of 30 hours or so before the extra body fluids are excreted, and aldosterone concentrations fall (Macfarlane, 1963; Follenius *et al.*, 1979). So the weight and body fluid of pigs or cattle during summer may be higher than those during a cold period. In the heat there will be more fluid and less tissue per kilogram live-weight.

Adrenal steroids in the heat

Stress reactions by animals are often equated with increases of cortisol or corticosterone in the plasma. If that were so, only rather extreme heat would be stressful, since there is a reduction in cortisol concentration and turnover in man and cattle during moderately hot weather (Robinson, Howard and Macfarlane, 1955; Macfarlane, 1963, 1964; Collins and Weiner, 1968; Christison and Johnson, 1972; Abilay, Johnson and Madan, 1975). The lowering of glucocorticoid secretion probably helps in the reduction of metabolic heat production from protein and carbohydrate turnover. This combines with the lessened food intake of hot animals, and the consequent reduction of thyroxine turnover (Good, Howard and Macfarlane, 1974; Macfarlane, 1976). There are large seasonal changes in the excretion of 17-ketosteroids (Macfarlane, 1963) and in plasma cortisol levels, with winter values reaching twice those in summer (Abilay, Johnson and Madan, 1975).

When unacclimatized animals are exposed to tropical heat, or for tropical animals when temperatures are unusually high, there may be heat strain, in the sense of adrenocorticotrophic hormone (ACTH) release and a rise in corticosteroid output (Robinson, Howard and Macfarlane, 1955; Abilay, Johnson and Madan, 1975; Follenius *et al.*, 1979). There may also be a rise of plasma ACTH in the heat, but less corticoid response from the adrenal than there is in cooler conditions. This has been suggested for cattle by Abilay, Johnson and Madan (1975) and for pigs by Marple *et al.* (1972) as the mechanism by which low plasma cortisol can come about in spite of greater amounts of ACTH in heated animals.

FERTILITY AND FECUNDITY

Some ecotypes are fecund in the heat, like the Kababish sheep of Nubia that annually produce 150 lambs per 100 ewes. But in the Australian Merino the lamb yield decreases steadily from an annual lambing rate of 85 per cent at 35° to 40 per cent at 20°S. Summer gestation in the subtropics (ambient temperatures of 35–40 °C) seems to reduce the size of lambs and the number of wool follicles (Yeates, 1958; Cartwright and Thwaites, 1976). The small lambs often fail to suckle and so die. The combined effects on the ewe of heat and reduced food intake also lead to poor gonadotrophin production and thus to fewer ovulations, implantations and fetal survivals (Macfarlane *et al.*, 1959). For comparison, in rats, lack of acclimatization and low levels of progesterone, B vitamins or protein reduce ova and implants but increase resorptions (Pennycuik, 1964).

Cattle studied in the mid-west of the USA ate less, grew more slowly and produced less milk at 33 °C and 55% RH than at cooler temperatures (18 °C). Guernsey cows also drank 19 per cent more at 33 °C than at 18 °C (Abilay, Johnson and Madan, 1975). Concomitantly there was a two-day lengthening of oestrous cycles, but the duration of oestrus was five hours less. At 33 °C the plasma progesterone level was increased by 1.1–2.6 ng ml^{-1}, mainly in the first 8 days of the cycle, but progesterone remained high till day 19. At the same time luteinizing hormone (LH) concentration in plasma decreased almost to half that at 18 °C (Madan and Johnson, 1971). This could arise from progesterone feedback on the hypothalamus, suppressing output of the LH-releasing hormone.

The reduced efficiency of oestrus in Guernseys is similar to the lower rates of ovulation and embryonic survival, with lengthened cycles, found in *Bos indicus* in the subtropics by Plasse, Warnick and Koger, 1970). Therefore, reduction of any aspect of heat strain is likely to increase reproductive output and to allow survival of more embryos. This is so in pigs (Edwards *et al.*, 1968). Shade, air movement and air cooling could all contribute. But the

use of tropical ecotypes, selection of animals and acclimatization are important for long-term production.

In males there is reduced fertility in the heat, partly because of pituitary
underfunction and partly because of direct scrotal heating (Waites and
Ortavant, 1967). The heat exchange between artery and vein in the blood
supply of the testis is a useful buffer, but once it is overcome, heat blocks the
seventh division of spermatocytes, and 2–3 weeks later there are few sperma-
tozoa for insemination. The sperm that are present may also be damaged,
since Howarth (1969) has observed fetal resorption in unheated ewes mated
with heat-stressed rams. Epididymitis and testicular atrophy are also common among tropical sheep (Moule and Waites, 1963).

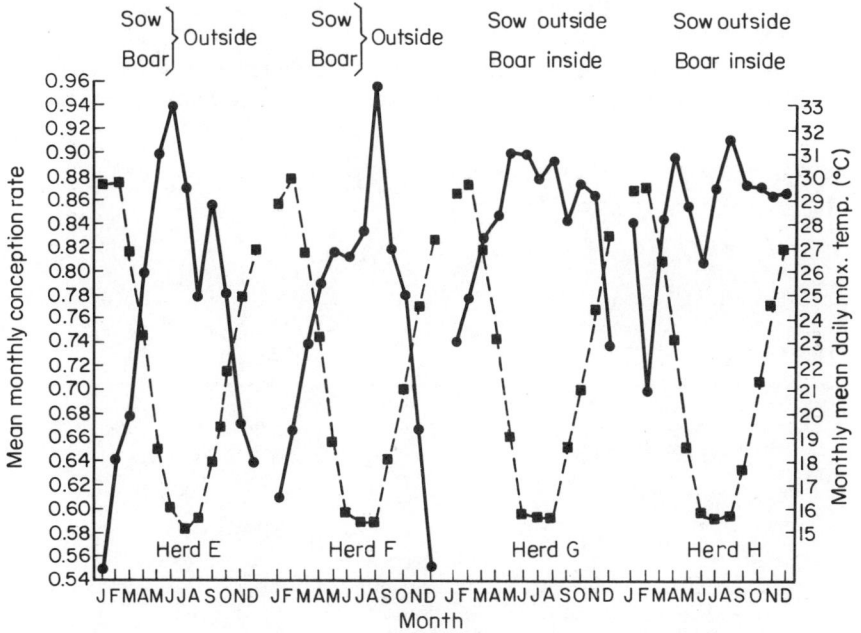

*Figure 15.3 Records of monthly conception rates (●) for four herds of Large White pigs,
showing the effects of exposure of the boar to the outdoor environment (lat. 35 °S) and of housing.
(■) Monthly mean daily maximum temperature. (From Stone, 1977, with permission)*

When Large White pigs were exposed to sun and heat during the summer
on 35°S (Stone, 1977) there was a 40 per cent reduction in conception rate
(*Figure 15.3*). However, housing the males more than halved the depression, even though the sows remained outdoors. Male infertility in the heat
therefore appears to be an important factor in the reproductive failure of
pigs in hot climates, as for sheep. The reduction in semen and sperm
production supports this interpretation. The rise of prolactin in the heat is
also associated with reduced libido and fertility.

COOLING ADJUVANTS AND PRODUCTIVITY

The inward flux of heat to buildings can be reduced by reflective, low-
emissivity roofing, insulation and shade.

In the dry subtropics of Biloela (lat. 24°S), wooden-walled pig houses with *reflective insulation* under galvanized-iron roofing, reduced average daily maximum air temperatures of 37 °C by 3–4 K, relative to half-tank iron shelters (Todd and Daniels, 1968). Large White pigs in the improved housing averaged a daily gain in weight of 636 g (compared with 558 g, without the reflective layer), an 11.4 per cent increment. Thus lessening long-wave radiation reduced strain. The pigs were panting on hot days, but survived 40–43 °C ambient. Since the optimum temperature for most 80 kg pigs is 24–25 °C, they benefit also from less fat, wet floors and good air movement (0.2–0.4 m s⁻¹) when air temperatures rise above the critical balance point.

Feral (black) pigs die if confined in the sun at 35 °C dry-bulb. Acclimatized domestic pigs can survive these conditions, but they lose appetite and become less fecund. Rannfelt and Kroeske (1974) recommended that piglets in the first week be kept at 24–30 °C, falling to 19–24 °C at 8 weeks. They stressed that shade is the essential protection, aided by an insulated, white-painted roof, and a 1 m wall to cut off ground radiation. Sprays are valuable

Figure 15.4 Shade in Ghana (lat. 6 °N) is chosen by growing pigs, rather than the wallow. There is good air movement through the galvanized-iron shelter with eaves to east and west. Pigs do not huddle in the heat (air temp. 30 °C), though most conform socially by sleeping—at the same time

(*Figure 15.4*), though in 10 hours' spraying, the amount of water used was equivalent to 20 litres for a 24 hour period, an appreciable cost in both water and pumps.

LACTATION

Synthesis of milk is metabolically costly. At peak production the turnover of water and energy by a lactating dairy cow is more than doubled (Brody, 1945). In less specialized milk producers, the extra cost in heat and water is about 50 per cent above the non-lactating state (Macfarlane and Howard,

1972). Virtually all hormones are mobilized for lactation: prolactin, oxytocin, growth hormone, gonadotrophins, oestrogens, progesterone, thyroxines, adrenal steroids, parathormone and vasopressin. So any reduction in food or water supply lowers the endocrine output and milk production falls precipitately.

Hot weather acts through the hypothalamic and limbic control of both temperature and neuroendocrine mechanisms. Hahn and Osburn (1970) and Hahn (1976a) have constructed maps predicting the likely reduction of milk output during summer. Hahn tested animals in controlled and natural environments, then predicted the fall in production likely at each high temperature. The predictions were followed in the field, and agreed with the values found to within 5–17 per cent. In the USA the depression of milk production due to heat ranged from 59 kg near lat. 40°N, to 45 kg on lat. 32°, while growth of pigs could be 30 per cent below optimum over the same geographical range (Hahn, 1969). The approach is useful for farmers, though predictions are not easily validated.

ADAPTATION AND BEHAVIOUR

Although it is possible to provide thermoneutral environments for animals, this is expensive. Instead there is a degree of elasticity in the functioning of animals which can be drawn upon as policy in the design of housing. With habituation and acclimatization the effects of heat are less, but cooling by convection (by fan), by evaporative coolers and showers all help sustain milk output (Hahn, 1976b; Folman *et al.*, 1979).

Habituation to hot environments occurs automatically, particularly in young animals. Repeated exposure to heat results in changes of the interneurones of the spinal cord, brain stem and the cortex, so that less discomfort from heat is transmitted through to the perceptive areas of the thalamus and cortex. This form of inhibition can be acquired early in life and is more difficult to implant as neurones become less plastic (Glaser, 1966; Kozak, Macfarlane and Westerman, 1962; Macfarlane, 1963). In addition to reduced perception, there is reduction of the reflex responses to heat, in terms of heart rate changes and vascular adjustments.

Acclimatization to heat is due to a complex of processes, partly nervous (by increased sweating and respiratory cooling), partly endocrine (in the depression of adrenal glucocorticoid output) and metabolic (reduced food intake, increased water turnover and lower rates of use of thyroxine). There are also cellular and enzymatic aspects of acclimatization. Some of these events occur within one week though the greater part of acclimatization takes a month. There is also slower adjustment, over years if achieved at all—tropical Merinos transferred to good, cool pastures did not reach the rates of growth, wool production or lactation of those from temperate zones (Macfarlane and Howard, 1972).

Within any breed or species there is a wide range of potential adaptability. Pennycuik (1964) bred laboratory mice at 22 °C and at 32 °C. At 32 °C the mice were mostly slow in growth, flaccid and infertile. She then selected those which grew, ate and reproduced well at 32 °C, over 12 generations. This resulted in a line of more active mice, with longer tails than at 22 °C, and at 32 °C their metabolic rates and reproductive performances were equal to those at 22 °C. Similar selection is in principle possible for large animals, but the time scale is longer.

Within each species different susceptibilities to environmental pressures are due to breed differences involving metabolic rate, sweating ability, size, coat colour, insulation and behaviour. Selection for tropical climates can be made from this polymorphism. Breeding for improved animal production in the heat should be directed to adaptable animals with lower metabolic rates (compatible with production), light colour, low levels of insulation from hair, and reflective surfaces, good evaporative cooling and relatively small size or large surface area. Endocrine status, both metabolic and reproductive, should be maintained.

The *behaviour* of individuals and groups of animals affects their reaction to housing (Hafez, 1962). Behaviour, such as shade-seeking, differs between individuals and species. For example, when goats, sheep, a donkey and a camel were together in a paddock with trees, behaviour patterns were interestingly diverse. At ambient temperatures of 30–32 °C, the sheep, black goats and donkey sought shade, but smooth-coated Friesian cattle, the camel and white goats remained in the sun. At 35–40 °C with radiation of 1.1 kW m^{-2} the cattle and white goats sought shade. The camel initially was not interested in shelter, but as the summer progressed, it learned the comfort of shade, and joined the others. Thus, learning is probably important as an adjunct to the instinctive and preprogrammed behaviours.

When the pig is hot it seeks any cool material on which to lie while panting, including rolling in its own urine. When pigs had the choice of solid shade, a fabric shelter or a wallow (Heitmann and Hughes, 1949; Bond, 1967), they used more than one cooling mode. When temperatures rose above 21 °C some pigs began to seek shade; by 30 °C over 80 per cent of the animals took some form of shelter; and in the hottest weather (up to 43 °C) there were 86 per cent under shade, 44 per cent under the fabric and 29 per cent in the wallow during the heat of the day. Solid shade was clearly the preferred environment, by day, as it is in the wild pig population, preferably with a damp substrate. But individual animals have different reaction patterns (*Figure 15.4*).

Animals exposed to heat also behave in relation to each other. Social hierarchies show in the takeover of shade or water by the dominant members of the group, while subdominants are denied access to shelter or water. There is segregation of species: horses like to be with horses, and goats with goats. When mixed they segregate within minutes, and the least aggressive species may be denied shade. Feeding behaviour and drinking also change in acclimatization. With acclimatization, food is taken in greater amounts. Sheep show parallel changes in eating and drinking patterns. The more food eaten, the more water is taken, and the less water or the less food, the less of the other components is ingested (Clark and Quin, 1949). In contrast, pigs tend to fill the stomach by drinking more when they eat little dry matter (Yang, 1979). Usually sheep do not drink beyond their water needs, but with long-term heating they may learn to overdrink. Sheep kept at 40 °C drank over 30 litres a day from the fifth to eighth month of exposure. Extracellular volumes expanded and plasma vasopressin rose above 200 μU ml^{-1}. On their return to a cool environment, water turnover dropped to 3 litres a day and vasopressin levels returned to normal. Possibly this was behavioural cooling with large quantities of water, but there could be a neurotic component in which drinking became a diversion. Other neuroses such as tail-biting in pigs and crib-biting by horses are largely signs of boredom. The provision of diversions such as chains or ropes to be played with, and some social interaction outside the shed, helps reduce these, as does lower social density.

Behaviour may also be induced by rewards. Conditioning of milking cows to move to individual stalls, using food as the reward, is standard practice in herring-bone milking sheds. Milking ewes can also be conditioned to move spontaneously to position in rotolactors. Food thus becomes the reinforcing agent in behaviour as well as the nutrient source for lactation.

SUMMARY

In housing animals for hot climates, the biological components are complex. Size, cooling mechanisms and insulation enter into the picture. Young animals produce relatively more heat and water than mature forms and lactation is always expensive in both water and energy. Habituation, adaptation, selective behaviour and learning all play their part in the adjustment which animals make to captivity. The animal house is a basic jail and the amelioration of its conditions should be the aim of good productive processes. In addition, sensitivity to the instincts and reaction patterns of the animals is essential to the successful raising and breeding of animals.

Building design for tropical environments

The principles of building for hot climates were well understood in Roman times. Vitruvius wrote, 'Quod spectat ad occidentam sole exorto tepexit meridie calet vespere fervet'. ('Buildings facing west are warm in the morning, hot at noon and burning by evening.')

In contrast to buildings for cold countries, tropical buildings for intensive production need higher ceilings (5–7 m at the apex) and a greater area of ventilation openings. Openings should be on both sides, while the roof should have sufficient overhang to prevent tropical downpours from entering the animal holding area. The north and south walls can be left open (*Figure 15.1d*). Two storeys are useful, the lower being cooler than the upper. In dry environments mechanical ventilation and cooling have been used for forty years, and may increase productivity (Kelly, Bond and Ittner, 1959). Where fans are used, openings should be near ground level to allow air movement to remove water vapour, carbon dioxide and pathogenic bacteria.

Traditional animal shelters have grown out of the needs, resources and ingenuity of farmers. Only for research and industrial buildings have architects and engineers been brought in; Vitruvius long ago pointed out that most building was vernacular and worked well without benefit of architects. Recently, limitations of resources have forced architects to think in terms of low energy costs and biological realities. The accommodation of large animals in hot climates may be characterized in terms of the levels of sophistication and diversity of purpose for which the shelter is provided.

VERNACULAR STRUCTURES

Sound traditions of housing construction have often been evolved in the field (*Figure 15.5*). For example, in the wet tropics, simple shade rectangles help cattle or buffalo; in Bali (*Figure 15.6*), a bamboo framework with a thatched roof provides shade from the sun, and transmits little long-wave radiation from the undersurfaces. The low-level walls to east and west give shelter from the rising and setting sun, while the open sides allow convective cooling. Trees also cool the air and lessen low-angle solar loads.

In monsoon latitudes, with hot wet and dry periods, solid-wall animal housing is made from mud bricks, wattle and daub or concrete. The roof may be thatch, or flat and made of solid materials. Though ventilation is often poor and little flow-through occurs, exclusion of hot winds is useful for

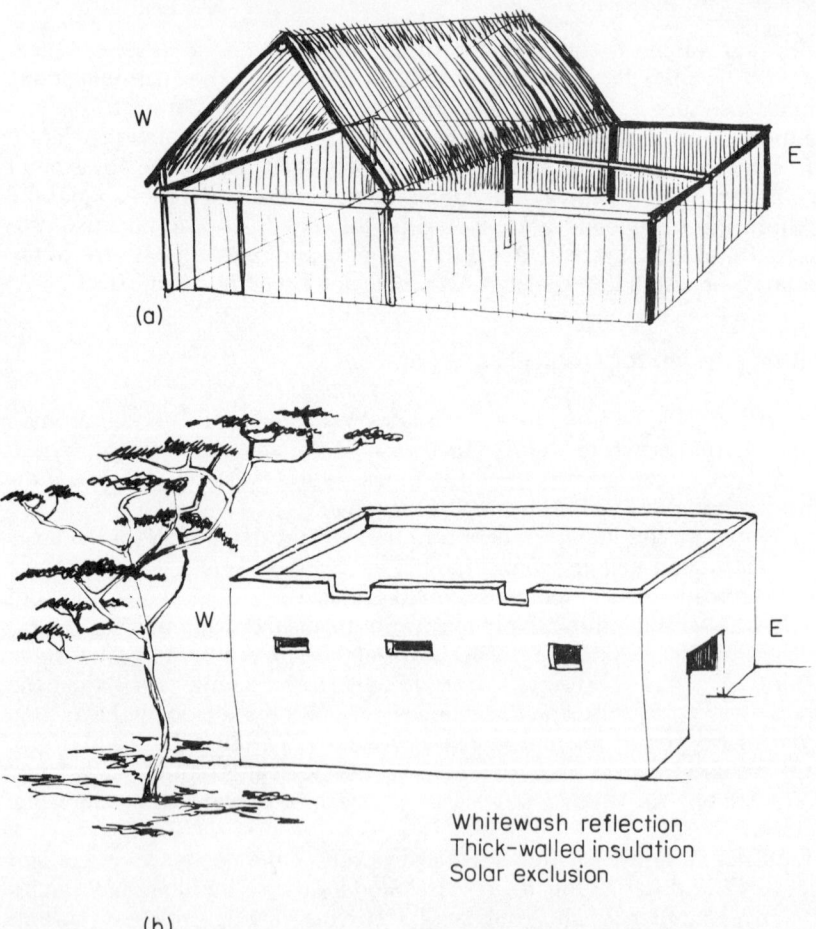

(a)

Whitewash reflection
Thick-walled insulation
Solar exclusion

(b)

Figure 15.5 (a) Indoor–outdoor cattle shed in Kenya. The cool, shaded area under the thatch has good air flow. The outdoor area is suitable for cool-season living. (b) Tropical desert building in Somalia, with small windows, thick walls and white reflective surfaces. Shade trees to the west. There is not much air movement, but if the air temperature is 40 °C or more, it is better outside than flowing through the building

Figure 15.6 Shelter for Bibos banteng *on Bali (lat. 9 °S) made from bamboo and thatch. There is a high woven wall to the west and a lower one on the east. Good shade and flow of air is obtained*

the dry heat, white walls are effective reflectors and narrow windows exclude sun (*Figure 15.5b*).

Stall feeding of pigs and cattle is increasing in both the dry and monsoonal areas (Köppen climates BS, Af and CW; *see* Smith, Chapter 14). Economies in animal housing can be obtained by thoughtful siting. As pointed out by Vitruvius and quantified by Givoni (1974), the morning and afternoon sun can be a powerful source of radiant energy, so that few windows should be built on the east and west walls. The main axis of the building should be east–west, with openings on the long axis. Trees can provide shade and cool the air by evapotranspiration. Tropical or arid-zone animal houses need good access for the winds. Advection may help cooling. Water surfaces, in particular, cool (but humidify) the air moving through the building, and sea breezes can be exploited several kilometres inland from the coast. Similarly, gulley winds moving downhill in the evening can provide cool air.

The removal of effluents by gravity is also important. In the wet tropics drainage may be a major problem and flooding or excess mud may occur in the wet seasons, so elevated sites are even more important in low latitudes.

The best housing design is based on a steep, insulated, white roof, with wide overhang on the east and west against low sun angles, combined with open sides to facilitate ventilation (Gupta, 1970; Starr, Neubauer and Melzer, 1978). Showers, turned on hourly in the summer, are a simple solution to cooling requirements in the management of pigs (*Figure 15.7*). For milk production in hot, arid conditions in the Rift Valley of the Negev, a high roof with free air flow works well (*Figure 15.8*)—there is always wind in the desert.

Traditional thatch is an excellent insulator and is still difficult to match economically except in high-wage areas. The main disadvantages of thatch are that its life is short and that it acts as a refuge for insects, snakes and rats. Polystyrene is effective insulation, but it can be attacked and removed by

Figure 15.7 Intensive pig production in Taiwan. Good air flow. Showers provided for each pen. High roof

Figure 15.8 Cattle barn for beef and dairy Friesians in the Palestine Rift Valley at Yotvata (lat. 30 °N). Air temperatures here reach 45 °C. A corrugated asbestos roof reflects some solar energy and its height (6 m) allows good air flow and reduces long-wave radiation. The floor was covered with straw litter and remained dry during the hot season. Showers were provided thrice daily to dairy cows yielding up to 9000 l per lactation

birds. A combination of corrugated iron or asbestos with thatch is cool and more lasting.

Iron roofing benefits from a reflective inner layer when used alone. The modern practice of bonding a thin sheet of aluminium to iron gives very low emissivities (in the order of 0.04 when the surface is clean), so that the amount of long-wave radiation passing down to the animals is greatly reduced. The inner foil is equivalent to abut 5 mm of insulation.

Where solar radiation is a major source of heat, white walls and roof aid the comfort of animals (*Figure 15.5b*). A white wall heats only to 55–60 °C in the sun, whereas a dark wall or roof heats to 70–80 °C, according to its reflectivity. White roofing is superior to either aluminium or galvanized iron in its raw state, both of which oxidize and lose reflectivity.

Conclusions

Animals find thermal comfort and function most efficiently near their thermoneutral zone. They grow better and produce more if they are selected for thermal tolerance, are habituated to heat from birth and are acclimatized. The basic ecophysiology of each species determines the amount of heat, water and egesta released to the built environment. Small size, young age group, lactation and tropical ecotype all reduce the thermal range of effective production because of high heat output and demands on water and food. Ecophysiological traits also determine the relative effects of thermal stress on growth, physiological strain and reproduction. Overall, the stress of hot environments reduces food intake, growth rate, reproductive output and lactation. Shelter can help lessen the strain which leads to losses of production.

Housing animals may have several advantages; in particular, some of the disabilities arising from thermal strain can be reduced. The advantages have not shown up consistently, however, and depend on whether the animals are adapted and on the magnitude of heat stress, though shelter can increase the efficiency of breeding and milk production. Therefore, each set of variables needs local experimental determination of the economic use of shelter. In addition to its direct effects on the animals, housing also allows better control of food intake and its quality. With this goes a reduction in the amount of work an animal must do in sweating, panting and feeding; but the costs of construction, labour and maintenance must be set against the gains. High prices for animal products and improved food use are needed in order for housing to be justified. The comfort of the animals is also sometimes forgotten, but a perceptive policy in providing comfort is likely to have economic yields also.

Disadvantages may also accrue. First, the cost of materials and energy, for construction of shelter, food preparation, cartage, maintenance and cleaning can be formidable. Convective or evaporative cooling and refrigerative cooling are even more costly. Secondly, not all animals are equally suited to the social contiguities and dense packing that animal houses demand. Infectious diseases spread readily and particular difficulties may occur with the feet of animals unaccustomed to hard and sometimes treacherous substrates. Thirdly, metabolic disturbances are readily induced by inapposite food; for example, dairy cattle suffer considerable acidotic and ketotic diseases together with disturbance of calcium metabolism.

References

ABILAY, T.A., JOHNSON, H.D. and MADAN, M. (1975). *J. Dairy Sci.* **58**, 1836–1840

ADOLPH, E.F. (1949). *Science, NY* **109**, 579–585

BLIGH, J. (1972). *Symp. zool. Soc. Lond.* **31**, 357–369

BOND, T.E. (1967). *Microclimate and Livestock Performance in Hot Climates. Ground Level Climatology Symposium*, pp. 207–220. Ed. by R.H. Shaw. American Association for the Advancement of Science, Washington, DC, Publication No. 86

BRAUDE, R. and ROWELL, J.G. (1967). *J. agric. Sci., Camb.* **68**, 325–330

BREWER, N.L. and MACFARLANE, W.V. (1967). *Aust. J. exp. Biol. med. Sci.* **45**, 37

BRODY, S. (1945). *Bioenergetics and Growth.* Reinhold, New York

BRUCE, J.M. (1979). *Fm Bldg Prog.* **55**, 1–4

BURTON, A.C. and EDHOLM, O.G. (1955). *Man in a Cold Environment.* Arnold, London

CARTWRIGHT, G.A. and THWAITES, C.J. (1976) *J. agric. Sci., Camb.* **86**, 573–585

CHRISTISON, G.L. and JOHNSON, H.D. (1972). *J. Anim. Sci.* **35**, 1005–1009

CLARK, R. and QUIN, J.I. (1949). *Onderstepoort J. vet. Sci. Anim. Ind.* **22**, 335–343

COLLINS, K.J. and WEINER, J.S. (1968). *Physiol. Rev.* **48**, 785–839

EDWARDS, R.L., OMTVEDT, E.J., TURMAN, D.E., STEPHENS, D.E. and MAHONEY, G.W.A. (1968). *J. Anim. Sci.* **27**, 1634–1637

FOLLENIUS, M., BRANDENBERGER, G., REINHARDT, B. and SIMEONI, M. (1979). *J. appl. Physiol.* **41**, 41–50

FOLMAN, Y., BERMAN, A., HERZ, Z., KAIN, M., ROSENBERG, M., MAMEN, M. and GORDIN, S. (1979). *J. Dairy Res.* **46**, 411–425

GIVONI, B. (1974). In *Progress in Biometeorology*, pp. 183–193. Ed. by S.W. Tromp. Swets & Zeitlinger, Amsterdam

GLASER, E.M. (1966). *The Physiological Basis of Habituation.* Oxford University Press, London

GOOD, B.F., HOWARD, B. and MACFARLANE, W.V. (1974). *Proc. Endocr. Soc. Aust.* **17**, 32

GRIBBLE, D.J. (1973). 'Dairy systems in the warmer climates'. *National Dairy Housing Conference, American Society for Agricultural Engineers, St Joseph, Michigan, Special Publication 01–73*, pp. 66–75

GUPTA, C.L. (1970). *Bldg Sci.* **5**, 165–173

HAFEZ, E.S.E. (1962). *The Behavior of Domestic Animals.* Lea & Febiger, Philadelphia

HAHN, G.L. (1969). *J. Dairy Sci.* **52**, 800–802

HAHN, G.L. (1976a). *Biometeorology* **6**, 106–114

HAHN, G.L. (1976b). In *Progress in Biometeorology*, pp. 496–503. Ed. by H.D. Johnson. Swets & Zeitlinger, Amsterdam

HAHN, L. and OSBURN, D.D. (1970). *Trans. Am. Soc. agric. Engrs* **13**, 289–294

HAINES, H., MACFARLANE, W.V., SETCHELL, C. and HOWARD, B. (1974). *Am. J. Physiol.* **277**, 958–963

HEITMAN, H., Jr and HUGHES, E.H. (1949). *J. Anim. Sci.* **6**, 171–181

HERRE, W. and RÖHRS, M. (1977). In *Origins of Agriculture*, pp. 245–280. Ed. by C.A. Reed. Mouton, The Hague

HOWARTH, B. (1969). *J. Reprod. Fert.* **19**, 179–183

INDIAN STANDARDS INSTITUTION (1978). 'Recommendations for farm cattle housing for arid areas.' IS : 8845, parts I, II, III. 'Recommendations for farm cattle housing for heavy rainfall and high humidity areas.' IS : 5605, 1970. Manak Bhavan, New Delhi

INGRAM, D.L. (1965). *Nature, Lond.* **207**, 415–416

JENKINSON, D.M. (1972). *Symp. Zool. Soc. Lond.* **31**, 345–356

KELLY, C.F., BOND, T.E. and ITTNER, N.R. (1959). *Trans. Am. Inst. elect. Engrs*, Part 2: *Applics Ind.* **40**, 512–517

KING, J.M. (1979). *J. agric. Sci., Camb.* **93**, 71–79

KOEPPE, C.E. and DE LONG, G.C. (1958). *Weather and Climate*. McGraw-Hill, New York

KOZAK, W., MACFARLANE, W.V. and WESTERMAN, R. (1962). *Nature, Lond.* **193**, 171–173

LEDGER, H.P. (1968). *Symp. Zool. Soc. Lond.* **21**, 289–310

MACFARLANE, W.V. (1963). In *Environmental Physiology and Psychology in Arid Conditions*, pp. 153–222. UNESCO, Paris

MACFARLANE, W.V. (1964). In *American Handbook of Physiology*, vol. 4, pp. 509–531. Ed. by D.B. Dill. American Physiological Society, Washington, DC

MACFARLANE, W.V. (1968). In *The Adaptation of Domestic Animals*, pp. 164–182. Ed. by E.S.E. Hafez. Lea & Febiger, Philadelphia

MACFARLANE, W.V. (1971). In *Salinity and Water Use*, pp. 161–178. Ed T. Talsma and J.R. Philip. Macmillan, London

MACFARLANE, W.V. (1976). In *Veterinary Physiology*, pp. 463–539. Ed. by J.W. Phillis. Wright-Scientechnica, Bristol

MACFARLANE, W.V. (1978). In *Water*, pp. 108–143. Ed. by A.K. McIntyre. Australian Academy of Science, Canberra

MACFARLANE, W.V. and HOWARD, B. (1972). *Symp. Zool. Soc. Lond.* **31**, 261–296

MACFARLANE, W.V., HOWARD, B. and GOOD, B.F. (1974). In *Tracer Techniques in Tropical Animal Production*, pp. 1–23. IAEA, Vienna

MACFARLANE, W.V., PENNYCUIK, P.R., YEATES, N.T.M. and THRIFT, E. (1959). In *Endocrinology of Reproduction*, pp. 81–96. Ed. by C.W. Lloyd. Academic Press, New York

MACFARLANE, W.V., ROBINSON, K., HOWARD, B. and KINNE, R. (1958). *Nature, Lond.* **182**, 672–673

McMAHON, T. (1973). *Science, NY* **179**, 1201–1204

MADAN, M.L. and JOHNSON, H.D. (1971). *J. Dairy Sci.* **54**, 793–797

MARPLE, D.N., ABERLE, E.D., FOREST, J.C., BLAKE, W.H. and JUDGE, M.D. (1972). *J. Anim. Sci.* **34**, 809–812

MOULE, G.R. and WAITES, G.M.H. (1963). *J. Reprod. Fert.* **5**, 433–446

MOUNT, L.E. (1979). *Adaptation to Thermal Environment*. Arnold, London

PENNYCUIK, P.R. (1964). *Aust. J. med. Res. biol. Sci.* **17** (1), 245–260

PLASSE, D., WARNICK, A.C. and KOGER, M. (1970). *J. Anim. Sci.* **30**, 63–72

PRIESTLEY, C.H.B. (1971). *Aust. Physicist* **8**, 9–15

RANNFELT, C.A. and KROESKE, D. (1974). *Wld Anim. Rev.* **10**, 24–30

ROBINSON, K.W., HOWARD, B. and MACFARLANE, W.V. (1955). *Med. J. Aust.* **2**, 756–760

SCHMIDT-NIELSEN, K. (1975). *How Animals Work*. Oxford University Press, Oxford

STARR, G., NEUBAUER, L. and MELZER, B. (1978). American Society of Agricultural Engineers, paper No. 78-4003, pp. 1–19

STONE, B. (1977). *Agric. Rec. (South Australia)* **4**, 22–25

TODD, A.C.E. and DANIELS, L.J. (1968). *Proc. Aust. Soc. Anim. Prod.* **7**, 285–288

UCKO, P.J. and DIMBLEBY, G.W. (1969). *The Domestication and Exploitation of Plants and Animals*. Duckworth, London

VITRUVIUS, M. (30 BC). *Decem Libri de Architectura*. Ed. by F. Granger. Heinemann, London (1931)

WAITES, G.M.H. and ORTAVANT, R. (1967). *Aust. J. exp. Biol. med. Sci.* **45**, 4

YANG, T.S. (1979). *PhD Thesis*, University of Adelaide

YEATES, N.T.M. (1958). *J. agric. Sci., Camb.* **51**, 84–89

ZEUNER, F.E. (1963). *A History of Domesticated Animals*. Hutchinson, London

16

EXTREME HEAT STRESS IN DAIRY CATTLE AND ITS ALLEVIATION: A CASE REPORT

R.H. ANSELL
World Bank Agricultural Development Service, Sudan

Introduction

There has been much controversy in the past regarding the suitability of *Bos taurus* dairy stock for use in the tropics. The proponents and opponents of such stock have variously blamed the unfamiliarity of local workers with the needs of such animals, poor management practices, poor nutrition or the effects of high ambient temperatures. Without doubt, it is the last factor which has caused the most controversy—are high ambient temperatures and humidities inimical to satisfactory milk production?

With the financial aid of the US Department of Agriculture, the University of Missouri initiated a comprehensive programme in 1948 to investigate this problem of heat stress in cattle and their work (referred to later in the text) has continued almost to the present day. Other workers in various parts of the world have, of course, also made valuable contributions to the subject, but Missouri seem to have been predominant in the field.

Unfortunately, most of the work that has been done by these workers and others has been carried out in climatic chambers. The term 'unfortunately' is used since there is mounting evidence that some of the results obtained from chamber experiments are at variance with results obtained under field conditions and it is possible that the close confinement consequent on climatic chamber experiments may have had psychosomatic effects.

It is to be regretted that very little work from the field has been published on this problem either to support or refute the laboratory findings, but this is understandable. Scientists are usually reluctant to carry out research in which the variables cannot be strictly controlled and field work with cattle is notoriously difficult in this respect. There is thus an enormous mass of data on the response of European dairy cattle to heat stress under laboratory conditions, but very little confirmatory work under more natural conditions. When, therefore, an opportunity presented itself to carry out such a study under field conditions of very low variability it was recognized as being unique and was promptly utilized.

Such a situation arose in 1970 in the United Arab Emirates (then called the Trucial States). Thirty pure-bred Friesian in-calf heifers were imported from the UK, together with two bulls, to the Agricultural Experimental Station of Digdaga in the state of Ras al Khaimah. The original intention was that these animals should constitute a demonstration herd for the Agricultural School students attached to the centre. However, it soon became apparent that the cost of this herd could not be justified on that count alone and therefore permission was sought and obtained for the herd to be used for research into the response of such animals to the very high ambient temperatures and humidities of the Arabian Gulf littoral. This area is one in which the climate is extremely severe for cattle, particularly of the *Bos taurus* type. It has very high temperatures, high humidities, low wind velocities and little diurnal variation; all the essentials for a high stress factor. In addition, the weather pattern is extremely stable. During the summer months the cloud cover is virtually negligible and the wind, temperature and humidity vary very little, making the site particularly suitable for research purposes. It could, in fact, be likened to a natural climatic chamber.

One other factor that was advantageous from the research aspect was that the system was one of zero grazing; all food was cut and carted, thus making a controlled diet possible.

A four-year study was therefore carried out on this herd and extensive and detailed records kept. It is the findings and conclusions from this study that are presented here in the hope that they can be of assistance to others who wish to raise *Bos taurus* stock under high-temperature regimes and, in addition, to resolve in some measure some of the controversies that have surrounded the introduction of such stock into the tropics.

Climate

Figure 16.1 gives an indication of the dry-bulb temperature and relative humidity during a fairly typical mid-summer day at Digdaga, and although

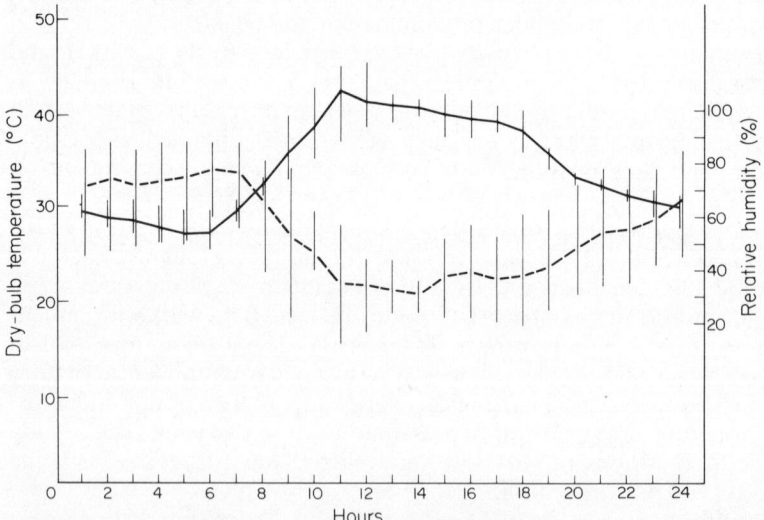

Figure 16.1 Mean diurnal variation of relative humidity (– – –) and ambient temperature (——) for July 1972. Hourly readings were taken during one day of each week and the curves are the arithmetic means of these four sets of readings for the month in question. The vertical bars represent the range of the readings constituting the mean. These readings are deep-shade readings and are much lower than those recorded in the standard Stevenson screen, which can be 7–8 K higher

somewhat cumbersome, this method of depicting a climate is necessary to be of any real use. It will be noted that the diurnal variation is depicted as recorded hourly. When considering physiological effects on stock, it is of little value quoting arithmetic means of maxima and minima in the ordinary meteorological manner, as in *Figure 16.2*, since it has been amply demonstrated that cattle are able to tolerate very high temperatures during the day

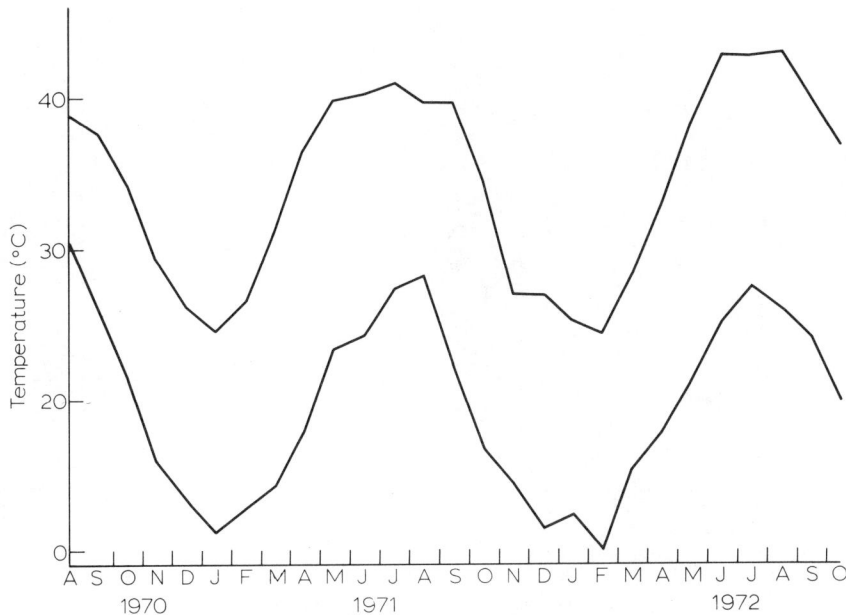

Figure 16.2 The monthly means of maximum and minimum deep-shade temperature in a covered yard for late 1970, 1971 and 1972. In 1972 temperatures exceeded the upper limit of tolerance of the lactating animals. Monthly means are of little significance, however, without the information in Figure 16.1

if there is adequate remission at night for them to eliminate their surplus body heat. Thus, a climate having very high day-time temperatures with low night temperatures (sometimes called a true desert climate) is far better tolerated than one in which the arithmetic mean is the same but which has little diurnal variation.

Care should be taken when considering these temperatures, since they were all taken in the deep shade of the cowshed away from high-radiation effects. They bear little relation to the temperature as normally recorded in the standard 'Stevenson screen' meteorological box. These boxes are inadequately shielded from solar radiation in the tropics, and record temperatures that are considerably higher than deep-shade ones—which reflect a truer air temperature (Ansell, 1974). The deep-shade system was chosen since, although the results are not comparable with standard meteorological

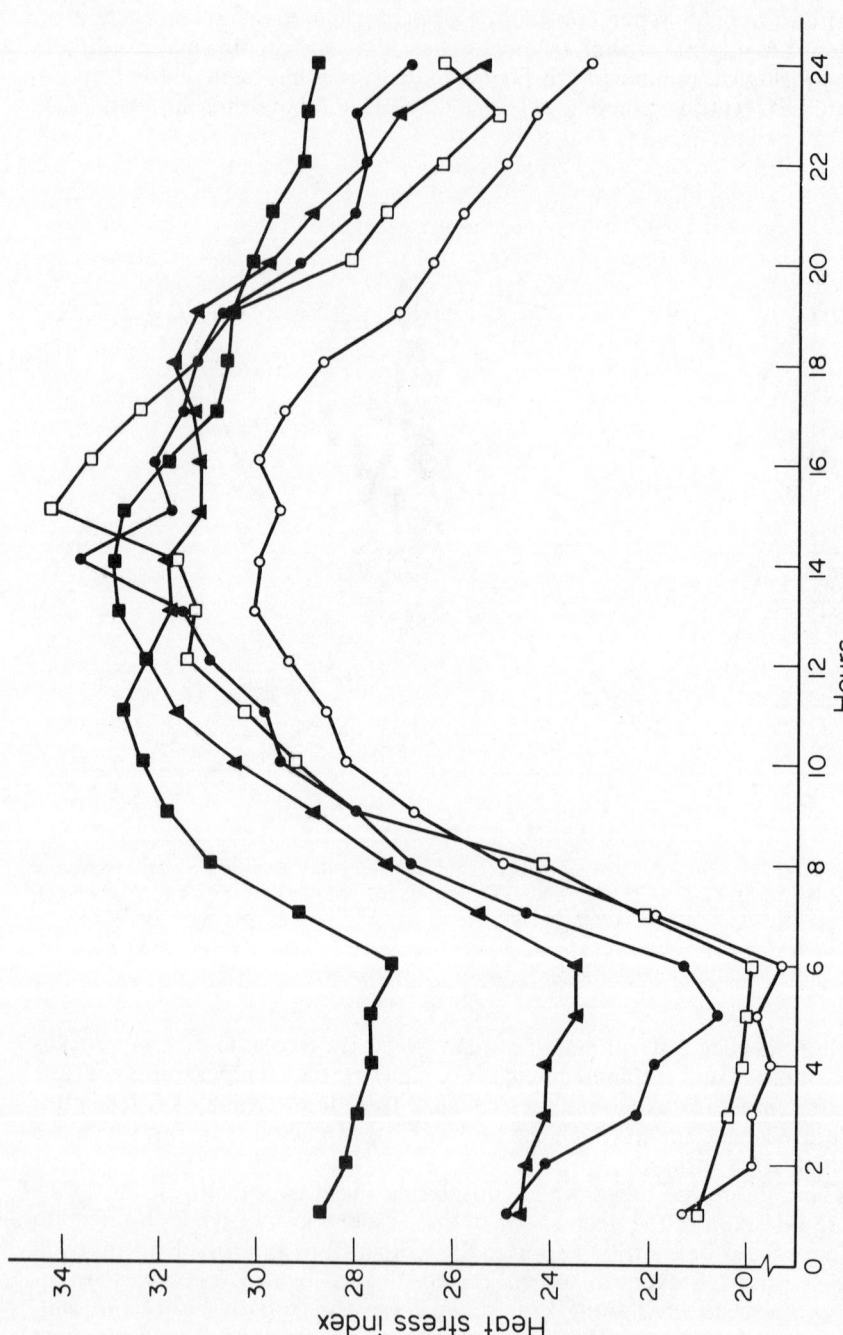

Figure 16.3 Hourly values of Bianca's Heat Stress Index (Bianca, 1962) for five days in 1972. (○) 1 June; (●) 8 June; (□) 15 June; (■) 22 June; (▲) 29 June. The index is calculated as a temperature equal to (0.35T + 0.65T) where T ... dry bulb and ... is the ...

records, they are more nearly comparable with climatic-chamber temperatures, with which most previous work was done.

In this work also, only sling or wet- and dry-bulb hygrometers were used
for relative-humidity measurements, since it was found that paper and hair
hygrometers were grossly and inconsistently inaccurate.

It is unfortunate that it is not yet possible to compare differing climates in
terms of their stressing effect on cattle. Each combination of wind speed,
temperature, humidity and time varies the effect. Attempts have been made
both in human and veterinary medicine to deduce a stress factor based on
relative humidity and temperature alone. Perhaps the most promising one
for cattle is that of Bianca (1962), in which the dry-bulb temperature
multiplied by 0.35 is added to the wet-bulb temperature multiplied by 0.65,
to give a stress factor. This factor was calculated over a 24 hour period during
the study (*Figure 16.3*). This represents a period during which the stress was
known to be approaching the upper limit of tolerance of the experimental
animals when in milk, and could be of some use to those wishing to ascertain
whether their own climate is more severe than that of Digdaga. It should be
remembered, however, that this and other formulae have not been fully
confirmed experimentally and in any case ignore wind, time and radiation
effects. They are thus only guides.

Feeding

After some experiments the diet of these animals was stabilized and that
chosen was then fed both summer and winter for two years. The dry-matter
intake and metabolizable energy are lower than would be recommended for
similar animals in a European climate, but this diet maintained the animals
in good and often fat condition at all times.

The feeding regimes were as follows:

Maintenance ration

0400 6.8 kg fresh-cut alfalfa
0700 1.8 kg bran + 14 g vitamin and mineral supplement (Coopavite) + 57 g
       dicalcium phosphate
1130 6.8 kg fresh-cut alfalfa
1630 3.2 kg Bronze Label Dairy Nuts (British Oil & Cake Mills)
1730 6.8 kg fresh-cut alfalfa

Production ration

Bronze Label Dairy Nuts (BOCM) (1.8 kg per 4.5 kg of milk), divided into
two feeds and fed during milking.

Calves

Calves were suckled for the first 4 days and then transferred to the calf house
and bucket-fed fresh or reconstituted milk for a period of 3 months with an
average of 4.5 kg d^{-1}. In addition, fresh-cut alfalfa and calf weaner pencils
were available *ad libitum* at all times.

Drinking water was freely available to all stock at all times.

The only variation to this regime was that during the height of the summer the cows in the milking herd took considerably longer to consume their ration of alfalfa at 1130 hours; it was found necessary to cut each animal's ration by 2.3 kg at this time and add it to the evening meal. At no time was the total ration not consumed, however, even at the height of summer.

Production

The main concern of this study was to establish whether or not reasonable milk yields could be obtained in the climate existing in the Arabian Gulf from a pure-bred dairy breed of European origin. For this purpose it was necessary to assess the depressant effect of the hot season on the lactation curve, and to what extent the yield would recover on the resumption of the cool season. These factors are also important in that they should indicate the pattern of breeding to be adopted for maximum milk yield. As will be seen later, the temperature effects on breeding efficiency prevent the lactation from being confined entirely to the cooler months.

The inevitable problem facing the worker in this field is devising a method for estimating what would have been the yield without the experimental conditions. Four main methods have been utilized in the past:

(1) the use of identical twins;
(2) the reversal or double reversal system;
(3) animals paired on the basis of age, previous lactation, etc.
(4) the Standard Lactation Curve method, whereby expected yield is calculated from the lactation curves compiled by various authorities.

All these methods have various disadvantages, but in this work it was decided to utilize the fourth system as being the only practical alternative.

Yields during the cool months prior to stress conformed remarkably well to the normal curve described in the UK Milk Marketing Board's Production Division Report for 1969/70 (Anon., 1969/70), whose formula has been utilized in this study (*Figure 16.4*). However, where an animal's lactation curve prior to stress differed markedly from the standard curve, so that no trend could be established, that lactation was eliminated when calculating deviations. It must be remembered, however, that whatever method is used to determine the effects of experimental conditions on expected milk yields, results can only indicate trends when dealing with such an inherently variable quantity, and these results should be viewed in this light.

Using the lactation curve method it was found that there was little depression of milk yield for the herd capable of being attributed solely to the climatic factors. In the first lactation (in 1970) it was calculated that the total loss of production from this cause during the five months of May to September inclusive, i.e. the whole of the summer, was 2.8 per cent of the expected 305 day lactation. In the second lactation the result was calculated by a slightly different statistical approach, but again produced a low figure, the total loss for five months amounting to 2.0 per cent of the total expected yield.

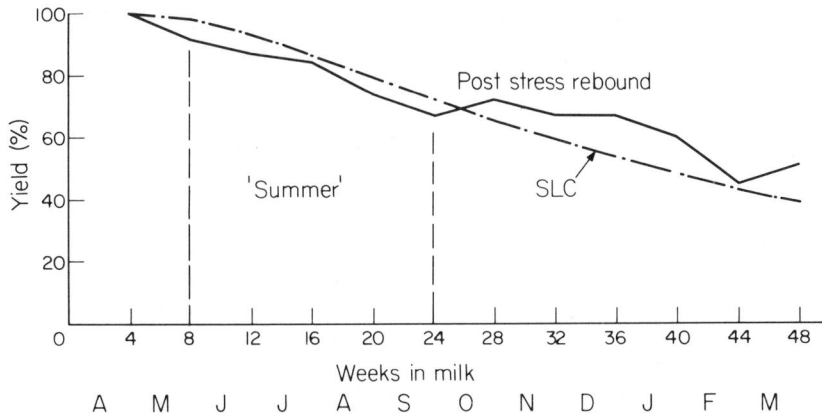

Figure 16.4 Comparison of the Standard Lactation Curve (SLC) (–·–·–) with a typical pattern of milk production from the Digdaga herd (————). The depression during the summer months was frequently followed by an increase, compared with the expected yield, at the onset of the cooler weather

It must be emphasized that these figures were produced whilst the animals were suffering the full rigours of the Gulf climate and often showed signs of advanced heat stress. The only mitigation of the climate was the provision of adequate shade. In 1972, however, the summer was particularly severe and it was apparent that the animals had reached the upper limit of their tolerance. Steps were necessary to protect the health or even the life of the herd and, as mentioned later under the heading 'Shelter engineering', a rotating lawn sprinkler was erected in the roof of the cowshed. The animals had free access to this during the period from 1130 to 1700 hours. Under this regime the estimated monthly depression in yield was calculated separately and amounted to 0.6 per cent of the total expected lactation.

Another effect noticed during this work was that during April and May, when temperatures were increasing and were already higher than normal European summer levels, milk production often rose above the expected yield. Although the rise was small it was a fairly consistent occurrence. This observation might well suggest that the ideal ambient temperature for these cows is much higher than that normally considered optimum for *Bos taurus*.

Yet another and very unexpected effect was a 'post-stress rebound'. At the onset of the cooler weather in October the lactation curves show a marked upward trend, attaining a level above the 'norm'. This was nearly enough to offset the previous heat stress depression (*Figure 16.4*). One implication of this phenomenon is that it may be misleading to connect, in a graph, the yield before and after exposure to heat stress to determine the shape of the lactation curve, a technique often used in climatic chamber experiments.

Although the animals at Digdaga were pure-bred British Friesians, they were not of a very high quality; in fact some were of a decidedly poor conformation. Despite this, the average yields of the herd were:

| 1st Lactation | 19 animals | 305 days | 3253 kg |
| 2nd Lactation | 7 animals | 305 days | 4569 kg |

Table 16.1 FIRST LACTATION

Cow No.	Calving date	Lactation length (d)	Total yield (kg)	305 day yield (kg)	Mean daily yield to 210 days (kg)
19	4. 3.70	525	4937	3023	10.35
121	13. 5.70	329	2551	2454	9.19
127	18. 4.70	413	4180	3347	11.54
171*	14. 1.70	Died in 39th week		–	12.90
223*	27. 1.70	368	4011	3454	12.10
227*	4. 2.70	354	3582	3368	13.02
254*	28. 1.70	332	3192	3061	10.54
255*	10. 1.70	343	3472	3231	11.84
308	26. 3.70	553	5168	3449	12.51
313	5. 4.70	427	5060	3749	12.52
314*	20. 1.70	480	4291	3156	11.27
315	30. 5.70	399	3014	2389	8.02 3 teats
401*	15. 2.70	293	3320	–	13.33
431	10. 5.70	462	4970	3449	11.59
535	20. 3.70	365	2465	2115	7.34
549	4. 4.70	448	4454	3489	11.86
757	28. 2.70	294	2897	–	10.97
764*	5. 1.70	292	2565	–	10.53
782	4. 3.70	504	5531	3615	12.46
791*	23.12.69	333	3353	3296	12.12
792*	26.12.69	476	4760	3521	12.80
811*	2. 1.70	331	3650	3547	12.49
907	8. 6.70	378	4427	3541	10.97
957*	30. 1.70	330	3181	3046	10.87

*Milk recording commenced 25 February 1970. Yields prior to this have been estimated on the basis of the Lactation Curve.

Table 16.2 SECOND LACTATION

Cow No.	Calving date	Lactation length (d)	Total yield (kg)	305 day yield (kg)	Mean daily yield to 210 days (kg)
19	22.10.71	266	3970	–	17.03
121	13. 8.71	154	1291	–	– 3 teats
127	15. 8.71	387	4495	4142	15.40
223	21. 3.71	245	3958	–	17.30
227	9.11.71	301	4288	–	17.76
254	3. 2.71	259	3419	–	14.55
255	5. 1.71	238	2505	–	11.61
308	10. 9.72	Not complete, aborted 2. 5.71 and 4.11.71			17.89
313	12. 8.71	336	5534	5387	19.32
314	20.11.71	329	5237	5151	20.43
315	21. 8.71	308	2848	2844	10.29 3 teats
401	13. 1.71	238	3667	–	16.26
431	6.11.71	322	4591	4537	17.94
535	15. 8.71	252	2177	–	9.78
549	18. 9.71	259	2851	–	12.90
757	21. 2.71	Died 7.5.71	–	–	–
764	30.11.70	259	3250	–	13.98
782	18.11.71	266	5367	–	23.03
791	23.12.70	238	2998	–	13.79
792	28. 8.71	392	3717	3291	11.97
811	17. 1.71	301	5189	–	19.30
907	6. 9.71	287	3945	–	15.50
957	27. 2.71	280	4125	–	16.93

NB. Cows 19, 227, 314, 431, 782 and 792 had access to the overhead sprinkling system as from 1.7.72.

Table 16.3   THIRD LACTATION

Cow No.	Calving date	Lactation length (d)	Total yield (kg)	305 day yield (kg)	Mean daily yield to 210 days (kg)
19	19. 7.72	322	2220	2169	8.52
121	8. 7.72	266	2747	–	12.12
127	26. 9.72	210	1402	–	6.68 3 teats
223	21. 1.72	301	5054	–	20.46
227	14.12.72	*	–	–	–
254	7.12.71	252	3745	–	16.64
255	24.11.71	266	3410	–	14.69 3 teats
313	17. 9.72	*	–	–	18.98
314	15.12.72	*	–	–	–
315	1. 7.72	231	1393	–	6.50 3 teats
401	16.11.71	273	4158	–	17.31
431	3.11.72	*	–	–	17.58
535	2. 7.72	301	2516	–	10.32
549	30. 7.72	273	3223	–	14.51
764	28.10.71	259	2738	–	11.94
782	31.10.72	*	–	–	18.57
791	5.11.71	273	3276	–	14.21
792	29.11.72	*	–	–	–
811	13. 1.72	294	4753	–	19.10
907	21. 7.72	*	–	4198	13.55
957	1. 3.72	252	3036	–	13.47

*Lactation continuing.
NB. All but Cow 764 had access to the overhead sprinkling system during part of these lactations.

A summary of the yields is depicted in *Tables 16.1–16.3*. These compare favourably with the average lactations of 859 Friesians which were used in Britain to compile the standard lactation curve. Their corresponding yields were 3842 kg and 4512 kg. In addition, no culling of any kind was carried out in the experimental herd, as would have been normal under commercial conditions.

These results would certainly not be expected from climatic-chamber work, since practically all workers have reported very substantial drops in yield when operating at 'Gulf' temperatures. For example, Ragsdale *et al.* (1953) found that milk production almost ceased at 37.7 °C and 47% relative humidity (RH). The same authors earlier noted a drop of 60 per cent when Brown Swiss cattle were subjected to an increase in chamber temperature from 10 °C to 40.5 °C. Unfortunately, they did not specify the relevant RH measurements, but they were probably between 40% and 60%. Again, Brody *et al.* (1955) found that a diurnal temperature cycle between 18 and 38 °C with RH varying between 40% and 60% produced a 20 per cent reduction during a three-week exposure but, as has been mentioned previously, it is possible that the close confinement in experimental chambers could itself produce depressant effects.

Attention should be drawn to cows 223, 957 and 811, whose second lactations extended throughout the summer. Calving on 21.3.71, 27.2.71 and 17.1.71 respectively, their mean daily yields to the 210th day were 17.30, 16.93 and 19.30 kg. These animals are fairly convincing proof that even pure-bred Friesians—which are probably not the most tolerant of European breeds—can produce remarkably well under these conditions and there is reason to suppose that with a certain amount of selection, herd results of this order might be achieved.

It appears from the Digdaga results, therefore, that pure-bred Friesians can produce satisfactorily in the climate of the Arabian Gulf although, by virtue of its high humidity–temperature combination, it is one of the most severe in the world in which dairy farming of any kind could be contemplated. Even on those occasions when the stress does exceed tolerable limits, the provision of a simple showering mechanism has been shown to control adverse effects (*see below*).

Reproduction

The reproductive performance of the herd was, however, less satisfactory, as shown in *Figure 16.5*. Between the end of May and the end of October 1970 only one service resulted in a successful pregnancy.

All the cows calved after arrival between December 1969 and June 1970. In general, the earlier calvers were able to conceive before the temperatures became excessive but the later calvers failed to do so. The herd thus became split into two groups; the latter group did not conceive until the autumn and thus had extended lactations.

Stott and Williams (1962) concluded from their work carried out in Arizona that, under their climatic conditions (which unfortunately are not described in sufficient detail to use for comparison with the present work), the effects of high ambient temperature on reproduction are temporary and that the possibility of an animal conceiving and maintaining the conceptus, having failed to do so previously owing to heat stress, is not thereby lessened. However, the pattern which emerged during the present study was that there was an excessively long period between the end of the hot season and the initiation of a successful pregnancy in those animals served during the summer months (*Figure 16.5*). Johnson, Naelapaa and Frye (1963) have shown that bulls exposed to only seven days at temperatures with a diurnal variation approximately equal to those at Digdaga, suffered semen abnormalities up to nine weeks after exposure. Thus, some of the difficulties experienced with this herd in 1970 could be attributed to high temperatures, but on the other hand, five of the ten animals served as early as November conceived, indicating some degree of fertility in the bull one and a half months after the (admittedly arbitrarily chosen) end of summer, despite the severity and duration of exposure. The severity of the exposure of this bull was such that towards the end of each summer it was regularly observed that the scrotum developed numerous ecchymoses and petechiae. This phenomenon was often accompanied by a generalized reddening of the skin of the body of the animal, an occurrence that was not observed in the cows, although the bull's rectal temperature never reached the levels experienced by the lactating animals.

These records suggest that, under environmental conditions of this type, successful matings are unlikely to occur in summer, here between June and October. Thus the breeding programme should be arranged to avoid the necessity for mating during these months. Evidence of milk production, birth weights of calves, premature births and homeothermic regulation of

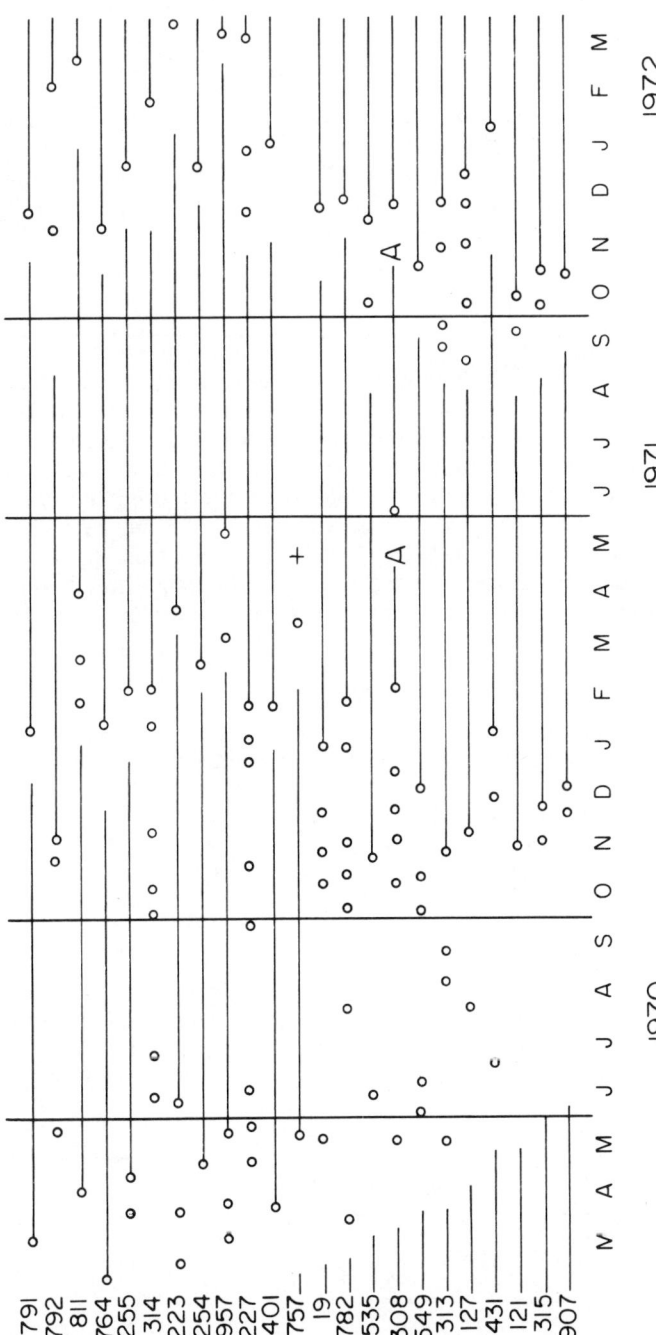

Figure 16.5 Reproductive pattern in temperature-stressed Bos taurus cattle; individual cattle numbers on the left. (○) Served; (———) pregnant; A, aborted; +, died. No successful conceptions occurred from the end of May until the beginning of November, even though October was cooler than May (when conceptions were occurring). This could be due to either impaired spermatogenesis or early fetal death

calves (*see below*) also suggests that it is undesirable to allow calving during the summer. One must therefore conclude that, for maximum performance of a herd under these climatic conditions, seasonal calving is the system of choice. With these two limitations the mating period will be short. A simple calculation will show that only February, March and April are suitable for service and it follows that the bull should be introduced into the herd in February in order that the poorer breeders will finally be settled by April. If seasonal calving is to be avoided in this climate then artificial insemination must be practised or the bull maintained under air-conditioning.

GESTATION LENGTHS

It also became apparent during this work that gestation lengths were consistently shorter than the commonly accepted 283 days, found in temperate climates. In the winter the mean for those animals with accurately known conception dates was 278 days, while the summer mean was 264; i.e. 16 days 'premature'. This, of course, has a bearing on the 'drying off' periods, birth weights and, as will be mentioned later, the thermoregulatory mechanisms of the calves. The 'drying off' period, in particular, must be adjusted to allow for this phenomenon. Some gestation periods were considerably less than the mean: the earliest were 56 and 35 days premature. This effect on the gestation period is yet another reason for avoiding summer calving.

ABORTIONS

Only three spontaneous abortions occurred between January 1970 and July 1973 and none of these could be fairly attributed to heat stress, although Brody *et al.* (1955) have reported what they believed to be temperature-induced abortions. If this phenomenon does exist, then the critical limits were not reached in the United Arab Emirates (UAE) study.

CALF BIRTH WEIGHTS

The birth weights of all calves born at Digdaga were recorded. These were consistently low and there was also a significant difference (at the $P < 0.05$ level) between summer and winter calves. The mean weights of all calves were as follows:

	Mean weight	*No. involved*
Summer born	29.2 kg	25
Winter born	39.3 kg	24

Normal birth weights of Friesian calves are about 41–43 kg.

Detailed analysis of the results, bearing in mind parity and sex, suggested that this was indeed a temperature effect. This could be related to tropical 'stunting', which has been debated now for many years. It is contended by some that all *Bos taurus* stock introduced into the tropics suffer a gradual decline in stature. If calves are consistently born lighter than in temperate climates and these offspring are served at a given age, according to the regrettable but common practice, then the stock are likely to decrease in stature. If on the other hand, the time of first service is controlled by body weight then this trend should not occur. This might explain why some authorities concede tropical stunting and others do not.

PARTURITION

One striking feature of this herd was the very high incidence of primary inertia experienced, amounting to 25 per cent of all births. There was, however, no statistically significant difference between winter and summer. No satisfactory explanation has been found for this phenomenon. A standard Compton Metabolic Profile showed all parameters to be within normal limits and no reference can be found in the literature to the subject. It does mean, however, that the veterinary supervision of such herds must be of a high order or perinatal deaths are likely to be excessive.

Animal health

RECTAL TEMPERATURES

Without doubt the most convenient method of assessing the heat stress on animals is by monitoring rectal temperatures. Pulse rates and respiratory rates are too erratic to be of much use. The Digdaga herd were therefore monitored twice daily—during the milking periods. Since high rectal temperatures can be expected under these climatic conditions it must be considered essential to carry this out as a routine, since it is the only way to differentiate between climatic and pathological pyrexia. It was found that the rectal temperatures rose steadily from approximately 38 to 39 °C over the spring period and then remained between 39 and 40 °C during most of the summer, although temperatures of 41 °C were not uncommon at 1600 hours. This latter temperature can be tolerated for considerable periods of each day in some animals, provided there is a nightly remission.

For calves the situation was different. Whereas calves over 2 weeks had no difficulty in tolerating the most extreme of the conditions encountered, those under that age showed little thermoregulatory ability and would run lethal pyrexias if unattended. Usually one or two soakings with water in mid to late afternoon were sufficient to control the situation. This phenomenon is, of course, of great practical significance, and the high perinatal mortality reported from some tropical herds may be due to deaths from hyperthermia. *Figure 16.6* shows the temperatures of some calves from the fourth day onwards. 'Normal' temperatures are not attained until 8–10 days *postpartum*. The temperatures recorded prior to the fourth day have been excluded, since wetting of the calves' coats was often necessary during this

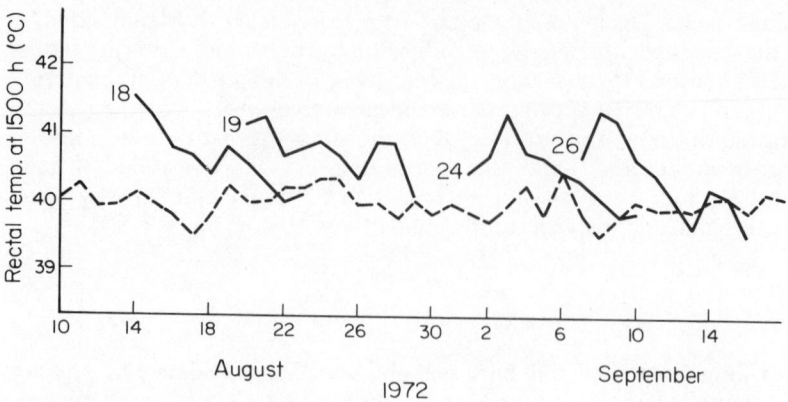

Figure 16.6 Evidence of thermal incompetence in newborn calves. The full lines are rectal temperatures of individual calves from day 5 to day 14 post-partum. The broken line represents the mean rectal temperature of calves aged more than 21 days. Temperatures recorded during the first four days post-partum are omitted, since artificial cooling was necessary during this period to avoid lethal hyperthermia

period, which invalidated the records. There was also evidence that premature animals needed a correspondingly longer period to attain adequate thermoregulation than those born at full term.

RESPIRATORY RATES

During the summer, respiratory rates of the order of $100–120$ min^{-1} were normal during the day. These were tolerated for long periods, without producing significant respiratory alkalosis or hypoxia. Night-time respiration rates were approximately $70–80$ min^{-1}. 'Second-phase' breathing, which is a sign of advanced heat stress, was observed from time to time. This respiratory phase was encountered when rectal temperatures reached approximately 41.8 °C. At this point there is a sudden drop in respiration rate, concurrent with an increase in tidal volume. As this phenomenon is a sign of severe hyperthermia, under no circumstances should it be allowed to continue. It was of uncommon occurrence in the adult stock but frequently observed in calves in the first 10 days or so of thermoregulatory incompetence.

GENERAL

After the initial difficulties of the first summer, the herd was maintained in remarkably good health. The incidence of parasitic diseases, which can create health problems in temperate climates, was low. No cases of helminthiasis occurred, the parasites apparently being unable to survive the ambient temperatures, but some coccidiosis was experienced. No case of calf scour occurred in the three and a half years of observation and out of 94 calves born only 5 died, all in the perinatal period. No significant troubles were experienced in the adult stock and the Standard Compton Metabolic Profile test indicated a blood chemistry well within the norm during both summer and winter. However, two cases of hypomagnesaemia were encountered in 1972. The incidence of mastitis was within acceptable limits.

One important health hazard was the pathogenicity of *Theileria annulata* for European stock. This was, however, brought under control with twice-weekly power spraying with an acaricide.

Only one other health hazard of significance was encountered, particularly in the first year: a marked lameness, mostly in the hind feet, which was probably a form of laminitis. Although not serious in itself, it became of great concern if it resulted in recumbency, since the animals were then unable to control their temperatures.

Some necrosis of the plantar surface of the hoof was also encountered in the first year when the animals were permanently housed, presumably resulting from standing on damp bedding at bacterial incubation temperatures. This was resolved when paddocks became available.

It should be remembered, however, that the remarkably good health record of this herd, both in the calves and the cows after the first year, was due in no small part to the very high degree of supervision given to the animals. The day to day care and management was in the hands of a veterinary surgeon with many years of tropical dairying experience. Without her meticulous care results might have been different.

The behaviour of cattle under heat stress

Since management is a major factor in the wellbeing of cattle in a hot climate, it is worth describing the symptoms that indicate heat stress in cattle. A number of these are valuable guides to those having care of such animals, as they can indicate when it is necessary to take steps to alleviate undue strain. In approximate order of ascending severity they are as follows:

(1) Refusal to lie down. This is a common phenomenon and is presumably caused by the need to expose as much of the body surface as possible to the atmosphere. It was noted that animals that became recumbent involuntarily had the greatest difficulty in avoiding hyperpyrexia and needed intermittent hosing or sprinkling.
(2) Huddling. Contrary to expectation (and the behaviour in man), cattle under high stress huddle together and will disperse only when the temperature drops. This is presumably an endeavour to create their own microclimate.
(3) Body splashing. Strenuous efforts may be made by animals to wet their coats either by pawing water back over themselves or by beating their heads into the water in their troughs. They do not, however, use their saliva for this purpose as do some animals.
(4) High respiratory rates. These have already been mentioned and become serious only when second-phase breathing occurs—which can be recognized by the deep inspiratory movements of the flanks. This condition must always be alleviated.
(5) High rectal temperature. Any persistent temperature higher than 41 °C must be promptly treated.
(6) Open-mouth breathing. This also is indicative of advanced stress and is

characterized by the head being extended, the tongue protruded and profuse salivation taking place. In addition the elbows are often ablated in this phase.

Shelter engineering

The importance of good shelter engineering was amply demonstrated in the first summer of the Digdaga work. The heifers were imported heavily in calf in late December 1969 but were not taken over by the author until May 1970. At that time the provision for their housing and shelter was found to be unsuitable, the design of the buildings having obviously been influenced by European requirements.

The cattle were housed 24 hours a day in a cowshed having concrete block walls and asbestos roofing. Bedding consisted of a few inches of wood shavings—the only material available. The walls were of solid construction up to a height of 1.75 m, thereby obstructing what little breeze was available, and the roof's apex was only 2.6 m. This height was much too low and resulted in high radiant heat loads on the animals. The wood shavings, being short 'staple', afforded little protection against the concrete flooring and resulted in extensive bedsores, which proved intractable.

These conditions proved intolerable to the cattle experiencing their first tropical summer and nearly resulted in a major disaster; three animals died during this period as a direct result of climatic conditions. It was found that any animal becoming recumbent and unable to rise for any reason was incapable of controlling its body temperature (presumably owing to the reduced area available for evaporation) and usually died despite strenuous efforts to save it.

It was not until near the end of the summer that the necessary alterations were completed and the situation brought under control. The walls were demolished and replaced by steel piping, the roof was raised by 1 m and outside paddocks were constructed complete with palm thatch shelters. Thereafter almost no trouble due to the climatic stress was experienced with the herd, except for the late summer of 1972, which has been dealt with separately.

This amply demonstrates the need for those importing animals into the tropics to be aware of the fundamentals of shelter engineering and of the great difference between the heat tolerance of European breeds and *Bos indicus*. As an example of the latter, during the study one recumbent Friesian cow became exposed to the afternoon sun for a few hours and during this time attained a lethal hyperthermia, whereas an indigenous animal could often be seen lying in the sun from choice under the same climatic conditions.

GENERAL CONSIDERATIONS

There is a common fallacy that even in the climates of the Middle and Near East formal-type cattle buildings are needed as protection against the weather. Cattle are particularly cold-tolerant, having many physiological adaptations to combat low temperature, and it is difficult to envisage temperatures in these regions being low enough to warrant such housing, except at very high altitudes. The main housing problem in hot climates is therefore to provide suitable shade.

Radiation loads on animals in shadow originate from five principal sources (Kelly, Bond and Ittner, 1950). These are: from the shade itself, the area of cool sky, the area of hot sky, the ground outside the shaded area and the shaded ground. All these factors must be taken into account when designing shades, thus there can be no hard and fast rules governing the construction of such buildings; each situation requires a thorough knowledge of the principles involved together with the radiation loads from the principal sources. Certain guidelines, however, can be proposed for simple buildings that do not employ artificial cooling.

SHADE CONSTRUCTION

Ideally, any construction casting shade should have one or more of the following properties: high reflectivity, low conductivity, low under-surface emissivity, correct profile and maximum practical height.

(1) The most commonly used materials for shelters are galvanized iron and aluminium sheet. Of the two, aluminium is usually the most highly reflective and thus superior, but this depends on the degree of corrosion and discoloration of the two materials (Nelson, Mahoney and Berousek, 1961). Other materials can, of course, have their reflectivity modified by the use of reflective paint.
(2) The effective conductivity of a roofing material may be lowered by the use of 'double skinning', leaving an air gap between the layers, or by covering it with a layer of hay, palm thatch or any other suitable insulating material. Double-skinned roofs should have an adequate slope to take advantage of the induced convection currents, an adequate apex opening and an inter-skin gap big enough to take advantage of any winds available. There appear to be no published data on the best dimensions, but in practice a 20–30 cm gap and a slope of at least 20–25° seem satisfactory. A venturi-type construction at the apex may also be an advantage in areas of little wind and is simple to construct.
(3) The roof profile needs to be designed for the circumstances. In the Northern Hemisphere, for instance, it is normally advantageous to have the main slope of the roof to the south, since the northern sky has a lower emissivity than the southern. In areas of very high heat stress even the late afternoon sun may produce enough radiation load to cause critical levels of stress. Vertical shade or a very low-eaves, high-apex configuration must therefore be provided. The desired profile of the roof of a shelter will also be affected by whether it is to stand free in a paddock, for optional use, or to have confining walls as in a cow house. In the former, a cheap, flat roof will suffice, since the cattle may follow the shade, but in the latter the traditional apex type is usually called for, since the shade of a high, flat roof may be cast outside the confines of the walls. For similar reasons a north–south orientation is more suitable

for a shade in a paddock since it allows more sun onto the 'poached' area below it for drying urine, etc., but the east–west configuration is more suitable for a confined area. A large leafy tree is often considered the ideal shade, but is not usually to be found in extreme climates.

(4) A suitable height for a roof is important. Normally, the higher the roof the lower will be the radiation load, as more of the 'cool sky' is exposed. The disadvantage of high roofs, on the other hand, is that the rate at which the shadow moves across the ground is directly proportional to the height of the roof. Thus cattle under high shade are constantly having to move onto hot ground, but this may not be significant if the

Figure 16.7 An example of a shelter with incorrect orientation. As can be seen from the shadows, the long axis of this building is orientated approximately north–south. Thus the shadow will be cast outside the confines of the shelter for a considerable period of the day; an east–west orientation would have been better

Figure 16.8 A cowshed designed to take maximum advantage of air movement. The arrangement at the apex provides a simple venturi effect to assist in the extraction of air heated by the roof. Protection from the late-afternoon sun was afforded by removable rush mats

surface in question rapidly cools when shaded. Rapid shade movement would be an insignificant factor on a grass surface but important on concrete. Again, very high roofs are impracticable with open-sided buildings as the shade may be cast outside the confines of the structure.

(5) Artificial cooling of the roof area with a water spray can be very effective in reducing radiant heat loads, but provision must be made for recycling the water or for adequate drainage.

Protection against the other significant sources of radiation, namely the 'hot sky' and the unshaded land surfaces, is not usually attempted, since the 'hot sky' area is confined to an area from the horizon to approximately 10° above it and radiation from unshaded ground is also usually confined to a narrow angle.

Some of these points are illustrated in *Figures 16.7* and *16.8*.

HOUSING

Although conventional housing is usually inappropriate in areas of high heat stress, under certain conditions it may be necessary: for example, when stock need to be protected from pests and disease vectors or there is insufficient water to allow showering. Under these circumstances the same general principles as those outlined for shade construction should be observed, but additional measures are needed to provide a suitable environment in closed houses.

(1) Air-conditioning is a very satisfactory method, since both temperature and humidity can be lowered and accurately controlled, but the cost is very high except in those countries where energy is cheap.

(2) Evaporative coolers are less satisfactory, since they may raise the humidity to unacceptable levels, but their energy consumption is low (Nelson, Berousek and Mahoney, 1956).

(3) Cold-air hoods may be considered a modified form of air-conditioning, but they have a much lower energy requirement. Cattle soon learn to inhale from these hoods from which a constant stream of cooled air flows. Since a considerable portion of heat is lost to the atmosphere from the respiratory system by humidifying and warming it, these hoods can be quite effective. Simple calculations can determine the potential heat loss available to the stock from this system (Hahn *et al.*, 1965).

(4) Attic ventilation. In a building with a ceiling the forced or natural ventilation of the attic space may produce worthwhile lessening of the radiation loads from the ceiling (Nelson, Mahoney and Berousek, 1961), since under high radiation loads attic space temperatures may be very high.

Insulation may also be used to keep out heat, following the well-documented principles for retaining heat originating from countries with temperate or cold climates.

CLIMATIC ALLEVIATION

Under very severe climatic conditions, such as those encountered in the UAE study, the provision of good shade and suitable buildings may be either inadequate on their own or, in the case of artificially cooled buildings, too costly. There are, however, other means of diminishing high temperature stress. It is obviously an advantage to stall-feed the cattle, since the energy cost of walking is considerable (Hall and Brody, 1934), but if grazing is obligatory then it should be confined to the hours of darkness. Similarly, a high-concentrate, low-roughage diet lessens the heat increment of feeding (Stott and Moody, 1960). The provision of fans, even in the open, has been shown to have beneficial effects in areas of low air velocities. Cooled drinking water may also assist in maintaining a thermal balance (Ittner, Kelly and Guilbert, 1951), but the long-term effects of this on ruminant digestion appear not to have been investigated. One of the most promising methods, however, is the provision of optional showering facilities for the stock. Since the great difference between heat-tolerant mammals such as man and the relatively heat-intolerant *Bos taurus* is in the sweating capacity of the species, it seems logical to provide cattle with an artificial sweating mechanism. Showering seems the simplest choice, and was adopted at Digdaga in 1972. During that summer the heat was particularly severe, and it became obvious that the limit of tolerance had been reached by those animals in milk, although non-lactating and young stock were still in no danger. Although from a scientific viewpoint it would have been informative to have followed the natural course of events to their sequel, neither on financial nor humanitarian grounds was this possible. It was therefore decided to take steps to alleviate the climate by this method.

After some preliminary trials the simple expedient was adopted of inverting an ordinary domestic lawn sprinkler in the roof of the cowshed. This threw a pattern of water some 15 m in diameter. The cattle were allowed free access to this 'shower' during the afternoon and all availed themselves of it immediately. They would form an outward-facing circle at the periphery of the wetted area, thus wetting only their hind-quarters. Although

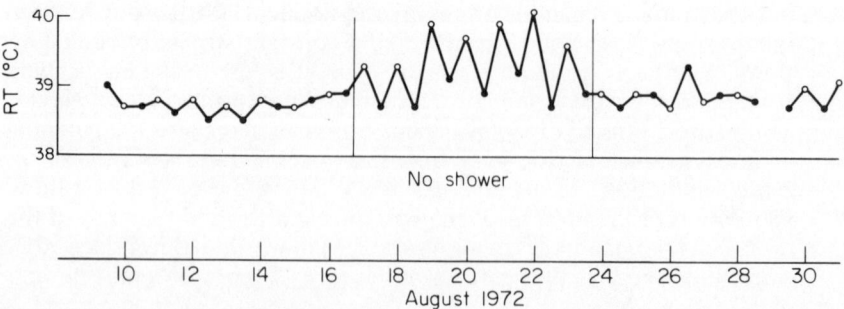

Figure 16.9 Mean rectal temperatures (RTs) of lactating cows, with and without free-choice showering available from 1130 to 1700 hours. (●) Mean a.m. RT; (○) mean p.m. RT. Temperatures in June, July and August of 1972 were frequently above the limit of tolerance of most of the lactating cows. The temperatures recorded show a diurnal variation of rectal temperature when the shower was not available, indicative of heat stress

there were doubts that this system would be effective in the high ambient humidities of the Gulf, *Figure 16.9* shows that it was entirely successful. Since the water temperature was frequently of the order of 35 °C it is apparent that evaporative cooling was the effective factor. One limitation to this system is that wetting of a coat for more than 4–5 hours is apt to lower its resistance to abrasion, which is of some concern if bedding is inadequate. However, this method is likely to be effective even when temperatures reach 50 °C or more, as occasionally occurs in the Gulf area.

Conclusion

The main conclusion drawn from the work at Digdaga is that given careful management, satisfactory buildings and sufficient water to operate a domestic garden sprinkler for 4–5 hours a day, European Friesian cattle can yield very satisfactorily in a climate as severe as the Arabian Gulf littoral, provided that seasonal calving is acceptable. Thus any lesser climate should prove no barrier to successful dairy farming.

Admittedly, the Digdaga herd had the advantage that they were on a zero-grazing system and thus the heat increment associated with grazing was negligible. Nevertheless, where a grazing system must be adopted it is unlikely that this will prove a serious factor, if grazing is restricted to night and the cooler hours of the day. It appears, therefore, that many of the poor results obtained from 'exotic' stock in the tropics in the past may have been due to factors other than temperature stress *per se*, and that management and nutritional factors must be scrutinized more closely in the future.

It is to be regretted that the work at Digdaga was terminated in 1974 for financial reasons, since the herd had by then reached a stage where almost every animal was in calf at the correct period, and the herd could have been used to solve some of the many problems that still remain. There is, for instance, little information on the effects of cross-breeding between *Bos indicus* and *Bos taurus* on early fetal death and spermatogenesis in such climates, or whether routine showering of both bull and cows eliminates these problems.

These and many other problems require answers in the near future, now that more reliance is being placed on exotic stock for increased yields in the tropics.

This chapter is based largely on a doctoral thesis submitted by the author to the University of Bern (Ansell, 1974).

References

ANON. (1969/70). English Milk Marketing Production Division report, pp. 105–116. Thames Ditton, Surrey.
ANSELL, R.H. (1974). *Thesis*, University of Bern
BIANCA, W. (1962). *Nature, Lond.* **195**, 251–252
BRODY, S., RAGSDALE, A.C., YECK, R.G. and WORSTELL, D.M. (1955). *Missouri agric. exp. Stn Res. Bull. 578*

HAHN, L.G., JOHNSON, H.D., SHANKLIN, M.D. and KIBLER, H.H. (1965a). *Trans. Am. Soc. agric. Engrs* **8**, 332–334

HAHN, L.G., JOHNSON, H.D., SHANKLIN, M.D. and KIBLER, H.H. (1965b). *Trans. Am. Soc. agric. Engrs* **8**, 337

HALL, W.C. and BRODY, S. (1934). *Missouri agric. exp. Stn Res. Bull. 208*

ITTNER, N.R., KELLY, C.F. and GUILBERT, H.R. (1951). *J. Anim. Sci.* **10**, 743–751

JOHNSON, J.E., NAELAPAA, H. and FRYE, J.B., Jr (1963). *J. Anim. Sci.* **22**, 432–436

KELLY, C.F., BOND, T.E. and ITTNER, N.R. (1950). *Agric. Engng, St Joseph, Mich.* **31**, 601–606

NELSON, G.L., BEROUSEK, E.R. and MAHONEY, G.W.A. (1956). *Agric. Engng, St Joseph, Mich.* **37**, 98–102

NELSON, G.L., MAHONEY, G.W.A. and BEROUSEK, E.R. (1961). *Tech. Bull. Okla. agric. Exp. Stn* (87), 5–30

RAGSDALE, A.C., THOMPSON, H.J., WORSTELL, D.M. and BRODY, S. (1953). *Missouri agric. exp. Stn Res. Bull. 521*

STOTT, G.H. and MOODY, E.G. (1960). *J. Dairy Sci.* **43**, 871

STOTT, G.H. and WILLIAMS, R.J. (1962). *J. Dairy Sci.* **45**, 1369–1375

VI

ENGINEERING AND CONTROL OF THE HOUSE ENVIRONMENT—1

17

MONITORING THE HOUSE ENVIRONMENT

J. A. CLARK
Department of Physiology and Environmental Science, University of Nottingham

K. CENA
Department of Environmental Physics, Technical University of Wrocław, Poland

Introduction

Only a decade ago, a review of the principles and techniques involved in measurements of the environment within animal houses would have been devoted largely to a discussion of the techniques of measurement (Wadsworth, 1968; Smith, 1970); for example, the construction and calibration of temperature sensors. However, those ten years have seen first the general adoption of the digital data logger for the recording and of the computer for analysis of repetitive measurements and, latterly, the advent of microprocessor- and minicomputer-based systems of measurement. Associated developments have also produced advances in both the sensitivity and reliability of digital electronics, as well as a drastic reduction in cost, and their routine application to control. Developments in this latter area are considered in Chapter 20 by Stenning.

These changes in the technology of recording have made possible two changes in our approach to measurements. First, digital data loggers and computer analysis make it possible to handle much greater quantities of data than before. At the research level this has allowed, for example, detailed mapping of temperature distributions in broiler houses showing the effects of brooders, etc. (Moulsley and Fryer, 1978). Secondly, the improved sensitivities of digital voltmeters, as compared with the majority of analogue recorders, and the ease of subsequent computer processing have reduced the requirement for high sensitivity in the *transducer*. Thus, in the past, thermistors were often employed for temperature measurements because of their high sensitivity, a valuable factor when used with analogue instruments despite concomitant problems of output linearization. However, electronic cold-junction compensation combined with digital electronics now allows the choice of thermocouples as a simpler alternative in most situations.

It is unnecessary to pursue the above theme too far, since our point is that modern technology makes the choice of the actual transducer employed to measure a particular variable comparatively unimportant. We can therefore concentrate our attention largely on the principles of measurement: on deciding what we should measure and where, on the desirable characteristics of the transducer and on how we wish to use the measurements. In addition, we shall pay particular attention to those measurements which are

309

or could be used for control purposes: temperature, humidity, the concentration of the contaminant gases such as carbon dioxide and ammonia, and on the measurement of ventilation, which is usually the factor altered in order to control other environmental variables. We shall also discuss the measurement of radiant energy in animal houses, a subject that has received inadequate attention.

General principles

Whether animals are housed in order to alleviate thermal strain or for reasons of management, the maintenance of an optimal environment depends on measurement and control. Whereas we wish to control our own environment for optimum comfort (Cena and Clark, 1981a), for our farm animals we require optimum production. We may regard the house environment and its controls as a feedback loop with five main parts (*Figure 17.1*). If

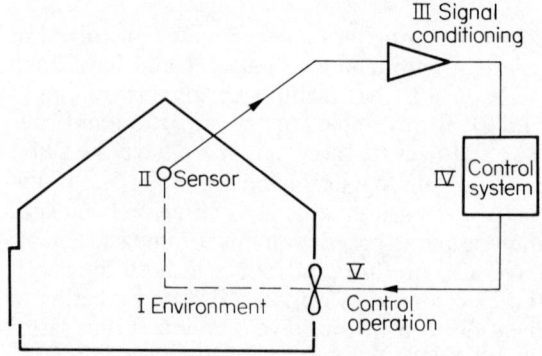

Figure 17.1 Diagram of the control loop for the environment in an animal house. The main components of the system are numbered

we accept that modern technology is capable of performing exactly to specification, the critical path in the loop becomes the interaction between I and II—between the environment and the sensor that provides the signal to the control system. Therefore, the whole control system can operate satisfactorily only if the sensor provides a signal that is both prompt and typical of the environment.

Siting

The first and simplest requirement is correct siting of the sensor, a problem that may require an arbitrary decision. For example Charles (Chapter 11) shows that in the corridor between layer cages there is usually a vertical gradient of temperature which results in a difference of about 3 K between the temperatures at the levels of the bottom and top cages. Differences of similar magnitude between cage and passage temperatures have also been reported. Therefore, a thermostat situated near floor level would need to be set 3 K lower than one at top cage level to produce the same *house* temperature. Trials have shown that a temperature difference of 3 K can produce significant differences in the economic returns from laying poultry—so all house temperatures must be specified relative to temperature measure-ments at a standard height, usually eye level. This must also be remembered when comparing the results of calorimeter studies with those of production trials. The calorimeter temperature is that experienced by the animal, but in trials the stock experience a temperature several degrees above the house set temperature, which is that recorded.

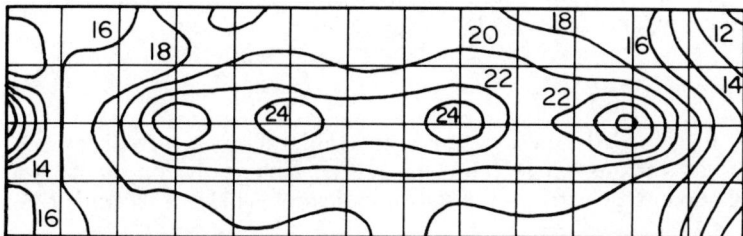

Figure 17.2 Part of a temperature distribution measured in a broiler house with four gas-fired brooder heaters in operation. The contours show intervals of 2 °C. (From Moulsley and Fryer, 1978, courtesy of the authors)

Horizontal temperature variations may be even larger, as shown in *Figure 17.2*, which presents measurements made by Moulsley and Fryer (1978). Incorrect siting of the controlling thermostat could obviously result in a significant departure from the optimal environment for the sensitive broiler chicks in the house. Even though a temperature distribution of this type may have advantages—it allows the animals scope to correct our mistakes by selecting their own preferred environment—with modern fuel costs it would be more sensible to provide the optimal environment under the brooders only. Again, the control sensor should be sited according to rational and consistent criteria rather than by chance.

Time constant

The second requirement, assuming that we have resolved where to site the sensor, is that it should give a true reading of the particular environmental variable. A major determinant of error is the time constant of the sensor. As in the preceding section we will use thermometry as the example, but the principles also apply to other sensors.

The recent text by Fritschen and Gay (1979) contains an excellent outline of the theory of the time constants of thermometers and their influence on the temperature indicated. The effects of thermometer time constant on measurements in animal environments have also been considered by Smith (1970) and in more detail by Cromarty (1974). In fact, Cromarty presented a careful analysis of the factors which determine the 'ideal' thermometer for various measurements in poultry houses and, with his permission, we present some of his results to illustrate the theory.

If there is a sudden step change in the temperature of an environment, the temperature of a thermometer used to measure the environment will approach the new true temperature according to Newton's law of cooling.

Thus, if T is the temperature indicated by the thermometer and T_a the air temperature, the rate of change of the *thermometer* temperature dT/dt is proportional to $(T - T_a)$. Hence

$$\frac{dT}{dt} = \frac{1}{\tau}(T - T_a) \tag{17.1}$$

where τ is the 'time constant' of the thermometer. (The determinants of τ for thermometers will be discussed later.) Integration then yields

$$T - T_a = \Delta T \exp(-t/\tau) \tag{17.2}$$

where t is the elapsed time and ΔT is the initial temperature difference. Therefore, at various times following a step change in conditions, the remaining percentage error of the sensor indication will be as in *Table 17.1*.

Table 17.1   PERCENTAGE ERROR OF SENSOR INDICATION (RELATIVE TO STEP AMPLITUDE) FOLLOWING A STEP CHANGE IN THE ENVIRONMENT

Elapsed time	Error (%)
τ	37
2.3τ	10
4.6τ	1

For the present purposes we need not concern ourselves with higher accuracies. Though unusual, step changes may occur; most often as the result of moving the sensor rather than changing the environment. *Table 17.1* tells us that the person who takes a 'typical' mercury-in-glass thermometer (which has a time constant of about one minute) into an animal house at $20\,°C$ from an outside environment at $0\,°C$ needs to wait several minutes before he can rely on the reading.

For control purposes we also need to consider the effect of sensor time constant on the relation between the indicated and true readings in oscillating environments. For simplicity (and because it is the only case that can be solved easily by analytical methods) we shall confine ourselves to sinusoidal changes. Again using temperature as our example: if the amplitude of the oscillation is T_1 and its angular frequency is ω (in rad s^{-1}), then at a time t the temperature of the environment is

$$T_a = \bar{T}_a + T_1 \sin \omega t \tag{17.3}$$

where $\bar{T}_a$ is the mean temperature. Though the mean value of the sensor temperature will be the same as that of the environment and it will follow a cycle with the same frequency, it will both lag behind and show a smaller amplitude, as shown diagrammatically in *Figure 17.3*. The sensor temperature is given by

$$T = \bar{T}_a + \alpha T_1 \sin(\omega t - \beta) \tag{17.4}$$

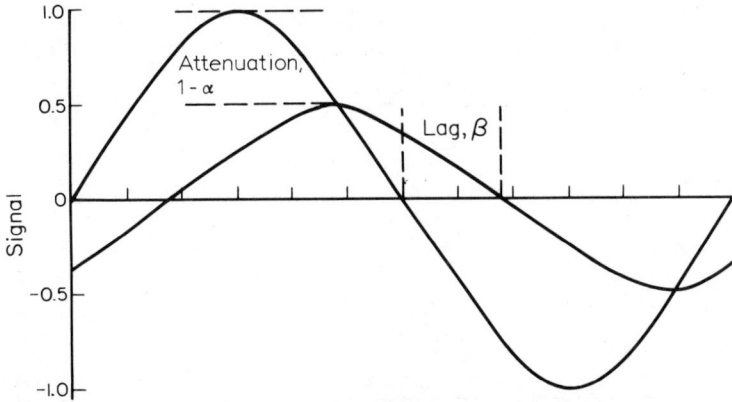

Figure 17.3 The phase lag, β, and attenuation coefficient, α, of a sensor output relative to environmental conditions. Sinusoidal variations are shown for simplicity

For regular sinusoidal oscillations the phase lag, β, is

$$\beta = \text{arc tan } \omega\tau \tag{17.5}$$

and the signal attenuation coefficient, α, is

$$\alpha = 1/[1 + (\omega\tau)^2]^{\frac{1}{2}} \tag{17.6}$$

Both factors may conspire to cause overshoot or hunting in control operations. The effects of instrument time constant on readings are presented in *Figures 17.4* and *17.5*. In each case the scale of the abscissa is the ratio of the cycle period (τ_p) to the sensor time constant (τ), and is therefore dimensionless. The ordinate scales are also dimensionless. In *Figure 17.4* the ordinate is the actual time lag divided by τ and in *Figure 17.5* the ordinate is α, the signal attenuation factor predicted by equation 17.6.

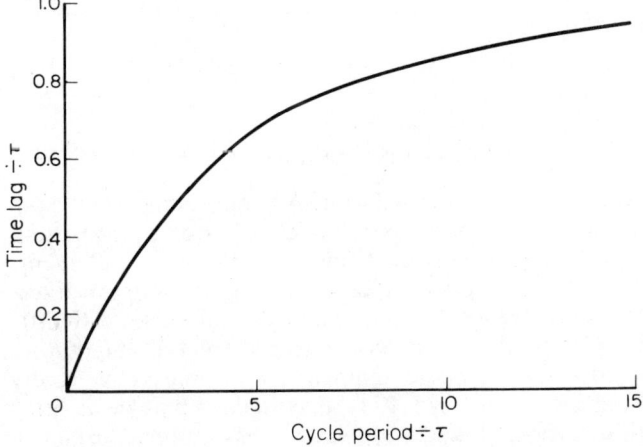

Figure 17.4 Graphical representation of the relationship between signal lag and cycle period for sinusoidal signal. Both axes are normalized with respect to the sensor time constant, τ

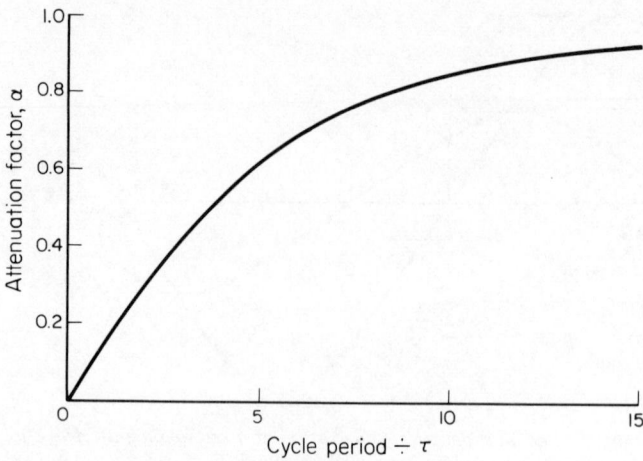

Figure 17.5 Graphical representation of the relationship between the signal attenuation, α, and cycle period. The cycle period axis is as in Figure 17.4

For periods much greater than τ the lag time approaches τ and the attenuation becomes small (i.e. $\alpha \to 1$). At progressively shorter periods (relative to τ) the lag becomes smaller, which is beneficial for control operation, but signal attenuation is increased (*Table 17.2*). Cromarty (1974) selected two criteria: in order to give a reading within 5 per cent of the true value (as a proportion of the true amplitude) the sensor time constant must be less than 1/22 of the cycle period; and for an indication within 1 per cent, τ/τ_p must be less than 1/44. For these τ must be about 5 per cent and 2.5 per cent of τ_p, respectively.

Table 17.2 LAG FACTOR AND SIGNAL ATTENUATION FOR SELECTED RATIOS OF CYCLE PERIOD TO SENSOR TIME CONSTANT

Ratio of period/τ	Time lag (β)	Signal attenuation ($1 - \alpha$)
∞	τ	0
22	$\simeq \tau$	0.0
2.8	0.5τ	0.6
1.0	0.2τ	0.85
0.4	0.1τ	0.95

What oscillations need we be concerned with? An important period for all animal houses is that of the diurnal temperature cycle driven by insolation. This presents few problems: almost all common thermometer elements (other than the water-filled black globe) have time constants of much less than an hour and can therefore follow diurnal temperature cycles with little error. The corollary also holds: if one wishes to construct a temperature sensor that will record a 'daily average' temperature this must have a very long time constant indeed. Cromarty (1974) suggested a 1 pint ($\cong$ ½ litre) vacuum flask for this purpose. Filled with water it has a time constant of about 18 h. For a 24 hour cycle this gives an attenuation factor of 0.2 and $\beta = 85°$, equivalent to about 5½ h. A time constant of 75 h would be required to reduce α to 0.05, which is impractical. Cromarty's technique for obtaining 'average' temperatures from the vacuum flask was to estimate the zero points of the cycle and make measurements at one of these, conveniently at about 5 p.m. each day.

The other cycling period of interest is that of the control system (*Figure 17.5*). However, there have been surprisingly few reports of the control cycle periods in animal houses. Cromarty (1974) gave one figure of half an hour, but from theoretical considerations we might expect the response time to be proportional to ventilation rate, at least in the cooling part of the cycle. Esmay (1969) also presented one example, with about 13 control cycles per hour (a period of 5 minutes) but this is from a computer simulation. We are, therefore, indebted to P. Hearn (personal communication) and his staff at Gleadthorpe, EHF for making some measurements for us at rather short notice. They found periods of between 3 and 23 minutes for the Gleadthorpe deep-pit house, when conditions demanded just above the minimum ventilation rate. This house is controlled by mercury contact thermometers. Comparable periods of between 6 and 20 minutes were measured for a house controlled by Danfoss thermostats, when cycling between its maximum ventilation rate and the next lower fan setting. Both houses contained laying hens. Oscillations of the longer of these time periods will be followed adequately by an unventilated mercury-in-glass thermometer— but periods of 5 minutes require a time constant of only 14 s to give an amplitude attenuation of less than 5 per cent. If we accept that Cromarty's (1974) criteria are unrealistically severe, a 10 per cent loss of signal occurs when the ratio of τ_p/τ is 13. Then for $\tau_p = 5$ min, τ must be 23 s, and the lag is 12 s. This order of time constant is obtainable with ventilated thermometers and thermocouples or thermistors, but it is likely to be exceeded considerably by mercury contact thermometers and conventional industrial thermostats. For example, Smith (1970) quoted a typical time constant for an *aspirated* mercury-in-steel thermometer as 280 s, while that for a single unencapsulated thermocouple may be as short as 2 s.

A thermometer time constant of 280 s is similar to the shorter control periods measured. For a 5 minute period it will result in a time lag of about 0.2τ (≈ 1 min) and, more important, in a signal loss of about 85 per cent ($\alpha = 0.15$). Hence, we might expect imperfect control operation and 'hunting' of the house temperature, even with an on/off system. If the thermostat dead band is ± 0.5 K, a house temperature span of almost 7 K could occur for this combination. The mathematics may exaggerate the problem, since animals weighing more than a few hundred grams themselves have long thermal time constants and will respond physiologically to the average temperature. However, poor control will certainly impose unnecessary loads on the control system, and the animals may also experience discomfort, as do humans exposed to rapid fluctuations in temperature. Another possible consequence is loss of control over the airflow pattern within the house, an important design factor in current ventilation systems (Carpenter, Chapter 18; Randall, Chapter 19). Poor control is therefore undesirable at the least.

Temperature measurement

The discussion of time constants, which used thermometers as its example, has already defined one desirable characteristic of a thermometer to be used for the control of temperature in animal houses: a short time constant relative to the shorter control cycle periods observed. For cycle times of 5 minutes τ should be about 20 s. What determines τ is the ratio between the thermal mass of the thermometer—the product of its mass, m, and specific heat, c—and the rate of heat transfer from its surface per unit temperature gradient—the product of the heat transfer coefficient, $1/r$, and surface area, A. Hence

$$\tau = amcr/A \qquad (17.7)$$

where a is a constant of proportionality and r is the resistance to heat transfer as defined by McArthur (Chapter 3). Assuming that heat transfer from the (cylindrical) thermometer is due to forced convection, r will be approximately inversely proportional to the square root of the Reynolds number (Monteith, 1973). Allowing for the increase of mass and surface area with size the thermometer time constant should therefore be proportional to $d^{3/2}$ and $u^{-1/2}$, where d and u are the thermometer diameter and the ventilating air speed respectively. To record true air temperature and have a short time constant, a thermometer should therefore be small, well ventilated and shielded from radiation. These are not common characteristics of the temperature sensors used to operate control systems in animal houses, but are they necessary and what are the desirable characteristics of a thermometer for this purpose? In Chapter 3, McArthur has discussed the factors that determine the sensible heat loss from animals, $\mathbf{G}$, which has two main components: radiation, $\mathbf{R}$, and convection, $\mathbf{C}$. Using the same definitions,

$$\mathbf{G} = \mathbf{R} + \mathbf{C}$$
$$= \varrho c_p \frac{(T - T_R)}{r_R} + \varrho c_p \frac{(T - T_a)}{r_a} \qquad (17.8)$$

We can rearrange this equation into a more convenient form with two terms, one representing the temperature of the animal and the other that of the environment.

$$\mathbf{G} = \varrho c_p T \left(\frac{r_R + r_a}{r_R r_a} \right) - \varrho c_p \left(\frac{T_R + T_a}{r_R + r_a} \right) \qquad (17.9)$$

We may regard the second term as the mean of the air and radiant temperatures, weighted according to the values of the respective transfer resistances. This can define an *environmental temperature* T_e, such that

$$\mathbf{G} = \varrho c_p \left(\frac{r_R + r_a}{r_R r_a} \right) (T_a - T_e) \qquad (17.10)$$

Hence

$$T_e = \frac{T_R r_a + T_a r_e}{r_a + r_R}$$

T_e is analogous to the *operative temperature* defined by Gagge (1981) as an index of dry heat exchange for humans.

How does this influence the choice of the ideal temperature sensor? T_e, which is the temperature experienced by the animals, can be measured by a temperature sensor that has the same ratio of radiant to convective transfer resistance as the actual animals. The standard black-globe thermometer is designed to satisfy this criterion for man (Humphreys, 1977), but has an inordinately long time constant.

The radiative resistance, r_R, is the same for all surfaces of unit emissivity (a reasonable approximation for animal integuments and animal houses) and is about 2.12 s cm^{-1} ($\equiv 6$ W m^{-2} K^{-1}) at 20 °C, but unfortunately our choice is complicated by the variation of r_a with animal size and air speed and by the need for a short time constant. We therefore need to estimate r_a, which is possible if the air speed and temperature and the typical body dimension are known.

Animals of agricultural interest cover a range of weights between about 50 g for a newly hatched chick and 1 tonne for mature cattle. If we make the crudest possible assumption, that the animals approximate to spheres of unit specific gravity (closer to the truth for small than for large animals), we need concern ourselves with a range of diameters from about 5 cm to at most 1 m. In practice the range is somewhat smaller, the typical diameter for convection even for large cattle being about 0.4 m only. A short thermometer time constant may be obtained by 'squashing' the black globe flat—using a thin plate instead of a sphere; all we require is the same heat transfer resistances as for the animal. *Figure 17.6* shows calculated boundary-layer resistances for convective heat transfer from plates and spheres with characteristic dimensions from 5 cm to 1 m. These are based on the Nusselt number–Reynolds number relationships presented by Monteith (1973). Curves A (for spheres) and B (for flat plates) are for an air velocity of 0.2 m s^{-1}—representative of conditions in well wind-proofed houses. These relationships suggest that the same weighting of resistances will be provided by a plate 1 m across as for a sphere of 0.45 m diameter—but this is rather impractical. More usefully, a disc of 0.5 m diameter is the median of the whole range required, while one about 10 cm across will simulate the resistances appropriate to poultry with an error of about ±20%. The size of disc required will be different when free convection is the dominant mode of heat transfer, but the predictions are not too dissimilar: *see* curve C in *Figure 17.6a* (the comparison is still with forced convection for the plate, because it will be closer to air temperature than an animal). The radiative resistance is also shown for comparison (line D). *Figure 17.6b* shows the corresponding combined resistances, which are always less than the radiative resistance alone.

Is a plate temperature sensor practicable? In the sense that we can clip any type of thermometer element to a plate heat sink the answer is yes; indeed, for resistance thermometers this procedure will have the additional advantage of almost eliminating self-heating errors. However, the time constant of a plate is related to its thickness, Z (m), by

$$\tau = \frac{Z \varrho' c'}{(\varrho c_p / r_e)} \tag{17.11}$$

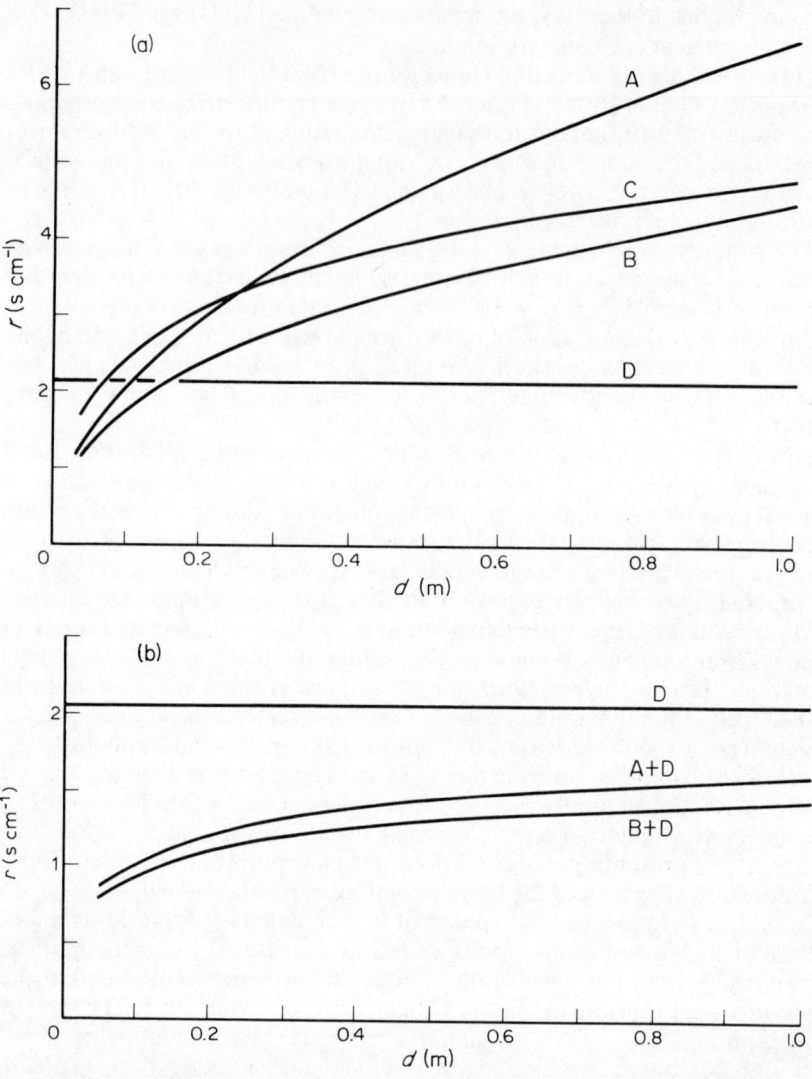

Figure 17.6 (a) Idealized variation of heat transfer resistances with size. Line A is the resistance for a sphere in forced convection at 0.2 m s⁻¹ air speed, and line B that for a flat plate in the same conditions. For comparison, line C is the free-convection resistance for spheres for a temperature excess of 5 K. Line D is the radiative resistance, which is independent of size. (Values calculated from relations given by Monteith, 1973.) (b) Variation with size of combined resistances for radiation + convection heat transfer. The radiative resistance is also shown for comparison. The labels A + D and B + D refer to combinations in parallel of the resistances shown in Figure 17.6a

where ϱ' is the density of the plate material. For a disc 10 cm in diameter, insertion of values appropriate to aluminium ($\varrho' = 2.7$ tonne m⁻³ and $c' = 880\,\mathrm{J\,kg^{-1}K^{-1}}$) unfortunately yields $Z = 0.1$ mm for a time constant of 20 s—too thin to be useful. $Z = 1$ mm gives a time constant of 200 s—and similar problems to those of existing thermostats. Aluminium sheet 0.5 mm thick should give a time constant of about 100 s, and would be worth trying.

This will produce a phase lag of about 50 s and a signal attenuation of approximately 50 per cent for sinusoidal control cycles of 5 minutes, but would respond well to radiant temperature.

In well-insulated animal houses a small ventilated thermometer is the best temperature sensor, but an 'environmental temperature' sensor could give better results where radiant loads are significant and for control of houses containing large animals, because at low air velocities these are coupled much more strongly to their radiant environment than to the air temperature. Perhaps the best approach would be to measure *two* temperatures: true air temperature, obtained using a small sensor of short time constant, and that of a flat plate, to give an indication of the radiation field, which will vary less rapidly than air temperature. The two readings could then be combined electronically to provide the control input, a procedure which is facilitated by the advent of microprocessor-based control systems.

Humidity measurement

The humidity in animal houses is rarely measured, except for the diagnosis of problems that have already arisen. There are two reasons for this: first, in temperate climates humidity has little direct effect on the stock for most of the year; secondly, traditional humidity sensors are either unreliable or inappropriate in animal house environments. Ventilated wet- and dry-bulb thermometers will quickly become contaminated by dust and pollutants in the house atmosphere, and consequently require regular and conscientious maintenance to make accurate readings possible, while hair hygrometers have time constants of between 10 and 60 min (Smith, 1970), depending on their preparation and the environmental temperature. Obviously, neither are suited to routine application for control purposes in animal houses, though wet-bulb thermometers are employed in the control of hatchery incubators, because water loss from the eggs is important.

Humidity control can (or could) be of value in animal housing in two circumstances. At low temperatures and at low ventilation rates the limited moisture capacity of cold air, which is apparent from the hygrometric chart (*Figure 17.7*), may be easily saturated by the vapour exhaled by stock or evaporated from bedding and faeces. Condensation in the atmosphere of the building may facilitate the transmission of disease between animals, indirectly causing a loss of production, and condensation on or within the fabric of the building may result in damage to the structure (Wathes, Chapter 21). Both could be avoided by an automatic increase of ventilation or by heating the environment when relative humidities become high. In contrast, at high temperatures the heat stress imposed on the stock by the environment may depress production, as shown in *Figure 17.8*. The possibility of evaporative cooling is also related more closely to the wet-bulb temperature (*Figure 17.7*) than to air temperature. Here the control variable should be a weighted mean of the wet- and dry-bulb (or environmental) temperatures (Ingram, 1974; Starr, Chapter 2), rather than dry-bulb temperature alone.

There are a number of methods for the measurement of atmospheric humidity that could be applied for control in animal houses. Automatic

320

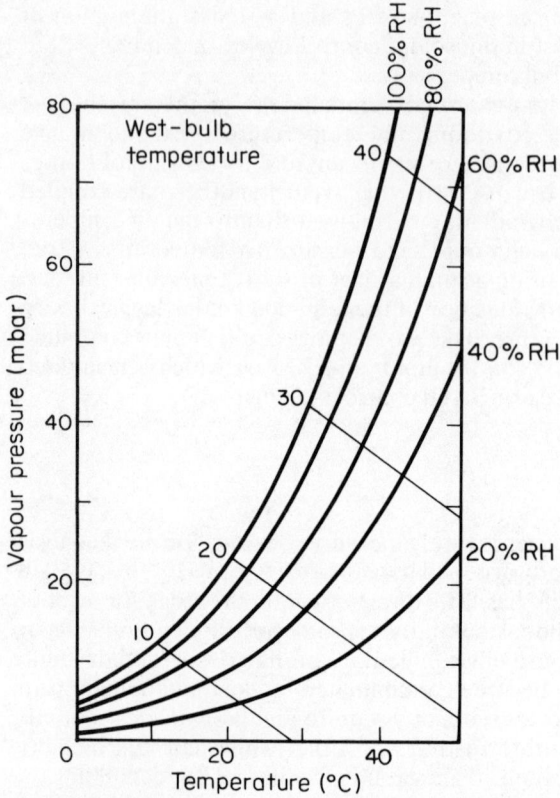

Figure 17.7 Psychrometric chart showing the relationship between air temperature, water vapour pressure and relative humidity. The diagonal lines are loci of constant wet-bulb temperature, as indicated by the figures beside the line of 100% relative humidity

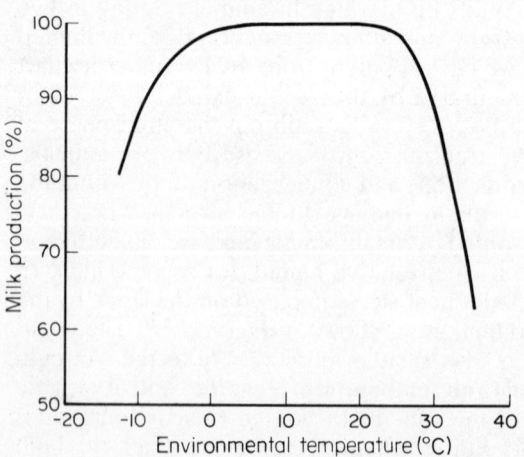

Figure 17.8 Variation of milk production, expressed as a percentage of normal production, with environmental temperature. This shows the depression of production at high temperature, which is exacerbated by high humidities. (Redrawn from Yeck and Stewart, 1959)

dewpoint hygrometry and infrared gas analysis would be ideal, but are too expensive to be warranted for other than research purposes. However, devices are now available which measure the resistance or capacitance changes of porous elements as they absorb or desorb water in response to changes of relative humidity. These could be applied for control purposes. Capacitative sensors (mostly based on oxides of silicon or aluminium) are the more recent development (to the extent that they are not mentioned by Fritschen and Gay (1979)) and are probably the most promising. Both resistance and capacitance elements essentially sense relative humidity, and the sensor output must therefore be combined with a temperature measurement in order to estimate water vapour concentrations or dewpoint temperatures (*Figure 17.7*). Early capacitance sensors were subject to excessive calibration drift due to contamination, but the manufacturers of the latest types claim that they may be handled and even washed or immersed in water without significant changes in sensitivity (e.g. Moisture Control and Measurement Ltd, Wetherby; Michell Instruments, Cambridge). Robustness of this order is required for use in animal houses. The sensors may also be small, of the order of millimetres, and they have time constants of the order of seconds, suitable for control applications. However, there is one potential problem. Because they work by the physical absorption of water molecules in fine pores within an evaporated oxide film, the absorption sites may also be occupied by any other polar molecule of comparable size to water. Ammonia, in particular, which is a common contaminant in animal houses, may compete with water for absorption sites and therefore affect the readings. The relative sensitivity of such humidity sensors to water and ammonia, and their application to animal house control, warrant further study.

Gaseous contaminants

The measurement of gaseous contaminants in animal houses presents few problems for research purposes. Carbon dioxide may be present in concentrations of several per cent without adverse effects on the stock, while the legal limit for ammonia is 25 volumes per million (vpm). Both are best measured using infrared gas analysis. This technique has been considered in detail by Hill and Powell (1968). However, IRGAs are unlikely to be used for control in animal houses. Their time constants are about 1 min at normal flow rates, but since a gas concentration sensor would be used only to override temperature control at minimum ventilation rates, this is not important. The problem is again price, which is too high to warrant their use for a control function required for little of the time.

Control for carbon dioxide is probably unnecessary, since the rate of emission from stock, q ($g\,s^{-1}$), can be estimated with reasonable accuracy from a knowledge of their feed intake (and therefore metabolic rate). The maximum allowable concentration of CO_2, ψ($g\,m^{-3}$), then determines the minimum design ventilation rate for the house, $\dot{V}$ ($m^3\,s^{-1}$), as

$$\psi = q/\dot{V} \tag{17.12}$$

This will suffice for most circumstances.

Ammonia presents a less tractable problem. In addition to legal limitations on the exposure of workers, experiments have shown that stock exposed continuously to concentrations of 50 vpm and above may show both reduced production and disease induced by the exposure (Charles and Payne, 1966a, b). For the advisory worker the solution is to use chemical detection methods, such as the 'Dreager tube' (*Figure 17.9*). Though simple, this can give an indication of ammonia concentration to about ±5 vpm, and at low capital cost.

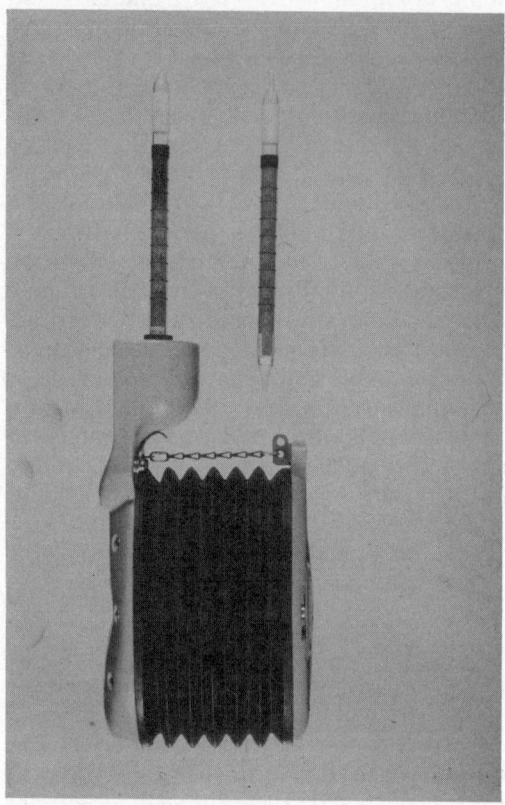

Figure 17.9 'Dreager tube' used for the detection of ammonia, showing darkening of the crystals, here indicating 15 vpm, and an unused tube for comparison

There would, however, be considerable potential for a low-cost ammonia sensor for control applications in animal houses, since ammonia may present a persistent problem, especially in winter in broiler houses and those for laying poultry. Conductimetric analysers have been successfully applied for the control of carbon dioxide concentrations in greenhouses (Bowman, 1968) and have proved economic. These appear the most likely candidates for ammonia control in animal houses, though undoubtedly the non-specific response of most conductivity cells would present difficulties: they would respond not only to ammonia, but also to CO_2 and other gases which dissociate on solution in water. Again, this is an area that warrants further work.

Ventilation

This is the last part of this chapter concerned with control inputs—because although ventilation is usually the controlled variable in the animal house environment, it is the most difficult to measure. Indeed, it is probably impracticable to consider routine monitoring of ventilation, other than via measurements of fan speed and pressure differentials. Both of these indicators may be unreliable, however, because wind will interfere with the ventilation of any practical animal house. *Figure 17.10* shows measurements of ventilation rate made on a 'high-speed jet' ventilated house at Gleadthorpe EHF by one of the authors (J.A.C.) and Cartledge (Cartledge, 1979). This system, as described by Carpenter (Chapter 18) and Randall (Chapter 19), is the most wind-proof currently available, but the measurements show changes by a factor of almost two between the ventilation rates measured on different days for the same fan settings. The only variables that could explain this were differences in wind speed and direction.

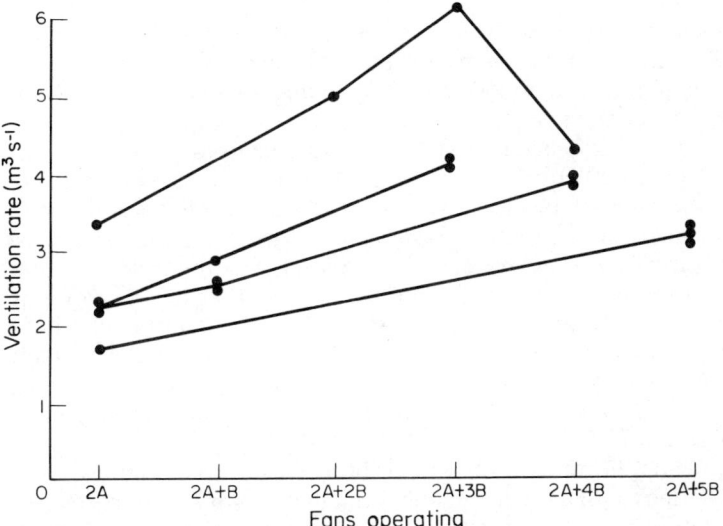

Figure 17.10 *Changes of ventilation rate with fan setting, measured using the tracer decay method. The four lines were obtained on different days for the same laying house at Gleadthorpe EHF. The differences may be ascribed to differences in wind speed and direction, even though the ventilation system employed (high-speed jet) is regarded as wind-resistant. (Measurements by J. A. Clark and V. Cartledge; Cartledge, 1979)*

Ventilation control of temperature or gas concentrations works by 'diluting' the air inside the house with air from outside containing a lower concentration of the particular entity. Thus to control temperature we add cold air (with a low 'heat' concentration) to the warm air in the house. The methods employed to measure the ventilation rate of animal houses work in the same way. If the concentration of a tracer in the house is ψ_i, and that outside is ψ_o (usually ≈ 0), then the quantity of material removed per unit time is $\dot{V}\psi_i$ (by the interchange of volumes containing $\dot{V}\psi_i$ and $\dot{V}\psi_o$), where $\dot{V}$ is the rate of ventilation air exchange in $m^3\,s^{-1}$. If the total volume of the house is V (m^3), and the tracer material is introduced into the atmosphere at a particular instant, then the subsequent dilution will result in a rate of decay of the concentration of

$$\frac{d\psi}{dt} = -\frac{\dot{V}\psi_i}{V}$$ (17.13)

Rearranging,

$$\frac{1}{\psi_i} d\psi = -\frac{\dot{V}}{V} dt$$ (17.14)

which on integration and the insertion of the appropriate initial boundary condition ($\psi_i = \psi_{(t=0)}$ when $t = 0$) yields

$$\psi_i = \psi_{(t=0)} \exp\left(-\frac{\dot{V}}{V}t\right)$$ (17.15)

Hence the locus of $\ln\psi_i$ versus t yields a slope of $(-\dot{V}/V)$. If the house volume is known, the ventilation rate in $m^3\,s^{-1}$ may be estimated or, more usually, multiplication of the slope by 3600 gives the number of air changes per hour (i.e. the ventilation rate in h^{-1}).

Various tracers have been used for measurements of this kind. For example, Evans and Webb (1971) have used radioactive gases to measure the ventilation rates of broiler (Webb, 1967) and turkey (Webb, 1966) houses, but radioisotopes are not suitable for routine use. The most convenient tracer is nitrous oxide, which is harmless in the concentrations required, but detection must be by infrared gas analysis, requiring comparatively delicate and bulky instruments. The recent introduction of a portable gas chromatograph detector for sulphur hexafluoride, originally developed for the detection of leaks in gas mains, has consequently made available a useful tool for the measurement of ventilation rates in commercial situations.

For research applications ventilation is best measured continuously, but this is rarely done though the technique is relatively simple. It requires only accurate calibration of a flowmeter and an infrared gas analyser. Rearranging equation 17.13 gives

$$\dot{V} = \frac{q}{\psi}$$ (17.16)

Therefore, the ventilation rate (in $m^3\,s^{-1}$) may be obtained by dividing the rate at which a tracer is released into the house by the concentration measured concurrently. *Figure 17.11* shows part of a 24 hour series of measurements of ventilation obtained by this method for the high speed jet ventilated house at Gleadthorpe EHF (Cartledge, 1979). The control of house temperature appears excellent, with no significant change when the ventilation rate increases by a factor of more than two in response to the morning commencement of the activity of the birds.

It is relatively easy to measure the ventilation rate of whole houses, by siting the tracer sensor at a known air exit. However, many ventilation

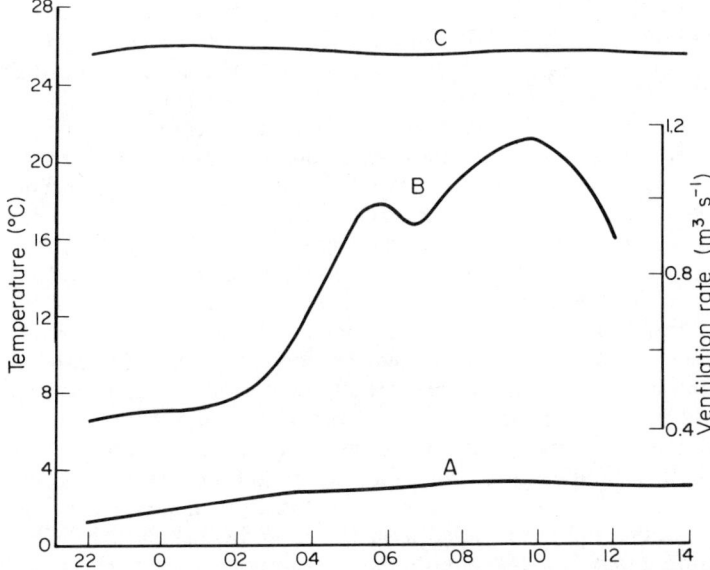

Figure 17.11 Part of a 24 hour series of measurements of ventilation and heat balance obtained from the same house as in Figure 17.10. The measurement of ventilation (line B) was made using the continuous tracer release method. The lower line (A) is the inlet air temperature and the upper line (C) the outlet air temperature. (Measurements by J. A. Clark and V. Cartledge; Cartledge, 1979)

problems are concerned more with differences in air exchange between various points within the enclosed space. Lidwell (1960), who was concerned with hygiene problems in operating theatres, and Smith (1972) suggested the use of a 'transfer index' to describe the local ventilation rate, equal to the inverse of the rate of air supply (i.e. to $\dot{V}^{-1}$). Hence, the index has units of $s\,m^{-3}$. A more unified treatment of heat and mass transfer in animal houses may be facilitated by using the treatment adopted by Monteith (1973). He expressed the ventilation rate per unit of *floor* area, A. The transfer index then becomes a resistance, r_v, in units of seconds per metre (Cena and Clark, 1978), as employed earlier in this volume by McArthur (Chapter 3).

$$r_v = A\,\dot{V}^{-1} \tag{17.17}$$

This resistance will act in series with the resistances between the animal and the house environment. It is therefore easily envisaged that transfer will be 'difficult' from locations with a high ventilation resistance and 'easy' where the resistance is small. The transfer resistances between particular positions in the house and the outside environment may be determined for each locality by the tracer decay method, while 'noise' on the decay curve offers a measure of the uniformity of mixing within the house air (Evans and Webb, 1971).

Radiative energy

It is implicit in the treatment of temperature measurements earlier in this chapter that radiative energy transfer plays an important part in the heat balance of housed animals. At typical house temperatures the radiative resistance is about 2.2 s cm^{-1}. Therefore, at the low air velocities typical inside animal houses in temperate climates, thermal radiative heat transfer is comparable with convection even for animals as small as newly hatched chicks. For an air speed of 0.2 m s^{-1}, heat transfer by convection is only about 50 per cent more than that by radiation for a sphere 5 cm in diameter; the two are almost equal for an animal the size of a mature hen or a piglet, and for fully grown cattle or pigs, radiation exchanges may exceed those due to convection by a factor of two (Mount, 1968; Monteith, 1973). The variation is evident in *Figure 17.6*. The thermal radiation environment is therefore important for all housed animals. However, in current practice almost all temperature sensors other than the black globe are intended to measure air temperature. The solution suggested earlier may be useful for control purposes, but the radiation environment is also neglected in much research and trials work, though appropriate measurements are not difficult.

Thermopiles are the most reliable basis for instruments intended to measure radiative temperatures and thermal radiation fluxes, and the radiative characteristics of the environments to which animals are exposed are best measured using a net radiometer (Cena and Clark, 1981b). This type of instrument may be adapted either to measure the net radiant flux or, by enclosing one face in a reference cavity of known temperature, to measure the radiant temperature of the environment. They have much shorter time constants than a black globe and do not require an air speed correction, because of the polyethylene film shielding their surfaces. The radiant temperatures of particular surfaces may be measured either with comparatively inexpensive radiometric thermometers, such as those manufactured by Barnes Engineering in the USA and Heimann in Europe, or by infrared thermal scanners. The latter present a visual analogue of the temperature field, which is probably of more potential value to the building engineer than to those concerned directly with husbandry. The temperature distributions on the surfaces of a building may indicate imperfections of construction and the presence of damage, which result in poor insulation, or the locations of air leaks. Therefore thermal scanners, though expensive to hire and run, can aid rapid diagnosis of building faults. The problem with all these devices is that they will not operate satisfactorily on prolonged exposure to dusty environments. Dust and other surface contaminants of biological origin have thermal emissivities close to unity. Such contamination of the protective surfaces of radiometers will therefore make them inoperative. For this reason radiative measurements can be applied only for research and diagnostic purposes in animal house environments.

Light

Both the intensity of light and the duration of 'day length' are already employed routinely to control the activity and reproduction of housed animals (Morris, Chapter 5). Of these two characteristics of the light regime, day length is the easiest to specify. It may be determined simply by a time switch. Indeed, in many experiments on the effects of light (for example, on the egg-laying performance of poultry) day length is the only measurement of light noted. However, though the contribution of light sources to the heat balance of the house is almost always trivial, the intensity of illumination is very important in stock management. In addition, very few publications reporting experiments on the effects of light on production by housed animals give any details of the spectrum of the light source used. Since different spectral regions can affect growth or development to a markedly different extent, this is a neglected area of research.

In the animal world, only the action spectra for vision have been determined in detail, in comparison to our much greater knowledge of the photochemistry of plant pigments (Bainbridge, Evans and Rackham, 1966; Evans, Bainbridge and Rackham, 1975). However, at least some stages of the development of animals can be remarkably sensitive to light. For example, *Figure 17.12* shows the response of chick embryos to ultraviolet

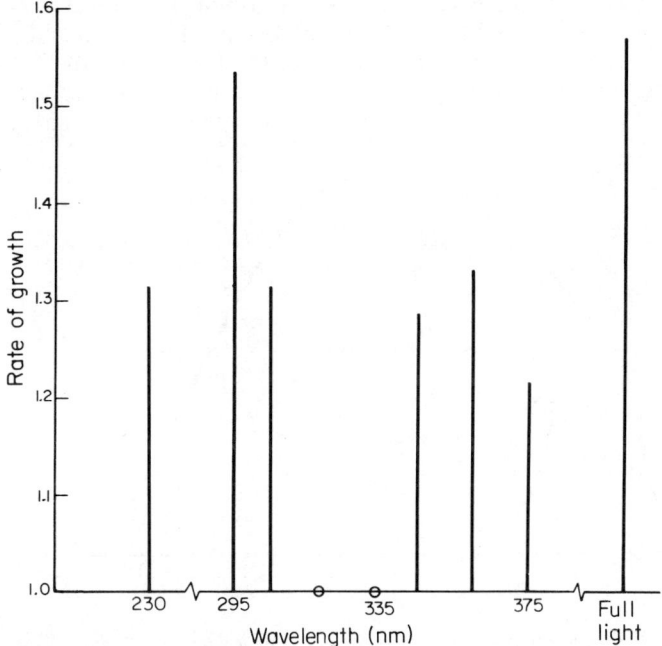

Figure 17.12 Bar chart showing the variation of the relative growth rates of chick embryos with the wavelength of light used to irradiate the eggs. The rates are normalized to that of embryos growing in the dark and the rate for full light is shown for comparison. (Redrawn from Coleman et al., 1977)

light, determined by Coleman *et al.* (1977). The relative growth rates may be regarded as a crude action spectrum, showing an increase in mass of up to 50 per cent in response to light of particular wavelengths. In these experiments the embryos were harvested at 3 days' incubation. It is interesting that eggshells have strong absorption bands at 345 and 385 nm whereas the maximal photoacceleration of embryonic weight occurred at 295 nm. Coleman *et al.* concluded that overheating of eggs should be avoided and that cool-white fluorescent tubes are to be preferred to incandescent bulbs. For the best embryonic weight the light sources should be placed no further than 12 cm from the eggs. Coleman *et al.* also concluded that it would be preferable to describe the type of light fully, including both wavelength and flux densities, rather than by illumination, which is not very useful in photobiology.

There is also another pressure for improved measurements of light intensities in animal houses: tungsten filament bulbs, though cheap, are energetically inefficient. Fluorescent lighting can provide the same illumination at a significantly lower cost. However, all parts of the visible spectrum are present in 'tungsten' light, while the most efficient fluorescent lamps have their light output concentrated in restricted parts of the spectrum. Therefore, substitution of fluorescent for tungsten lighting either presupposes a knowledge of the photobiological needs of the animals or requires extensive trials to show satisfactory equivalent performance. Both paths require adequate means for the measurement of light.

The 'traditional' instruments employed for measuring light intensities in this industry have been based on silicon photocells. These have the advantage that they can be self-powered and robust, but their sensitivity is limited at low light levels and the response of the basic cells peaks in the near infrared (*Figure 17.13*). Most instruments of this type are designed to

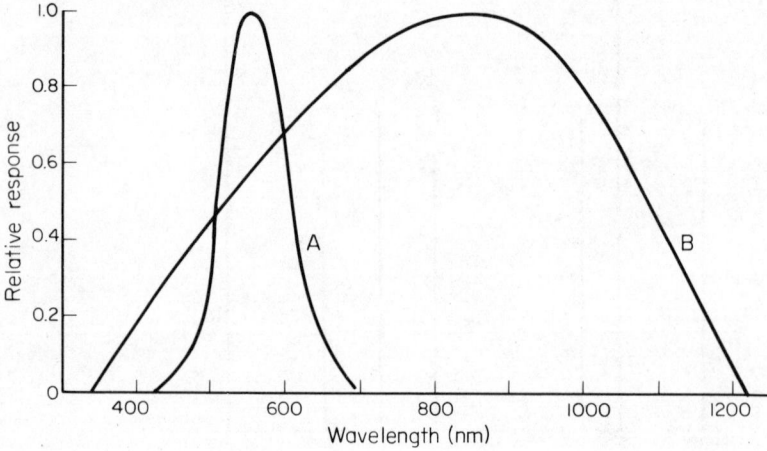

Figure 17.13 Variation of relative response with radiation wavelength, line A for the human eye and line B for silicon photocells. Both curves are normalized to the peak response

measure 'illumination', and the best are equipped with filters that give a very good match to the spectral sensitivity of the human eye (*Figure 17.13*). Measurements of perceived illumination have their place in animal environments—because what the *stockman* can see may be an important variable in the standard of husbandry—but they may bear little relation to the response of the stock when the spectrum emitted by the light differs markedly from that of daylight. Sophisticated spectrophotometers are unnecessary in this context. Comparatively simple but sensitive quantum sensors are now available at comparable prices and in a form no bulkier than photometers; indeed, some may also be employed as photometers using interchangeable sensing heads. Band filters enable the measurement either of total photon flux (*Figure 17.14*), or that in particular spectral bands. The units of measurement should be $\mu mol\,m^{-2}\,s^{-1}$. Though the response of

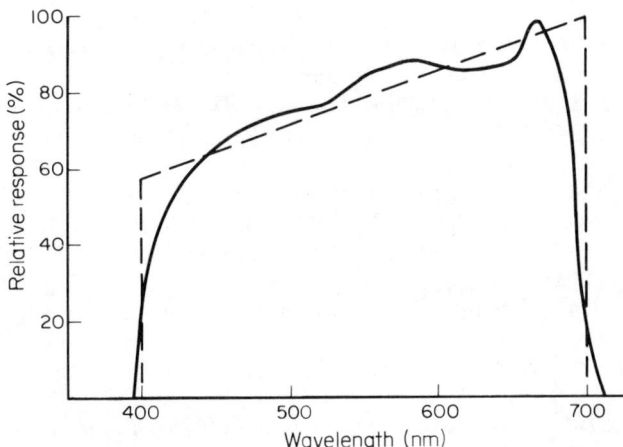

Figure 17.14 Variation of relative response with radiation wavelength for an 'ideal' quantum sensor (dotted line), and for a typical current instrument designed for the measurement of photosynthetically active radiation

current instruments has been tailored for work on photosynthesis rather than to animal pigments, they should be adequate for the present purpose. Commercial instruments of this type are currently produced, for example, by Macam in the UK and by Li-cor in the USA.

Conclusions

Though measurements of the animal house environment for research purposes need present few problems, the design and selection of sensors for control applications could make better use of existing knowledge. We suggest that thermometers for use in animal houses should ideally have a time constant of about 20 s, and that the optimum thermometer for temperature control in animal houses should be a black disc of thin metal. Unfortunately these requirements are difficult to reconcile unless combined with a separate sensor for air temperature. Humidity control may be possible using recently developed capacitative sensors, while control for ammonia contamination could be based on existing conductimetric devices.

Measurements of radiant energy in animal houses are often neglected, and quantum sensors should be used to specify light levels and spectral distribution in animal experiments.

References

BAINBRIDGE, R., EVANS, G. C. and RACKHAM, O. (Eds) (1966). *Light as an Ecological Factor*. Blackwell, Oxford

BOWMAN, G. E. (1968). In *The Measurement of Environmental Factors in Terrestrial Ecology*, pp. 131–140. Ed. by R. M. Wadsworth. Blackwell, Oxford

CARTLEDGE, V. (1979). *BSc Dissertation*, University of Nottingham

CENA, K. and CLARK, J. A. (1978). *J. thermal biol.* **3,** 173–174

CENA, K. and CLARK, J. A. (Eds) (1981a). *Bioengineering, Thermal Physiology and Comfort.* Elsevier, Amsterdam

CENA, K. and CLARK, J. A. (1981b). In *Bioengineering, Thermal Physiology and Comfort,* pp. 271–284. Ed. by K. Cena and J. A. Clark. Elsevier, Amsterdam

CHARLES, D. R. and PAYNE, G. C. (1966a). *Br. Poult. Sci.* **7,** 177–187

CHARLES, D. R. and PAYNE, G. C. (1966b). *Br. Poult. Sci.* **7,** 189–198

COLEMAN, M. A., McDANIEL, G. R., NEELEY, W. C. and IVEY, W. D. (1977). *Poult. Sci.* **56,** 1421–1425

CROMARTY, A. S. (1974). *Reading University Agricultural Engineering Report TEN-7401*

ESMAY, M. L. (1969). *Principles of Animal Environment.* Avi, Westport, Connecticut

EVANS, G. C., BAINBRIDGE, R. and RACKHAM, O. (Eds) (1975). *Light as an Ecological Factor,* 2nd edn. Blackwell, Oxford

EVANS, G. V. and WEBB, J. W. (1971). *UKAEA Report, AERE R6709.* HMSO, London

FRITSCHEN, L. J. and GAY, L. W. (1979). *Environmental Instrumentation.* Springer-Verlag, New York

GAGGE, A. P. (1981). *Bioengineering, Thermal Physiology and Comfort*, pp. 79–98. Ed. by K. Cena and J. A. Clark. Elsevier, Amsterdam

HILL, D. W. and POWELL, T. (1968). *Gas Analysis.* Adam Hilger, London

HUMPHREYS, M. A. (1977). *Ann. occup. Hyg.* **20,** 135–140

INGRAM, D. L. (1974). In *Heat Loss from Animals and Man,* pp. 233–254. Ed. by J. L. Monteith and L. E. Mount. Butterworths, London

LIDWELL, O. M. (1960). *J. Hyg., Camb.* **58,** 297–305

MONTEITH, J. L. (1973). *Principles of Environmental Physics.* Arnold, London

MOULSLEY, L. J. and FRYER, J. T. (1978). *National Institute of Agricultural Engineering, Departmental Note DN/EN/927/1003*

MOUNT, L. E. (1968). *The Environmental Physiology of the Pig.* Arnold, London

SMITH, C. V. (1970). 'Meteorological observations in animal experiments.' *WMO Tech. Note 107.* World Meteorological Organization, Geneva

SMITH, C. V. (1972). 'Some environmental problems of livestock housing.' *WMO Tech. Note 122.* World Meteorological Organization, Geneva

WADSWORTH, R. M. (Ed.) (1968). *The Measurement of Environmental Factors in Terrestrial Ecology.* Blackwell, Oxford

WEBB, J. W. (1966). *Wantage Research Laboratory (AERE) Physics Group Report 154. E/JWW*

WEBB, J. W. (1967). *Wantage Research Laboratory (AERE) Physics Group Report 161. E/JWW*

YECK, R. G. and STEWART, R. E. (1959). *Trans. ASAE* **2,** 71–77

18

VENTILATION SYSTEMS

G. A. CARPENTER
National Institute of Agricultural Engineering, Silsoe, Bedford

Introduction

GENERAL OBJECTIVES

For human subjects, ventilation is mainly concerned with comfort. For livestock it is also concerned with comfort interpreted through welfare, behaviour and health. In addition, and most importantly, it is concerned with quantifiable factors such as conversion ratio, growth rate and mortality. In designing a ventilation system, the engineer has to come to terms with, first, the external interfering influences, namely climatic and environmental factors, airborne pollutants and failures of electricity and fuel supplies, and secondly, with the building design, the stock requirements and the stock enclosure system. The resulting internal environment, even if temperature-controlled in time, can have considerable spatial variations in temperature and also large variations in other factors such as air speed, and gas and dust concentrations. This complex, variable environment also interacts with the health and nutrition of the stock. Although economic production on the farm is the ultimate measure of the success or failure of a controlled environment, the engineer can make physical measurements that are essential for an understanding of how a ventilation system functions. As this understanding increases, a progressive improvement in the level of control is possible.

Later in this chapter it will be shown that both the nature of internal air movement and the principles of ventilation systems are now well understood, and this means that much greater significance can now be attached to

Table 18.1 POPULATIONS OF CLASSES OF LIVESTOCK IN THE UK, DEC. 1975 (THOUSANDS)

Finishing pigs (above 50 kg)	2 597	Layers	48 499
Weaners (20–50 kg)	2 281	Broilers	50 111
Weaners (below 20 kg)	1 905	Rearers	16 228
Sows and gilts in pig	618	Breeders	5 568
Boars	41	Turkeys	5 407
Farrowing and dry sows	226	Ducks and geese	1 227
Calves up to 6 months old	1 528		
Calves 6 months to 1 yr old	2 140		
Cows, milking	2 574		
Other cows and heifers	3 244		
Other cattle over 1 yr old	4 429		

From HMSO (1978)

the requirements of the different classes of housed livestock and associated husbandry practices (*Table 18.1*).

Although certain technical aspects are common to most of these classes (for example propeller fans are often used to ventilate houses for all except adult cattle), there are differences between them that are sufficiently large to preclude any one design of ventilation system being applied to all. Thus, an important practical objective is to find the optimum design of ventilation for each class of livestock.

THE BUILDING CHARACTERISTICS

The designs of livestock buildings have evolved from a combination of the needs to control the aerial environment, to enclose, feed and inspect the stock and to collect waste. Of the five aspects considered below, all but the last, namely thermal insulation, apply to all livestock buildings.

Table 18.2 APPROXIMATE SENSIBLE HEAT OUTPUTS AND OTHER DATA FOR COMMON CLASSES OF LIVESTOCK

Class of stock	Animal wt (kg)	Heat output per animal (W)	Total area of house floor per animal (m²)	Heat output per unit floor area (W m⁻²)
Weaners, 8 weeks old	23	55	0.25	220
Finishing pigs	64	100	0.77	130
Finishing pigs	90	125	0.77	162
Layers in cages	1.8	8.5	0.47	183
Layers in cages	1.8	8.5	0.67	122
Broilers, 1 day old	0.04	0.3	0.047	6
Broilers, 8 weeks old	1.8	8.5	0.047	183
Turkeys (fattening)	4.5	20	0.093	215
Adult cattle	500	890	3.58	250
Calves, 12 weeks old	136	250	1.95	128
Calves, newborn	45	120	1.95	62

(1) Generation of sensible heat by the stock. This is always reasonably high for adult stock, but can be lower by a factor of over 40 for young stock (*Table 18.2*).
(2) Building shape and its aerodynamic implications. The majority of buildings are single-storey, detached and of length greater than their width. Unless there are partitions, the resistance to air movement within the building is negligible. Also, as the fans and vents are usually mounted directly in the walls or roof, the ventilation of the buildings is potentially vulnerable to wind interference.
(3) Stock distribution. Stock are usually distributed throughout the building and all parts require ventilating.
(4) Direction of air movement. In order to reduce end-to-end temperature gradients at low ventilation rates and to avoid high air velocities longitudinally at high ventilation rates, most buildings are ventilated with transverse rather than longitudinal air movement.

(5) Insulation. Buildings for all classes of stock, except adult cattle and most calves, are insulated to a high standard, with U values of the order of 0.6 W m^{-2} K^{-1}.

TEMPERATURE CONTROL—THE HEAT BALANCE EQUATION

The heat balance equation for a building is well known. The interrelated variables of ventilation rate, external ambient and inside temperature for a fully stocked insulated building can be shown graphically in two ways. *Figure 18.1* illustrates how changes in ventilation rate affect the inside temperature. *Figure 18.2* shows the ventilation rate required at each external ambient temperature in order to maintain a particular internal temperature, here 18 °C. The form of this curve is only slightly modified by moderate differences in building dimensions and heat release (and therefore stock density), and this curve can be used as the design basis for the control requirement for most ventilation systems.

Air distribution—the significance of internal airflow patterns

TYPES OF AIR MIXING

Although for a given situation the quantity of the ventilating air may be correct for the desired mean internal air temperature, the actual environment close to the stock will depend also on how the fresh air mixes with the

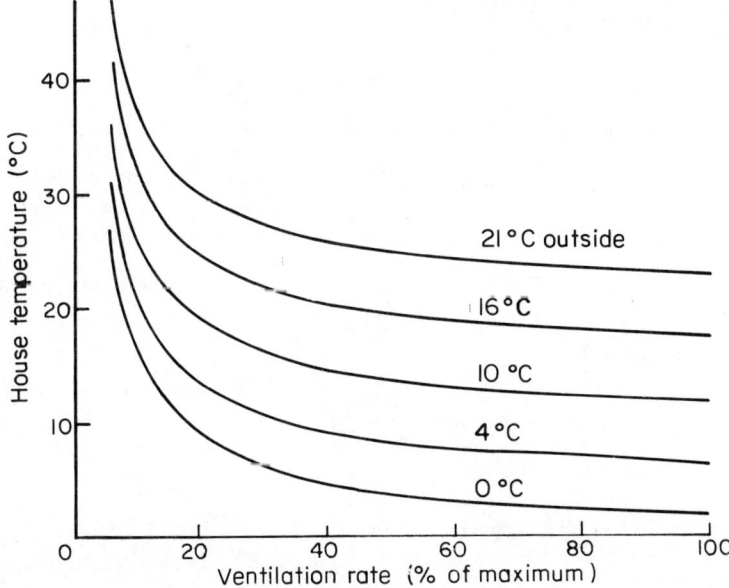

Figure 18.1 Effect of ventilation rate on inside temperature at various outside temperatures for a typical insulated building for intensive animal production

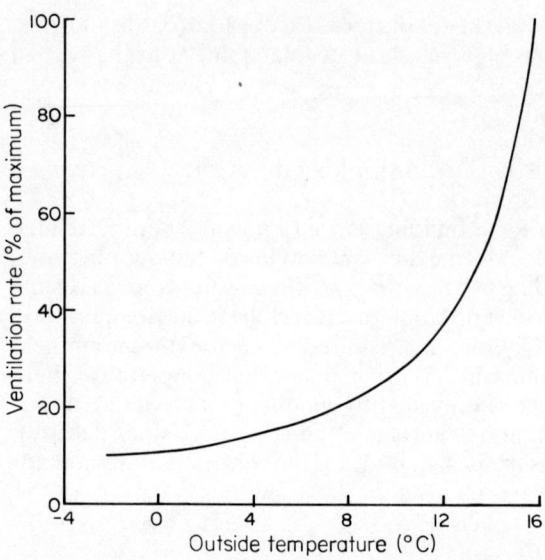

Figure 18.2 Effect of outside temperature on the ventilation rate required to maintain 18 °C inside a typical insulated building for intensive animal production

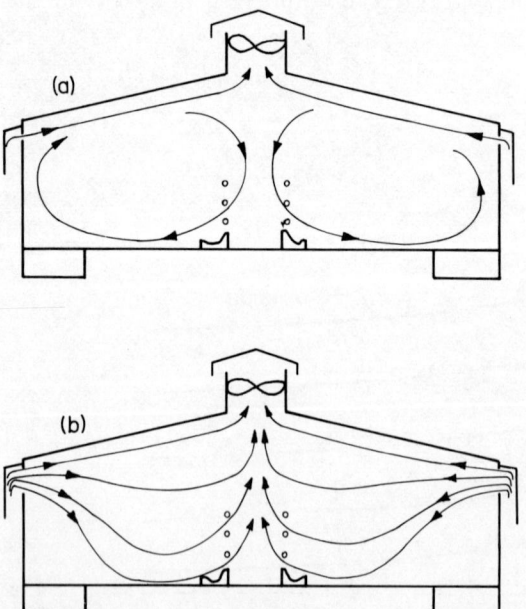

Figure 18.3 Typical airflow pattern showing (a) rotary movement and (b) the (often presumed) air movement which is never found under normal circumstances

air in the building. At one extreme, unidirectional airflow can occur, consisting of parallel paths throughout the building. To achieve this in a horizontal direction, air speeds of the order of 0.5 m s^{-1} are necessary to prevent natural convection from interfering with the flow. At the other, a high degree of mixing can be obtained using mechanical stirring, giving highly turbulent flow with no readily identifiable air paths. Real situations in livestock buildings lie between these extremes, and flow consists of one or more rotary movements (*Figure 18.3*). The pattern of mixing, or airflow pattern, is usually depicted as a map of the dominant air paths. It constitutes the link between the point of entry of the fresh air, the environment around the stock and the environment at any controlling sensor.

THE PREDICTION OF AIRFLOW PATTERNS

The various factors in a ventilation system that affect the airflow pattern have been summarized by Carpenter (1974). The entry conditions determine the initial air direction. For example, the air speeds in the vicinities of an inlet and an outlet for a vent velocity of 10 m s^{-1} are shown in *Figure 18.4*.

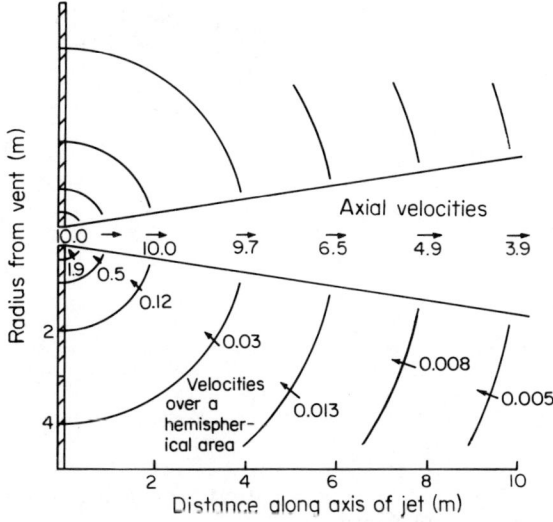

Figure 18.4 Air speeds in the proximity of a vent of diameter 0.62 m when used respectively for supply or exhaust

The flow due to the inlet is also modified by buoyancy effects and physical obstructions.

Airflow patterns have been studied in detail in a full-scale section of a building containing simulated pigs (Randall, 1975), and real pigs (Boon, 1978). The position of the outlet was shown to have a negligible effect on the flow pattern, while inlet jets and convection currents have a large effect. Obstacles also can readily deflect air currents and so modify the airflow pattern. When situated close to a jet orifice or slot, even a small obstacle can have a pronounced effect, which can be detrimental to the environment or, if exploited correctly, can help to produce the required airflow pattern (Randall and Battams, 1976).

Randall (1975) also observed that if air inlets are left open too wide at night when ventilation rates are reduced, the inlet jet is consequently both of low temperature and low speed and is readily overcome by convection currents from the stock. In one commercial layout common in the UK, the night-time temperature is sufficiently low to cause the direction of airflow across the stock to reverse on an average of one day in three, taking the year as a whole, as shown in *Figure 18.5*. This effect is most pronounced in spring and autumn. Control of a ventilation system should be thought of in terms of hot or cold ambients on a *diurnal* rather than a seasonal basis.

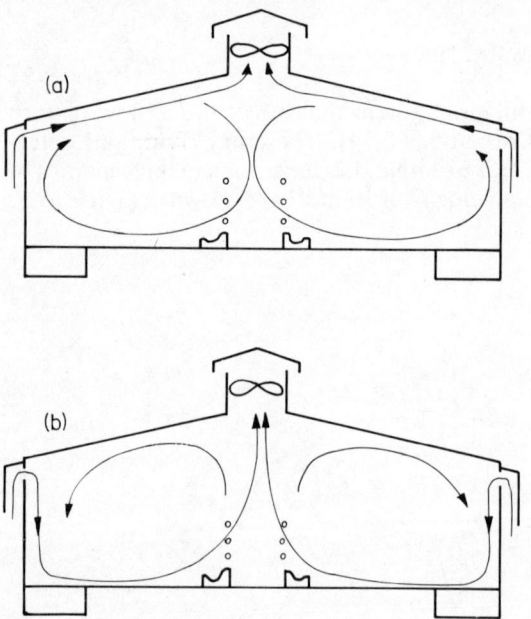

Figure 18.5 Diagram of the reversal of direction of airflow pattern due to changes in external temperature: (a) hot, (b) cold

Randall and Battams (1979) used the concept of the Archimedes number to characterize the initial trajectory of ventilating jets in an experimental section of a livestock building. They found that an air jet projected horizontally from a bluff, long, narrow slot remained horizontal if a corrected Archimedes number was <30 and fell when it was >75. The Archimedes number (Ar) is the ratio of buoyancy to dynamic pressures, which determines motion in non-viscous, non-isothermal fluids:

$$Ar = \frac{g\,d\,(\varrho_1 - \varrho_2)}{u^2\varrho_2} \qquad (18.1)$$

where g is the acceleration due to gravity (in $m\,s^{-2}$); d is the characteristic dimension (in m); ϱ the air density (in $kg\,m^{-3}$), and u is the characteristic air speed (in $m\,s^{-1}$). The subscripts 1 and 2 refer to separate elements in the fluid. When applied to a horizontally projected air jet inside an enclosure, the path of the jet is a function of Ar in the form

$$Ar = \frac{g\,d\,(T_w - T_a)}{T u^2} \tag{18.2}$$

Temperatures replace densities in equation 18.2 because density is inversely proportional to absolute temperature, T. T_w is the temperature of the heated (wall) surface, T_a is the temperature of the incoming air and T is the mean temperature, $(T_w + T_a)/2$, all in kelvins. The characteristic dimension here is the hydraulic diameter of the enclosure. This is given by

$$d = \frac{2\,B\,H}{B + H} \tag{18.3}$$

where B is the width of the enclosure (the side containing the vent) and H the height of the enclosure, both in metres. In equation 18.2 u is a hypothetical mean air speed (in m s^{-1}), given by

$$u = \frac{\dot{V}}{BH} \tag{18.4}$$

where $\dot{V}$ is the volume rate of airflow through the inlet (in m^3 s^{-1}). In order to take account of the effects of the vent's being long and narrow and also having a bluff entry, corrections give rise to a corrected Archimedes number, Ar_c, where

$$Ar_c = Ar\,\frac{0.6ab}{d^2} \tag{18.5}$$

where a and b are the vertical (short) and horizontal (long) dimensions of the vent, respectively, both in metres. Substituting in equation 18.5,

$$Ar_c = \frac{5.89ab\,BH(B + H)\,(T_w - T_a)}{\dot{V}^2(T_w + T_a)} \tag{18.6}$$

When values are appropriate to a typical modern pig-fattening house, using an inlet velocity of 5 m s^{-1},

$$Ar_c = \frac{39(303 - T_a)\,(290 - T_a)}{[1.7 - 0.016(290 - T_a)]\,[303 + T_a]}$$

from which it can be shown that for all values of inlet air temperature above 0 °C, $Ar_c < 30$.

As an alternative to introducing air into an enclosure by means of high-speed inlets of small area, it is possible to use permeable material of an area so large that the incoming face velocity is of the order of one hundredth of that from high-speed inlets. Under these conditions, convection currents readily dominate the incoming air and determine the patterns of airflow

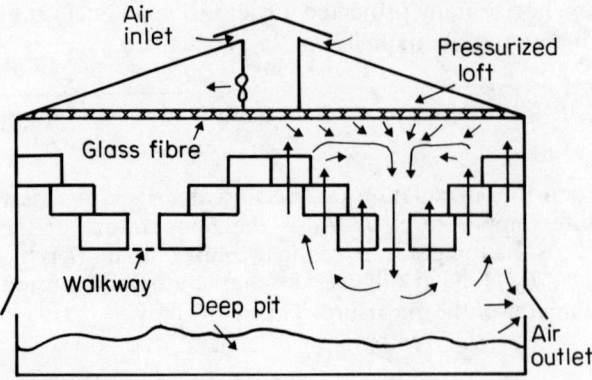

Figure 18.6 Example of the airflow patterns produced in a deep-pit house by natural convection due to thermal buoyancy

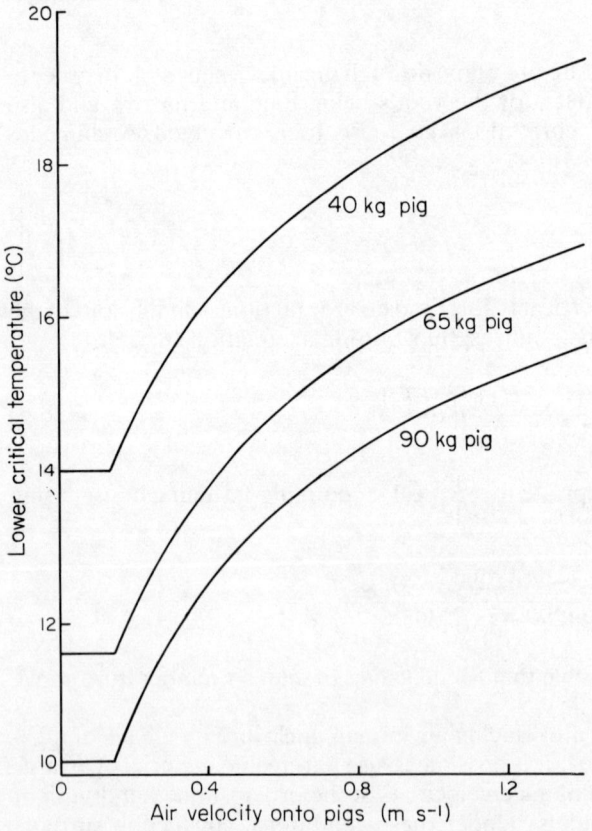

Figure 18.7 The effect of incident air speed on the lower critical temperature of the stock

(*Figure 18.6*). Currents of the order of 0.4 m s⁻¹, for example, arise from laying birds in tiered cages.

In general terms, it can be assumed that inlet velocities in the range 0.5–2.0 m s⁻¹ will result in unstable airflow patterns and should therefore be avoided or used with great care. There is one important case when inlet air speed is not critical, and that is for a discrete jet blowing vertically downwards; under these circumstances, the initial part of the flow pattern will be maintained over a wide range of inlet air temperatures and velocities. However, the instances when this approach can be adopted are likely to be limited to those building layouts that give protection to the stock by solid-fronted enclosures.

THE COOLING EFFECT OF AIR VELOCITY

For a given weight of livestock, husbandry regime and pen design, the heat loss is determined by a combination of air temperature and air velocity.

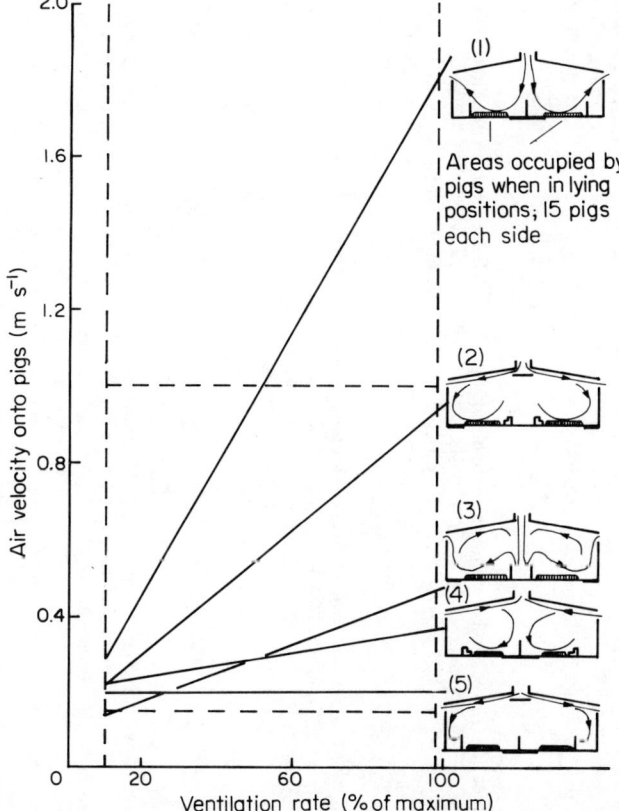

Figure 18.8 The effect of the building ventilation rate (expressed as a percentage of the maximum) on the air speed at stock level, for the five piggery layouts indicated on the right of the figure

Table 18.3 CLASSIFICATION OF VENTILATION SYSTEMS HAVING A CONTROLLED AIRFLOW PATTERN

Type	How airflow pattern controlled	Way in which air is mixed in enclosure	Way in which air is discharged into enclosure	Commercial system			Characteristics
				Name	Description	Applications when used alone	
(a)	Convection currents	Natural entrainment	Large area source, variable temperature, very low air speed (<0.1 m s⁻¹)	Glass-fibre ceiling	Pressurized loft space with entire ceiling of air-permeable glass-fibre	For intensive stock in fixed positions, e.g. layers in cages	Stable patterns over 10 : 1 change in fresh air rate in presence of fixed strong convection currents
(b)	Forced jets	Natural entrainment	High-speed jet (4–5 m s⁻¹), variable temperature	High-speed jet (automatic vents/stepped fans)	Air inlets of automatically controlled area, linked to fans switched on/off in groups	Large finishing piggeries	Stable patterns over 10 : 1 change in fresh air rate. Inlet speed constant
(c)	Forced jets	Mechanical mixing by fan and ducts	Medium-/high-speed jet (2–5 m s⁻¹), constant temperature	Recirculation	(a) Single point discharge (b) Ducted discharge	(a) Symmetrical buildings and layouts with a single fan (b) Normal shaped buildings with a single fan. Calves, sows	Stable conditions over > 10 : 1 change in fresh air rate. Inlet temperature constant. Inlet speed adjustable
(d)	Forced flow	No large-scale mixing inside enclosure	Air source over entire area of end. Medium air speed (0.4 m s⁻¹)	Uni-flow tunnel	Tunnel with uni-directional air flow, heated and filtered	Early weaners	Defined environment at all times

Bruce and Clark (1979) have quantified these factors for pigs and predict that in a real building, in which spatial variations of both velocity and temperature occur, there are optimum combinations of these factors. The effect of air speed on the lower critical temperature is shown in *Figure 18.7*. Randall (1980) used a building section containing simulated pigs to study the relationship between building and ventilation layout and the temperature and velocity distributions close to the stock. In particular, he was able to relate air speed at stock level to ventilation rate (*Figure 18.8*) and to select systems that give (a) good air mixing, (b) a stable airflow pattern, (c) a lying area warmer than the dunging area, and (d) a tendency to increase air speed at stock level at the highest ventilation rate when outside temperatures are high and so increase the range of heat loss control. In *Figure 18.8*, the preferred system for a side-dunging finishing piggery is (2).

Classification of ventilation systems

DESCRIPTION OF SYSTEMS

Most attempts to classify forced-ventilation systems have been based on such considerations as pressurization or suction in the building, the use of ducts or false ceilings, the position of the vents and fans, and of how the fans are controlled. Most of these aspects are of considerable significance in design, but they do not necessarily form a basis for specifying the fundamental characteristics of a system. A more rational approach is to start with the basic premise that any sound system is likely to maintain a defined and desirable airflow pattern and then classify these systems on the basis of how this is achieved. *Table 18.3* identifies four types that are in commercial use, although the unidirectional tunnel is as yet used only on a very limited scale. *Figure 18.9* illustrates the corresponding patterns of airflow.

Type (a) is the conventional 'glass-fibre ceiling' system in which the loft space is pressurized and the entire ceiling made permeable to air by a suspended 50–100 mm thickness of glass fibre. Essentially, the airflow pattern is determined by the convection currents from the livestock, regardless of the quantity and temperature of the incoming fresh air. Because the building space itself is used to mix the fresh air with air already in the building and no mechanical recirculation or mixing by fans is involved, the system can be termed 'straight-through' It is suitable for multi-fan buildings and for controlling ventilation rates down to one-tenth of maximum.

Type (b) is the conventional layout in which the fresh air enters directly from the outside of the building through slot inlets and is exhausted by fans. However, in the best practice the inlet vents are adjusted to maintain an air speed through the vents of 5 m s^{-1} and the interior of the building is designed to ensure that the required flow pattern is obtained. Like type (a), it is straight-through. This system has had several names: 'constant-velocity inlet', 'adjustable inlet/stepped fan', or 'high-speed jet'. It is suitable for multi-fan-exhausted buildings and like (a), will give ventilation rates varying by a factor of ten.

Type (c) is a recirculation system. Its most important difference from the first two is that fresh air is mechanically mixed with house air *before* being

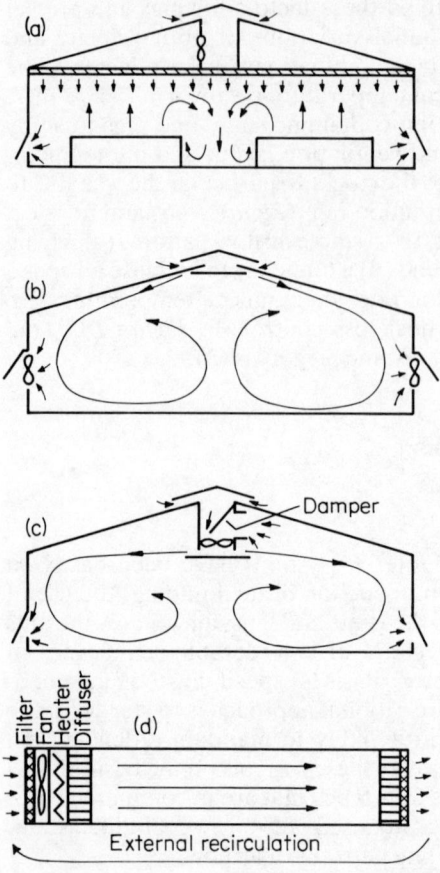

Figure 18.9 Airflow diagrams for four systems of forced ventilation that give controlled and predictable airflow patterns. (a) The pressurized diffusing plenum ceiling; (b) the constant-velocity inlet; (c) the recirculation unit; (d) the unidirectional-flow tunnel

blown into the building space, and that the ventilation rate is controlled by the position of a flap as opposed to regulation of switching or speed of fans. Several points follow from this: the temperature of the air blown into the building is almost constant and only just below that of the air in the building; the air velocity is also constant; this velocity need not be as high as 5 m s^{-1} in order to produce a stable airflow pattern; ventilation rate can be controlled down to below one-tenth of maximum and is more precise at these low rates than for other systems; it is a suitable system for a single fan. Although most designs usually produce a positive pressure in the building, the pressure can be balanced or even give slight suction.

Type (d) is derived from the concept of laminar-flow cabinets (Butler and Egan, 1974). The airflow pattern is unidirectional and the velocity (about 0.4 m s^{-1}) is sufficient for this flow not to be deflected by convection currents from the livestock. Because this velocity has to be sustained over the cross-section of the tunnel, it cannot be a 'straight-through system' in any but the warmest weather if either overcooling or the alternative of gross

artificial heating loads is to be avoided. It therefore has an external recirculation loop, which is normally the room that the tunnel is placed in; this loop
could, however, be an integral part of the unit. It is therefore a special case
of recirculation in which stock are housed in filtered, heated air downstream
of the fan.

Application of ventilation systems

The key to the successful application of ventilation systems is to match the
design to the special requirements of each class of livestock, as dictated
by the optimum environment, husbandry methods and house design and
layout. Even within a single class, different husbandry and layout
considerations may have led to the evolution of different housing systems.
For example, for finishing pigs there are continuous stocking and all-in,
all-out systems, or side-dunging and fully slatted systems, or solid-sided pens
and open-sided pens. Laying poultry may be in cages with scraped manure
belts or in cages that are above a collecting pit. These different housing
systems must be related to the ventilation design, and this will be illustrated
by considering the various factors involved.

TEMPERATURE

First, classes of livestock are divided into cattle on the one hand and pigs and
poultry on the other. Cattle generally require no control of temperature,
and hence are usually housed in naturally ventilated and uninsulated buildings, though young calves require straw bedding and some protection from
very low temperatures. In contrast, pigs and poultry require close control of
temperature and are invariably housed in well-insulated buildings with
control of ventilation rate. Although a significant proportion of pigs are
naturally ventilated, most poultry are fan-ventilated.

Secondly, the required temperature for young pigs and poultry can be
achieved only by using heating, even in well-insulated buildings. Heating
systems interact with ventilation in several ways. When propane burners are
exhausted directly into the building space, as is most common, there is a
critical minimum ventilation rate, required to ensure satisfactory combustion and avoid production of carbon monoxide. Temperature stratification
can also occur, causing very high temperatures close to the roof in the
absence of forced air circulation, particularly with radiant heating systems.
Ventilation systems using recirculation can help to reduce this stratification
(Moulsley and Fryer, 1979). A further interaction between ventilation and
heating is due to the need to interlock the controls: the heat supplied must
not immediately be removed from the building by an increase in ventilation
rate.

Thirdly, a spatial temperature difference is sometimes required within the
livestock building. For example, in a farrowing house, the creep area for the
piglets needs to be 10 K warmer than the general environment around the
sow. For an open-type creep, the effectiveness of the local heating provided
to maintain the higher temperature is greatly reduced if the ventilation

system, owing to poor design, causes excessive air movement in this region. The remedy is to prevent cold air from falling onto the creep area and provide shelter from air currents along the floor (*Figure 18.10a*). Another example is for finishing pigs. It is often believed that these animals do not defaecate in the lying area if the dunging passage is cooler than the lying area. Using a 'straight-through' jet system, this can readily be provided (*Figure 18.10b*), and experiments to investigate this idea are in progress.

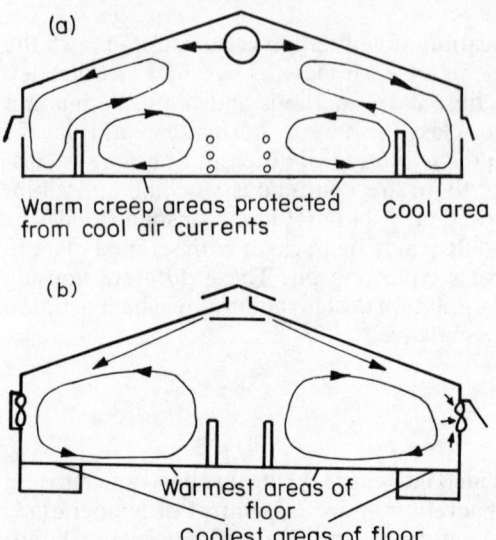

Figure 18.10 Ventilation rates v. fan operation for all-in, all-out systems combining a recircu-differences in (a) a farrowing house and (b) a finishing piggery

Fourthly, enhanced cooling is sometimes required under conditions of exceptionally high outside temperatures. While it may not be justified to use mechanical refrigeration or even evaporative cooling, the cooling effect of air speed can be exploited at modest cost either by switching on an additional fan system or by changing the direction of existing air jets so that the air impinging on the stock has travelled a shorter distance and is therefore cooler and faster (*Figure 18.11*). This is not to be confused with increasing the ventilation rate of the building, which is relatively ineffectual, as can be seen from *Figure 18.1*. A secondary effect of an increase in velocity at stock level may be to increase evaporation from any source of moisture.

BUILDING WIDTH

In any system that employs air jets it is necessary to permit the jet velocity to dissipate and the jet to entrain sufficient air from within the building for it to have become relatively warm by the time it reaches the stock. This is difficult in a narrow building, especially one containing young stock, as the distance between the air inlet and the stock is insufficient. The problem could occur even with a recirculation system operating at reduced fan speed. The

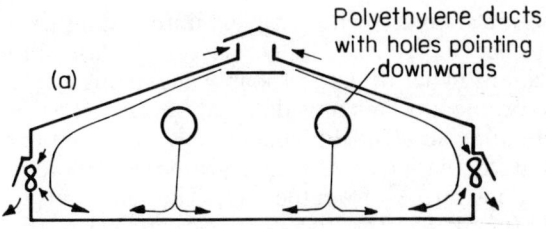

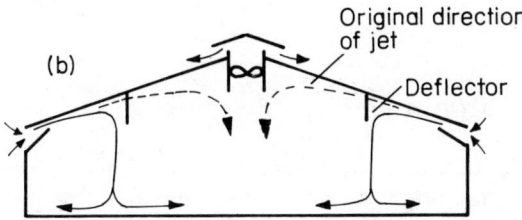

Figure 18.11 Examples of layouts for use in hot weather. These give airflow patterns that cause increased cooling of stock (a) by switching on additional fans and ducts or (b) by changing the original direction of the air jets using a deflector

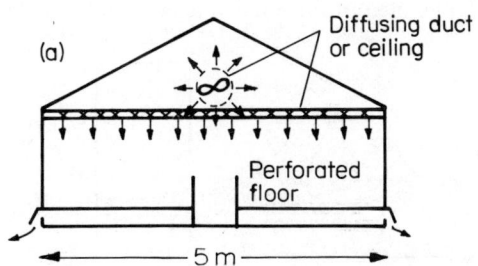

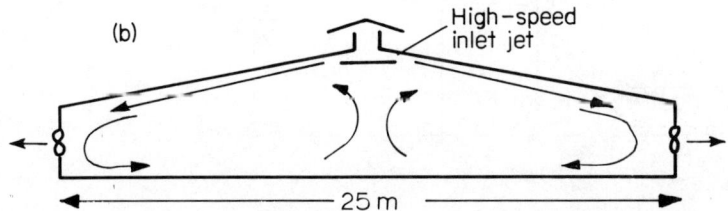

Figure 18.12 Diagrams of optimal ventilation systems for (a) a narrow early-weaner building and (b) a wide turkey fattening house

solution is to use a diffuse air source, such as a glass-fibre ceiling or a semi-diffusing polyethylene duct (*Figure 18.12a*). The success of glass-fibre ceiling systems for early weaners on flat-decks is probably due to this.

Conversely, the opposite problem occurs in wide buildings (in excess of 20 m), especially when containing adult stock, namely, how to sustain the jet over a long distance at low ventilation rates in cold weather and thereby avoid stagnation in those regions furthest from the inlet. The solution is to not only maintain an inlet air speed of 5 m s^{-1}, but also to exploit the Coanda effect, by causing the jet to cling to a smooth roof or ceiling surface. The jet throw can also be increased by increasing the depth and decreasing the width of the inlet slot to ensure the maximum dynamic energy in the jet (*Figure 18.12b*).

HUSBANDRY: THE AGE DISTRIBUTION OF STOCK

Figure 18.2 shows that for adult stock of constant mean weight in an insulated building, a typical temperature of 18 °C can be maintained against external temperatures from –2 to 16 °C by control of the ventilation rate within a range factor of 10. For adult stock such as laying birds, this is the variation normally used. For growing stock, however, two approaches are possible: continuous stocking with animals regularly entering and leaving, or an all-in, all-out system. When the former is applied to finishing pigs, a 10 : 1 ratio in ventilation rate is also employed. However, all-in, all-out systems require a higher ratio, depending on how much the stock grow whilst they are in the building. Thus, early weaners kept from 2 to 8 weeks typically require a range of ventilation of at least 20 : 1, and broilers as much

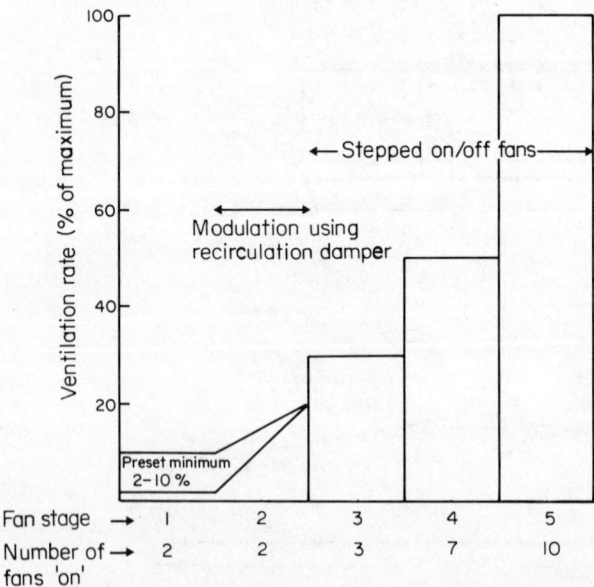

Figure 18.13 Ventilation rates v. fan operation for all-in all-out systems combining a recirculation and a stepped 10 fan system to give a wide range of ventilation rates

as 30 : 1. This full range is most likely to be encountered in the spring, when the increase in temperature between the coldest nights in March and the warmest days in May can be of the order of 15–20 K and coincides with an increase in the weight and metabolic heat production of the livestock.

Low minimum ventilation rates may be achieved without resorting to operating fans at very low speeds by using composite systems, cycle-timing or a range of fan sizes. A good example of a composite system is to combine a recirculation unit powered by a single fan with a step system operating several fans (*Figure 18.13*). The step system controls ventilation down to 20 per cent of the maximum, and the control damper of the recirculation unit provides the range from 20 per cent down to a minimum determined by the current age of the stock. The use of a full-speed fan in the recirculation unit also ensures reasonable air circulation even at very low fresh air rates. At any age, when the number of stock and consequently the size of the building and the number of fans are small, control of ventilation by step control is difficult and a recirculation system is preferable.

HUSBANDRY AND LAYOUT: WASTE COLLECTION SYSTEM

For stock housed above stored slurry or manure, it is desirable to minimize the air exchange between the stock environment and the waste. When waste is collected by gravity, the cross-sectional area of the space separating the stock from the waste is invariably very large, hence the pressure drop in the air is too small to ensure that air flows in the direction from the stock to the waste. For laying birds in cages over a deep pit, at all times there are likely to

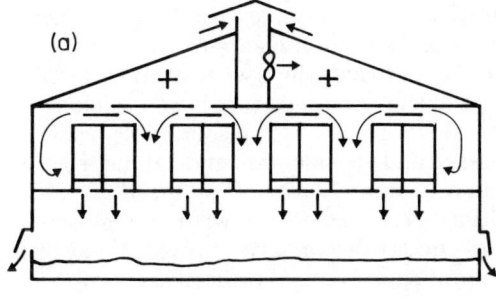

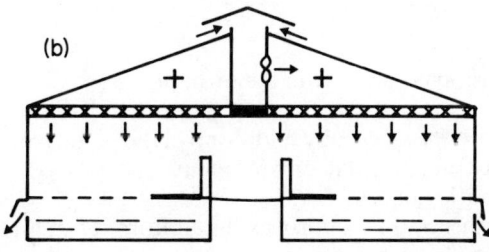

Figure 18.14 Diagrams of optimal ventilation systems for in-house storage of waste for (a) caged laying birds over a deep pit and (b) early weaners on mesh floors

be strong convection currents up through the cages and down into the pit space, particularly at low ventilation rates. This interchange can probably be eliminated only by using an almost solid floor, with slots of limited area through which the manure is scraped (*Figure 18.14a*).

For pigs on slats, there is no way of reducing further the free area between the slats. If air is to be extracted beneath the slats, a separate air duct beneath the slats is therefore needed to give uniform extraction. However, Bruce (1975) has shown that air exchange through a slatted floor can be minimized if jets of air are prevented from blowing vertically downward onto the slats. For early weaners kept on wire floors, the use of a glass-fibre ceiling ventilation system should help to minimize this air exchange (*Figure 18.14b*).

HUSBANDRY AND LAYOUT: POSITION OF STOCK AND DESIGN OF ENCLOSURE

A common ventilation design is the 'straight-through' jet system shown in *Figure 18.9b*. In this, by maintaining a high inlet speed, the jet can be made to throw a long distance and so entrain warm air and produce good mixing. However, even if this is done, in very cold weather this jet is still considerably colder than the air in the building and may constitute a draught for stock that have little scope for movement, or are very young or have no overhead protection. For example, if stock are housed adjacent to walls in a house with a centre access passage and eaves inlets, it is essential that the jet is directed across the ceiling and that its speed is high. With ceiling jets from a ridge inlet, however, the jet will fall on stock whether it is slow or fast, and the solution is to direct the jet into the centre passage and provide protection at the fronts of the enclosures. Both of these examples are shown in *Figure 18.15*. With stock housed beneath the ridge, as in *Figure 18.16*, the reverse solutions apply: it is essential that the jet speed is high when using ridge inlet jets. Eaves inlets directed across the ceiling on the other hand are likely to cause draughts whether the jet is slow or fast and the solution is again to direct the jet directly into the passage and provide protection at the front of the enclosures. When stock are housed in three rows, slow-speed jets from either ridge or eaves will cause draughts (*Figure 18.17*), whereas high-speed ceiling jets can be brought in from either ridge or eaves, provided that they are deliberately deflected down into the passages. An alternative approach is to use a diffuse air source such as a perforated duct or glass-fibre ceiling.

Conclusions

We may present the main conclusions succinctly in the form of a list:

(1) Soundly designed ventilation systems usually both control the pattern of internal airflow and enable the internal environment to be largely independent of external climatic factors.
(2) Control of the airflow pattern not only improves the uniformity of air distribution and temperature, but also introduces some control of air velocity at stock level, so improving the control of heat loss from the stock.

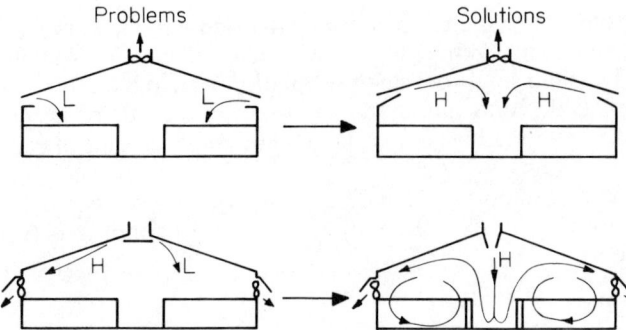

Figure 18.15 Airflow diagrams and ventilation designs when using jets with stock constrained against outer walls. H denotes high speed, L low speed

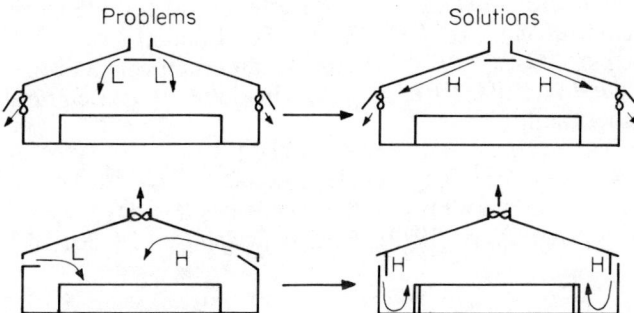

Figure 18.16 Airflow diagrams and ventilation designs when using jets with stock constrained beneath the ridge. H denotes high speed, L low speed

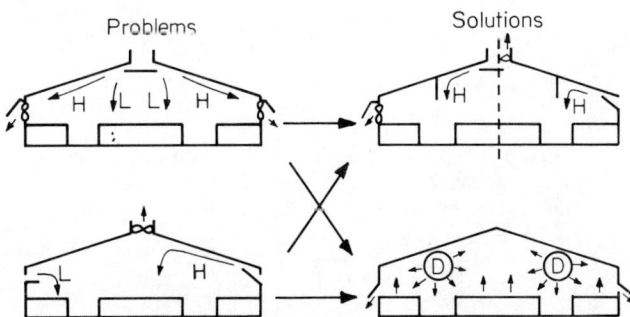

Figure 18.17 Airflow diagrams and ventilation designs when using jets with stock constrained in three rows. H denotes high speed, L low speed, D diffusing air source

(3) The optimum solution to the control of an environment will vary with the class of livestock and even with the husbandry and housing systems within a class. Thus, no single design of ventilation system is likely to be universally applicable, and any system must be matched in design, complexity and cost to the husbandry, layout and requirements of each class of stock.

(4) An important new development could be the use of composite systems for livestock requiring an exceptionally wide range of conditions (for example, broilers).

References

BOON, C. R. (1978). *J. agric. Engng Res.* **23,** 129–139

BRUCE, J. M. (1975). *Farm Building R & D Studies*, p. 3. SFBIU, Aberdeen

BRUCE, J. M. and CLARK, J. J. (1979). *Anim. Prod.* **28,** 353–369

BUTLER, E. J. and EGAN, B. J. (1974). *Wld's Poult. Sci. J.* **30,** 32

CARPENTER, G. A. (1974). In *Heat Loss from Animals and Man*, pp. 389–404. Ed. by J. L. Monteith and L. E. Mount. Butterworths, London

HMSO (1978). *Agricultural Statistics, UK 1975.* HMSO, London

MOULSLEY, L. J. and FRYER, J. T. (1979). 'Air recirculation units in a broiler house.' *Dep. Note EN/929/10003.* National Institute of Agricultural Engineering, Silsoe, Bedford

RANDALL, J. M. (1975). *J. agric. Engng Res.* **20,** 199–215

RANDALL, J. M. (1980). *J. agric. Engng Res.* **25,** 169–187

RANDALL, J. M. and BATTAMS, V. A. (1976). *J. agric. Engng Res.* **21,** 33–39

RANDALL, J. M. and BATTAMS, V. A. (1979). *J. agric. Engng Res.* **24,** 361–374

19

VENTILATION SYSTEM DESIGN

J.M. RANDALL
National Institute of Agricultural Engineering, Silsoe, Bedford

Introduction

The purpose of a ventilation system is to provide the optimum conditions for the livestock by controlling the thermal environment. To achieve this, the two basic functions of a ventilation system are to provide effective control over the ventilation rate and efficient control over the airflow pattern within the ventilated structure. These should be attained simultaneously. It is therefore necessary to pay careful attention to the design and siting of the ventilating equipment. In general, any one component of a ventilation system is designed to influence only one of the basic functions. For example, a fan extracting air effects control only over the ventilation rate, and a polyethylene distribution duct effects control only over the airflow pattern. Nevertheless, it is important to design each of these components correctly, because otherwise they are likely adversely to affect each other and the performance of the ventilation system as a whole; for example, a distribution duct with insufficient holes will reduce the throughput of a fan and thus the ventilation rate.

Essentially then, a ventilation system should be designed by combining components selected from two groups: those which influence ventilation rate and those which influence the airflow pattern. In addition to these functions, which are essential to all systems, there may be other special requirements to be incorporated, such as the exclusion of external influences (such as light, rain, snow and noise) and the provision of adequate ventilation in the event of power or equipment failure.

Control of ventilation rate

FANS

Except in naturally ventilated buildings, propeller fans are the most common means of providing ventilation and air exchange in livestock buildings. These low-pressure fans can be used because the ventilating air is not normally passed through long lengths of restrictive ducting, which would call for a high-pressure source of air. The building itself often acts as the distribution or collection duct, and, having a large cross-section in relation to the quantity of air flow, it does not cause significant pressure losses. However, in future designs there may be special circumstances for fans to work against high pressures, such as when air may require filtering or bringing to one place so that it can be passed through a heat exchanger or through an air or waste treatment plant.

Apart from their relatively high throughput at low pressures, propeller fans have been popular because their speed can be controlled by varying the applied a.c. voltage. Thus, by using a simple thermostat the quantity of ventilating air could be automatically and continuously varied, the consequences of which are discussed later. It is standard practice to install fans of similar rotational speed and of adjacent sizes in one particular building; because when fans of widely differing characteristics exhaust from the same space, the more powerful ones are likely to override and cause damage to the smaller ones. Fans may be used either to pressurize or to exhaust a building. In very cold climates, when many days are well below freezing, pressurizing should be avoided as the fan may become laden with ice, causing imbalance and subsequent damage to bearings and motor. In these circumstances exhausting ensures that warm air always passes through the fan. Propeller fans designed for agricultural use are available with fireproof motors for use in dusty conditions and with waterproof motors suitable for outdoor use.

Speed control of fans

Traditionally, in livestock buildings the rate of ventilation has been controlled by continuously varying the speed and hence the throughput of all the fans installed in a building, although in some installations some of the fans are turned off during cold weather to allow the remaining ones to operate at higher speeds.

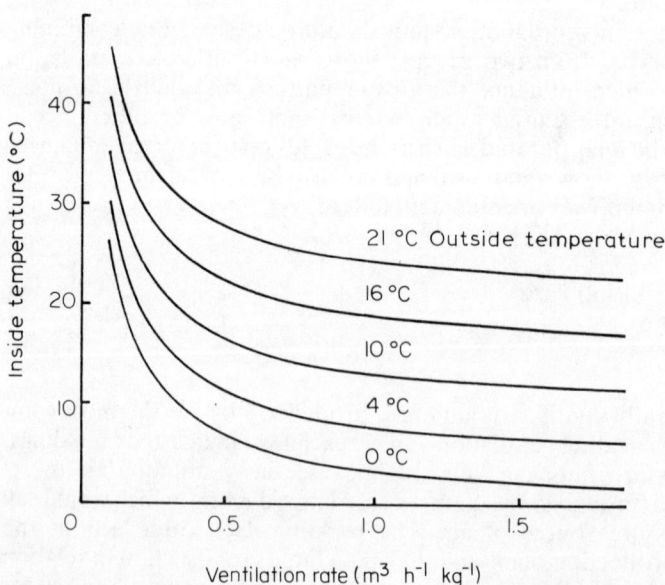

Figure 19.1 *The effect of ventilation rate on house temperature for various outside temperatures.* (*After Carpenter, 1972a*)

The effect of small changes in ventilation rate on the internal temperature is shown in *Figure 19.1*, where it may be seen that in hot weather (i.e. at high ventilation rates), small changes have little effect on inside temperature whilst in cold weather (i.e. at low ventilation rates) they have a large and significant effect. Thus a ventilation system is required which provides close control, particularly over low ventilation rates. This is not obtained with speed control, because fans running at low speeds are readily influenced by the wind (Carpenter, 1972a). This is illustrated in *Figure 19.2*, which shows

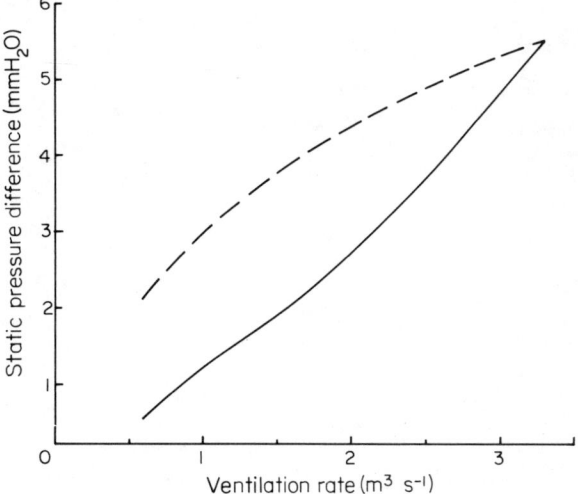

Figure 19.2 The effects of vent adjustment on fan performance; (– – –) adjusted vents, (—) fixed vents. The curves apply both to full fan speeds and variable fan speeds. (After Boon, 1980)

that with vents set permanently at a fixed gap (that gap which permits an air speed through it of 5 m s^{-1} at maximum ventilation rate), as the fan throughput is reduced so the pressure generated falls off dramatically and consequently the air speed through the vent falls (Boon, 1980). This is so whether the appropriate number of fans are running at full speed or all the fans are operating at an appropriate reduced speed (as in the conventional speed control system). These low pressures are negligible when compared to a moderate wind of about 6 m s^{-1}, which would generate pressures around 2 mmH$_2$O gauge. The only way to maintain an air speed of 5 m s^{-1} in the inlet at low ventilation rates is to use the appropriate number of fans at full speed. If speed control is attempted, then with the vents set at the appropriate gaps all the fans need to run at almost full speed in order to obtain the necessary pressure to give sufficient air speed in the inlet; in effect, control of ventilation rate is being obtained by fan throttling. Thus, there is no point in attempting to achieve 5 m s^{-1} in the inlet with speed-controlled fans as they would all be running at nearly full speed and be consuming a disproportionate amount of power.

Sequential or step control of fans

Ventilation rate may be better controlled by switching groups of fans on and off in predetermined steps. The cooling capacity of each group should be similar and, in order to obtain the required ventilation rate, it is usually necessary to have different numbers of fans in each group. Thus the first group which operates in the coldest weather may contain only one fan, whereas the group sizes increase with increasing outside temperature until the final group may contain half or more of the total number of fans. Four, five or six stages are suitable for most livestock buildings (Randall, 1977a). As an example, *Figure 19.3* gives the ventilation rate for each stage, corresponding to equal steps of outside temperature, required to produce an

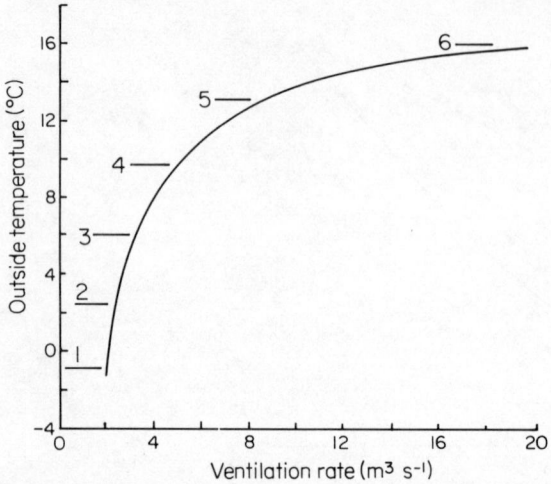

Figure 19.3 An example of the required ventilation rates in a six-stage system, for equal steps of outside temperature, to produce an internal temperature of 18°C

internal temperature of 18 °C. These steps were calculated from the known characteristics of a building and the heat released by the stock. The curve is reasonably typical of most modern intensive livestock buildings and may be used to derive *Table 19.1*, giving the proportion of the maximum ventilation rate required at each stage in a four-, five- or six-stage system when the minimum required rate is 10 per cent of the maximum.

Because of the difficulties in attaining a required minimum ventilation rate, this system is particularly suited to large buildings requiring at least 10 standard fans. Obviously, a building with 10 fans requires one of these only to achieve a minimum 10 per cent of the maximum. Therefore, when there are fewer than 10 fans some other approach is required to obtain the desired

Table 19.1   VENTILATION RATES FOR EACH STAGE IN FOUR-, FIVE- AND SIX-STAGE SYSTEMS

Stage	Percentage of maximum ventilation rate		
1	10	10	10
2	15	14	13
3	28	19	17
4	100	35	25
5		100	44
6			100

minimum ventilation rate. An ideal system has not yet been developed for this case, but one method is to have one fan on speed control for minimum ventilation; however, doing this negates the entire purpose of step control systems, which is to maintain relatively high pressure differences, particularly at low ventilation rates. A second method, which is likely to be used in the future, is speed control of one special fan, which has been selected to maintain a higher pressure at low speeds. A third method that may be satisfactory is to provide 'time proportioning' of one fan; during a five- or ten-minute cycle the fan operates for the appropriate time to achieve, on average, the minimum ventilation required. However, during the off period there is no control over the airflow pattern. It is unwise, therefore, to have the fan off for periods in excess of four or five minutes, and this means that any requirement less than about 50 per cent of the fan rating is likely to give rise to unsatisfactory conditions. Another approach, perhaps the best presently available, is to install an additional *small* fan to give the minimum ventilation rate; but if too small it will not generate sufficient pressure to overcome wind even at full speed and it is likely that it will need to be switched off when the larger fans operate.

Time switching of fans

This system, pioneered by Cromarty (Dias, 1976), is designed to switch all the fans on or off for a proportion of the time which depends on the currently required mean ventilation rate. Thus, at the maximum ventilation rate all the fans are on full time and as less ventilation is required the off periods are increased. This method may be combined with permanently switching off some of the fans to reduce the length of the 'off' period in cold weather.

Time proportioning of fans has the same problem at all ventilation rates as does group switching of fans near the minimum ventilation rate, namely loss of control over the airflow pattern when the fans are off. In addition, the overall power consumption is higher than for staged switching, because of the high starting currents each time the fans are switched on.

DAMPER PROPORTIONING (RECIRCULATION)

One of the most satisfactory means of controlling ventilation rate, particularly for small buildings and those requiring only a few fans, is to run one or more fans at full speed and use them in combination with a damper. This enables mixing of varying proportions of outside air with air recirculated from within the building. The mixture is then discharged into the building. The method has been studied by Carpenter (1972b) and Owen (1977), and *Figure 19.4* represents a typical damper-controlled recirculation unit. The units may be controlled automatically by sensing the temperature inside the building, so causing the damper to open or close depending on whether more or less outside air is required to maintain a selected internal temperature. Adjustment of the damper may be made by a linear push–pull motor or by a rotary motor connected to a rod passing through the hinged edge. The motor may operate in a continuous or step mode depending on the controller design. The minimum required ventilation rate is guaranteed by providing a stop that prevents complete closure of the fresh-air route.

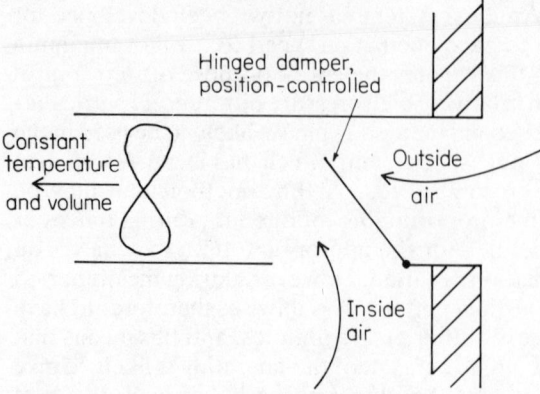

Figure 19.4 The essentials of a damper-controlled air recirculation unit for animal house ventilation

Some commercial recirculation units incorporate an exit air vent and, although they are generally expensive, they have the advantage that they are self-contained and are easily installed. Most units of this type employ single-speed fans, but some installations are designed to allow the system to operate normally at half speed, so that in hot weather additional cooling may be obtained from an increased air speed and capacity on switching to full fan speed.

The main assets of this system are that at all times and over the normal range of temperature control the temperature and quantity of the air discharged into the building both remain constant. These constant conditions allow the pattern of airflow distribution to be maintained without adjustment of the areas of the air inlets.

NATURAL VENTILATION

The term *natural ventilation* refers to those systems not using mechanical power to produce air movement, the motive power being provided from two sources, wind and thermal buoyancy. Both of these have been studied in detail (Bruce, 1975c, 1978). With natural ventilation it is not usually possible to maintain close control over the ventilation rate but it is essential to be able to achieve the minimum required rate under any external conditions.

Natural ventilation is particularly suited to cattle buildings, where the prime requirement is to ensure adequate minimum ventilation to reduce the incidence of disease, and not necessarily to maintain a particular temperature (Bruce, 1975b). Poor ventilation is most likely to occur during calm weather. Thus in cattle buildings sufficient motive power for air movement is required from thermal buoyancy alone. Design criteria have been established for buildings with open ridges (Bruce, 1975a) and slotted roofs (Bruce, 1977, 1978).

For open-ridge buildings, ventilation due to thermal buoyancy depends on the areas of the air inlets and outlets, the difference in their height, the heat produced by the stock, the outside temperature and the thermal conductance of the building. The outlet area required for a vertical separation of

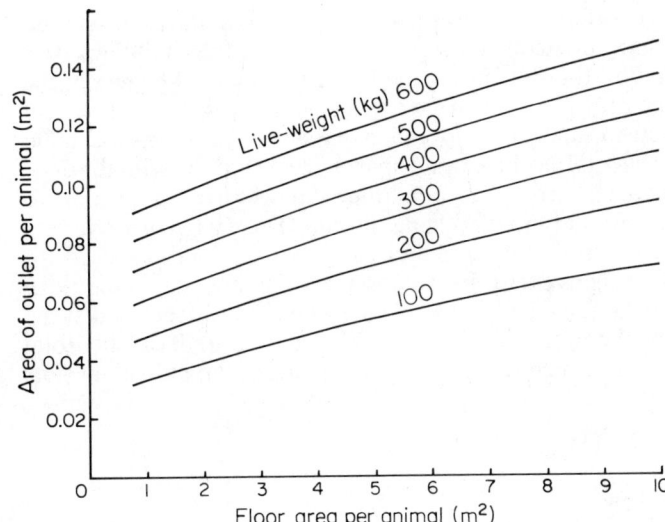

Figure 19.5 Ventilator outlet area required for cattle housing, calculated for a height difference between the inlet and outlet of 1 m and for specified animal stocking rates. (After Bruce, 1977)

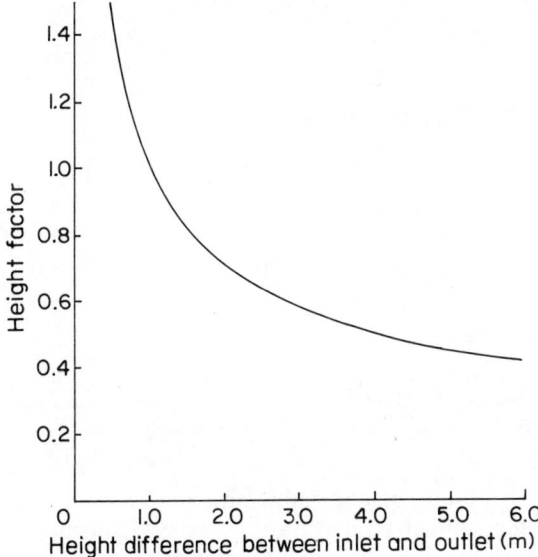

Figure 19.6 Variation of the 'height factor' for ventilator outlet areas with the difference in height between inlets and outlets. The factor is expressed as the ratio to the area required for a 1 m height difference. (After Bruce, 1977)

1 m between inlets and outlets can be determined from *Figure 19.5*, when the floor area per animal and their live-weight are known. It is generally assumed that the area of the inlets should be twice the area of the outlets. The outlet area required for inlet–outlet separations other than 1 m apart may be determined by multiplying the area obtained from *Figure 19.5* by the height factor given in *Figure 19.6*.

Where it is not possible to provide sufficient openings at the ridge, slotted roofs have provided an excellent alternative. On existing buildings with inadequate ventilation the slots are made by cutting along the crest of the corrugations on the roof sheets from eaves to ridge, leaving the sheets uncut at the horizontal laps. Usually not more than one slot is cut in each sheet. On a new building the sheets may be spaced apart during erection with the edge corrugations turned upwards. The minimum slot width is 15–18 mm, to reduce the risk of frost and snow bridging the gap. The design procedure to establish the open areas of the slots and eaves is based on that used for designing the open-ridge system. In the absence of an opening at the ridge the ratio of the area at the eaves to that which would have been used at the ridge outlets is calculated. *Figure 19.7* can then be used to derive the total area of the slots, from which the number and width of the individual slots may be determined.

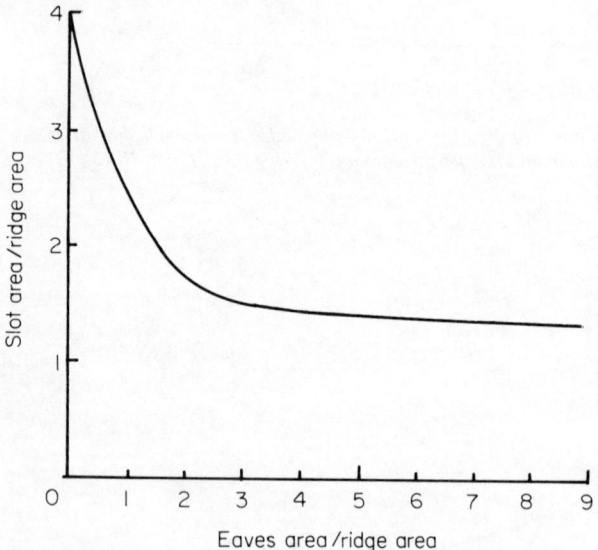

Figure 19.7 Ventilation by slotted roofs. The ratio of the slotted area on a roof to the area which would otherwise be used for a ridge opening, is plotted against the ratio between the total area of the eaves openings and the area which would be necessary for a ridge opening. (After Bruce, 1977)

The areas of the inlets and outlets of naturally ventilated systems are usually fixed, but because the critical temperature of cattle is normally below ambient temperatures in the UK there is no need to make adjustments to them even in windy or cold weather. However, the lower critical temperature of pigs is often above ambient temperature and when they are housed in naturally ventilated buildings their performance would be adversely affected if some degree of control over the ventilation were not available. It is likely that adequate conditions may be provided by straw bedding, in which the pig can insulate itself and so have control over its lower critical temperature, or by manually adjusted air inlets or outlets. The adjustment of these vents depends on the stockman, who usually achieves an acceptable standard of temperature control by 'trial and error', but when the outside temperature varies considerably from day to night much time and effort is required to achieve the desired environmental conditions.

Ventilation rate may be varied automatically in naturally ventilated houses using control systems that are being studied by Bruce (1979). It is likely that the internal temperature will be detected by a thermostat which activates an electric motor that opens or closes a flap. Limit switches would control the maximum and minimum positions of the flap. In the first design the opening of the eaves inlets was automatically adjusted and the opening of the outlet, an insulated chimney passing through the ridge, was adjusted manually. This method has rarely been used in the UK, but is likely to become more widely accepted in the future.

Control of air distribution

AIR JETS AND CONVECTION CURRENTS

Control over air distribution is achieved by selecting the correct balance between the buoyancy and dynamic forces in the air. Because the air outlet has little effect on the air distribution (Randall, 1975), control of the air distribution depends almost entirely upon the design of the air inlet. Air speed at the inlet may be comparatively slow or comparatively fast. When comparatively slow the convection currents arising from the stock overcome the inlet currents and so dominate the airflow pattern. An example is the use of a large area of diffusing ceiling for stock confined in cages or pens.

When the speed of the air is comparatively fast then the stability of an airflow pattern is characterized by the value of the Archimedes number, defined as the ratio of the buoyancy to dynamic pressures in the air jet (Randall and Battams, 1979). A horizontally projected jet does not fall when the Archimedes number is less than 30, but falls when it is greater than 75, thus causing stable air flows with opposite directions of rotation. Intermediate values should be avoided to prevent unstable patterns and thus unstable environmental conditions near to the stock. In practice, an Archimedes number of less than 30 can usually be obtained by adjusting the air inlet to maintain an air speed of about 5 m s^{-1} through it, irrespective of the temperature of the external air and the ventilation rate.

BUILDING LAYOUT

Obstructions to airflow considerably modify the airflow pattern that would exist in a smooth-surfaced empty building, by deflecting the dominant air current from its path (Randall, 1975, 1980). Thus, control of the airflow distribution cannot be isolated from the building layout, as was discussed in the previous chapter by Carpenter.

When the building is ventilated with air jets, the inlets are usually fitted between structural members. The inlets should be as short as possible compatible with the air jets spreading to the entire length of the building by the time they reach stock level. Making the inlets as short as possible ensures that for a given inlet area the inlet gap is as wide as possible. It is easier to control the aperture of a wide gap in practice, and this produces an air jet that is more likely to remain coherent when projected across the width of the building.

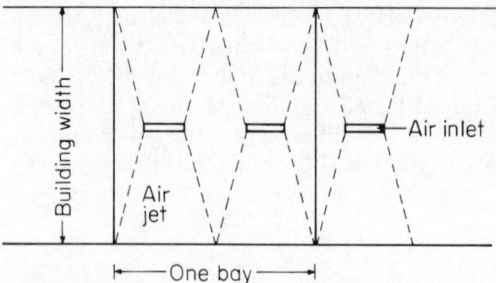

Figure 19.8 Siting of air inlets within a bay of a livestock building to ensure that the air jets cover the total length of the building before reaching the stock

The possible layout of inlets is influenced by both the length and width of the structural bays, as shown in *Figure 19.8*, where it is assumed that the individual air jets must overlap when they reach halfway across the building; this layout is applicable whether the inlets are at the eaves or at the ridge.

ADJUSTABLE INLETS

Because of the need to maintain the inlet air speed of 5 m s^{-1} at all times, because access to air inlets is difficult in many buildings and because there is normally a large number of inlets, the vents need to be linked together and operated remotely. In addition, they have to be adjusted frequently, that is, whenever the ventilation rate changes significantly (which is at least diurnally), and it is therefore desirable to control the operating mechanism automatically in conjunction with the control of ventilation rate.

Several designs of baffles and flaps are amenable to automatic adjustment. They may be used for air entry either at the ridge or at the eaves. The close tolerances on the size of the inlet gap that are required dictate a baffle construction which is unlikely to warp. The inlet must also be made of an insulating material to prevent the formation of condensation, caused by the warm, moist air inside the house contacting the baffle, which is cooled by the incoming air. A satisfactory baffle may be constructed of a metal frame, infilled with an insulation board. In farm conditions plywood and insulation board without a metal frame will warp, as does a wooden frame.

Ridge inlets with horizontal swinging baffles

When air is brought in at the ridge it is usually deflected by a horizontal baffle so that it discharges beneath and along the roof or ceiling, towards both eaves. Most designs consist of a series of horizontal baffles (*Figure 19.9*) each supported from four suspension cables of fixed length. Each of the baffles is attached to a connecting cable running the length of the building. The inlet gap formed between the baffle and the ceiling is reduced by pulling this cable horizontally and thereby causing the baffles to rise. Releasing the cable allows the baffles to fall back and increase the inlet gap. The tension in the main connecting cable increases as the inlet gap is reduced, and to

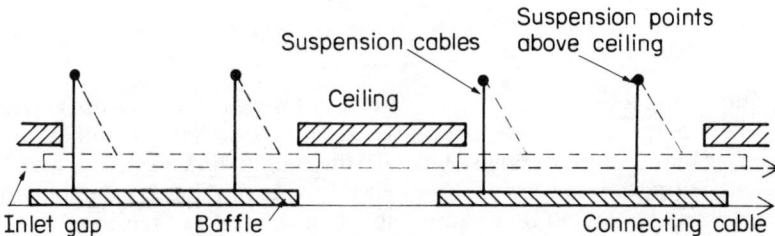

Figure 19.9 Diagram of a swinging-baffle suspension system for the automatic adjustment of the air inlet gap for high-speed jet ventilation, where suspension points are available above ceiling level

prevent it from becoming excessive at the minimum inlet gap the four suspension cables must be hung from a level well above the ceiling. If this is physically impossible, down-turned 'legs' are required on the baffles, to lower the attachment point as shown in *Figure 19.10*. Increasing the length of the suspension cables reduces the tension in the connecting cable, but it also causes an increase in the horizontal movement of the baffle. In practice,

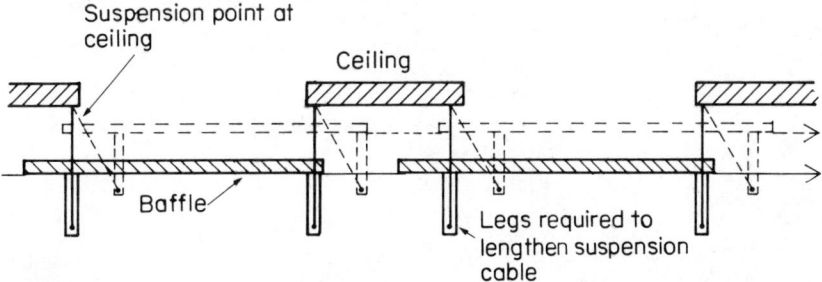

Figure 19.10 Diagram of a swinging-baffle suspension system for the automatic adjustment of the air inlet gap for high-speed jet ventilation, where suspension points are not available above ceiling level

it is convenient to restrict this forward movement and a compromise has to be reached between the cable tension and the forward movement (Randall, 1977a). A major advantage of this method of operation is that the total pull required in the connecting cable can be designed to be less than the dead weight of the baffles, by selection of an appropriate length for the suspension cables.

Ridge inlets with lifted baffles

When the baffle is too long to swing within the length of a bay, the cables supporting the baffle can be passed over pulleys and all the cables attached

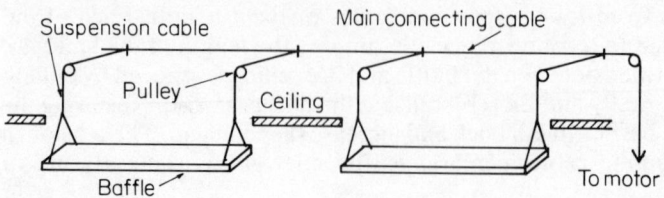

Figure 19.11 Diagram of a lifted-baffle suspension system for the automatic adjustment of the air inlet gap for high-speed jet ventilation

to the main connecting cable (*Figure 19.11*). In this case the aperture above the baffle is equal in length to the baffle. In some installations pulleys are replaced by metal rings, but these increase the friction unnecessarily, which increases the operational forces; they cause fraying of the cable, leading to premature breakage; and they cause uneven distribution of tensions, leading to differential stretching of the connecting cable and uneven inlet sizes.

Eaves inlets with hinged flaps for horizontal air discharge

For air entry at the eaves the air may be discharged either beneath the roof or down the outer wall and here hinged flaps are satisfactory (Randall, 1977a). The design and construction requirements for eaves inlet flaps are similar to those for ridge baffles, in that they must not warp and must be free from condensation. The hinge along the lower edge (*Figure 19.12*) should be

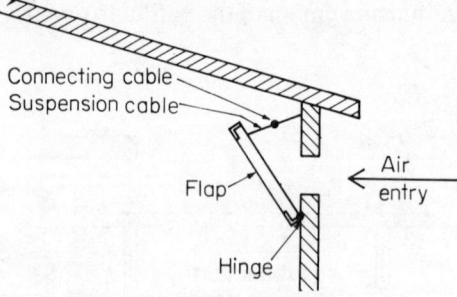

Figure 19.12 Diagram of an automatically adjustable inlet flap for high-speed jet ventilation, to discharge air from the eaves beneath a ceiling

designed to prevent air leaks and be almost frictionless. Metal hinges are always unsatisfactory, and perhaps the most serviceable hinge is to use pins on each end of the flap that turn in appropriate plain bearings or eyes. Air sealing is then provided separately along the lower edge with a strip of butyl rubber attached with battens to the flap and wall. The ends of the flaps also require sealing.

The mobile edge of the flap is suspended from the wall by a cable, and the size of the inlet is varied by pulling or releasing the main connecting cable, attached to the midpoint of each of the suspension cables.

Eaves inlets with hinged flaps for downward air discharge

The design of eaves inlets with hinged flaps for downward air discharge is similar to that for horizontal air discharge and is shown in *Figure 19.13*.

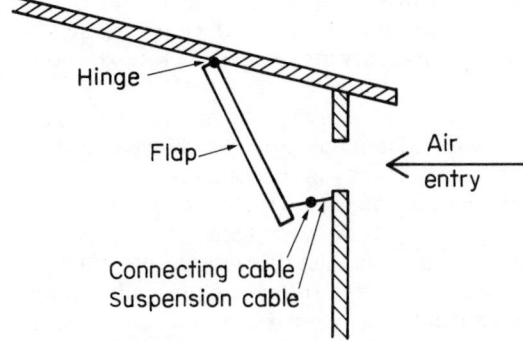

Figure 19.13 Diagram of an automatically adjustable inlet flap for high-speed jet ventilation, to discharge air from the eaves down the outer wall

However, because the distance in which the air can fan out before reaching the stock is less than for horizontal air discharge, the total length of the eaves inlet must be greater to ensure that air reaches the stock uniformly along the length of the building.

DUCTS

Ducts provide a convenient means of distributing air uniformly along the length of a building. In practice, inflatable ducts made of plastic film with holes for air discharge distributed along their length provide a cheap and convenient means of air distribution, particularly in the conversion of old buildings. They are usually inflated by a fan of a cross-section similar to that of the duct.

Different hole sizes and positions can be used in a single duct. Ducts of circular cross-section may have the holes sited at any point of the circumference, allowing air to be discharged in one predetermined direction, in several directions or in many directions. The last of these gives a diffusing duct, where the air discharged is not required to have a specific direction.

The normal requirement is to discharge an even quantity of air per unit length of duct at a constant speed, and special design precautions are required to achieve this (Carpenter, 1972c; Zamir, 1973). When using standard propeller fans the total area of holes in a duct should be 1.5–2.0 times the duct cross-sectional area (aperture ratio). Smaller ratios prevent the fan from developing its full throughput and larger ratios allow the duct to collapse. The size of the individual holes is selected to allow the required throw of the air, larger holes for long distances and small holes for a diffusing duct. Holes of diameter 50 mm are a practical size to give a good throw of the air jet, but holes less than 10 mm in diameter should be avoided, because they are likely to clog with dust.

The distribution of the holes is important, for a parallel-sided duct with uniform hole spacing discharges more air from the holes *furthest away* from the fan than from those close to it. For even air distribution the average distance apart of the holes is calculated and the spacing of the holes at the closed end is made half as big again, whilst the spacing at the fan end of the duct should be half the average. The spacings along the duct are graduated between these two extremes. The figures apply for an aperture ratio of about 1.85 and for ducts less than 45 m long.

The disadvantage of all parallel-sided ducts is that the air velocity inside the duct decreases towards its closed end and the corresponding increase in static pressure increases the discharge velocity in this section. Uniform air distribution can be obtained by tapering the duct (Carpenter, 1972c) and maintaining uniform hole spacing—easy to do with ducts made of plastic film. In general, it is satisfactory to taper the duct diameter to one-third of the full diameter at the closed end; the degree of taper being reduced for very long ducts. Further, the duct can be suspended from the surplus web attached to the duct as shown in *Figure 19.14*. Air is discharged from the

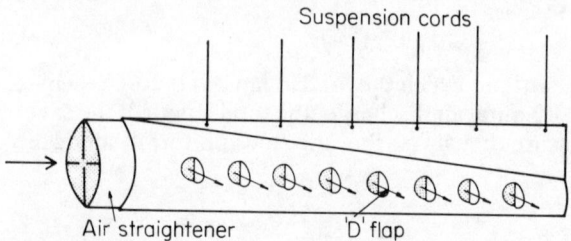

Figure 19.14 Diagram of a tapered polyethylene film duct used for distributing air from a fan uniformly along the length of a livestock building

holes in both a downstream and a tangential direction, and can result in poor control over the airflow pattern in a building. The downstream movement can be counteracted by making the holes of a semicircular D shape, instead of circular, and leaving the D-shaped flap fixed to the duct downstream of the hole. For holes of diameter less than 20 mm the flap does not open readily unless cut by a punch designed to remove a small piece around the circular edge of the flap. In addition, the tangential movement makes it impossible to give the discharged air a symmetrical pattern. This is easily overcome by fitting an air straightener between the fan and the duct inlet. The length of the straightener should be twice the duct diameter and it should be fitted with at least four radial fins that are not joined at the centre.

DIFFUSING SURFACES

Passing air through diffusing surfaces allows the convection currents from the stock to control the airflow pattern. Thus stable flow patterns exist only in buildings in which the stock are confined and at a high density. The traditional use of diffusing surfaces is in houses for caged laying hens (King *et al.*, 1968), where the air passes through a false ceiling of permeable material. This material is usually a glass fibre mat of the type used for insulation. To obtain a uniform distribution of air over the entire ceiling it is necessary to create a considerable pressure drop through the material by selecting an appropriate thickness. *Figure 19.15* shows the approximate relationship between the pressure drop and air speed through glass fibre (Randall,

1977b; Carpenter and Moulsley, 1978). At the maximum ventilation rate when using propeller fans, a satisfactory pressure drop is about 3 mmH$_2$O gauge, giving an air speed of 0.2 m s^{-1} or 0.1 m s^{-1} through a thickness of 50 mm or 100 mm respectively. At minimum ventilation rates of about one-tenth of the maximum rate, the mean air speeds fall to 0.02 m s^{-1} and 0.01 m s^{-1} and the pressure drops are correspondingly less.

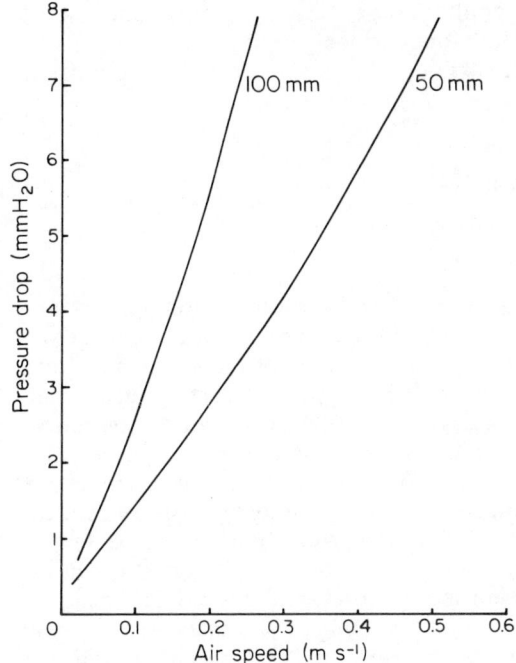

Figure 19.15 The pressure drop through glass fibre mats of two thicknesses, plotted against the air speed through the mats in m s^{-1}

For satisfactory air distribution, a material is required with a back-pressure at least as large as that for 50 mm of glass fibre. Very few of the commercially available air filter materials have such characteristics, since they are designed to have as low a resistance as possible. Nevertheless, a material that allows three-dimensional flow through it is required, because permeable cloths such as hessian or woven polypropylene rapidly clog with dust. In Norway a combination of glass fibre and a smooth-woven nylon cloth has been shown to be satisfactory in a relatively clean atmosphere (Graee, 1976).

Apart from caged poultry, the most frequent use of permeable ceilings is in flat-deck weaning houses for pigs. A similar pressure drop is required across the material, but because the rooms used are much smaller the total area of ceiling is much less. Part or all of the ceiling may be covered in a permeable material where the floors are totally slatted or perforated, because there is no need to encourage the pigs to lie on, or defaecate on, any particular part of the floor. Care has to be taken where solid lying areas are separated from a perforated dunging channel; and in order to encourage the young pigs to lie on the solid floors, these should be kept warmer than the dunging area. This might be achieved by making sure that incoming fresh air cannot fall directly into the lying area. Thus, only the area of ceiling above the dung channels should be permeable to allow the cold air to fall there, to be warmed by the stock and to rise above the lying area. This arrangement allows a stable flow pattern to be developed, which encourages the required stock behaviour. Because the areas of permeable material are much less in weaning houses than in caged-poultry houses, more expensive alternatives to glass fibre, which allow three-dimensional air flow, are acceptable. Multiple thicknesses of air filter material or foamed plastics are possible alternatives.

Control of external influences

AIR LEAKAGE

The performance of the equipment designed to control ventilation rate of a building can be degraded by air leaking through the structure. Leakage will adversely affect both the quantity and distribution of the air. In a building from which the air is extracted, any hole in the structure acts as an air inlet, causing local draughts and cold areas and so upsetting the airflow pattern or even the stock behaviour. Also, at low ventilation rates, when the designed area of the air inlet is small, air leaks into the building may cause a reduction in the air speed through the inlets and consequent loss of control over the airflow pattern. In pressurized buildings the problems are in part different. Warm, moist air is forced through holes in the building material, so causing condensation within the insulation and eventual breakdown of the insulating properties of the material. This may be combated by installing a vapour barrier between the inside of the building and the insulation which is designed to prevent air movement into the wall.

Air gaps in the building structure also allow adventitious ventilation due to wind. For example, Moulsley, Fryer and Pike (1976a) showed that the ventilation rate per unit wind speed of a good-quality broiler house with all the vents closed was 0.63 changes per hour per m s^{-1}. Moulsley, Fryer and Pike (1976b) have also observed that in a wooden building shell with a leakage rate of 0.2 changes per hour per m s^{-1} of wind speed, the addition of a layer of insulating slabs provides little improvement, and that considerable trouble has to be taken before a worthwhile improvement can be achieved. Possibilities include lining the building with polyethylene sheet or spraying the internal surface with an insulating foam. Although these measures may not always be practical, great care should be taken to prevent excessive air leakage, by sealing around cladding sheets, around doors and at the ends of slurry channels.

BACK-DRAUGHT SHUTTERS

Wind may interfere with ventilation not only by infiltration through cracks in the structure of a building but also by interfering with the functioning of the inlet and outlet vents. The most certain means of preventing wind infiltration is to fix back-draught shutters on the air outlets of a building, whether or not they are powered by a fan. They are particularly useful on systems employing group switching of fans, as the fans that are switched off are then completely sealed by the shutters, owing to the reduced pressure inside the building created by the fans that are on. In a pressurized system, back-draught shutters on the air outlets may be closed or partly closed by the action of wind, causing a degree of isolation. If vents are located on both sides of a building, then the partial closing of shutters on one side is usually accompanied by an increased opening on the other side, preventing a significant influence of the wind on the ventilation rate.

There are a variety of back-draught shutters in use. Commercially available types are mainly of multi-louvre design and made of aluminium or plastics. Some designs provide an adequate seal on closing when new, but many have narrow gaps between and at the ends of the louvres. In some installations the total area of these gaps may approach the required minimum inlet area and thus destroy the efficiency of the inlet. The accumulation of dust and corrosion can impair the performance of louvre shutters, calling for frequent cleaning to maintain efficiency, and such shutters significantly reduce the throughput of fans, by up to 20 per cent. A single-vane shutter is usually preferable to the multivane types, but requires equally careful attention at both the design and construction stage.

Randall (1977a) has described a simple back-draught shutter for fans. This consists of a lightweight wooden frame, slightly larger than the aperture to be sealed, covered with a sheet of heavy-gauge polyethylene or reinforced plastic film that projects beyond the frame. The upper projecting edge is attached above the building aperture with a wooden batten, to form a frictionless and corrosion-free hinge. It is essential that the frame is light enough for the entire frame to be raised almost to the horizontal when the fan is switched on. In windy situations it is advantageous to provide a short length of plywood trunking surrounding the baffle, to prevent opening by strong winds blowing along the length of the building. Lightweight aluminium sheet with a flexible nylon hinge has also been used for shutters, but does not provide such a good air seal against the face of the building. Outlets which are long and narrow do not require a full-framed shutter. A flexible plastic sheet with an appropriate weighting along the lower edge is usually adequate.

VENT DESIGN

Whereas air outlets can be effectively isolated from the wind, it is not so easy to do the same for air inlets, and the traditional approaches of installing cowls, chimneys, fixed baffles and hoods do little to reduce the effects of wind interference. Only a relatively expensive structure surrounding a building, in the form of a windbreak combined with hoods on each inlet, has been shown to reduce wind interference (Hearn and Charles, 1978). There is no evidence that any simpler form of structure is at all satisfactory. In the absence of efficient cowls, wind interference may be minimized by maintaining a high inlet air speed of the order of 5 m s^{-1}, as discussed earlier. Where hoods, baffles or cowls are used they are likely to perform best with two external openings, designed to prevent high wind pressures from being created inside them.

Continuous ridge inlets usually allow the ingress of precipitation. Graee (1972) proposed several designs to overcome this problem, and also suggested that attic-ventilated systems are less subject to wind if both ridge and soffit eaves inlets are employed.

FAIL-SAFE SYSTEMS

A well-designed fan-ventilated building is almost free from the influence of wind, thus a power failure results in an almost total lack of ventilation. It is therefore necessary to incorporate some form of fail-safe device against power failure, overheating or overcooling.

Failure of the ventilating system may be indicated to the stockman by an automatic warning signal; he can then make manual adjustments to the ventilation. Alternatively, emergency ventilation may be provided automatically. Electromagnets may be used to hold vents, doors or special flaps in position, which open when the power fails, allowing a degree of natural ventilation. A time delay of about five minutes before the release of the magnets is usually advisable, to prevent unnecessary responses to momentary interruptions of the power supply. A standby generator is an alternative means of maintaining ventilation.

Inlet vents that are operated automatically by a cable can be fitted with a device which holds a loop in the cable. This is released when an electromagnet is de-energized. Similarly, back-draught shutters may be raised by cables attached to a master cable, which is pulled by a weight normally suspended by an electromagnet.

Conclusions

This chapter stresses the need to select and design components of ventilation systems which provide effective control over ventilation rate and air distribution. Most components influence only one of these requirements and the appropriate equipment for each has been presented. Obviously, not all combinations of the equipment to control ventilation rate and air distribution are compatible. It was the purpose of the previous chapter by Carpenter to show which systems are acceptable in this and other respects. However, the most common systems which are satisfactory from the aspects of design and efficient functioning are:

(1) group switching of fans combined with adjustable inlets;
(2) mechanical air mixing combined with diffusing surfaces and air filtration;
(3) mechanical air mixing combined with polyethylene ducts;
(4) group switching of fans combined with diffusing surfaces.

In naturally ventilated buildings there is rarely an effective means of control over the airflow pattern.

References

BOON, C.R. (1980). *Report DN/En/1009/10003*. National Institute of Agricultural Engineering. Silsoe, Bedford

BRUCE, J.M. (1975a). *Fm Bldg R & D Stud.* **6**, 1–8

BRUCE, J.M. (1975b). *Fm Bldg Prog.* **57**, 17–20

BRUCE, J.M. (1975c). *Fm Bldg R & D Stud.* **7**, 1–7

BRUCE, J.M. (1977). 'Slotted roofs for cattle buildings, design aid.' Scottish Farm Buildings Investigation Unit, Aberdeen

BRUCE, J.M. (1978). *J. agric. Engng Res.* **23**, 151–167

BRUCE, J.M. (1979). *Fm Bldg Prog.* **58**, 1–2

CARPENTER, G.A. (1972a). *Agric. Engr* **27**, 92–100

CARPENTER, G.A. (1972b). *Report DN/FB/218/3020*. National Institute of Agricultural Engineering, Silsoe, Bedford

CARPENTER, G.A. (1972c). *J. agric. Engng Res.* **17**, 219–230

CARPENTER, G.A. and MOULSLEY, L.J. (1978). *J. agric. Engng Res.* **23**, 441–451

DIAS, B. (1976). *Poult. Ind., Godalm.* **40**, 13

GRAEE, T. (1972). 'Åpne møner.' *Saertrykk nr. 157*. Institutt for Bygningsteknikk, Norges Landbrukshøgskole, Ås

GRAEE, T. (1976). 'Foreldrehus for høner.' *Saertrykk nr. 184*. Institutt for Bygningsteknikk, Norges Landbrukshøgskole, Ås

HEARN, P. and CHARLES, D. (1978). *Poultry Booklet 1978*, pp. 113–119. Gleadthorpe Experimental Husbandry Farm, Mansfield

KING, A.W.M., CHARLES, D., SPENCER, P.G., WALKER, G. and BENHAM, C.L. (1968). *Wld's Poult. Sci. J.* **24**, 319

MOULSLEY, L.J., FRYER, J.T. and PIKE, K.S. (1976a). *Report SN/FB/18/3020*. National Institute of Agricultural Engineering, Silsoe, Bedford

MOULSLEY, L.J., FRYER, J.T. and PIKE, K.S. (1976b). *Report SN/FB/9/3020*. National Institute of Agricultural Engineering, Silsoe, Bedford

OWEN, J.E. (1977). *Pig Fmg* **25**, 94–95

RANDALL, J.M. (1975). *J. agric. Engng Res.* **20**, 199–215

RANDALL, J.M. (1977a). 'A handbook on the design of a venilation system for livestock buildings using step control and automatic vents.' *Report 28*. National Institute of Agricultural Engineering, Silsoe, Bedford

RANDALL, J.M. (1977b). *Report SN/En/29/10003*. National Institute of Agricultural Engineering, Silsoe, Bedford

RANDALL, J.M. (1980). *J. agric. Engng Res.* **25**, 169–187

RANDALL, J.M. and BATTAMS, V.A. (1979). *J. agric. Engng Res.* **24**, 361–374

ZAMIR, N. (1973). *J. agric. Engng Res.* **18**, 397–406

20

CONTROL SYSTEMS FOR ANIMAL HOUSES—THE WAY AHEAD

B.C. STENNING

National College of Agricultural Engineering, Silsoe, Bedford

Introduction

Animal production has undergone many significant changes in the past thirty-five years. The movement has been towards large-scale enterprises, often involving intensive housing and a closely controlled environment. Each step along the way has been taken with care, but economic viability has, of necessity, been the major consideration and few producers would claim that the ideal production environment has yet been attained.

In any attempt to anticipate future trends in environmental control, consideration must be given to:

(1) the factors that have led to the present situation;
(2) the limitations and drawbacks of present methods;
(3) current trends in people's eating habits;
(4) the social (and legal?) pressures that might influence the intensive producer in future;
(5) engineering developments that may influence his methods;
(6) expected improvement in our knowledge of the animal, its health and its reaction to non-ideal environments.

The demand for food

In real terms the 'standard of living' in the UK continues to rise. Attendant upon this is a desire to eat tasty, high-quality foods and, according to present-day fashions, this implies increased demand for animal products and expensively produced salad crops. *Table 20.1* summarizes the production of certain items over a ten-year period, potatoes being quoted as a basis for comparison (HMSO, 1974; Toland, 1980). At the same time, the spending power per family has increased (Toland, 1980) (*see Table 20.2*), and whilst this trend continues it is unlikely that the purchase of luxury items will fall. On economic grounds, then, it would appear likely that the demand for favoured animal products will persist for at least a few years and that intensive production will continue to be a necessity.

Public awareness of the implications of intensive animal production was sharpened by Ruth Harrison's book in 1964, and the recommendations of the Brambell Committee (HMSO, 1965) are well known. More recently the

Table 20.1a   PRODUCTION OF SELECTED CROPS IN THE UK OVER THE PERIOD 1965–1974

	Livestock on agricultural holdings			Vegetables for human consumption		
	Total cattle (thousands)	Pigs (thousands)	Poultry (thousands)	Tomatoes (thousands of tonnes)	Lettuce (thousands of tonnes)	Potatoes (thousands of tonnes)
1964/65	11 943	7 979	118 141	89	12	5 085
1969/70	12 581	8 088	143 430	105	18	5 729
1974/75	15 203	8 544	139 672	120	32	5 357

From *MAFF Agricultural Statistics 1974*

Table 20.1b   CONSUMPTION OF CERTAIN FOODS IN THE UK OVER THE PERIOD 1957–1978 (1957 = 100)

Year	Poultry	Pork	Potato	Eggs	Mutton and lamb
1957	100	100	100	100	100
1963	320	130	95	100	100
1969	590	140	90	100	90
1975	700	140	80	100	70
1978	750	170	75	100	70

From Toland (1980)

Table 20.2   REAL DISPOSABLE INCOME PER HEAD AT 1975 PRICES (£ per annum)

1963	1969	1975	1978
1000	1100	1350	1400

From Toland (1980)

UK's press and broadcasting authorities have kept the matter before the nation by a variety of documentary articles and programmes. It is also significant that comments on proposed revisions to the welfare codes for domestic fowls and turkeys are currently being invited not only from farming and research organizations, but also from the general public. It should not go unnoticed, however, that public reaction to the alleged cruelty that attends intensive housing has not been sufficiently high to bring about a decline in sales of intensively produced animal products.

The effect of such publicity on the producer is difficult to judge, but some examples of its influence may be seen: for instance, the number of 'free-range' eggs offered for sale has shown a significant rise in recent years; in a different field, considerable attention is being paid to the production of veal calves by means that do not involve housing them in crates in windowless houses. Moreover, within the rest of the EEC, too, there is a move away from some intensive techniques—battery egg production in West Germany must shortly cease and the Council of Ministers is being urged to bring other countries into line.

In addition to humanitarian pressures being brought upon the farmer, there is the trend towards public acceptance of 'analogue foods'. This suggests that animal products must remain highly competitive in order to attract custom, particularly from the food processing industry.

Accordingly, a compromise is necessary. A moderate solution is required to the matter of animal welfare, with the introduction of generally acceptable codes of practice. Current investigations into the detection of stress conditions in animals and the development of modified housing systems will help in this direction. But the real cost of production cannot be allowed to rise significantly.

Trends in animal environmental control

If intensive animal production is to continue for the foreseeable future, it is worth while to review some of the recent developments and some of the remaining deficiencies in our methods of control.

In the early 1970s, artificial heating of poultry laying houses was regarded as a viable practice (HMSO, 1970) the relative costs of food and oil fuel being favourable under the conditions of ventilation that were employed at that time (0.7 cubic feet per minute per 4 lb bird: 0.34 m^3 s^{-1} per 1000 birds). Subsequent recommendations allowed the rate to be reduced to half of this value, with the proviso that uniformity of ventilation within a house could be achieved. With the consequent proportionate increase in use of animal metabolic heat, the economic picture immediately changed and heating is now rarely encountered. The heating of fattening piggeries, using fossil fuels, has been shown to be uneconomic for similar reasons (Stenning, 1974; Park, 1978). Have we reached the final and correct conclusion?

The introduction of fan speed controllers sensitive to air temperature within the stock building at first appeared to herald a new era in animal house control. Certainly, great improvement was noted in houses that would otherwise have been poorly ventilated, but problems of draught and of non-uniformity of air temperature were quickly observed (Owen, 1973). Poor speed control was noted (Barrett, 1977) and the resulting over- or underventilation accounted for undue food consumption and pneumonia outbreaks respectively (Looker, 1977). Do we now know the answer to the problem?

C.R. Boon (private communication) observes, for pigs in the 30–75 kg range, that when air temperature is within -1 to $+2$ K of the lower critical temperature, the behaviour or adopted lying position does not indicate discomfort. He suggests, however, that if more than 40 per cent of the pigs in a pen are huddling, the building is too cold for maximum food conversion. It is evident, then, that attention must be paid to good-quality temperature and air speed control. Randall (1979) has shown how this requirement may be met using step control of fans coupled with automatic vent adjustment, but it is not claimed that perfect spatial uniformity has been achieved.

The loss of heat from animals to the floors on which they lie is also of significance and Bruce (1979), in developing a method for classifying the thermal characteristics of floors, has shown that the thermal loss to concrete can be equivalent to a drop in air temperature of 7 K for 40 kg pigs. Present control systems take no account of this situation.

Environmental control in other fields

We can learn much from current advances in related fields. Fruit storage and greenhouse crop production both rely on control of the aerial environment, including temperature, humidity and gas concentration. However, it was necessary for satisfactory control of air temperature to be attainable before adjustment of the other variables could be considered. Indeed, only in 1962 was the enrichment of glasshouse air by carbon dioxide introduced into commercial practice. By this time it had become possible to control temperature, the 'primary' variable in the house, to a standard that allowed the exploitation of the hitherto uncontrolled gaseous balance.

In the drying of deep beds of grain, suitable control of the air stream is essential if uneven drying or condensation of airborne moisture is to be avoided. In this case the required air temperature depends both upon the water vapour pressure of the air and on the grain moisture content. Only recently has a generation of satisfactory hygrometers emerged which has made possible the development of comprehensive control equipment.

The lesson here is that the variables in a control system may be ranked in order of importance and the beneficial control of any one may depend upon those that are higher in rank already having been mastered. But the lower-ranked variables may still be very influential.

Whither automatic control?

Without question, environmental control in stock buildings implies the need for automatic control. Analogue controllers have been used in the adjustment of the environment of agricultural buildings (particularly glasshouses) for almost twenty years, but have always suffered from the disadvantage of being able to process only a limited amount of information. Modern digital electronic equipment, in the form of the microcomputer, is capable of far greater versatility and must surely take its place at the head of future control systems. All control systems, however, require reliable transducers to produce relevant information for the central processor, and reliable actuators to bring about the adjustment of the measured variable.

Enormous advances have been made in microprocessor technology in recent years—but almost without exception, the development of peripheral equipment, particularly transducers, has lagged behind. For situations where effective sensors are available, however, large quantities of data may be obtained and processed and appropriate corrective action initiated.

Reliability is an essential attribute of any control system, and an added advantage of the microprocessor-based system over its analogue counterpart is its ability to take decisions about the accuracy of its own operation. Where several such processors are interconnected in any one control system, a 'majority voting' scheme may be introduced, with a consequent improvement in system reliability. Whilst this facet of the microprocessor's ability has not yet been developed to agricultural advantage, work is in hand (Weaving, 1980) on a distributed-processor system for glasshouse control by which, in the event of a failure of one device in a chain of perhaps three, then one of the sound processors can be programmed to take over emergency control duties.

Looking ahead

Furtherance of our knowledge of the environmental conditions required by the animal relies to a large extent upon study of the animal when housed under closely controlled conditions. As engineering technology advances, ever better quality of control can be attained. Specification of target environmental conditions for use in commercial practice, however, demands a thorough knowledge of the performance of the animal and of its reaction to slight deviations from the ideal. Modern agricultural research recognizes the interdependence of engineers and animal physiologists in this work and the resulting scientific advances are well known.

As more information becomes available about the influence of the major environmental variables, it is not unreasonable to suppose that some less well recognized influences may be examined (or new ones discovered). For example, in the realms of human biology, experiments have been made to examine the effect of negative air ions upon psychomotor tasks (Hawkins and Barker, 1978). Elsewhere, ion generation has been reported as 'producing an improvement in cheerfulness and alertness' of staff whilst at the same time 'decreasing errors and complaints of headache and lassitude'. It is believed that work is currently in progress to examine possible beneficial effects of negative ions upon people suffering from respiratory disorders. Perhaps from this there may fall a crumb of useful information to help in reducing virus pneumonia in pigs?

In the more immediate future one looks forward to more and better transducers which may be so sited in stock buildings as to allow improved assessment of animal wellbeing. In the dairy industry the possibility is being discussed of implanting transponders into cattle for purposes of electronic identification (Grant, 1980). Already in the USA an ingestible temperature transmitter is being used in animal trials (Stermer, Camp and Smith, 1979). There is a growing need for inexpensive, robust and accurate transducers capable of transmitting those data which are of environmental importance, such as air temperature and air speed, and perhaps radiant temperature, atmospheric vapour pressure, atmospheric gas content, animal heart rate and animal respiration rate. With the versatile microprocessor at our elbow it should then be possible to monitor, and subsequently to control, the animal's environment to a degree that is unknown today, at least in commercial practice.

Another virtue of the microcomputer is its ability to adopt a routine managerial role. Given appropriate software and, for example, information about market trends and animal weight, there emerges the possibility of recommendations being made as to market readiness, nutritional levels and anticipated financial rewards!

Should the social or legal pressures necessitate the lowering of stocking density, with the possible renewed justification of supplementary heating systems, it is likely that use will be made of 'waste' heat from other sources, or from the exhaust air of the stock building itself instead of from fossil fuels.

A wide range of heat-exchanging devices exists (MacCormack and Bruce, 1975; Woods, 1979) and their introduction into intensive animal housing systems may be only a matter of time.

Conclusions

Progress occurs in steps. Improvement of the control of animal environment will depend upon the interaction of workers from a range of disciplines: animal physiology, electronics, heating and ventilation engineering and others. Effort is being directed towards a better understanding of the animal and better control of its environmental conditions. For the future we should be looking towards reinforcement of both aspects and their extension to commercial practice. This, as always, implies advancement for the leaders in the field—but more important is the need to improve the standards of the *average* producer. Effective and reliable equipment and techniques are essential for this purpose, and the introduction of microcomputer devices into animal house control is expected to lead towards this end.

But the role of the stockman will remain, even though he will be relieved of a number of routine duties. The extra time made available to him should be used to advantage, to allow him to exercise his professional skills rather than to impose upon him any extra routine task that may be just outside the capability of his microcomputer.

References

BARRETT, M. (1977). *NAC News* (Sept. 1977). National Agricultural Centre

BRUCE, J.M. (1979). *Fm Bldg Prog.* **55**, 1–4

GRANT, A.J. (1980). *Agric. Engr* **35**, 27–29

HARRISON, R. (1964). *Animal Machines.* Vincent Stuart, London

HAWKINS, L.H. and BARKER, T. (1978). *Ergonomics* **21**, 273–278

LOOKER, M. (1977). *Pig Fmg* (Suppl.) (July 1977), 48–51

HMSO (1965). *Report of the Technical Committee to Enquire into the Welfare of Animals Kept under Intensive Livestock Husbandry Systems. Cmnd 2836*

HMSO (1970). 'Heating of laying houses.' *MAFF Short Term Leaflet 102*

HMSO (1974). *Agricultural Statistics, UK 1974.* HMSO, London

MacCORMACK, J.A.D. and BRUCE, J.M. (1975). *Fm Bldg Prog.* **56**, 17–20

OWEN, J.E. (1973). *Pwr Fmg* **52**, 38–39

PARK, R.J.D. (1978). *BSc Dissertation*, National College of Agricultural Engineering, Silsoe, Bedford

RANDALL, J.M. (1979). *Fm Bldg Prog.* **57**, 1–5

STENNING, B.C. (1974). In *Heat Loss from Animals and Man*, pp. 367–388. Ed. by J.L. Monteith and L.E. Mount. Butterworths, London

STERMER, R.A., CAMP, T.H. and SMITH, L.R. (1979). *Trans. ASAE* **22**, 375–376

TOLAND, S. (1980). In *Social Trends 10*, pp. 13–38. Ed. by E.J. Thompson. HMSO, London

WEAVING, G.S. (1980). *Agric. Engr* **35**, 44–46

WOODS, J.L. (1979). *Agric. Engr* **34**, 20–22

VII

ENGINEERING AND CONTROL OF THE HOUSE ENVIRONMENT—2

21

INSULATION OF ANIMAL HOUSES

C. M. WATHES*

Agricultural Development and Advisory Service, Shardlow, Derby

Introduction

The primary function of animal house insulation is the reduction of structural heat flow. In fulfilling this function there are three consequences. First, the desired house temperature is achieved more readily; secondly, condensation on the interior surfaces is eliminated; and thirdly, minimum ventilation rates can be exceeded, thus lowering concentrations of airborne pollutants while still maintaining a satisfactory temperature lift.

The provision of insulation in a livestock building must be determined from considerations of the physiological needs of the animal, the economic cost of the insulation, the thermal behaviour of the building and the financial penalty incurred if the optimum climate is not preserved. In temperate climates pig and poultry buildings are usually insulated; calf houses are not (Mitchell, 1976).

In this chapter the insulation of animal houses is discussed under four headings: the physical properties of insulation materials; the energy and moisture balance of a wall or roof structure; the measurements of and design values for the thermal transmittances of a building structure; and the economic level of insulation in livestock buildings.

Physical properties of insulation materials

The physical properties of insulation materials that determine their suitability for agricultural use include their thermal conductivity, water vapour resistivity, bulk density and loading characteristics. Both the thermal conductivity, k_H, and the bulk density, ϱ, are dependent upon the moisture content of the material, which should therefore be specified in measurements of k_H and ϱ.

THERMAL PROPERTIES

The thermal conductivity of a material reflects its resistance to heat flow and is the flow per unit area per unit temperature gradient between two faces of

*Present address: Department of Animal Husbandry, University of Bristol. School of Veterinary Science, Langford, Bristol

the material. Similarly the vapour resistivity, ϱ_v, describes a material's resistance to water vapour movement. It is the reciprocal of the vapour diffusivity and, if the vapour pressure difference driving the flux is measured in pascals, has units of $N\,s\,kg^{-1}\,m^{-1}$.

Heat transfer within a material occurs by convection, conduction and radiation. The transfer processes within building materials, animal pelts and human clothing are similar and have been reviewed by Cena and Clark (1978). If water evaporates from or condenses within the insulant, the latent heat of evaporation acts as either a sink or a source of energy respectively. The temperature and water vapour gradient are modified and the heat and vapour fluxes altered accordingly. The thermal conductivity of the material, k_H, therefore does not have a single value but is affected by the material's temperature and moisture content. It is the sum of three components: $k_H = k_C + k_G + k_R$, where k_C, k_G and k_R are the thermal conductivities due to convection, conduction and radiation respectively. Evidence for this treatment includes the work of Skochdopole (1961), Lao and Skochdopole (1976), Pelanne (1977) and Paljak (1973), and was extensively reviewed by Pratt (1969). These workers observed that the thermal conductivity of a homogeneous material depends on its thickness. This observation, together with the report by Cena and Monteith (1975b) that the conductivity of an animal pelt is influenced by the temperature gradient across it, suggests that the magnitude of k_H is determined by the transfer processes within the material. For example, free convection usually dominates the heat transfer within animal pelts (Cena and Monteith, 1975b) and is important in fibrous insulants (Pelanne, 1977).

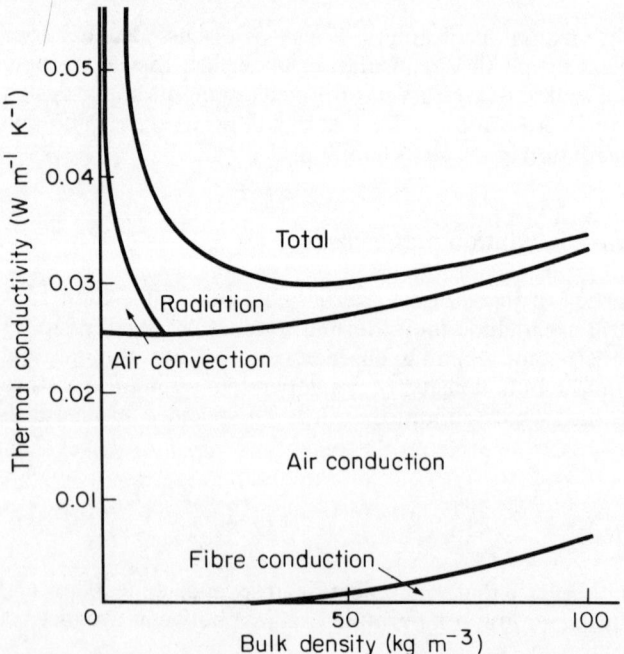

Figure 21.1 The contributions of the four component mechanisms to the total thermal conductivity of a glass fibre mat as a function of bulk density. (Redrawn from Pelanne, 1977)

The effect of density upon the thermal conductivity has been investigated by several workers. *Figure 21.1*, taken from Pelanne (1977), shows the effect of ϱ on k_H for glass fibre. The minimum thermal conductivity occurs at about 40 kg m^{-3}. Below this density, radiant transfer between the fibres causes a sharp increase in k_H, while above about 40 kg m^{-3} conduction along the fibres produces a gradual rise in k_H. Skochdopole (1961) found that polystyrene foams were influenced in a similar manner. As the density increases, transfer by gas convection and radiation is reduced, whilst that by conduction along the solid fibres becomes larger. Pelanne suggests that free convection ceases to be an important transfer mechanism in glass fibre once the bulk density exceeds 12 kg m^{-3}, because the solid fibres impede the convection currents. Within glass fibre the largest contribution to k_H is that due to gas conduction.

Conduction along the solid fibres or cells accounts for little of the apparent conductivity. Cena and Monteith readily dismissed the importance of this mechanism and this was confirmed by Pelanne for fibrous materials. If another gas of greater molecular weight is substituted for air within the insulant (such as Freon in polyurethane foams) then k_H can be markedly reduced. The reverse is true for lighter gases. However, even in closed-cell materials the substitute gas eventually diffuses out of the insulant and is replaced with air. For example, the captive blowing agents used in the manufacture of polyurethane foams diffuse from the foam within about one year, and this phenomenon is responsible for a change in k_H with age (Skochdopole, 1961). *Table 21.1* shows the effect of different gases on the

Table 21.1   EFFECT OF PORE GAS COMPOSITION ON THE THERMAL CONDUCTIVITY OF GLASS FIBRE INSULATION MATS AS A FUNCTION OF BULK DENSITY

Density (kg m^{-3})	Thermal conductivity (W m^{-1} K^{-1})			
	Helium	*Air*	*CO$_2$*	*Freon-12*
8.73	0.219	0.057	0.047	0.038
24.0	0.192	0.043	0.033	0.023
74.0	0.181	0.036	0.026	0.017

Based on data from Pratt (1969)

thermal conductivity of glass fibre (Pratt, 1969). Pratt gave a theoretical analysis that related the conductivity due to gas conduction to the molecular mean free path. Decreasing the ratio of the cell or fibre spacing to the mean free path reduces gas conduction; for air at room temperature the mean free path is of the order of 0.1 µm compared with an average pore size of 1 mm and mean fibre diameter of 4 µm for glass fibre insulation of bulk density 8.0 kg m^{-3} (Pelanne, 1977).

The second largest component of k_H is radiative transfer. Cena and Monteith (1975a) and Cena and Clark (1978) gave details of a theory of short- and long-wave radiative transfer within animal pelts and human clothing that could, in principle, be applied to other fibrous insulants. A theory of radiant heat transfer in fibrous insulating materials was also presented by Hager and Steere (1967). The rate of transfer depends on the coat structure and is influenced by the orientation and density of the fibres. Lao and Skochdopole (1976) have considered the radiative transfer within a cellular plastic foam, adopting a theoretical model similar to that of Cena and Monteith involving transfer in a homogeneous material in the steady state. Lao and Skochdopole derived both numerical and analytical solutions and showed that the theoretical temperature gradient within a homogeneous foam is non-linear, being maximal at the boundaries. This indicates that the insulation near the boundary provides a higher resistance than that of the bulk. In general their expression for k_R was accurate to $\pm 10\%$. They

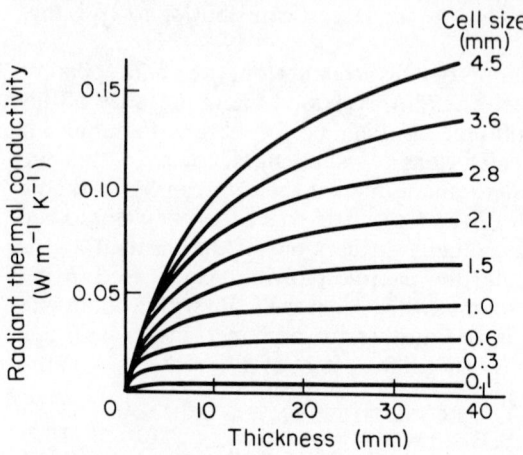

Figure 21.2 The theoretical effect of cell size and thickness of an extruded polystyrene sheet on the radiant component of the thermal conductivity, k_R. The surface emissivity is assumed equal to 0.85. (Based on Lao and Skochdopole, 1976)

concluded that radiative transfer accounts for an increase in k_H with thickness, and that the magnitude of the increase depends on pore size. Their predictions for extruded polystyrene foam of boundary emissivity equal to 0.85 are shown in *Figure 21.2*.

   Table 21.2 shows published values of k_H and ϱ_v for some common building materials (IHVE, 1971). Prangnell (1971) has reviewed values of ϱ_v published between 1964 and 1970 and part of his survey is summarized in this table. *Table 21.3* (IHVE, 1971) shows the effect of moisture content on the thermal conductivity of masonry materials, such as concrete and brickwork. Values are given for brickwork exposed to or protected from rain, as might be appropriate for the outer and inner walls of a farm building, respectively. Further information on the effect of moisture on k_H is given by Novels and Clegg (1977). Ball (1968) presented detailed measurements of k_H for many different concretes and mortars.

   Interstitial condensation within insulants increases their thermal conductivity in addition to causing physical deterioration. Movement of water vapour in insulants is considered in the next section, but *Figure 21.3*, taken from Kelly (1973), shows the effect of moisture on k_H for the more common agricultural insulants. For example, for urea formaldehyde foam k_H changes from 0.032 W m^{-1}K^{-1} at 0% moisture content to 0.066 W m^{-1}K^{-1} at 20% moisture content.

The rate at which materials absorb surface water also affects their suitability for agricultural use. Paljak (1973) exposed dry extruded polystyrene, expanded polystyrene and polyurethane foams, of bulk densities 34.8, 18.0 and 40.5 kg m^{-3} respectively, to a vapour pressure gradient of 27.8 kPa and measured the changes in moisture content with time. After 148 days the moisture contents (% by volume) were 6.3, 24.0 and 28.8% for extruded and expanded polystyrene and polyurethane foam respectively. At 340 days the moisture contents of the extruded and expanded polystyrene had increased

Table 21.2   BULK DENSITY, ϱ, THERMAL CONDUCTIVITY, k_H, AND VAPOUR RESISTIVITY, ϱ_v, OF SOME COMMON BUILDING MATERIALS

Material	Condition*	ϱ $(kg\ m^{-3})$	k_H $(W\ m^{-1}\ K^{-1})$†	ϱ_v $(GN\ s\ kg^{-1}\ m^{-1})$‡
Aluminium foil				175–10 000
Asbestos cement sheet	C	1600	0.400	1.60–3.50
Asbestos insulating board	C	750	0.120	–
Asphalt roofing	D	1600–2325	0.430–1.150	4.35–100.0
Brickwork	§	§	§	25–167
Concrete hollow blockwork	§	§	§	32.5
Fibre insulating board	C	260	0.050	20–40
Glass wool, mat or quilt	D	25	0.040	–
Hardboard, medium		600	0.080	1450
standard		900	0.130	1000
Metals, aluminium alloy		2800	160.0	–
steel, carbon		7800	50.0	–
Mineral wool, felted	D	50	0.039 ⎫	
semi-rigid felted mat	D	130	0.036 ⎬ 6.0	
loose felted mat	D	180	0.042 ⎭	
Perlite, loose granules	D	65	0.042	–
Plasterboard, gypsum		950	0.160 ⎫ 60.0	
perlite		800	0.180 ⎭	
Plastics, cellular:				
expanded polystyrene sheet	D	15	0.037	100–600
extruded polystyrene sheet	D	33	0.029	1000
polyurethane, new	D	30	0.017–0.023 ⎫ 1000	
aged		30	⎭	
Plastics, solid:				
epoxy fibre glass	D	1500	0.230	–
polystyrene	D	1050	0.170	
Polyethylene films				
0.05 mm				125–220¶
0.10 mm				250–350¶
0.15 mm				440¶
Snow, fresh		190	0.170	–
compacted		400	0.430	–
Timber, across grain				
softwood	C		0.130	500
hardwood			0.150	–
plywood	C	530	0.140	1500–6000
Urea formaldehyde foam		9	0.033	20–30
Vermiculite loose granules		100	0.065	–
Wood chipboard	C	800	0.150	–
Woodwool slab	C	500	0.085	15–40

*Refers to moisture content: D—dry; C—conditioned to constant weight at 20 °C and 65% relative humidity
†From IHVE (1971)
‡ From Prangnell (1971)
§ Values in *Table 21.3*
¶ Vapour resistance, not resistivity

Table 21.3   EFFECT OF MOISTURE CONTENT ON THE THERMAL
CONDUCTIVITY OF BRICKWORK AND CONCRETE AS A FUNCTION OF BULK
DENSITY

Density (kg m⁻³)	Thermal conductivity (W m⁻¹ K⁻¹)		
	Brickwork protected from rain (1% m.c.)	*Concrete protected from rain (3% m.c.)*	*Brickwork or concrete exposed to rain (5% m.c.)*
200	0.09	0.11	0.12
400	0.12	0.15	0.16
600	0.15	0.19	0.20
800	0.19	0.23	0.26
1000	0.24	0.30	0.33
1200	0.31	0.38	0.42
1400	0.42	0.51	0.57
1600	0.54	0.66	0.73
1800	0.71	0.87	0.96
2000	0.92	1.13	1.24
2400	1.49	1.83	2.00

m.c.—moisture content
Adapted from data given in IHVE (1971)

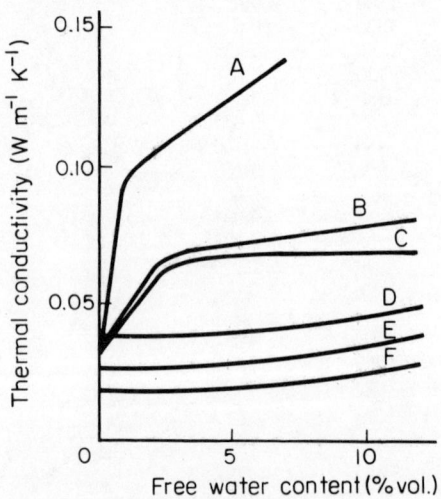

Figure 21.3 The effect of free water content (expressed percentage by volume) on the total thermal conductivity of A, mineral wool; B, glass wool; C, urea formaldehyde; D, extruded polystyrene; E, expanded polystyrene; and F, polyurethane. (From experimental results cited by Kelly, 1973)

to 11.6 and 34.6% respectively and were still increasing, whereas the poly-urethane had reached equilibrium at a moisture content of 29%. Substrates for animals are also of interest. The insulation of bedding has been studied by Gatenby (1977). She measured values of 0.12, 0.04, 0.11 and 0.72 W m⁻¹ K⁻¹ for the thermal conductivity of straw, wood shavings and dry and wet sawdust respectively at compressed bulk densities of 43, 370, 150 and 1100 kg m⁻³. The sixfold increase in k_H for wet as compared with dry sawdust demonstrates the considerable importance of maintaining a dry litter. It is also desirable that structural insulants should be exposed neither to wetting nor high vapour pressure gradients.

Published estimates of the absorptivity, α, over the entire solar spectrum, and the long-wave emissivity, ε, over the temperature range 0–100 °C, were surveyed by Holden and Greenland (1951). *Table 21.4* is taken mainly from their survey but includes some values cited by Diamant (1977). Apart from

Table 21.4 SOLAR ABSORPTIVITY AND LONG-WAVE EMISSIVITY OVER THE TEMPERATURE RANGE 0–100 °C FOR SOME COMMON BUILDING MATERIALS

Material		Solar absorptivity, α	Long-wave emissivity, ε
Aluminium:	metal	0.15–0.26	0.08–0.14
	foil	–	0.03–0.09
	paint	0.54	0.29–0.55
Asbestos:	cement new/aged	0.61–0.75	0.95
	insulating board	–	0.93–0.96
Asphalt:	new	0.91	–
	weathered	0.85	0.96
Bitumen felt		0.88	0.91
Bricks:	colour-dependent	0.40–0.89	0.94
Concrete:	rough	0.65	0.94
Glass:	smooth	0.83	0.92–0.95
Iron:	galvanized, new	0.64–0.66	0.22–0.28
	aged	0.90	–
Slates		0.86–0.93	–
Tiles:	colour-dependent	0.43–0.91	–

From Holden and Greenland (1951) and Diamant (1977)

polished metals such as aluminium, the absorptivity of most materials ranges between 0.60 and 0.90 while a long-wave emissivity of 1.0 represents an acceptable approximation for normal use.

VERMIN, INSECTS AND OTHER HAZARDS

Rats, mice and some insects, in particular the lesser mealworm beetle (*Alphitobius diaperinus*) and the hide beetle (*Dermestes maculatus*), may cause damage to the fabric and insulation of livestock buildings, in part because the environmental conditions promote rapid breeding. Eradication is the only sure cure but damage to insulants could be almost eliminated if laminated materials, such as extruded polystyrene bonded to steel panels, were used. Such composite insulants are now available. In addition, a hazard can arise when PVC-covered electric cables come into contact with expanded polystyrene insulation and other thermosetting plastics. Migration of the plasticizer from PVC can cause cracking of the cable insulation, thereby reducing its life. In general, if electric cables are installed in cavity insulants the current rating must be reduced, as there is a danger that the cable will overheat.

Insulants may also be degraded during routine operations, such as cleaning between crops. The ease with which building materials can be cleaned is difficult to quantify but Sundahl (1975) attempted an assessment of the problems. The factors she considered included the tendency to collect dirt, accumulation of smell, wear and tear with washing and the number of organisms remaining on washed surfaces. She found that wood and wood products and materials with smooth hard surfaces, such as glazed tiles and laminated plastics, were the most difficult to clean; the easiest were materials coated with resins and bitumen paints. Cleaning produced the most wear on hard-to-wash, softer and more porous materials.

TEST METHODS

It is desirable that measurements of k_H, ϱ_v and other material parameters should be completed under standard conditions, so that valid comparisons can be made. Indeed, the more common test methods are the subject of a British Standard.

Descriptions of methods for measuring k_H are numerous and have been reviewed by Pratt (1969). The most common method involves a guarded hotplate; other steady-state methods include the calibrated hotbox and the heat-flow meter. However, steady-state procedures can lead to errors when moist samples are used owing to the migration of the water; thermal equilibrium may require many hours and the tests are costly. Novels and Clegg (1977) describe a transient technique that overcomes these disadvantages. The measurements are completed within 30–40 minutes and the temperature gradient across the material is at most 2 K. For moist materials the error is ±2%, double that for dry samples.

The vapour resistivity of a material, ϱ_v, is the reciprocal of the water vapour flow per unit area per unit vapour pressure gradient between two faces of the material. The measurement of vapour resistivity has been described by Pratt (1958). The method is simple but errors can arise from a change in the absorptive power of the desiccant and the non-establishment of a moisture equilibrium in thick materials (Prangnell, 1971). Paljak (1973) gave details of a method for measuring the water absorption of cellular plastics.

BS 2972 (BSI, 1975) is a standard describing brief details of test methods for most pertinent properties of inorganic insulants (bulk density; thermal properties; the assessment of fire hazard; resistance to compression; vibration settlement; covering capacity; moisture content; water absorption; water vapour permeance; odour; emissivity of metal foils, and flexural strength). Rigid cellular materials are the subject of BS 4370 Part 1 (BSI, 1968) and Part 2 (BSI, 1973a), whilst urea formaldehyde foam is covered in BS 5617 (BSI, 1978). In the context of this chapter, the measurement of k_H is described in BS 874 (BSI, 1973b) whilst the method for measuring ϱ_v for sheet materials is considered in BS 3177 (BSI, 1959). The latter standard suggests that materials should be conditioned in either a temperate climate, $T_a = 25 \pm 0.5\,°C$ and $75 \pm 2\%$ relative humidity (RH), or a tropical climate, $T_a = 38 \pm 0.5\,°C$ and $90 \pm 2\%$ RH. By contrast, most methods for measuring k_H specify that the sample should be conditioned to $65 \pm 5\%$ RH at $T_a = 20 \pm 2\,°C$.

BEHAVIOUR IN FIRE

The behaviour of an insulant or of a roof or wall structure in a fire also affects the choice of building materials. There have been few studies on the behaviour in fire of large agricultural buildings, but Kelly and Ross (1975) made a qualitative assessment of extruded polystyrene and polyurethane foam sheets for this application. They had observed that insurance premiums for both the building's contents and structure were raised when cellular plastics were employed in agricultural construction and queried whether this was justified in terms of their performance in fire. When the boards were placed in a small building (6.5 × 14.7 × 1.8 m) they ignited a wooden crib, of known heat output, and recorded the material behaviour with a camera. The temperature was measured with thermocouples. They concluded that whilst large-scale tests gave more realistic information on board performance, other factors to consider were 'the type of facing or core, the design and sealing of joints and the location and size of their gaps'. Extruded polystyrene retracted from the heat source but formed dripping molten globules and streamers, which were not, however, a secondary fire source. Both foams yielded prodigious quantities of dense black smoke when the foam core ignited. Nevertheless, Kelly and Ross concluded that these foams should not be excluded from agricultural use since fire performance and insulation value should be considered together and not in isolation.

Fire tests on building materials and structures are the subject of BS 476, which describes amongst other tests those for surface spread of flames (Part 7; BSI, 1971), fire resistance (Part 8; BSI, 1972) and ignitability (Part 5; BSI, 1979). Surface spread-of-flame tests are small-scale and are performed on six samples, each measuring 230 × 900 mm; the fire conditions that prevail in large agricultural buildings are unlikely to replicate those of this test. Materials are classified into four classes: Class 1 materials have the lowest surface spread of flame and are recommended for agricultural buildings; Class 4 the highest. Whereas Part 7 is concerned with individual elements of the structure, such as the inner cladding, the fire resistance tests are on whole structures and use the criteria of stability, integrity and thermal insulation for their assessment. In the latter tests sample sizes are larger; for example, wall samples should be at least 2.5 × 2.5 m and of service strength and moisture content. The ignitability test is straightforward and materials for structures are designated as either ignitable (Class X) or non-ignitable (Class P).

Energy and moisture balance

Though the transmission of heat through a building structure is largely governed by the normal processes of conduction, convection and radiation, condensation or evaporation of water within the structure or on either surface may also contribute to the energy balance. It is customary to specify an overall transmittance of a structure, but this procedure neglects the detailed transfer processes that operate.

Figure 21.4 shows the equivalent electrical analogue for the combined heat and mass transfer through a building structure. Condensation or

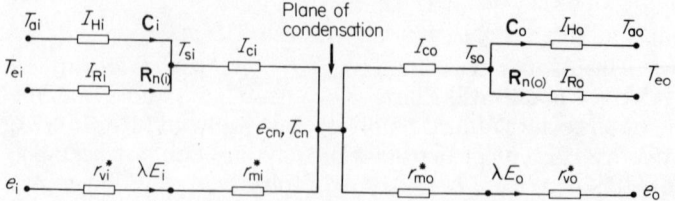

Figure 21.4 The equivalent electrical analogue of the two-dimensional heat and moisture balance of the wall or roof structure under steady-state conditions with condensation occurring at some plane within the structure. (C, convective flux density; R_n, net radiative flux density, λE, latent heat flux density; T_a, dry-bulb temperature; T_e, effective environmental temperature; T_s, structure temperature; e, vapour pressure; I_H, boundary-layer thermal resistance; I_R, equivalent radiative resistance; I_c, roof materials' thermal resistance; r^*_v, boundary-layer resistance to water vapour; r^*_m, roof materials' vapour resistance; T_{cn}, temperature at plane of condensation; e_{cn}, vapour pressure at plane of condensation. Subscripts i, interior; o, exterior)

evaporation of water vapour occurs at the plane of condensation at a rate ε_{cn} ($kg\,m^{-2}\,s^{-1}$); at this point the temperature is T_{cn} and the absolute vapour pressure is e_{cn} (Pa). The heat balance equation is

$$\mathbf{J} + \mathbf{C}_i + \mathbf{R}_{n(i)} + \lambda\varepsilon_{cn} + \mathbf{R}_{n(o)} + \mathbf{C}_o = 0 \qquad (21.1)$$

where $\mathbf{C}$ and $\mathbf{R}_n$ are the mean convective and net radiative fluxes ($W\,m^{-2}$) respectively and subscripts i and o denote those incident from the interior and exterior respectively (used throughout this chapter). $\mathbf{J}$ is the rate of change of heat storage in the structure and equals zero in the steady state, λ is the latent heat of vaporization of water. If $T_{cn} < 0\,°C$, hoar frost is formed and λ is replaced by $\lambda + \beta$, where β is the latent heat of fusion of water. The sign convention adopted is that fluxes of heat towards the plane of condensation, from either direction, are negative and those away from it are positive. The resistances to the transfer of water vapour and for heat transfer by convection and radiation are r^*_v, I_H and I_R respectively; I_c is the resistance to sensible heat exchange within the structure.

CONVECTIVE AND RADIATIVE TRANSFER

The rate of transfer of heat via convection is related to the temperature of the structure, T_s, and the air, T_a, by

$$\mathbf{C} = (T_a - T_s)/I_H$$

where I_H is the boundary-layer insulation resistance ($m^2\,K\,W^{-1}$). The value of I_H depends on whether forced or natural convection is the dominant mode of transfer; both McAdams (1954) and Monteith (1973) described a procedure for determining the appropriate regime. For a flat plate, similar to a roof or wall, of characteristic dimension 2 m and $(T_s - T_a) = 2\,K$, forced convection is dominant if the mean air speed over the surface is greater than $1\,m\,s^{-1}$; for speeds of less than $0.1\,m\,s^{-1}$, natural convection controls the transfer. Between 0.1 and 1.0 m s^{-1} neither mode is dominant and the lowest resistance should be used to estimate the convective heat exchange. For livestock buildings in temperate climates therefore, forced convection occurs at the exterior surfaces; transfer to the interior surfaces is via natural convection.

Many authors have measured I_H in both natural and forced convection for small, smooth plates in wind tunnels, but few have examined the boundary-layer resistance of full-scale buildings (Fishenden and Dufton, 1929; Loudon, 1963, 1968; Diamant, 1977). However, I_H is a sufficiently small proportion of the total resistance that its precise evaluation is unnecessary. Indeed the Institution of Heating and Ventilating Engineers Guide Book A (IHVE, 1971) specifies a fixed value of I_H irrespective of $(T_s - T_a)$ and the convective transfer regime. *Table 21.5* shows a comparison of values of I_H

Table 21.5 A COMPARISON OF PREDICTED BOUNDARY-LAYER RESISTANCES UNDER NATURAL AND FORCED CONVECTION FOR SMOOTH VERTICAL AND HORIZONTAL SURFACES, BASED ON RELATIONS BY THE AUTHORS LISTED

Author(s)	*Thermal resistance, I_H* *($m^2 K W^{-1}$)*				
	Natural convection			*Forced convection*	
	Vertical	*Horizontal upwards*	*Horizontal downwards*	*$u = 0.15\ m\ s^{-1}$*	*$u = 3.0\ m\ s^{-1}$*
Fishenden and Dufton (1929)	0.426	0.336	0.639	0.157	0.057
Loudon (1963)	0.352	–	–	0.159	0.056
IHVE (1971)	0.333	0.233	0.667	0.156	0.055
Monteith (1973)	0.607	0.482	1.532	0.917	0.080

u is the wind speed over the surface

predicted from relations suggested by four authors, based on a characteristic dimension of 2 m and $(T_s - T_a) = 2$ K. The resistances cited by Monteith are higher than those of the other authors, probably because the former are derived from empirical relations for smooth, flat plates whereas the IHVE resistances are based on measured values for small buildings. The values stated in the IHVE Guide Book are probably adequate for design purposes; more detailed calculations should be based on empirical engineering relationships.

The net exchange of radiation at the surface, $\mathbf{R}_n$, is the algebraic sum of the incident solar and absorbed long-wave radiation and the short-wave reflected and transmitted and the emitted long-wave radiation. Hence $\mathbf{R}_n$ is given by

$$\mathbf{R}_n = \alpha(\mathbf{S}_t + \mathbf{S}_e) + \varepsilon(\mathbf{L}_d + \mathbf{L}_e - \mathbf{L}_s)$$

where $\mathbf{S}_t$ is the sum of the direct and diffuse solar radiation, $\mathbf{S}_e$ is the sunlight reflected from the environment; $\mathbf{L}_d$ is the long-wave radiation from the atmosphere, $\mathbf{L}_e$ is that from the environment, and $\mathbf{L}_s$ is that emitted from the surface of the building (Monteith, 1973). A detailed account of radiation fluxes and geometry was given by Monteith. Values of α and ε are shown in

Table 21.4, but Loudon (1963) observed that ε is approximately 0.9–1.0 for most natural building materials, but is less for reflective insulations such as aluminium foil (for which ε is about 0.2).

STEADY-STATE HEAT BALANCE

Monteith (1973) described a procedure for combining the convective and radiative heat fluxes. If there is a net loss of radiation at the surface the sensible heat exchange becomes

$$\mathbf{C} + \mathbf{R} = (T_s - T_e)/I_{HR} \tag{21.2}$$

where T_e is an effective temperature of the environment and I_{HR} is the sum of the boundary-layer and radiative resistances in parallel and is given by

$$I_{HR} = (I_H I_R)/(I_H + I_R)$$

T_e is given by

$$T_e = T_a + \mathbf{R}_{ni} I_{HR} \tag{21.3}$$

where $\mathbf{R}_{ni}$ is the net isothermal radiation that would be received by an identical surface at air temperature in the same environment. Equations 21.1 and 21.2 can be combined with due regard to sign. Assuming $T_{si} > T_{so}$, then

$$(T_{ei} - T_{si})/I_{HRi} + (T_{si} - T_{cn})/I_{ci} + \lambda\varepsilon_{cn} = (T_{cn} - T_{so})/I_{co} +$$
$$(T_{so} - T_{eo})/I_{HRo} \tag{21.4}$$

where I_c is the thermal resistance of the structure from the plane of condensation to the interior or exterior surface. Equation 21.4 can be simplified to

$$(T_{ei} - T_{cn})/I_{si} + \lambda\varepsilon_{cn} = (T_{cn} - T_{eo})/I_{so} \tag{21.5}$$

where I_s $(= I_{HR} + I_c)$ is the combined resistance of the structure and the surfaces to sensible heat transfer. The reciprocal of $(I_{si} + I_{so})$ is the conventional definition of the thermal transmittance or U value of the structure.

A building structure normally comprises several layers or elements, each made of different materials. The thermal transmittance is therefore the sum of the resistances of the individual elements and the surfaces. Hence

$$I_s = I_{HR} + \Sigma d_j/k_{Hj}$$

where d_j and k_{Hj} are the thickness and thermal conductivity of the *j*th element of the structure respectively.

In practice, the steady-state heat balance is affected by several factors. First, heat bridges may form across the structure, in which case the surfaces become non-isothermal. Loudon (1963) suggested that the resistance of such bridges can be assumed to be proportional to their area. Similarly, building materials may be inhomogeneous owing to the ravages of rodents or simply faults originating during construction. Edge effects at the boundaries of the structure mean that the size and shape of the wall or roof will influence I_{HRo} and I_{HRi}. Secondly, the effect of moisture on the thermal conductivity produces a fall in I_s as the moisture content rises. If the condensate does not drain away and stays within the building fabric, positive feedback can occur in which condensation reduces the resistance of the structure, alters the temperature and vapour pressure profiles and leads to an increase in the rate of condensation. Finally, building structures are often porous and the effective value of $(I_{si} + I_{so})$ will depend on the rate of air infiltration through the structure and, in more favourable conditions, on the rate of loss of condensate by drainage or evaporation.

The justification for using a steady-state heat balance equation for agricultural buildings is twofold. First, the largest heat loss from the building is via the ventilation; even in winter only 30 per cent at most of the total heat loss is through the structure of a well-insulated building. Secondly, the response time of either the animals, in the case of adults, or the heating system, in the case of juveniles, to a sudden fall or rise in the external conditions is much less than the period of daily temperature cycles (Clark and Cena, Chapter 17). In a poorly insulated building, with a value of $(I_{si} + I_{so}) = 0.4$ m^2 K W^{-1} or less, the lack of insulation has a much greater effect on T_{si} and large fluctuations may result from rapid changes in outside conditions. For such a building a study of the non-steady-state heat balance may be appropriate and is analogous to heat flows in soil. A complete formal analysis of the non-steady heat flow equation was presented by Carslaw and Jaeger (1959); Monteith (1973) gave a solution to the one-dimensional heat flow equation for a homogeneous medium subject to a sinusoidal variation in surface temperature. Whilst not offering a formal solution to the heat flow equation, Diamant (1977) states that the velocity of propagation of a temperature sine wave through a wall or roof is proportional to the square root of the structure diffusivity and the angular frequency of the wave. In extreme climates, therefore, the building structure should be constructed from materials of large thermal mass; lightweight materials will suffice in temperate climates.

CONDENSATION

Condensation occurs within or on the surface of a structure whenever the temperature in some layer is less than or equal to the dewpoint temperature of the air. If T_{cn} is less than 0 °C hoar frost is formed. The prediction of interstitial condensation was first discussed by Rowley (1939) and more recently by Chang and Hutcheon (1956), Pratt (1958), Diamant (1967), Davies (1973) and Kelly (1973). The diffusion of water vapour through a structure is governed by Fick's law. Expressed in the units usually employed in building science,

$$\varepsilon = (e_i - e_o)/r^*_v$$

where ε is the mass flux density (kg m^{-2} s^{-1}) and e is the vapour pressure (Pa). r^*_v is the resistance to vapour transfer in N s kg^{-1} ($=$ m s^{-1}). (Unfortunately, if alternatively the mass concentration (kg m^{-3}) is considered as the driving force, then the resistance is in units of s m^{-1}.) The resistance to water vapour is the sum of the individual resistances of the elements of the structure. Hence

$$r^*_v = \Sigma d_j \varrho_j$$

where ϱ_j is the water vapour resistivity of the jth element. When heat transfer at a surface is mainly by forced convection I_H and r^*_v are related by the Lewis relationship (Monteith, 1973; Davies, 1973), which is

$$r^*_v = I_H(\varkappa/D)^{0.67} R^* T_m \varrho c_p/M_w$$

where M_w is the kilogram molecular weight of water, R^* is the universal gas constant, ϱc_p is the volumetric specific heat of air, T_m is the mean film temperature, $\varkappa$ is the thermal diffusivity of dry air and D is the diffusion coefficient of water vapour in air.

The simplest procedure for estimating the risk of and the position at which condensation occurs within the structure is to examine the temperature and water vapour pressure profiles. However, this procedure does not take account of the significant contribution of latent heat to the energy flow. The yield of latent heat raises T_{cn} and alters both the temperature and water vapour pressure gradients (Davies, 1973).

The following method to resolve this problem was suggested by Davies and has been extended considerably by Huang, Sian and Best (1979) in their study of heat and moisture transfer in concrete slabs. Conservation of mass gives

$$(e_i - e_{cn})/(r^*_{vi} + r^*_{mi}) = \varepsilon_{cn} + (e_{cn} - e_o)/(r^*_{vo} + r^*_{mo}) \qquad (21.6)$$

where r^*_v and r^*_m are the resistance to water vapour transfer of the boundary layer and structure respectively. r^*_m is analogous to r_c in heat transfer. Equation 21.6 can be simplified to

$$f_i(e_i - e_{cn}) = \varepsilon_{cn} + f_o(e_{cn} - e_o) \qquad (21.7)$$

where f_i and f_o are the interior and exterior vapour conductances and equal the reciprocals of $(r^*_{vi} + r^*_{mi})$ and $(r^*_{vo} + r^*_{mo})$ respectively. f_i and f_o are analogous to the reciprocals of I_{si} and I_{so} respectively. Rearranging equations 21.5 and 21.7 we have

$$\varepsilon_{cn} = \{T_{cn}(I_{so} + I_{si}) - (I_{si} T_{eo} + I_{so} T_{ei})\} \{I_{so} I_{si} (\lambda + \beta)\}^{-1} \qquad (21.8)$$

and

$$\varepsilon_{cn} = (e_i f_i + e_o f_o) - e_{cn} (f_i + f_o) \qquad (21.9)$$

where the term β is the latent heat of fusion of water and has been introduced to allow for the formation of hoar frost. If both sides of equations 21.8 and 21.9 are divided by $(f_i + f_o)$, then from equation 21.8

$$Z_1 = \varepsilon_{cn}/(f_i + f_o) =$$
$$\{T_{cn}(I_{so} + I_{si}) - (I_{si} T_{eo} + I_{so} T_{ei})\} \{I_{so} I_{si}(\lambda + \beta) (f_i + f_o)\}^{-1} \qquad (21.10)$$

and from equation 21.9

$$Z_2 = \varepsilon_{cn}/(f_i + f_o) = (e_i f_i + e_o f_o)/(f_i + f_o) - e_{cn} \qquad (21.11)$$

Equations 21.10 and 21.11 are simultaneous and have units of pressure. They can be solved for e_{cn} and T_{cn} by a graphical method (Davies, 1973). Equation 21.10 is a linear equation involving Z_1 and T_{cn}, of slope s:

$$s = (I_{so} + I_{si}) \{I_{so} I_{si} (\lambda + \beta) (f_i + f_o)\}^{-1}$$

and intercept T_{cn} at $Z_1 = 0$.

$$T_{cn} = (I_{si} T_{eo} + I_{so} T_{ei})/(I_{so} + I_{si})$$

Equation 21.11 relates the condensation rate to the saturation vapour pressure at the plane of condensation given e_i, e_o, f_i and f_o. By definition e_{cn} must equal the saturation vapour pressure at the temperature T_{cn}.

Two representative examples will now be considered: a roof structure at night during winter and during the day in summer. The roof structure is shown in *Figure 21.5*, which also lists the thermal resistances for the winter

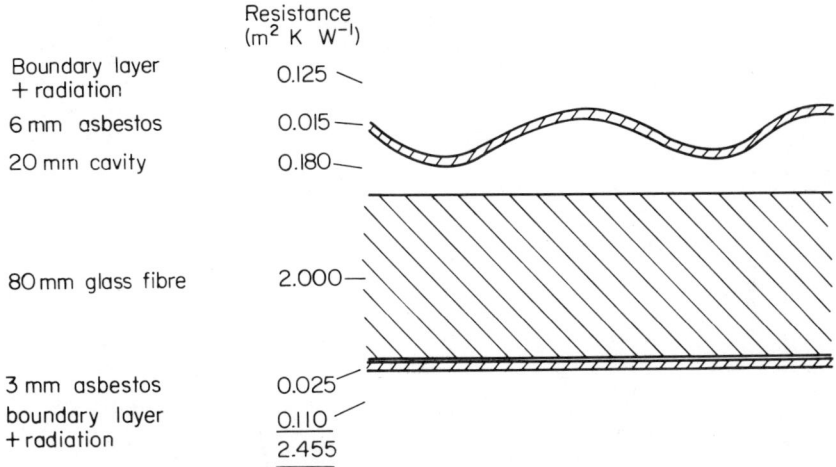

	Resistance (m² K W⁻¹)
Boundary layer + radiation	0.125
6 mm asbestos	0.015
20 mm cavity	0.180
80 mm glass fibre	2.000
3 mm asbestos	0.025
boundary layer + radiation	0.110
	2.455

Figure 21.5 The thermal resistance of a typical roof structure in winter. $T_{ai} = T_{ei} - 20\,^\circ C$, $T_{ao} = 0\,^\circ C$, $T_{eo} = -14\,^\circ C$, $(I_{si} + I_{so}) = 2.455$ m² K W⁻¹

case. The structure comprises a corrugated asbestos sheet 6 mm thick of corrugation height 30 mm, a 20 mm air cavity, an 80 mm glass fibre mat and an inner lining of 3 mm compressed asbestos board. Allowing for the corrugation, the mean air space thickness is 35 mm. The largest thermal resistance resides in the glass fibre mat, which accounts for 80 per cent of the total.

In the winter example the internal temperatures are $T_{ei} = T_{ai} = 20\,^\circ C$. Air speeds over the ceiling are low, less than 0.15 m s⁻¹, and convective transfer is in the mixed mode, giving a boundary-layer resistance $I_{Hi} = 0.293$ m² K W⁻¹. Allowing for an equivalent radiative resistance of $I_{Ri} =$

0.175 m² K W⁻¹ (Monteith, 1973), the combined surface resistance is I_{HRi} = 0.110 m² K W⁻¹. In this example two vapour pressures are considered: case 1 is e_i = 800 Pa (34% RH) and case 2 is e_i = 1440 Pa (62% RH). On the outside of the building the skies are cloudless and T_{ao} = 0 °C. The atmospheric long-wave radiation flux, L_d, can be calculated using the formula given by Monteith. At night time $S_t = S_e = 0$. Then, neglecting L_e and with

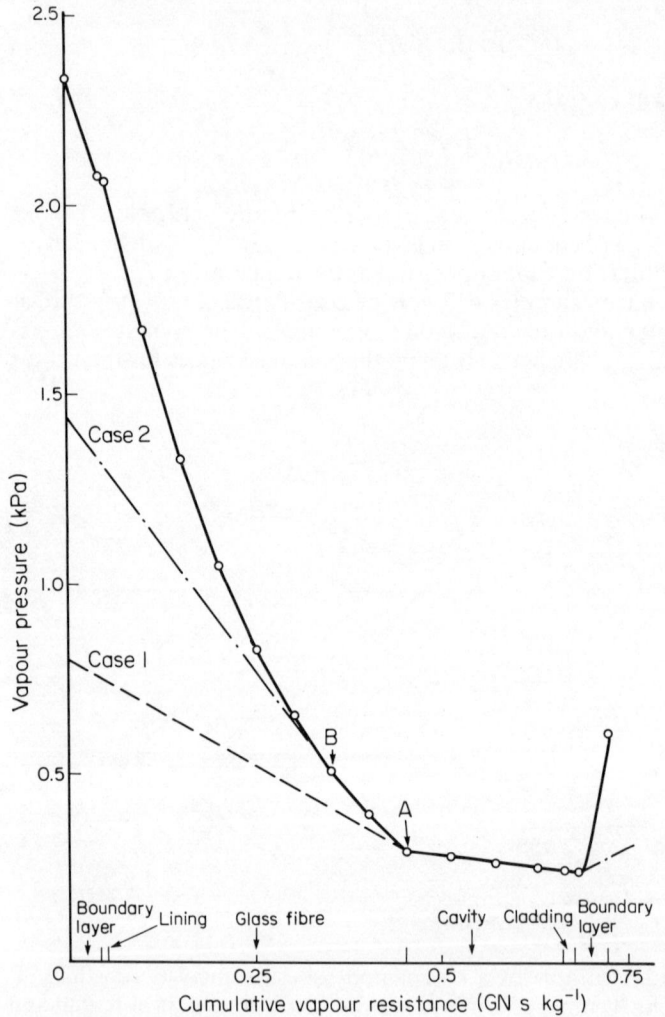

Figure 21.6 The saturation vapour pressure profile across the roof structure illustrated in Figure 21.5 (again in winter), without a polyethylene vapour check, as a function of the cumulative vapour resistance of the structure. Interstitial condensation occurs where the dashed and solid lines intersect at A and B for the two cases respectively; case 1, at the interface of the glass fibre and air cavity (– –); case 2, 20 mm from this interface within the glass fibre (– · –). Hence the thermal and vapour pressure resistances up to and beyond the plane of condensation are: case 1, I_{si} = 2.135 m² K W⁻¹, I_{so} = 0.320 m² K W⁻¹, $(r*_{vi} + r*_{mi})$ = 0.452 GN s kg⁻¹ and $(r*_{vo} + r*_{mo})$ = 0.272 GN s kg⁻¹; case 2, I_{si} = 1.635 m² K W⁻¹, I_{so} = 0.820 m² K W⁻¹, $(r*_{vi} + r*_{mi})$ = 0.352 GN s kg⁻¹ and $(r*_{vo} + r*_{mo})$ = 0.372 GN s kg⁻¹

$L_d = 207\ \mathrm{W\,m^{-2}}$, we calculate $R_{ni} = -108\ \mathrm{W\,m^{-2}}$. The effective environmental temperature is determined from equation 21.3 and is $T_{eo} = -14\,°C$. The resistances to vapour transfer are calculated from the material properties and from the Lewis relation for the boundary layer and cavity resistances.

Figure 21.6 shows the saturation vapour pressure as a function of the cumulative vapour resistance of the structure. The saturation vapour pressure (s.v.p.) line is calculated from the temperature profile, which in turn is determined from the thermal resistances shown in *Figure 21.5* and the temperature difference across the structure, $(T_{ei} - T_{eo}) = 34\ \mathrm{K}$. Interstitial condensation occurs within the structure at any point at which the vapour pressure equals the saturation vapour pressure. If no condensation occurs within or on the structure, the vapour pressure profile is a straight line joining e_i and e_o. If the profile and s.v.p. line intersect, condensation or evaporation will occur at some point within the structure. A decrease in the vapour pressure gradient at the point of intersection implies condensation; an increase in the gradient suggests evaporation. The intersection of the vapour pressure profile and the s.v.p. line must be made tangentially (points A and B in *Figure 21.6*), otherwise there would be an inference that more vapour diffuses away from the plane of condensation than arrives at it, an impossible condition whilst condensation is taking place or a steady state has been attained. Davies (1973) discussed this phenomenon in greater detail.

For the first case the vapour pressure profile and the s.v.p. line meet at point A. This is at the interface of the glass fibre mat and the air cavity. Inserting the appropriate values into equations 21.10 and 21.11 we find that the slope estimated from equation 21.10 is $214.5\ \mathrm{Pa\,K^{-1}}$ with an intercept at $-9.6\,°C$ and that $Z_2 = 491 - e_{cn}$ (Pa). The intercept of Z_1 is the temperature at point A if no condensation had occurred, whilst Z_2 can be calculated given the values of the saturation vapour pressure at various temperatures and e_i, e_o, f_i and f_o. *Figure 21.7* shows a plot of Z_1 and Z_2 as a function of temperature, calculated from equations 21.10 and 21.11. The intersection of Z_1 and Z_2 defines the temperature and vapour pressure gradient at point A and is at $(-8.8\,°C, 177\ \mathrm{Pa})$, e_{cn} is calculated from equation 21.11 and is 314 Pa. In this case therefore hoar frost is formed and its rate of formation, calculated from equation 21.7, is $\varepsilon_{cn} = 1.04\ \mathrm{mg\,m^{-2}\,s^{-1}}$ or $3.8\ \mathrm{g\,m^{-2}\,h^{-1}}$. The latent-heat yield is $(\lambda + \beta)\varepsilon_{cn}$, equal to $2.8\ \mathrm{W\,m^{-2}}$, compared with the sensible heat loss of the structure of $13.8\ \mathrm{W\,m^{-2}}$, and is responsible for raising the temperature of the plane of condensation from -9.6 to $-8.8\,°C$. The time taken to deposit a 0.1 mm layer of hoar frost may be calculated from the rate of deposition and the density of ice $(\varrho = 920\ \mathrm{kg\,m^{-3}})$ and is 26 hours. Although in theory the frost is deposited uniformly throughout the structure from the glass fibre/cavity face outwards, in practice the hoar frost will be formed preferentially at the interface. One possible explanation is that the ice forms a vapour check which modifies the vapour pressure profile.

In the second case, the vapour pressure profile and the s.v.p. line meet at B (*Figure 21.6*), which is 20 mm from the glass fibre/cavity interface. Following the same procedure as above, we estimate the slope of equation 21.10 to be $116\ \mathrm{Pa\,K^{-1}}$, the intercept $-1.6\,°C$ and $Z_2 = 888 - e_{cn}$ (Pa). The intersection of Z_1 and Z_2, shown in *Figure 21.7*, is at $(0.3\,°C, 260\ \mathrm{Pa})$; freezing does not take place. The rate of condensation, E_{cn}, is $1.4\ \mathrm{mg\,m^{-2}\,s^{-1}}$,

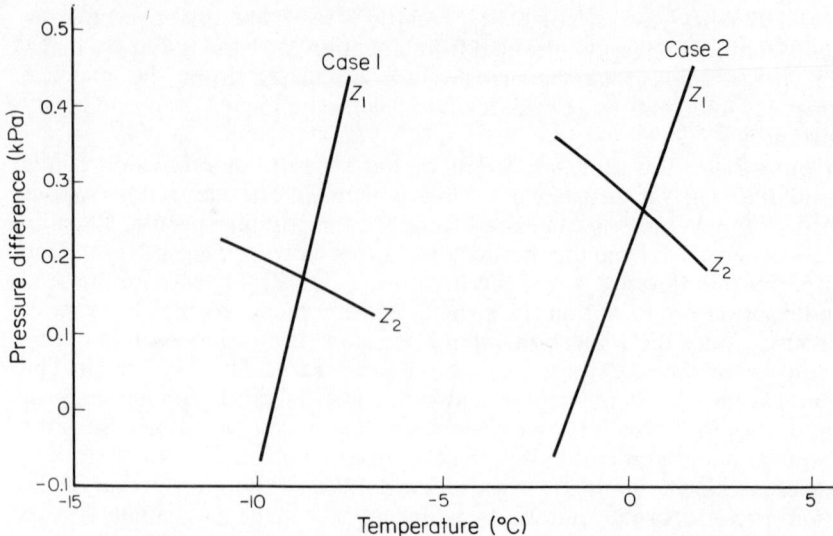

Figure 21.7 A graphical solution of the heat and moisture transfer equation for the roof structure shown in Figure 21.5 using the method of Davies (1973). Z_1 and Z_2 are defined by equations 21.10 and 21.11 respectively. Z_1 is a linear function of T_{cn} given I_{so}, I_{si}, T_{eo}, T_{ei}, f_i and f_o, while Z_2 relates E_{cn} to the saturation vapour pressure at the plane of condensation (at temperature T_{cn}) given e_o, e_i, f_o and f_i. The temperature and vapour pressure at the plane of condensation (T_{cn}, e_{cn}) are: case 1 (−8.8 °C, 177 Pa), freezing of the condensate occurs; case 2 (0.3 °C, 260 Pa), no freezing. The rates of condensation are 1.04 and 1.40 mg $m^{-2} s^{-1}$ respectively

or 5.0 g m⁻² h⁻¹...

or $5.0 \text{ g m}^{-2}\text{h}^{-1}$ and the latent heat yield is greater in this case at $\lambda E_{cn} = 3.5 \text{ W m}^{-2}$, which is larger in proportion to the same fabric heat loss of 13.8 W m^{-2}. Condensation raises the temperature at point B from −1.6 to $0.3 °C$ and the time taken to deposit a 0.1 mm layer of water is 20 hours.

In both cases the time taken to deposit a 0.1 mm layer of water or ice is large. These calculations refer to steady-state conditions alone and it can be seen that there would be a continuous deposition of interstitial condensation only if the internal and external environments remained constant. Even in winter the daily cycle of solar insolation will produce corresponding cycles of thawing and freezing, evaporation and condensation, within the structure. During the thawing period the condensate will drain through the insulation and produce visible evidence, by pattern staining on the ceiling (similar to that caused by condensation on the interior surfaces). Such condensation could lead to wet rot in the roof timbers. Freezing of liquid condensation may also lead to frost damage within the insulant and roof timbers; it will certainly facilitate larger structural heat losses. Whitehorne (1975) has described how interstitial condensation within the lightly compacted snow walls of their igloos causes the Inuit Indians to move house. When the porous snow is turned to ice its insulation is lost. These examples show that a vapour check is highly desirable. A 0.1 mm thick polyethylene sheet ($r^*_v = 300 \text{ GN s kg}^{-1}$ or greater), well sealed and placed on the 'warm' side of the structure, provides a useful preventative against the risk of interstitial condensation.

For contrast, *Figure 21.8* shows the s.v.p. profiles for the same structure in summer. In this example there is a reversal of the direction of heat flux and heat flows into the building. The conditions assumed are cloudless skies with $S_t = 400$ W m^{-2} and $T_{ao} = 30\,°$C. Assuming $\alpha = 0.68$, $\varepsilon = 1.0$, $S_e = L_e = 0$ and $T_{ao} = 30\,°$C we have $L_d = 403$ W m^{-2}, from which $R_{ni} = +197$ W m^{-2}. If $I_{Ho} = 0.23$ m^2 K W^{-1}, from Monteith's relation for flat plates in free convection, and if $I_{Ro} = 0.159$ m^2 K W^{-1} then $I_{HRo} = 0.094$ m^2 K W^{-1}. T_{eo} is then 48.5 °C. Inside the building, assuming a modest temperature lift

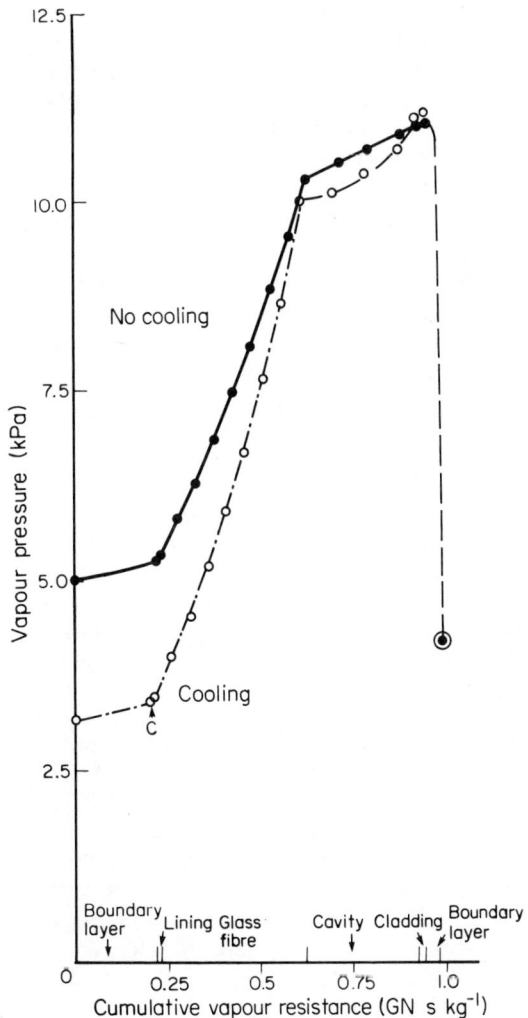

Figure 21.8 *The saturation vapour pressure profile across the roof structure of Figure 21.5, again without a polyethylene vapour check, but under summer conditions. No interstitial condensation occurs when the building is not cooled (●).* $S_t = 400$ W m^{-2}, $T_{ao} = 30\,°$C, $T_{eo} = 48.5\,°$C, $T_{ai} = T_{ei} = 33\,°$C. *If the air entering the building is cooled (○),* $T_{ai} = T_{ei} = 25\,°$C, *there is a risk of interstitial condensation at C, the interface of the asbestos inner lining board and the glass fibre*

of 3 K at maximum ventilation rate, $T_{ei} = T_{ai} = 33\,°C$ and $\mathbf{R}_{Hi} = 1.30\,m^2\,K\,W^{-1}$ which, if $I_{Ri} = 0.159\,m^2\,K\,W^{-1}$, gives $I_{HRi} = 0.140\,m^2\,K^{-1}\,W^{-1}$. The temperature gradient across the structure $(T_{ei} - T_{eo}) = -15.5\,K$. The boundary-layer vapour resistances are calculated from the Lewis relation and are $r^*_{vi} = 0.22\,GN\,s\,kg^{-1}$ and $r^*_{vo} = 0.038\,GN\,s\,kg^{-1}$.

It is apparent from *Figure 21.8* (solid symbols) that interstitial condensation will not occur under any normal combinations of e_i and e_o. If, however, the incoming air is cooled to (say) 25 °C with evaporative coolers, there *is* a danger of interstitial condensation. The open symbols in *Figure 21.8* show the saturated vapour pressure profile within the structure when the incoming air is cooled. Interstitial condensation would occur at point C if both e_i and e_o were close to saturation. Point C corresponds to the interface between the inner lining sheet and the glass fibre mat. Therefore in hot, humid climates a vapour check would be essential, but it should be placed close to the *exterior* of the structure, in contrast to its position in temperate climates. Where hot, humid days can be followed by cold, dry nights two vapour checks should be installed, to prevent condensation on the inner lining during the day and the outer cladding at night. In practice, the insulant could be encased in a polyethylene sandwich. There would be a small risk of saturation of the insulation if the polyethylene were punctured, because the condensate would remain trapped within the sandwich.

Measurements of and design values for thermal transmittances

The greatest proportion of the thermal resistance of light agricultural structures resides in the insulant, which can often provide 85 per cent of the total in a well-insulated building. In the 1950s a tradition was established in which design transmittances of a roof or wall were specified according to the building's orientation. Whilst not denying the influences of wind, solar radiation and rain on the precise value, Loudon (1968) proposed that design U values should be independent of the orientation and should instead be referred to the degree of exposure, classified as sheltered, normal and severe, corresponding to mean external wind speeds of 1.0, 3.0 and 9.0 m s⁻¹ respectively. The IHVE Guide Book A (1971) proposed further standardization with respect to 'surface resistance, moisture content and related thermal resistances of structural components'. Indeed, current building practice employs standard values not only for the resistance of both the interior and exterior surfaces (as noted earlier) but also for ventilated and unventilated air spaces. These are listed in the IHVE Guides and are summarized in the *Building Research Establishment Digest 108* (Building Research Station, 1975). The adoption of this procedure therefore assumes steady-state conditions.

MEASUREMENT TECHNIQUES

Most methods for measuring the thermal transmittance of a structure *in situ* are based on a heat balance. Thermal energy, usually via electric heaters, is supplied to the building at a known rate and the structural heat loss can be readily calculated if the ventilation rate is known. Ventilation heat losses are usually greater than those via the structure, so an accurate knowledge of the ventilation rate is essential. Continuous measurements of the ventilation rate over a long period (e.g. of the order of a month) are not normally available, because the tracer gases used in a gas dilution technique are too expensive and anemometer methods require a constant verification of the anemometer calibration. Small test chambers can usually be sealed, but air infiltration to larger buildings is greatly affected by external winds and is the largest source of error in the measurement of ventilation rates.

The most comprehensive study on the agreement between measured and calculated transmittances is that reported by Loudon (1963), based on the measurements of Pratt and Daws (1958). Loudon gave a full description of the effects of solar radiation and wind on the heat lost through the structure. During the winters of 1953–1957 Pratt and Daws measured the transmittances of six roof structures with total resistances ranging between 0.35 and 1.10 $m^2 K W^{-1}$ ($U = 2.86$ and 0.91 $W m^{-2} K^{-1}$ respectively). Most of the structures had a double-skin cavity, but with different inner lining boards. Each measured approximately 2.75×0.75 m. The roofs were inclined at $22.5°$ and faced south. During the first two winters, heat was supplied to the rooms via electrical heaters and the power consumption recorded. Heat losses through the walls and floor were measured with flux plates. In the first winter the roof transmittance was determined by difference, but in the final winter it was derived directly from flux plates attached to the roof panels. Ventilation rates, measured with katharometers, were 0.5 air changes per hour at most, although this was sufficient to cause an apparent increase of up to 50 per cent in the transmittance if its effects were neglected.

Loudon's analysis showed that measured transmittances were on average 13 per cent higher than calculated values after due allowance for all structural and climatic factors. The main errors were not instrumental but were due to factors such as heat bridging, edge heat losses, moisture within the insulant and lining boards and ventilation of the cavity air space and chamber. He concluded, 'It is feasible to base a scheme of computing heat losses or gains through building structures on calculations from the appropriate measured thermal properties of building materials.'

A similar but less detailed approach was adopted by Whiteside (1974), who investigated the benefits of injecting urea formaldehyde foam into the wall cavities of an occupied domestic house. Weekly measurements of gas and electricity consumption were combined with records of the mean house temperature and external climate to give estimates of the air-to-air transmittance of the walls. Prior to injection the transmittance was 1.6 ± 0.24 $W m^{-2} K^{-1}$, which was reduced to $0.46 \pm 0.17 W m^{-2} K^{-1}$ after injection of the foam. The major sources of error in this determination were in the calculation of the solar and long-wave radiation, the ventilation rate and the heat losses through the floor. Although the absolute values of the wall transmittance differed from those measured in small test panels, the estimate of the difference in the transmittance before and after injection was more accurate ($\pm 10\%$). Nevertheless these results indicate that a technique based on this method is feasible and reasonably accurate.

Measurements of thermal transmittance on a larger scale were reported by Borgnes (1979), who examined the cargo holds of refrigerated ships, each with a volume of 1.2×10^5 m³. His method, which was essentially a heat balance on the refrigerator, gave a mean ($\pm$ s.d.) value of 0.465 ± 0.055 W m⁻² K⁻¹ for the overall transmittance of the holds of eight ships, compared with a design value of 0.52 W m⁻² K⁻¹. Thermal equilibrium was achieved after 24 hours and readings were taken at hourly intervals for 6–8 hours, in which time the hold temperature varied by no more than 0.1 K h⁻¹. He attributed the 12 per cent range in the measured transmittance to solar and long-wave radiation and uneven workmanship during construction.

An alternative approach is to estimate I_{Hi} and I_{Ri} and determine $(I_{ci} + I_{co})$ by difference. Measurements are made of T_{so} and T_{si} and R_n. The convective transfer is determined from the wind speed and T_{ai}. Assuming that the structure is in thermal equilibrium the transmittance is given by

$$I_{ci} + I_{co} = I_{HRi}(T_{si} - T_{so})/(T_{ei} - T_{si})$$

For the roof structure shown in *Figure 21.5*, if $I_{HRi} = 0.11$ m² K W⁻¹ and $(T_{si} - T_{so}) = 25$ K then if $(T_{ei} - T_{si}) = 1.2$ K we calculate $(I_{ci} + I_{co}) = 2.22$ m² K W⁻¹. The accuracy of the estimate of $(I_{ci} + I_{co})$ is therefore determined by the assessment of I_{HRi}; precise evaluation of $(T_{ei} - T_{si})$ is also difficult, but necessary for the calculations, when the building is large and non-homogeneous.

DESIGN TRANSMITTANCES

Table 21.6 shows standard thermal resistances of the interior and exterior surfaces of roofs, floors and walls and unventilated cavities (IHVE, 1971).

Table 21.6 STANDARD SURFACE RESISTANCES OF WALLS, FLOORS, ROOFS AND CAVITIES WITH HIGH AND LOW LONG-WAVE EMISSIVITY

Structure		Heat flow direction	I (m² K W⁻¹)	
			Surface emissivity	
			High	Low
I_{si}	walls	Horizontal	0.123	0.304
	floors and roofs	Upwards	0.106	0.218
	floors and roofs	Downwards	0.150	0.562
I_{so}	walls sheltered[a]	Horizontal	0.080	0.110
	normal[b]	Horizontal	0.055	0.067
	severe[c]	Horizontal	0.030	0.030
	roofs sheltered[a]	Either	0.070	0.090
	normal[b]	Either	0.045	0.053
	severe[c]	Either	0.020	0.020
I	cavity 5 mm	Horizontal or upwards	0.110	0.180
		Downwards	0.110	0.110
	20 mm	Horizontal or upwards	0.180	0.350
		Downwards	0.210	1.060

Wind speeds over the surface are: a—sheltered, 1.0 m s⁻¹; b—normal, 3.0 m s⁻¹; c—severe, 9.0 m s⁻¹
From IHVE (1971)

Low-emissivity surfaces are mainly untreated or unpainted metals, such as aluminium or galvanized steel; all other surfaces are deemed to be of high emissivity, $\varepsilon \geqslant 0.8$. Most cavities in agricultural buildings are unventilated, although the definition of 'unventilated' allows for a low ventilation rate, sufficient to remove small amounts of cavity condensation. The inclination of the cavity and the cladding corrugations have little effect on the cavity resistance, which remains approximately constant if the thickness is 20 mm or greater. However, lining a cavity with aluminium foil increases its resistance, owing to the reduction of long-wave exchange. Further explanation of the values shown in *Table 21.6* can be found in Loudon (1968) and IHVE (1971).

The largest heat fluxes through solid floors in contact with the ground occur within 1 m of their perimeter (Wiseman, 1979). The heat flows horizontally close to the building's walls, and therefore the greatest benefit of floor insulation arises when it is installed around the boundary. It can either be laid as a horizontal strip 1 m wide together with a vertical strip throughout the full thickness of the floor or as a vertical strip alone to a depth of not less than 250 mm below floor level (Building Research Station, 1972). In each case the insulant should be protected from moisture. *Table 21.7* shows the standard resistances of solid floors (e.g. of concrete), with or without a hardcore bed and with four exposed edges (IHVE, 1971). The resistance depends upon the perimeter length. It is conventional to specify

Table 21.7   STANDARD RESISTANCES OF SOLID FLOORS IN CONTACT WITH THE GROUND, HAVING FOUR EXPOSED EDGES, TOGETHER WITH THE PERCENTAGE INCREASES DUE TO EDGE INSULATION INSTALLED TO DEPTHS OF 0.25, 0.5 AND 1.0 m

Dimensions (m)	Resistance ($m^2\ K\ W^{-1}$)	Percentage increase for edge insulation to depth:		
		0.25 m	0.5 m	1.0 m
Very long × 30	6.25	3	7	11
× 15	3.57	3	8	13
× 7.5	2.08	4	9	15
150× 60	9.09			
× 30	5.55			
60× 60	6.66	4	11	17
× 30	4.76			
× 15	3.12			
30× 30	3.84	4	12	18
× 15	2.77			
× 7.5	1.82			
15× 15	2.22	5	12	20
× 7.5	1.61			
7.5× 7.5	1.32	6	15	25
3× 3	0.68	10	20	35

From IHVE (1971)

the floor resistance in terms of an air-to-air temperature difference since the vertical heat flux through the ground is normally constant; heat losses through the floor are mainly determined by the external temperature close to the perimeter. The table also shows the effect of a vertical strip of insulation, installed to a depth of 0.25, 0.5 or 1.0 m below ground level, on the floor resistance. The thermal resistance of a suspended floor can be found in the IHVE Guide Book A (1971).

Economic levels of insulation

EFFECT OF INSULATION ON TEMPERATURE LIFT

The amount and type of insulation of a building is usually determined during the design stage. Structural heat losses are therefore fixed at the beginning of

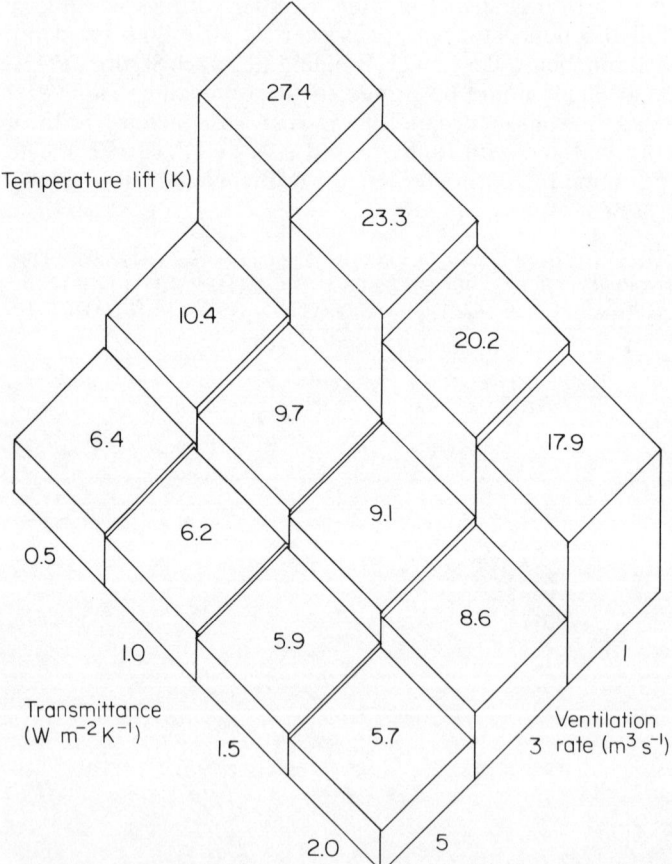

Figure 21.9 The effect of building ventilation rate and roof and wall transmittance on the temperature lift of a fattening piggery of length 46 m, width 7.5 m, eaves height 2 m and ridge height 2.5 m. The calculation was based on a house containing 450 pigs each of weight 64 kg and producing 100 W of sensible heat with $T_{ei} = 18\,°C$. Structural heat losses were calculated using the method of Carpenter and Randall (1975). The lowest ventilation rate corresponds to the current recommended minimum (Armstrong, 1974)

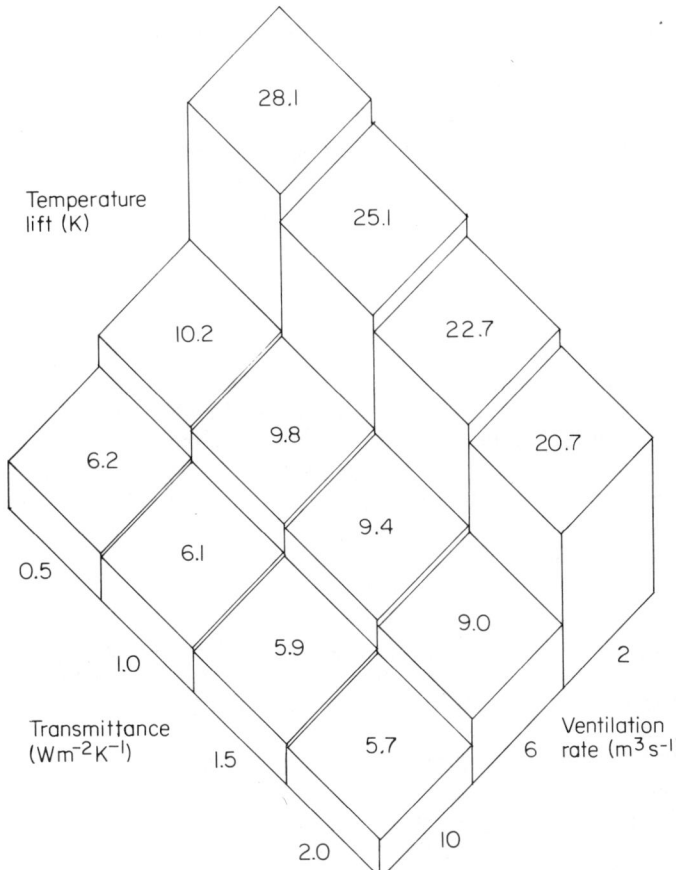

Figure 21.10 The effect of building ventilation rate and roof and wall transmittance on the temperature lift of a laying hen house, of length 33.3 m, width 15 m, eaves height 2 m and ridge height 3 m. The calculation was based on a house containing 10 000 hens each of weight 2 kg and producing 8.5 W of sensible heat with $T_{ei} = 21$ °C. Structural heat losses were calculated using the method of Carpenter and Randall (1975). The lowest ventilation rate corresponds to the current recommended minimum (Charles, 1979)

the building's life, apart from minor changes in the structural temperature gradient that result from alterations to the ventilation system. *Figures 21.9* and *21.10* represent the effect of ventilation rates and of the thermal transmittance of the roof and walls on the temperature lift above ambient, ($T_{ai} - T_{ao}$), for fattening pigs and laying hens respectively (after Carpenter and Randall, 1975). A ground temperature of 10 °C has been assumed and the transmittance of the floor has been calculated from *Table 21.7*. In each example the lowest ventilation rate is the minimum that is currently recommended (Armstrong, 1974; Charles, 1979).

The effect of insulation on the temperature lift is small at the higher ventilation rates. At five times the recommended minimum rate, reducing the transmittance from 2 to 0.5 W m^{-2}K^{-1} results in an increase in the lift attainable in both fattening pig and laying houses of about 0.5 K only.

Increasing the stocking rate would also have a slight effect only on the lift, because the ventilation rate would be increased *pro rata*. Insulation therefore has a significant effect on the building's temperature lift only when the ventilation rate is low. With this observation in mind, the optimum economic levels of building insulation can now be considered.

METHODOLOGY

The choice of the optimum economic thickness of a given insulant depends on knowledge of the responses of both the animals and the buildings to the external climate. At an earlier Easter School, Smith (1974) separated these responses into five factors: the thermoneutral zone of the animals under a particular feeding regime; the internal climate, preferably expressed as a single environmental temperature; the relationship between the animal's metabolic rate and this environmental temperature; the monetary values associated with a unit of feed, shelter and production; and 'probability distributions of the frequency and magnitude of the departures of the natural thermal environment, and the house environment, below nominated levels for a typical sample of seasons'. He then combined these responses and, using a Games Theory approach in which a matrix of outcomes is calculated from the different biological responses and possible weather alternatives, showed that the maximum economic level of expenditure on the building structure is determined by the consequences of not modifying the external climate—provided that shelter is the sole reason for housing the animals.

In essence this Games Theory approach to decisions on insulating and heating animal houses has been adopted by several authors (e.g. Stenning, 1974; Carpenter and Randall, 1975; Riley and Redfern, 1977; Sallvik, Nilsson and Nimmermark, 1978). Although a wealth of meteorological data are available, most studies have considered the effects of air temperature alone, probably because of the difficulties in estimating the influences of wind speed and solar radiation on building heat balances. Aceituno (1979) has shown that monthly heating and cooling degree-days can be readily calculated from the mean daily temperature, using statistical criteria, if the daily means are assumed to be normally distributed. If studies are confined to air temperature, this method is more convenient than direct calculation of degree-days, without loss of accuracy.

In the design of building insulation, knowledge of the First Law of Thermodynamics is used to minimize the structural heat flow. Bejan (1979) has proposed a novel approach based on the Second Law, which is more suitable for those building designs in which the ratio of the absolute surface temperatures, T_{si}/T_{so}, is large and the thermal conductivity of the structure increases with temperature. He showed that the rate of entropy generation (in essence the heat losses) was minimized when the temperature gradient within the structure was modified such that the structural heat flow increased linearly with decreasing structural temperature. The modification could include local cooling or even heating of the fabric. This novel design method may have applications for livestock buildings in extreme climates.

CASE STUDIES

The nutritional responses of animals to climate are well known (e.g. *see* Haresign, Swan and Lewis, 1977), and the economic consequences of departure from the optimum climate can be calculated from these known responses. Similarly, the heating requirement of a building can be derived from the building's ventilation rate, design temperature, external wind speed and other climatic variables. Changes in the biological perform-ance, such as increases in feed intake or mortality, can be accommodated by the financial penalty of not maintaining the optimum temperature (Carpenter and Randall, 1975). This penalty or production deficit is expressed per unit of temperature below the current economic optimum (in the UK as $£\,K^{-1}\,yr^{-1}$); this concept underlies all the case studies considered below.

On the basis of fuel usage alone, Riley and Redfern (1977) determined the optimum amount of insulation for 60 broiler houses in Arkansas. They estimated fuel usage from degree-day data using a simple equation describing the building heat balance. A survey of the actual fuel con-sumption showed no difference from the predicted usage. After allowing for the costs of insulation and fuel and other building charges, they concluded that the optimum transmittance of extruded polystyrene was about $1.0\,W\,m^{-2}\,K^{-1}$ for propane fuel costing £0.45 per litre and insulation at £1.05 per square metre (at 1977 prices). For their calculations they employed a standard economic method in which the marginal rate of substitution of fuel for insulation is equated to the price ratio of fuel to insulation. The optimum thickness depends on this price ratio and hence future optima can be readily predicted, if fuel costs alone are the primary consideration.

Sallvik, Nilsson and Nimmermark (1978) describe a similar analysis for fattening pigs, sows and broilers. However, for fattening pigs they found that the predicted fuel usage was between 40 and 75 per cent of the actual consumption; this large discrepancy was probably due to unknown variations in the ventilation rate. Assuming a 20 year lifespan for the building, a 12% annual interest rate and a 6 per cent annual increase in the cost of energy, they calculated that the transmittances of glass fibre which gave the highest margin over fuel costs were 0.36, 0.35 and 0.27 $W\,m^{-2}\,K^{-1}$ for sows, fattening pigs and broilers respectively. Their values are one-third of those of Riley and Redfern and reflect the greater severity of the Scandinavian climate and the higher cost of animal food.

Carpenter and Randall (1975) employed a higher degree of temperature resolution than is customary by choosing a 4 hour period as the base for their degree-day calculations. They derived the temperature during 4 hour periods from records of the maximum and minimum daily temperatures and related the data to the cost of the insulant and the production deficits. In their example of a fattening piggery they considered only the rise in feed intake with falling temperature but, as was discussed earlier, other factors that depend on temperature in the building such as carcass composition, mortality or heating costs could be incorporated into the production deficit. Two examples are given below of the calculation of the economic optimum transmittance (for fattening pigs and laying hens), using Carpenter and Randall's method. *Figures 21.11* and *21.12* show the effect of production

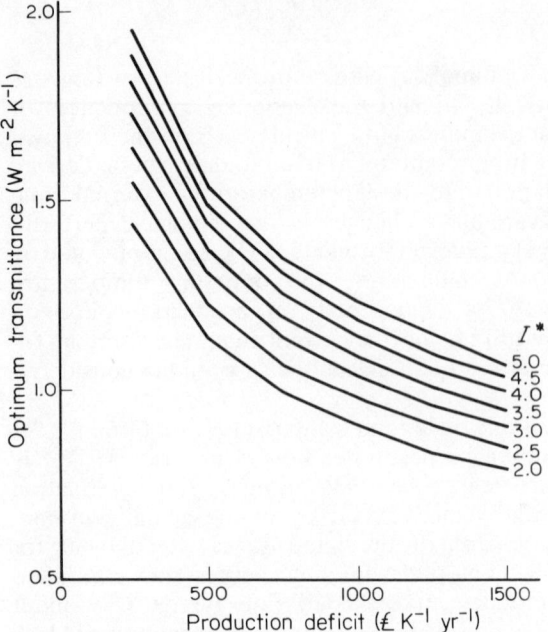

Figure 21.11 The effect of production deficit and the insulation cost index, I* (£ W m⁻⁴ K⁻¹), on the optimum transmittance for the roof and walls of the same 450 pig fattening house as considered in Figure 21.9. (After Carpenter and Randall, 1975. See text for further explanation)

deficit ($£ \, K^{-1} \, yr^{-1}$) and insulation cost index, I^* ($£ \, W \, m^{-4} \, K^{-1}$), on the optimum transmittance of the roof and walls of the 450-pig fattening house and 10000-bird laying house respectively, described on p. 403. The insulation cost index is equal to the product of the insulant's thermal conductivity and cost per unit area divided by the insulation thickness. It is a true comparative measure for different materials since it represents the cost per unit resistance. In these examples the degree-day data are from the records at Rothamsted, England, although Carpenter and Randall also considered two other sites. Other assumptions include a building life of 15 years, an interest rate of 10% on capital employed, full occupation of the building and minimum ventilation rates of $0.16 \, m^3 \, h^{-1} \, kg^{-1}$ and $0.4 \, m^3 \, h^{-1} \, kg^{-1}$ for the pigs and hens, respectively. As the production deficit rises the optimum transmittance falls. For example, when $I^* = 3.0 \, £ \, W \, m^{-4} \, K^{-1}$ the optimum transmittance for the laying hens decreases from $1.5 \, W \, m^{-2} \, K^{-1}$ at a deficit of $500 \, £ \, K^{-1} \, yr^{-1}$ to about $1.0 \, W \, m^{-2} \, K^{-1}$ at $1500 \, £ \, K^{-1} \, yr^{-1}$. The optimum transmittance for pigs follows the same trend as that for laying hens but is about $0.15 \, W \, m^{-2} \, K^{-1}$ less.

Table 21.8 shows the range of insulation materials suitable for agricultural use. The notional thickness refers to that commercially available thickness of insulation necessary to give a roof transmittance of $0.5 \, W \, m^{-2} \, K^{-1}$. The cost, at March 1980 prices, is for insulating a simple roof structure, comprising the outer cladding and timber support, and includes an allowance for breakages, fixing, contractors' overheads and profits. Where necessary, a 4.5 mm asbestos inner lining board and a 0.15 mm polyethylene vapour

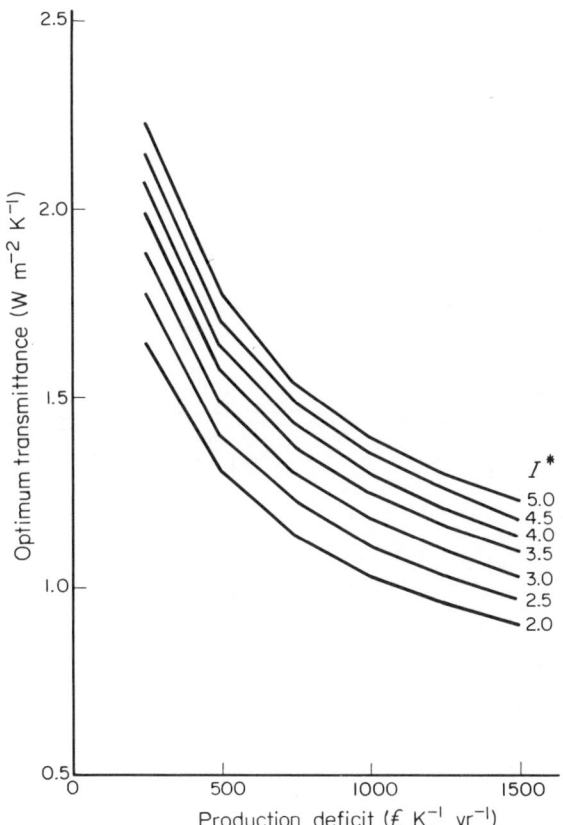

Figure 21.12 The effect of production deficit and the insulation cost index, I* (£ W m⁻⁴ K⁻¹), on the optimum transmittance of the roof and walls of the same 10 000 bird laying house as considered in Figure 21.10. (After Carpenter and Randall, 1975. See text for further explanation)

check has been included, at a cost of £4.10 per square metre and £0.30 per square metre respectively. At the time of writing, the material with the lowest insulation cost index is expanded polystyrene sheet, whilst that with the highest is blown mineral wool fibre. The last column in the table shows a cost performance index based on I^*. This performance index is referenced to a base of 100 when $I^* = 2.10$ £ W m⁻⁴ K⁻¹ for expanded polystyrene sheets. It is unlikely that the relative costs will alter unless there are substantial changes in the manufacturing processes of the different materials. Further information on comparative costs of building materials is given by Wight and Clark (1981), in an annual publication of the Scottish Farm Buildings Investigation Unit, Aberdeen.

Figures 21.13 and *21.14* show the relationship between the thickness of insulant and the total annual cost, which is equal to the depreciated cost of the insulation added to the product of the degree-day data and the production deficit, 1450 and 750 £ K⁻¹ yr⁻¹ for the fattening pigs and laying hens

Table 21.8 INSULATION MATERIALS SUITABLE FOR LIVESTOCK BUILDINGS

	k_H ($W\,m^{-1}\,K^{-1}$)	Notional thickness (mm)	Vapour check needed	Cost ($£\,m^{-2}$)	I^* ($£\,W\,m^{-4}\,K^{-1}$)	Cost performance index
Mineral wool fibres	0.045	75	Yes	7.90*	4.74	226
Mineral wool mat	0.042	70	Yes	7.30*	4.38	209
Glass fibre mat	0.040	75	Yes	5.85*	3.12	149
Expanded polystyrene sheet	0.037	60	Foil	3.40	2.10	100
Expanded polystyrene beads	0.035	75	Yes	8.10	3.78	180
Urea formaldehyde foam	0.033	75	Yes	6.15*	2.71	129
Cellulose fibres	0.032	75	Yes	5.60*	3.58	170
Extruded polystyrene sheet	0.029	50	No	5.50	3.19	152
Polyurethane sheet	0.023	35	Foil	5.00	3.29	157
Polyurethane foam	0.023	42	No	5.00	2.74	130

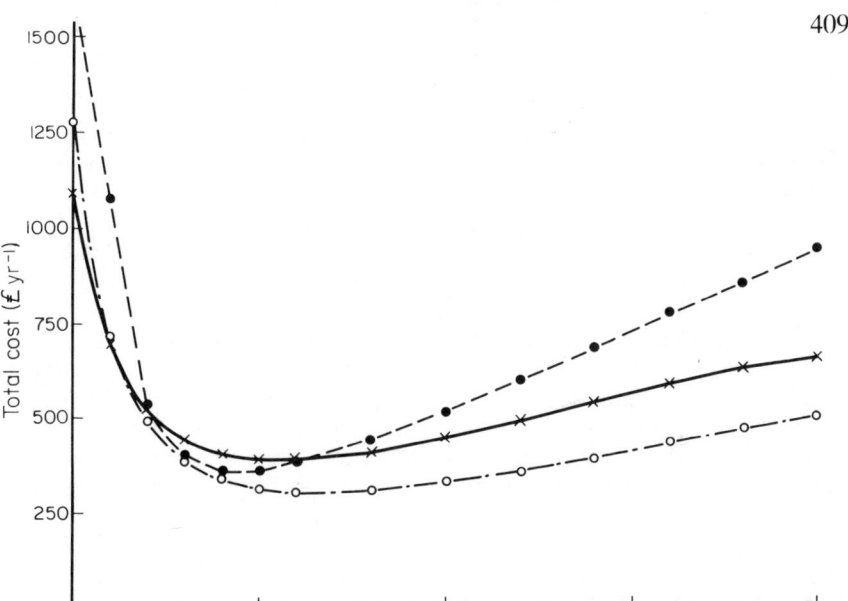

Figure 21.13 The effect of the thickness of insulation material on the total annual cost for the 450 pig fattening house considered in Figure 21.9. The total cost is equal to the sum of the depreciated cost of the insulation and the product of the degree-day data and the production deficit, = 1450 £ K^{-1} yr^{-1}. The optimum thicknesses of expanded polystyrene sheet (O), polyurethane foam (●) and glass fibre mat (X) are 32.5, 25 and 30 mm respectively. (After Carpenter and Randall, 1975)

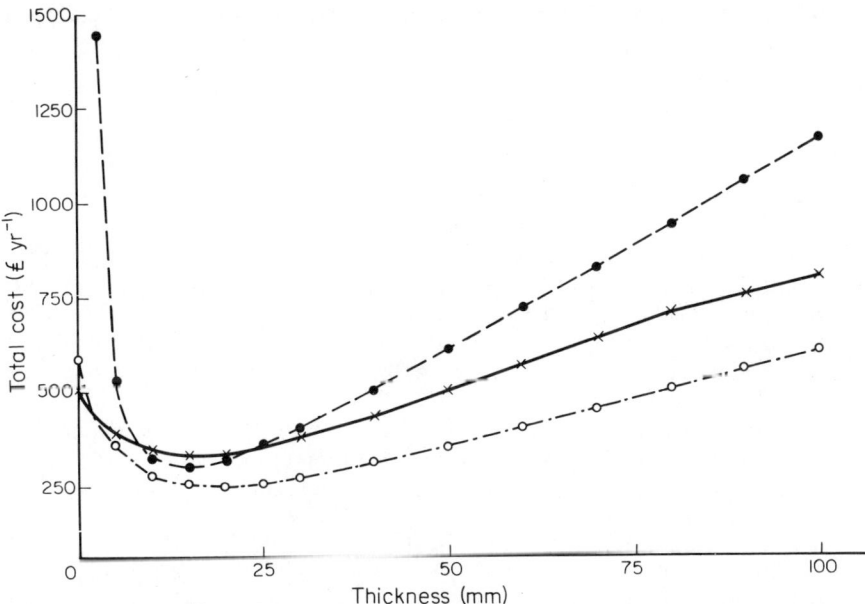

Figure 21.14 The effect of the thickness of insulation material on the total annual cost for the 10 000 bird laying house considered in Figure 21.10. The total cost is equal to the sum of the depreciated cost of the insulation and the product of the degree-day data and the production deficit, = 750 £ K^{-1} yr^{-1}. The optimum thicknesses of expanded polystyrene sheet (O), poly-urethane foam (●) and glass fibre mat (X) are 20, 15 and 17.5 mm respectively. (After Carpenter and Randall, 1975)

respectively. In deriving these relationships, allowances have been made for the thermal resistances of the interior and exterior surfaces, for a 6 mm corrugated asbestos cement cladding, for a 100 mm cavity and a 4.5 mm inner lining board where necessary. Three common insulating materials are shown in these figures: expanded polystyrene sheets, polyurethane foam and glass fibre mats. The optimum thicknesses for fattening pigs are higher than for hens, equal to 32.5 mm, 25 mm and 30 mm for the former and 20 mm, 15 mm and 17.5 mm for the latter. The change in slope for the glass fibre mat at 80 mm is due to the decrease in cavity resistance as the cavity thickness is reduced. The cheapest material, expanded polystyrene, shows a small rise in total cost above the optimum thickness; for example, doubling the thickness from 20 mm to 40 mm raises the annual cost from £256 to £312 for laying hens, but the financial penalties of *not* insulating a building can be severe. The optimum thickness of insulation therefore depends on the material and the production deficit associated with the departure from the optimum temperature. For a fattening piggery the optima given here correspond to roof transmittances of 0.82, 0.80 and 0.88 W m⁻² K⁻¹ for expanded polystyrene sheets, polyurethane foam and glass fibre mats respectively. For a laying hen house the transmittances are 1.13, 1.22 and 1.22 W m⁻² K⁻¹ respectively. These transmittances are higher than those suggested by Carpenter and Randall because of the recent rise in the cost of insulation relative to the production deficit, based in their case on feed. In addition the costs of the insulation assumed in the calculations presented here include full fitting and other building costs, and thus represent a true comparison between different materials. However, the conclusions drawn by Carpenter and Randall in 1975 are still valid today: 'for low-cost materials it would be prudent to increase the calculated optimum thickness of insulation to guard against increase in food price'.

Conclusions

The differences between measured and calculated thermal transmittances are small and there is little error if standard values are adopted for the resistances of air cavities and interior and exterior surfaces. Provided that the resistance of the insulant is related to its moisture content, calculated transmittances can therefore be used in the prediction of the heat balance of livestock buildings.

A vapour check, of minimum resistance 100 GN s kg⁻¹, is essential if interstitial condensation is to be avoided. In temperate climates the check should be sited as close to the interior as possible; in tropical countries the insulant should be sandwiched between two checks.

The current economic level of insulation corresponds to a roof or wall transmittance of about 0.8 W m⁻² K⁻¹ for fattening pigs and 1.2 W m⁻² K⁻¹ for laying hens. However, if in winter low ventilation rates are to be used and house temperatures of 20 °C or greater are to be attained, the structural resistance should be even higher. The economic optima represent an upper limit and a transmittance of 0.5 W m⁻² K⁻¹ is recommended for the roof and walls of livestock buildings, equivalent to 75 mm of glass fibre mat.

References

ACEITUNO, P. (1979). *Agric. Met.* **20**, 227–232
ARMSTRONG, B. (1974). In *Heat Loss from Animals and Man*, pp. 405–424. Ed. by J. L. Monteith and L. E. Mount. Butterworths, London
BALL, E. F. (1968). *J. Instn Heat. Vent. Engrs* **36**, 51–56
BEJAN, A. (1979). *Int. J. Heat Mass Transfer*, **22**, 219–228
BORGNES, O. (1979). *15th Int. Congr. of Refrigeration, Venezia, 23–29 September 1979*
BRITISH STANDARDS INSTITUTION (1959). BS : 3177. British Standards Institution, London
BRITISH STANDARDS INSTITUTION (1968). BS : 4370 : Part 1. British Standards Institution, London
BRITISH STANDARDS INSTITUTION (1971). BS : 476 : Part 7. British Standards Institution, London
BRITISH STANDARDS INSTITUTION (1972). BS : 476 : Part 8. British Standards Institution, London
BRITISH STANDARDS INSTITUTION (1973a). BS : 4370 : Part 2. British Standards Institution, London
BRITISH STANDARDS INSTITUTION (1973b). BS : 874. British Standards Institution, London
BRITISH STANDARDS INSTITUTION (1975). BS 2972. British Standards Institution, London
BRITISH STANDARDS INSTITUTION (1978). BS : 5617. British Standards Institution, London
BRITISH STANDARDS INSTITUTION (1979). BS : 476 : Part 5. British Standards Institution, London
BUILDING RESEARCH STATION (1972). *Building Research Station Digest 145, Watford*. HMSO, London
BUILDING RESEARCH STATION (1975). *Building Research Establishment Digest 108, Watford*. HMSO, London
CARPENTER, G. A. and RANDALL, J. M. (1975). *Agric. Met.* **15**, 245–255
CARSLAW, H. S. and JAEGER, J. C. (1959). *Conduction of Heat in Solids*, 2nd edn. Oxford University Press, London
CENA, K. and CLARK, J. A. (1978). *Physics Med. Biol.* **23**, 565–591
CENA, K. and MONTEITH, J. L. (1975a). *Proc. R. Soc., Ser. B* **188**, 377–393
CENA, K. and MONTEITH, J. L. (1975b). *Proc. R. Soc., Ser. B* **188**, 395–411
CHANG, S. C. and HUTCHEON, N. B. (1956). *Heat. Pip. Air Condit.* **28**, 149–155
CHARLES, D. R. (1979). *Massey University Poultry Convention, Palmerston North, New Zealand, May 1979*
DAVIES, N. G. (1973). *Bldg Sci.* **8**, 97–104
DIAMANT, R. M. E. (1967). *Heat. Vent. Engr* **40**, 423–426
DIAMANT, R. M. E. (1977). *Insulation Deskbook, 1977*. Heating and Ventilating Publications, Croydon
FISHENDEN, M. and DUFTON, A. F. (1929). 'Heat transmission.' *Building Research Spec. Rep. No. 11*. HMSO, London
GATENBY, R. M. (1977). *Agric. Met.* **18**, 387–400
HAGER, N. E. and STEERE, R. C. (1967). *J. appl. Phys.* **38**, 4663–4668
HARESIGN, W., SWAN, H. and LEWIS, D. (Eds) (1977). *Nutrition and the Climatic Environment*. Butterworths, London

HOLDEN, T. S. and GREENLAND, J. J. (1951). *Report R.6.* CSIRO Division of Building Research, Melbourne

HUANG, C. L. D., SIANG, H. H. and BEST, C. B. (1979). *Int. J. Heat Mass Transfer* **22**, 257–266

INSTITUTION OF HEATING AND VENTILATING ENGINEERS (1971). *IHVE Guide Book A.* Institution of Heating and Ventilating Engineers, London

KELLY, M. (1973). *Fm Bldg Prog.* **31**, 23–25

KELLY, M. and ROSS, P. A. (1975). *Fm Bldg Prog.* **40**, 23–27

LAO, B. Y. and SKOCHDOPOLE, R. E. (1976). *4th Int. Cellular Plastics Conference, Society of the Plastics Industry, Montreal, Canada*

LOUDON, A. G. (1963). *J. Instn Heat. Vent. Engrs* **31**, 273–298

LOUDON, A. G. (1968). *J. Instn Heat. Vent. Engrs* **36**, 167–174

McADAMS, W. H. (1954). *Heat Transmission*, 3rd edn. McGraw-Hill, New York

MITCHELL, C. D. (1976). *Calf Housing Handbook.* Scottish Farm Buildings Investigation Unit, Aberdeen

MONTEITH, J. L. (1973). *Principles of Environmental Physics.* Arnold, London

NOVELS, M. D. and CLEGG, J. T. (1977). *Bldg Serv. Engr* **45**, 170–173

PALJAK, I. (1973). *Matér. Constr.* **31**, 53–56

PELANNE, C. M. (1977). *J. therm. Insul.* **1**, 48–80

PRANGNELL, R. D. (1971). *Matér. Constr.* **24**, 1–7

PRATT, A. W. (1958). 'Condensation in sheeted roofs.' *National Building Studies Research Paper No. 23.* Department of Scientific and Industrial Research, HMSO, London

PRATT, A. W. (1969). In *Thermal Conductivity,* vol. 1. Ed. by R. P. Tye. Academic Press, London

PRATT, A. W. and DAWS, L. F. (1958). *J. Instn Heat. Vent. Engrs* **26**, 73–79

RILEY, W. R. and REDFERN, J. M. (1977). *Bull. Ark. agric. Exp. Stn,* No. 817

ROWLEY, F. B. (1939). *Trans. Am. Soc. Heat. Vent. Engrs* **45**, 545–560

SALLVIK, K., NILSSON, C. and NIMMERMARK, S. (1978). *American Society of Agricultural Engineers Paper No. 78–4508.* ASAE, Chicago

SKOCHDOPOLE, R. E. (1961). *Chem. Engng Prog.* **57**, 55–59

SMITH, C. V. (1974). In *Heat Loss from Animals and Man,* pp. 345–366. Ed. by J. L. Monteith and L. E. Mount. Butterworths, London

STENNING, B. C. (1974). In *Heat Loss from Animals and Man,* pp. 367–388. Ed. by J. L. Monteith and L. E. Mount. Butterworths, London

SUNDAHL, A.-M. (1975). *Fm Bldg Prog.* **40**, 19–21

WHITEHORNE, A. E. (1975). *Trans. J. Br. Ceram. Soc.* **74**, 1–4

WHITESIDE, D. (1974). *Building Research Establishment Current Paper CP20 74, Building Research Establishment, Watford.* HMSO, London

WIGHT, H. J. and CLARK, J. J. (1981). *Farm Building Cost Guide 1981.* Scottish Farm Buildings Investigation Unit, Aberdeen

WISEMAN, J. (1979). *Fm Bldg Prog.* **57**, 13–16

22

ASPECTS OF HEATING ANIMAL HOUSES

P. I. HAARTSEN
Institute of Agricultural Engineering, Wageningen, The Netherlands

Introduction

In the Western European climate only a limited number of animals need supplementary heating; it is needed when the animals' heat production is insufficient to keep the inside temperature at a level where gain, production and health are unaffected. There is clear evidence that for young animals, such as chicks and piglets, additional heating is indispensable in Western Europe, even during the summer. As to older animals, growing–finishing pigs in the range 20–60 kg may also show a decrease in average daily gain and feed efficiency during cold weather.

Generally speaking, in pig houses heating is necessary in the farrowing pen, in the piglet rearing pen and to a certain extent in the fattening house. In the farrowing pen a space temperature of 17–20 °C is wanted, depending on the floor type, while in the piglet creep the starting temperature should be about 30 °C. The piglet rearing pen needs a temperature of 18–24 °C, depending on the floor type (straw, concrete or metal slats). The temperature in fattening houses depends on various factors and is described in the following sections. The starting temperature for space-heated chickens is generally accepted as 30–32 °C but by 3–4 weeks, the temperature can be reduced to about 20 °C for broiler chicks and to about 15 °C for replacement pullets. For turkey rearing a starting temperature of 36 °C is wanted, reducing to about 18 °C.

In practical farming the heating costs can be considerable, even with adequate ventilation and insulation. *Table 22.1* gives some climatic data for the Netherlands. They show that even during the summer months additional heating may be needed in broiler houses and in farrowing pens. *Table 22.2* gives the mean fuel consumption for poultry and pig heating in the Netherlands, calculated per delivered animal the year around. About 40 per cent of the animal houses concerned are oil-heated, about 30 per cent heated by natural gas and the rest using propane. Owing to its high cost, electricity is used only for infrared lamps in piglet creeps.

Heating systems

Heating systems in animal houses can be divided into two main groups, localized heating and space heating. With localized heating, in contrast to

414

Table 22.1 CLIMATIC DATA FOR DE BILT (ROYAL NETHERLANDS METEOROLOGICAL INSTITUTE)

	Jan.	Feb.	Mar.	Apr.	May	June	July	Aug.	Sept.	Oct.	Nov.	Dec.
Mean daily temperature (°C)	1.7	2.0	5.0	8.5	12.4	15.5	17.0	16.8	14.3	10.0	5.9	3.0
Number of 'summer days' (max. temperature $\geq 25\,°C$)	0	0	0	0	2	6	7	6	2	0	0	0
Number of freezing days (min. temperature $< 0\,°C$)	15	14	12	3	1	0	0	0	0	2	5	12
Number of 'ice days' (max. temperature $< 0\,°C$)	5	4	0	0	0	0	0	0	0	0	0	3

Table 22.2 MEAN FUEL CONSUMPTION IN THE NETHERLANDS PER DELIVERED ANIMAL, FOR POULTRY AND PIGS

	Oil (l)	Natural gas (m³)	Propane (kg)
Piglets	8	10	4.5
Fattening pigs	10	13	–
Broilers and replacement pullets	0.35	0.45	0.25
Turkeys	1	1.3	0.6

space heating, it is possible to create variations of temperature at the level where the animals are living. This is important with broiler and turkey chicks and in pig breeding houses, where with spot heating fewer piglets are lain upon by the sow. Space-heating systems are easier to control but on the whole, fuel consumption is lower with localized heating than with space heating.

In broiler houses the most commonly applied localized heating units are infrared gas heaters; these produce 70 per cent of their heat as radiation and 30 per cent as convection. While the radiation is emitted directly to the floor and to the animal, the convection component brings the whole room temperature to an adequate level. Radiant heating of piglet creeps usually employs infrared lamps. However, though the investment cost is low, temperature control is poor and lamp lifetimes are limited. Therefore, infrared lamps are mostly used in combination with space heating. In strawless piglet creeps, conduction systems such as floor heating are also applied. Floor heating, whether electric or by hot-water pipe, has lower operating costs than infrared lamps and gas heaters.

Space heating in broiler houses usually employs convection systems such as gas- and oil-fired heaters, but air speeds around the animals can be very high, while the heating efficiency is lower than with radiant heating. As far as pig houses in the Netherlands are concerned, hot-water central heating units are preferred. The investment needed for this partly radiant, partly convective heating system is high, but control of the environmental temperature is easier and more accurate than with other heating systems. A combination with hot-water floor heating is attractive. Because of high fuel prices there is a strong incentive to reduce heating costs. In this chapter an attempt is made to evaluate the effect of changes in the house temperature and in feeding level on heating costs, especially with pigs.

Effects of heat loss

In practical farming the animals are often kept in conditions which vary in time and which may be below thermoneutrality. In the zone of thermoneutrality heat loss is constant, depending mainly on the animal's weight, activity and feed intake, and within certain limits it is independent of temperature. However, the ratio of heat loss by the various channels (radiation, convection, evaporation) is not constant. In this zone animals can regulate their heat loss by the various channels using physical means. When temperature drops to a certain level, heat loss cannot be further diminished and the animal will have to increase its heat production in order to maintain homeothermy. As mentioned in other chapters, this temperature is called the *lower critical temperature* (T_{cl}) (Mount, 1974). Below this zone animal heat production is regulated by chemical means. In such conditions the proportion of food energy used for productive purposes is diminished. Poor climatic conditions may also, apart from direct effects on production, have a considerable effect on animal health. For pigs, Tielen (1974) has shown that frequencies of infected lungs and livers, noticed at slaughter, are correlated with climatic conditions during the finishing period. Even in healthy animals climatic variables may affect production through the rate of heat loss from the animal to the environment (Close, 1970; Fuller and Boyne, 1971; Verstegen, 1971).

In order to meet these disadvantages there are several alternatives:

(1) acceptance of a reduction in the rate of weight gain and a loss in production, besides possible interference with the animal's health;
(2) supplying extra feed to compensate for reduced weight gain or loss in production;
(3) technical measures to ensure that the inside temperature stays within the thermoneutral zone.

These measures mainly involve a lower ventilation rate and better house insulation. Only if the desired indoor temperature cannot be maintained in spite of these provisions is supplementary heating needed. There are two ways of controlling the energy balance: manipulating the feed intake and manipulating the thermal environment. To obtain optimal farm results both the heating costs and the extra amount of feed to compensate for the disadvantages of lower temperatures must be calculated.

In the first place, it is important to establish separately the direct effects of climatic conditions on heat loss, and thus on weight gain, and the indirect effects by way of reductions in gain owing to infections. From the literature it can be stated that the critical temperature depends on the feeding level of the pig (Close, 1970); that the critical temperature decreases by about 1 K per 10 kg weight increase (Close, 1970; Verstegen, 1971) and that the floor type (straw bedding, insulated floor, concrete slats) affects the critical temperature (Close, 1970). For example, Verstegen and van der Hel (1974) found that concrete slats increased the critical temperature by about 7 K, as compared with straw bedding. *Table 22.3* gives some data concerning critical temperatures of fattening pigs on various floor types. The feeding level is

Table 22.3 LOWER CRITICAL TEMPERATURES (°C) OF FATTENING PIGS ON VARIOUS FLOOR TYPES

Weight (kg)	Feed (g d⁻¹)	Lower critical temperature (°C)			
		Straw bedding	Insulated floor	Partly slatted floor	Fully slatted floor
20	880	15	17	20	22
60	2160	11	13	16	18
100	2950	8	10	13	15

93 g (1000 kJ metabolizable energy) per $kg^{0.75}$ per pig. For each 5–6 per cent extra feed intake the critical temperature falls by 1 K. If pigs fed *ad libitum* eat about 400 g per day more than restricted-fed pigs, the critical temperature falls by 3–4 K (Verstegen, Brascamp and van der Hel, 1978). These critical temperatures are estimations only, as variation in conditions, such as the presence or absence of draughts and floor wetness, may affect the animals' heat loss significantly.

Trial measurements

Most of the rest of this chapter will be based on the results of one of the few attempts to determine the relative costs of heating and of the extra feed required if heating is not provided.

In the Netherlands a series of experiments with fattening pigs was set up to determine the reduction in rate of gain due to increased heat loss in the cold, the amount of feed required to compensate for reduced rate of gain in the cold, and the effect of fluctuating low temperatures compared with that due to constant cold. The experiments were carried out in two large climatic rooms at the Institute of Animal Husbandry at Zeist, in co-operation with the Agricultural University and the Institute of Agricultural Engineering, both at Wageningen. In order to make a correct estimate of weight gain depression possible, all treatments were performed below the lower critical temperature, for one group slightly so, for the other at somewhat lower temperatures.

Verstegen *et al.* (1979) have reported on the first four experiments, two with animals in the first half and two in the second half of the fattening period. In each room 32 pigs were housed in groups of 8, separated by sex. All animals were fed individually and received a similar amount of feed in each room. The results showed that on average the weight gain was 9 $g\,d^{-1}$ less for every 1 K below T_{cl} for the young pigs (25–60 kg) and 17 $g\,d^{-1}$ less for the older pigs (60–100 kg). For each 1 K the feed conversion ratio decreased by 0.04 in the 25–60 kg range and by 0.13 in the range 60–100 kg.

The additional feed required to compensate for reduced weight gain in conditions below thermoneutrality was investigated in a second series of experiments. Two additional groups of pigs were housed at the same temperature as the ones in the first series. However, they received sufficient extra feed to bring about the same gain as expected in thermoneutral conditions. From the results it has been calculated that 20 g of extra feed per day was required for young pigs and 40 $g\,d^{-1}$ for older pigs for every 1 K below T_{cl}. In practical conditions inside temperatures will not be constant, especially when below thermoneutrality, therefore two experiments have been performed to determine whether depression in body gain is similar in constant and in fluctuating temperatures. The results show a linear reduction in rate of gain in both groups. However, pigs in the range 80–100 kg showed a greater reduction in fluctuating temperatures, calculated for the whole fattening period, though it was not statistically significant.

The trials described above were performed with restricted-fed pigs. In the last two experiments ad-libitum-fed and restricted-fed castrated male pigs have been compared below their critical temperature. The feed intake of the ad-libitum-fed pigs was considerably higher, as expected. Weight gain in the range 20–60 kg was about 15 g d^{-1} higher, and in the range 60–100 kg about 25 g d^{-1} higher per °C than with restricted-fed pigs. The lean to fat ratio was clearly affected by feeding *ad libitum*.

Equations for environmental calculations

The experiments just mentioned show that feed conversion and weight gain are positively affected by a higher feeding level during cold periods. In what conditions is supplying extra feed more economic than supplying extra heat?

To answer this question, it is necessary to calculate the heating demand of an animal house. Van Ouwerkerk (1980) describes a standard method of simplifying the calculations. It is well known that the heating load is in balance when heat losses from the house are compensated by heat demand in the house:

> Required supplementary heating = transmission heat loss + ventilation
>
> heat loss – animal heat production

The transmission and ventilation sensible heat losses of a house may be calculated from the formula

$$q = UA + 0.35\dot{V} \tag{22.1}$$

where q is the heat loss per unit temperature difference between the inside and outside of the house (W K^{-1}), U is the thermal transmittance of the house in W m^{-2} K^{-1}, $\dot{V}$ is the ventilation rate in m^3 h^{-1} and A is the transmission area in m^2.

The outside air temperature at which heat losses are in balance with heat production is called the 'critical temperature of the building' (T_{cb}). When temperature falls below this level, supplementary heating is needed. This critical temperature depends on the house temperature chosen, the ventilation rate, the standard of insulation and the animal heat production:

$$T_{cb} = T_i - \frac{Q_{sb}}{q} \tag{22.2}$$

where T_{cb} is in °C, T_i is the desired inside air temperature, also in °C, and Q_{sb} is the sensible heat production of the animals in the building, in watts.

In order to estimate heating requirements some measure of mean outside temperature is also required. Meteorological data can be converted into accumulated temperature (degree hours or degree days). These are a measure of the heating required when the temperature falls below the acceptable outside level. It is possible to calculate the amount of heat that must be applied from the number of degree hours and the heating load:

$$Q_d = t_d q/\eta \tag{22.3}$$

where Q_d is the demand for extra energy in W h yr^{-1}, t_d is the number of degree hours, calculated from the critical temperature of the building and η is the heating unit efficiency.

It is possible to calculate the fuel costs, Ω, from the formula

$$\Omega = (Q_d/Q_f)\,\omega \qquad\qquad (22.4)$$

where ω is the fuel price per unit of supply and Q_f is the heating value of the same unit in watt-hours. The amount of feed needed to compensate for low-temperature effects can be calculated from the formula

$$F = 1000 \times t_d \times F_c \qquad\qquad (22.5)$$

where F is the total amount of compensation (extra) feed required in kg yr^{-1} per animal and F_c is the amount of compensation feed required in grams per °C drop in temperature below T_{cl}.

EXAMPLE

An economic analysis of extra fuel costs versus feed costs for a pig fattening house is presented as an example. The calculations assume the building and stock parameters given in *Table 22.4*.

Table 22.4   BUILDING AND STOCK PARAMETERS (*see text*)

Building dimensions: 20×12 m
Volume: 800 m^3
Transmission area: 425 m^2
U value: 1 W m^{-2} K^{-1}
Minimum ventilation per pig: 20 m^3 h^{-1}
Stocking: 180 fattening pigs (average stocking density: 90 per cent of 180 pigs)
Average weight: 55 kg (all in–all out system)
Average sensible heat production: 115 W per pig
Target inside temperature (average T_{cl}): 17 °C
Extra feed needed per °C below T_{cl} per pig per day: 30 g

The average animal heat loss into the house, assuming a loss of about 10 per cent of the animal heat by conduction to the floor, is

$$180 \times 0.9 \times 115 \times 0.9 = 16\,767 \text{ W}$$

The structural heat loss per 1 K difference between inside and outside is

$$q = 1 \times 425 + 0.35 \times 3240 = 1559 \text{ W K}^{-1}$$

The critical temperature of the building is therefore

$$(17 - 16\,767/1559), \text{ or } 6.2\,°C$$

In Dutch weather conditions the number of calculated degree hours for a T_{cb}

of 6.2 °C is 12 696. If the efficiency of the (oil-fired) hot-water central heating unit is 0.7, the extra amount of energy needed in the house per year is

$$(12\,696 \times 1559)/0.7 = 28\,275 \text{ W h}$$

The oil needed per year (heating value $= 9800 \text{ W h l}^{-1}$) is therefore 2885 litres, or 16 litres per pig-place.

For comparison, the extra feed needed to compensate for reduced weight gain when there is no extra heating in the house is

$$F = 529 \times 0.030$$

or about 16 kg of feed per pig place. The decision whether to feed or to heat therefore depends almost entirely on whether oil or feed is more expensive. In Dutch conditions, with oil costing Dfl 0.70 per litre and feed costing Dfl 0.55 per kg (1980 prices), it seems to be more profitable to supply extra feed than to supply extra heat.

Conclusions

When the house temperature falls below an acceptable level, artificial heating is required. This critical temperature of the building depends on the house temperature chosen, the ventilation rate, the standard of insulation and the animal heat production. As there are many different housing systems and conditions, calculations of heat demand and of eventual loss in production are needed for each situation before a decision is made whether or not to heat. The calculation systems described may be used for poultry and calf houses as well as those for pigs and are intended to indicate when extra heating would be economically justified.

The rapid rise in fuel prices makes it necessary to revise and refine the basic calculations needed to design a heating strategy. As the ultimate aim is economic production, a comparison must be made between the heating costs and the extra amount of feed to compensate for the disadvantages of lower temperatures; this suggests that feeding is usually cheaper. Despite the substantial increases in both fuel and feed costs since 1973, when Stenning (1974) considered the problem at a previous Easter School, the conclusion is basically the same. Presumably because food and fuel costs are linked via the chain of agricultural production, their relative costs appear relatively stable.

References

CLOSE, W. H. (1970). *PhD Thesis,* Queen's University, Belfast
FULLER, M. F. and BOYNE, A. W. (1971). *Br. J. Nutr.* **25**, 259–272
MOUNT, L. E. (1974). In *Heat Loss from Animals and Man,* pp. 425–440. Ed. by J. L. Monteith and L. E. Mount. Butterworths, London
STENNING, B. C. (1974). In *Heat Loss from Animals and Man,* pp. 367–388. Ed. by J. L. Monteith and L. E. Mount. Butterworths, London

TIELEN, M. J. M. (1974). *Thesis,* Med. Landb. Hogesch., Wageningen

VAN OUWERKERK, E. J. N. (1980). *IMAG-rapport 25*

VERSTEGEN, M. W. A. (1971). *Thesis,* Med. Landb. Hogesch., Wageningen

VERSTEGEN, M. W. A. and VAN DER HEL, W. (1974). *Anim. Prod.* **18,** 1–11

VERSTEGEN, M. W. A., BRASCAMP, E. W. and VAN DER HEL, W. (1978). *Can. J. Anim. Sci.* **58,** 1–13

VERSTEGEN, M. W. A., MATEMAN, G., BRANDSMA, H. A. and HAARTSEN, P. I. (1979). *Livestk Prod. Sci.* **6,** 51–60

23

HOUSING SYSTEMS AND THEIR INFLUENCE ON THE ENVIRONMENT

A.A. JONGEBREUR
Institute of Agricultural Engineering, Wageningen, The Netherlands

Introduction

Recent years have seen considerable changes in husbandry practices in the industrialized nations in most areas of animal production. The Netherlands has been no exception. For example, *Table 23.1* shows statistics indicating the recent increase in the practice of housing dairy cows. Approximately 53 per cent of dairy cows are now kept in loose house systems. The figures in *Table 23.1* also indicate that concurrently an increase in the average number of dairy cows per farm has taken place.

Table 23.1 RECENT CHANGES IN THE NETHERLANDS IN THE NUMBER OF CUBICLE HOUSES AND HOUSES WITH FEEDING CUBICLES FOR DAIRY COWS

Year	Number of houses with:		Number of dairy cows	Number of dairy farms	Average number of cows per farm
	Cubicles	Feeding cubicles			
1970	812	22	1.896×10^6	116 332	16
1975	8 379	675	2.218×10^6	91 560	24
1978	12 727	1 013	2.247×10^6	75 113	30
1980	16 674	1 269	–	–	–

A serious problem on quite a number of dairy farms is the rearing of the young stock. A rather high average mortality rate, of about 12%, indicates that the rearing of young calves must be improved. The prevention of diseases is the important factor (Webster, Chapter 13) and housing may have a considerable influence on the health of the calves and on the incidence of disease. To illustrate this, the preliminary results of an experiment on an open housing system for dairy calves will be discussed later in this chapter.

Table 23.2 presents further data showing the development of intensive animal production in the Netherlands between 1970 and 1978. As with cattle, there has been a considerable increase in the number of animals on each unit, while at the same time the number of farm units has fallen. Both are symptoms of increasing intensification.

The development of intensive livestock production has some disadvantages. In a densely populated country complaints about odours are common and there are manure surpluses in several regions of the Netherlands. This

Table 23.2 STATISTICS OF INTENSIVE LIVESTOCK PRODUCTION IN THE NETHERLANDS, GIVING THE NUMBER OF ANIMALS PER FARM AND THE NUMBER OF FARMS IN 1970 AND IN THE MOST RECENT YEAR FOR WHICH DATA ARE AVAILABLE

	Year	
	1970	*1978*
Number of pigs	5.5×10^6	9.2×10^6
Farms with pigs	76×10^3	50×10^3
Average number of pigs per farm	73	183
Farms with fattening pigs	42×10^3 (1972)	30×10^3
Average number of fattening pigs per farm	79	158
Farms with sows	46×10^3	29×10^3
Average number of sows per farm	16	39
Number of laying hens	18×10^6	22×10^6
Farms with laying hens	49×10^3	7×10^3
Average number of laying hens per farm	365	3×10^3
Number of broiler chickens	30×10^6	38×10^6
Farms with broiler chickens		
Average number of broilers per farm	10×10^3	18×10^3
Number of fattening calves	434×10^3	556×10^3
Farms with fattening calves	5.2×10^3	3.2×10^3
Average number of fattening calves per unit	83	172

intensification, with the development of large-scale operations, also has drawbacks for the welfare of the animals. This chapter will therefore review recent research in the Netherlands, not only on housing and its effects on the animals, such as open-house systems for fattening pigs and piglets, but also work on problems that result from intensive housing, for example odour emission from pig houses and laying hen houses and manure handling (in fattening pig units). Most of the results presented originate from the Institute in which the author works.

Storage of manure

GENERAL ASPECTS

In the Netherlands the use of straw as litter is an exception in all types of house, partly because of limited supply. Therefore, most dairy and pig farms use a liquid manure system. This brings savings both of labour and in the investment costs of the building. For fattening pigs the calculated yearly costs of a Danish-type house with a solid manure system are substantially higher than those of a totally slatted-floor house with liquid manure system. In the case of laying hens, however, the yearly costs per bird are only a little higher with an in-house drying system, in comparison with the 'slurry' system. The use of slatted floors in animal housing systems also implies the storage of slurry underneath the slats and the building. Despite its cost advantages there are drawbacks to the slurry system. The following aspects may be mentioned; odour emission in the ventilation air and during spreading of slurry; because the slurry has a high water content its transport over long distances is not financially attractive (in relation to the value of the fertilizer); and finally, the microclimate experienced by both animals and workers may be contaminated by emissions from the slurry—mixing of slurry in the channels under the house, in particular, demands a careful and intensive ventilation of the house because of the possible emission of large quantities of H_2S, which is highly toxic (Jongebreur, Kroodsma and Poelma, 1979). Poor ventilation may in any case depress the health of the animals and lower productivity. What, therefore, are the alternatives to the handling and storage of slurry for the disposal of manure? This problem will be approached from the different points of view of the various types of stock and housing systems.

CUBICLE HOUSES

As already shown in *Table 23.1*, an average of about 1700 new cubicle houses per year have been built in the Netherlands over the past ten years. It is estimated that about 95 per cent of the modern dairy houses are equipped with slatted floors (for reasons of reliability, to reduce problems with slippery floors and to give a dry walking area) though the investment costs are lower in the case of solid floors and tractor scraping of slurry. *Table 23.3* shows the relative investment costs and yearly costs of the two systems (Swierstra, 1976).

Table 23.3 DIFFERENCES IN INVESTMENT COSTS AND YEARLY COSTS FOR THE BASEMENT CONSTRUCTION OF A FOUR-ROW CUBICLE HOUSE WITH SOLID FLOOR AND SCRAPERS IN COMPARISON WITH SLATTED FLOORS AND CELLARS (TAKEN AS 100)

No. of dairy cows	Differences (%)	
	Investment costs	Yearly costs
80	0	8
120	−15	0
180	−22	−14

One problem is the physical size of the storage required; it is important to have enough storage capacity for at least three months and sometimes for six months. This is less of a problem when a new cubicle house is built, and it is possible to make chambers under the cubicles and even under the feeding passage. The required storage capacity per cow for three months is between 4.5 m³ and 5.4 m³ according to the type of housing. Depending on the area of storage chamber relative to the total area of the house, the storage capacity is adequate for between about 30 and 70 days when 1.2 m deep, and for 50 to 120 days when 1.8 m deep. Obviously, only the latter will make three months' storage possible and six months is difficult to achieve economically in this type of house.

Prefabricated elements (channel elements) are available for constructing the chambers under the cubicles and the feeding passage. The investment costs per cubic metre of storage decrease with increasing capacity. Using prefabricated channels, the cost per cubic metre for a 2 + 2 row cubicle house and a chamber depth of 1.50 m is Dfl 200, and for the same cubicle house with storage capacity also under the cubicles, about Dfl 160 per cubic metre (Swierstra, 1976). Further storage may be provided outside the house.

The investment costs of an above-ground concrete silo are lower: Dfl 40–50 per cubic metre in the size range 250–1000 m³, and a pit in the ground lined with plastics sheet is the cheapest storage (Jongebreur, Kroodsma and Poelma, 1979). A combination of underground storage in the house and above-ground storage outside it also requires a pump and stirring devices for the silo, the cost of which must be included in that of the storage facility. Mixing of the slurry is essential (Jongebreur, Kroodsma and Poelma, 1979). Without mixing it is impossible to empty the dung channels, so mixing must be carried out regularly and at least every 14 days. Special attention must be paid to the ventilation of the house during the homogenization because of the risk of emission of toxic gases. In addition, the design and construction of the slurry channels must be suited to the mixer employed and the construction of the walls of the chambers must be sufficiently strong to allow for changes in pressure due to differences in level of the slurry.

HOUSES FOR FATTENING PIGS

It is well known that bad odours in and around pig houses result from anaerobic processes in the slurry which is stored underneath the slatted floors. The formation of odour components was described by Spoelstra (1978). The basic explanation for the formation of volatile carboxylic acids and phenols is the absence of balance between the acid-forming and the methane-forming bacteria in the liquid manure. The formation of these components in pig houses may be considerably influenced by the climate, because the volume per animal is relatively small; typically about 3 m³, compared with 25–35 m³ per cow in dairy units. The total surface of slatted floors in pig houses influences the rate of odour emission into the ventilation air, because the exchange of air between the slurry passage and the house air depends on the relative areas of the slats and the gaps between them.

Some measured rates of odour emission are presented in *Table 23.4*. In this table the odour emission values are presented in arbitrary *odour units* (OU), determined from the reactions of a testing panel. The concentration of the most important and detectable odour components in the ventilation air is low. Some measurements of these concentrations are presented in *Table 23.5*.

It is certainly advantageous, if the climate in a house for fattening pigs fails to prevent anaerobic decomposition in the slurry, to store the manure outside the building, since measurements carried out in a respiration chamber with only a very little slurry indicated clearly that the concentrations of compounds giving rise to an odour are then very low. With this in mind a new system for handling of the manure in pig houses has been developed in our

Table 23.4 RELATIVE RATES OF ODOUR EMISSION IN THE VENTILATION AIR FROM DIFFERENT TYPES OF PIG HOUSE

Type of pig fattening house	OU h⁻¹ per kg weight	Calculated ratio		
		On the basis of surface of slatted floors	Real number	Number of houses
Danish	79	100	100	5
Half-slatted floor	95	126	120	6
Fully slatted floor	109	174	138	6

OU–odour units (arbitrary)

Table 23.5 AVERAGE CONCENTRATIONS (in mg m⁻³) OF A RANGE OF ODOUR COMPONENTS FOUND IN THE ATMOSPHERE OF PIG FATTENING HOUSES

Components	Summer			Winter		
	Average	Range	Number of measurements	Average	Range	Number of measurements
Carboxylic acid	0.46	0.13–1.37	26	0.65	0.22–1.36	22
Phenols and indoles	0.035	0.003–0.12	27		0.02	22
NH₃	3.8	<0.5–12	27	–	–	–

Institute of Agricultural Engineering. In this, faeces and urine are separated with the help of synthetic netting placed underneath the slats. The nets, which have a mesh of 0.85 × 1.15 mm, are connected with rollers which can be driven by a motor. To remove the solid manure from the net the driving units are switched on and the solid manure is scraped from the net and transported to a conveyor. The urine passes through the net and can be stored underneath the slats, or outside the pig house in a tank. Research is currently being carried out to investigate the practical application of the system. Very rough calculations give the investment costs of this manure handling system as Dfl 50–100 per pig place, depending on the nature of the fattening house and the number of pig places (which decides the length of the stall). Kroodsma (1980) points out that the system has the advantage that litter can be used in pens with half-slatted floors without causing trouble with manure handling. It is recognized that the supply of straw litter can save on food energy and improve the 'living environment' of the pig, but this is usually incompatible with mechanized manure-handling systems.

The separation of the faeces and urine with this device gives good results, as is illustrated by the measurements presented in *Table 23.6*. The solid material is more nutrient-rich than normal slurry, and is therefore more valuable. Thus in a region with slurry surpluses the solid manure can economically be transported to arable farms some distance away, while the urine can either be used for fertilizer in the immediate vicinity or can be purified. Another advantage is that the microclimate in a building with such a separation device underneath the slatted floors is much fresher. Measurements of the concentrations of odour compounds in the atmosphere confirm this. There are reductions in the amounts of volatile fatty acids and phenols,

Table 23.6 WASTE-SEPARATION EFFICIENCY WITH PIG MANURE OF A
SYNTHETIC NET WITH MESHES OF 0.85×1.15 mm

	Percentage
Average dry-matter (DM) content of the solid manure	35
Removed DM as proportion of total DM	83
Reduction of biological oxygen demand in the remaining liquid relative to total slurry	67
Percentage of the total available plant nutrients retained in the solid manure	
N_{tot}	62
CaO	93
MgO	87
Cl	22
K_2O	40
Cu	94

respectively, of 53 per cent and 55 per cent. This system can therefore aid
odour control on farms in difficult locations.

POULTRY HOUSES

Poultry farmers do have a real interest in the in-house manure-drying
system, for two main reasons: first, when using the in-house drying system
farmers are allowed to keep more birds at the same distance from dwelling-
houses and other buildings. In the Netherlands the official inspectors in
charge of environmental protection consider this manure handling method
as an odour control technique (Scheltinga, Jongebreur and Klarenbeek,
1981); and secondly, the sale of liquid manure to arable farms is becoming
more and more difficult, in spite of the provision of 'manure banks'
(Jongebreur, 1977).

The practical application of in-house drying in laying hen houses is restricted in the Netherlands to about 110 farms. The system can be used in the
so-called 'deep-pit' or 'high-rise' houses and also in 'traditional' houses with
dung channels below the cages. With laying hens, one of two systems can be
used for the in-house drying of manure. Both employ fan ventilation to drive
the airflow for manure drying. The first employs comparatively big fans
(capacity 11000 m^3 h^{-1}, power 60 W) fixed in the roof of the manure pit,
which extends under the whole of the house. For good results one fan is
necessary per 650–750 laying hens, at intervals of about 8 m. The electricity
consumption of these fans is relatively low, about 0.1 W per bird, but the
investment costs are high. An alternative system employs fans connected to
plastic tubes fixed under the walking paths in deep-pit or high-rise houses, or
under the cages in those with dung channels. Openings in the tubes provide
optimal air distribution over the manure. The installed power capacity per
laying hen is 0.25–0.30 W in this case (Kroodsma, 1976).

Good results can be achieved with either system of in-house drying, but
extra ventilation is essential under the conditions in the Netherlands, where
most of the laying hen houses are naturally ventilated and poorly insulated.
The evaporation of water from the manure is improved by good climatic

conditions in the house (high temperature and low relative humidity). Obviously, the heat production of the birds influences the in-house temperature and the ventilation rate required. The higher the stocking rate the higher the probability of good results in drying manure, as this enables a high house temperature to be maintained. On average the results for in-house drying of manure are better in traditional hen houses with dung channels than in deep-pit or high-rise houses. In the latter the climatic conditions in winter are mostly rather poor for drying manure. If in-house drying is successful a dry-matter content of 45–50% can be reached. This means a yearly production of waste of 25–30 litres (or 15–20 kg), compared with the amount of liquid slurry produced annually, which is about 80–100 litres. There is therefore quite a substantial reduction in the volume and weight of the waste (Kroodsma, 1981). Drying costs per bird per year are about Dfl 0.30, but these can be set against a saving on the costs of dehydration by other methods of about Dfl 0.80 per hen per year. However, about 50 per cent of the nitrogen and organic matter is lost during the drying and composting process.

Practical tests of a system with a centrifugal fan and small perforated tubes have been made in laying hen houses with flat floors and cages with manure belts (Kroodsma, 1981). Of these systems, the building costs of a manure pit with a concrete floor are about 27 per cent higher than those for a similar pit without the concrete floor. Because of higher building costs the yearly costs per laying hen for storage of dry manure during the whole production period are about Dfl 0.20 higher in comparison with liquid slurry.

ODOUR EMISSION RELATED TO THE MANURE HANDLING SYSTEM

Research on odour emission has been carried out on the three types of laying hen house shown in *Figure 23.1*, namely, a house with liquid slurry under the cages; a house with dry manure under the cages in channels (in-house drying system), and a house with daily removal of manure to a closed manure pit behind the house. The average emission rates were measured and different measuring techniques (three sensory, one chemical) compared. This part of the research has been described by Frijters *et al.* (1970). The results are presented in *Tables 23.7–23.9*. The liquid slurry systems gave the poorest results.

Open housing systems

DAIRY CALVES

Published work on the influence of climatic conditions on young calves shows clearly that it is possible to keep these animals under a wide range of temperatures (Mitchell, 1976). Young calves are little affected by low environmental temperatures, as long as they are in good health and the housing provides for a dry lying area and relatively little air movement (less than 0.2 m s^{-1}). One of the keys to reducing the mortality rate of young stock is the availability of sufficient space. Care of the newborn calves during the first 14 days is also important.

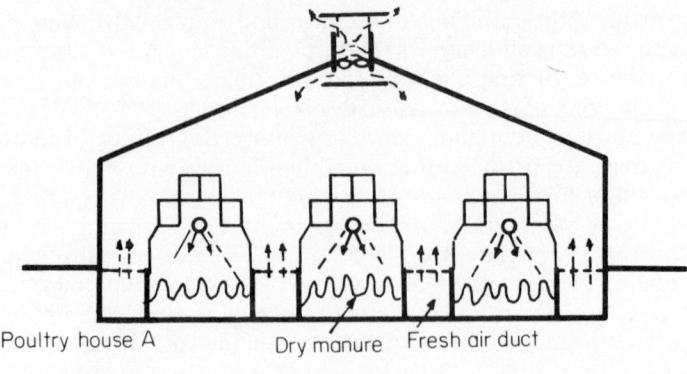

Poultry house A   Dry manure   Fresh air duct

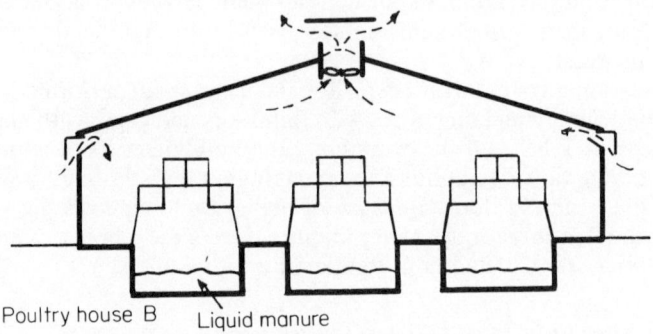

Poultry house B   Liquid manure

Curtain

Poultry house C

Figure 23.1 Diagrams of poultry houses from which the odour emission rates were determined

Our experience suggests that newborn calves are best housed in individual boxes in a separate part of the house. Over the past five years very favourable results have been achieved in an experiment with an open-front house (*Figure 23.2*), the young calves being brought in directly after birth. The results of rearing in this way have been compared with those achieved in the closed-house system on the same experimental farm in previous years. The calves were kept in the open-front house in individual pens for the first 14 days, then in group pens until weaning and after weaning in cubicles. In these five years the temperature and relative humidity followed the outside

Table 23.7 DETAILS OF POULTRY HOUSES FROM WHICH THE ODOUR CONSTITUENTS WERE SAMPLED

	*Poultry house**		
	A	B	C
Manure handling system	Under cage, drying with forced interior air	Liquid, with open interior storage up to three months	Daily removal by scraping into closed storage tank
Ventilation system	Roof-mounted exhaust fans	Roof-mounted exhaust fans	Natural ventilation
Number of birds	3100	3100	17 500
Average live-weight (kg per bird)	2.25	2.25	1.75
Type of cages	Fully stepped (3 rows)	Fully stepped (3 rows)	Vertically tiered batteries (5 rows)
Mean ventilation rate ($m^3 h^{-1}$ per kg live-weight at 20 °C)†	3.28	2.28	0.68
Mean internal temperature (°C)†	21	21	25
Mean internal humidity (% RH)†	44	57	40

*See Figure 23.1
†Averaged over 4 occasions

Courtesy of Information Retrieval Ltd

Table 23.8 EMISSIONS OF ODOUR COMPONENTS IN THE VENTILATION AIR FROM LAYING BIRD HOUSES WITH DIFFERENT MANURE HANDLING SYSTEMS

	Manure handling system					
	Daily removal of manure		In-house drying		Liquid slurry	
	Average	Range	Average	Range	Average	Range
Organic fatty acids C_2–C_5 ($mg m^{-3}$)	35	15–77	103	19–209	177	30–420
Phenols ($\mu g m^{-3}$)	6	3–10	7	3–13	58	27–216
Ammonia ($mg m^{-3}$)	2	1–4	12	1–24	4	3–8
Dust ($mg m^{-3}$)	1.5	0.1–7	0.9	0.1–6	0.8	0.1–3
OU (m^{-3} in ventilation air)	46	11–118	39	11–76	258	94–400

OU—odour units (arbitrary)
Means of 24 measurements

conditions with a minimum temperature of –16 °C and relative humidities of up to 95%. Some of the results obtained with this open-front house are presented in *Table 23.10*. They show a low mortality rate in the 'open'-house system. The weight gains obtained in open-front houses were at a normal level of 500–566 g d^{-1} from birth to weaning.

These results, especially with regard to health, have increased the interest of practical dairy farmers in this system, which has been adopted on six

Table 23.9 AVERAGE ODOUR EMISSION RATES FROM POULTRY HOUSES WITH DIFFERENT MANURE HANDLING SYSTEMS

Manure system	Odour emission ($OU\,h^{-1}$)	
	Per laying hen	Per kg live-weight
Daily removal	90	50
In-house drying	264	120
Liquid slurry A	1300	580
Liquid slurry B	352	210

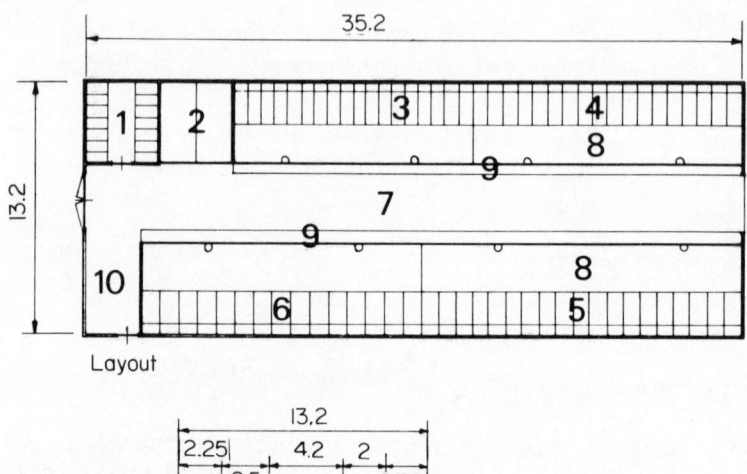

Layout

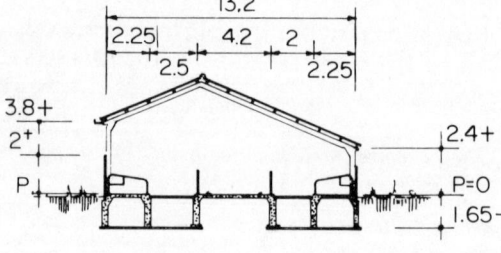

Cross section

Figure 23.2 Layout and cross-section of an open-front house for dairy stock. 1, Individual boxes 0–0.5 months; 2, group boxes 0.5–2 months; 3, cubicles width 0.7 m, 2–5 months; 4, cubicles width 0.8 m, 5–12 months; 5, cubicles width 0.9 m, 12–18 months; 6, cubicles width 1.0 m, 18–22 months; 7, feeding passage; 8, slatted floor; 9, manger; 10, feed storage and preparation

Table 23.10 COMPARISON OF RESULTS OF REARING DAIRY STOCK IN AN OPEN-FRONT HOUSE IN COMPARISON WITH THOSE FROM A CLOSED HOUSE

	Period	
	1971–1975	1975–1979
Housing system	Closed	Open
Number of births	423	429
Perinatal deaths (%)	4.0	3.2
Deaths after 1 day (%)	2.6	0.4
Total mortality rate (%)	6.6	3.6

practical dairy farms to date, and will come into use in a further four farms in the near future. Data on daily gain, health, mortality and climatic circumstances will be collected on these farms for several years. As a control we will also follow the rearing of the young stock on 10 farms with closed houses/ stalls for dairy calves, and collect the corresponding data. In addition, we plan to set up a comparison between open and closed housing systems on experimental farms with dairy calves and with young bulls.

Table 23.11   BUILDING COSTS FOR ALTERNATIVE
HOUSING SYSTEMS FOR YOUNG STOCK

Type of housing	Ratio
In cubicle house	100
By expansion of cubicle house	120–130
Separated young stock house, closed	140–150
Space boarding stall	150–160
Open-front stall	150–160
Outside feeding	110–120

There is a difference in building costs for the two types of system, as can be seen from *Table 23.11* (D. Swierstra, private communication). Surprisingly, the open-front house is one of the most expensive types, primarily because of the greater space allowance per animal.

PIGLETS AND FATTENING PIGS

For rearing piglets directly after weaning (5–6 weeks of age) we can distinguish between the various systems mentioned in *Table 23.12*. Systems 2, 4 and 5 are the so-called 'cold' systems. Trial results show clearly that there is

Table 23.12   HOUSING SYSTEMS EMPLOYED IN REARING PIGLETS

System	Space (m² per piglet)	Required temperature (°C)	Lying area
(1) Flat-deck cage	0.18–0.20	24–22	Totally slatted
(2) Cage with covered lying area and solid floor	0.3	–	Solid floor
(3) Normal pen with half slatted floor	0.3	22 20	Solid floor
(4) Open-house system with straw	0.5	–	Straw
(5) Housing system with covered lying area and solid floor with straw	0.35	–	Straw/sawdust

hardly any difference in daily gain, feed conversion ratio and mortality rate between the open-house system with straw and the flat-deck cage system. The results from the cage with a covered lying area and solid floor are slightly lower. Some of these results are given in *Table 23.13* (Buré and Koomans, 1980). They are surprising considering the well-established sensitivity of young pigs to low temperatures.

434

Table 23.13 RESULTS OF REARING PIGLETS IN OPEN-HOUSE SYSTEMS WITH STRAW IN COMPARISON WITH FLAT-DECK CAGES. DATA FROM TWO EXPERIMENTAL FARMS

	Farm A		Farm B	
	Open house with straw	Flat-deck cage	Open house with straw	Flat-deck cage
Number of animals	1479	1468	1824	1827
Starting weight (kg)	9.3	9.4	9.0	9.0
Final weight (kg)	23.9	24.1	21.3	21.0
Average daily gain (g d^{-1})	493	496	439	432
Average feed conversion ratio	1.79	1.75	1.49	1.47

From Koomans and Buré (1980)

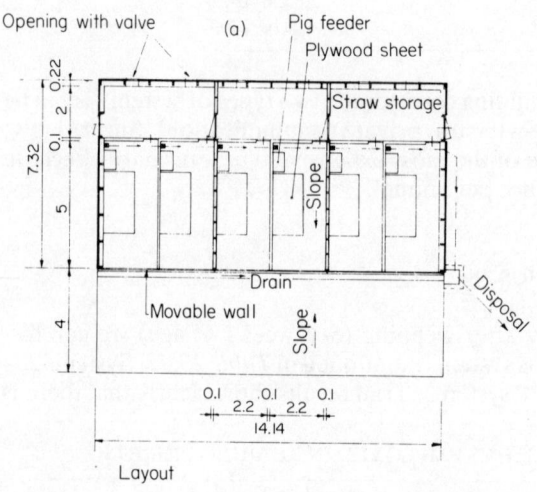

Layout

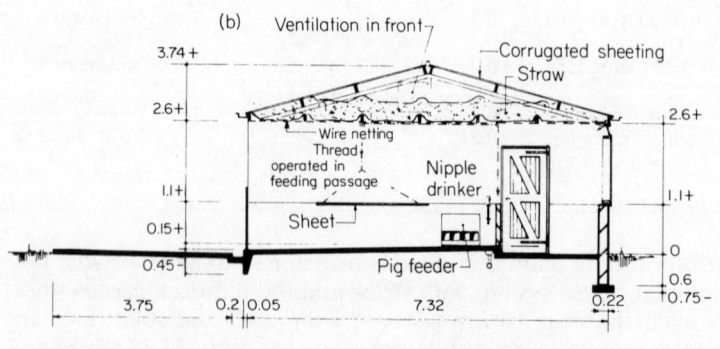

Cross-section

Figure 23.3 Layout and cross-section of an open-house system with straw for piglets

Figure 23.4 View of open house for piglets

All the piglets were placed in a closed housing system during the fattening period. It was observed that the piglets on the flat-deck cage showed more abnormal behaviour (suckling each other, tail biting, rooting, nibbling, etc.) than those in the open house with litter. This behaviour suggests that something is wrong with the living environment of the piglet in the flat-deck system. Open housing systems for fattening pigs were also investigated, and results of production in open houses with straw litter have been published by Koomans (1977). The house is shown in *Figure 23.3* and *23.4*.

In the open-house system 10–12 kg of straw was provided per piglet over the period. To summarize, in comparison with a closed housing system with a fully slatted floor, this system gave a favourable daily gain (approx. +8%), the same feed conversion ratio and mortality rate, but lower carcass quality (more fat) and lower profits (approx. Dfl 16 less per finished pig). The differences in daily gain and carcass quality were significant (respectively at $P < 0.05$ and $P < 0.01$). The investigations were continued by comparing open housing systems with a half-slatted floor and a covered lying area equipped with solid floor. The results indicated the same tendency, namely, the open-house system with straw gave a better daily gain (of +7.5%; significant at $P < 0.01$), a lower carcass quality (thickness of fat 8.7% higher; $P < 0.01$) and profits per finished pig lower by Dfl 8.50. The lower profits per finished pig can be explained by the cost of straw and the higher costs of handling solid manure.

Conclusions and summary

Development and research on housing systems in the Netherlands during the last ten years gives rise to the following conclusions.

First, the manure handling system may have considerable influence on the living environment of the animal. For example, odour levels determine the need for ventilation, which is therefore dependent on the handling system for manure. Storage of liquid manure outside the building or the use of alternative systems with dry manure can improve the environment and the climate in the house as well as odour emission nuisance. We also have to take into account the possibility of using straw or sawdust bedding; these contribute to the saving of energy by the animal, and to its wellbeing.

Secondly, open-house systems may provide a healthy and productive environment, especially for rearing dairy calves and piglets after weaning.

References

BURÉ, R.G. and KOOMANS, P. (1980). *De Boerderij/Varkenshouderij* **64**, 32–34

FRIJTERS, J.E.R., BEUMER, S.C.C., KLARENBEEK, J.V. and JONGEBREUR, A.A. (1979). *Chem. Senses Flavor* **4**, 327–340

JONGEBREUR, A.A. (1977). *Utilization of Manure by Land Spreading*, pp. 329–335. ECSC, EAEC, Luxembourg, EUR 5672 e

JONGEBREUR, A.A., KROODSMA, W. and POELMA, H.R. (1979). *Engineering Problems with Effluents from Livestock*, pp. 30–43. ECSC–EEC–EAEC, Brussels–Luxembourg, EUR 6248 EN

KOOMANS, P. (1977). *Landbouwmechanisatie* **28**, 1001–1003

KROODSMA, W. (1976). *Het Drogen van Kippemest met Behulp van Stallucht in Verschillende Staltypen*. IMAG, Wageningen. Publikatie 73

KROODSMA, W. (1980). *Landbouwmechanisatie* **31**, 137–142

KROODSMA, W. (1981). *Proc. 4th Int. Symposium on Livestock Wastes, Amarillo, Texas*, pp. 419–421. ASAE, St Joseph, Michigan

MITCHELL, C.D. (1976). *Calf Housing Handbook*. Scottish Farm Buildings Investigation Unit, Aberdeen

SCHELTINGA, H.J.M., JONGEBREUR, A.A. and KLARENBEEK, J.V. (1981). *Proc. 4th Int. Symposium on Livestock Wastes, Amarillo, Texas*, pp. 419–421. ASAE, St Joseph, Michigan

SPOELSTRA, S.F. (1978). *Dissertation*, University of Wageningen, the Netherlands

SWIERSTRA, D. (1976). *Landbouwmechanisatie* **27**, 45–50

VIII

THE CONSTRAINTS OF WELFARE AND DISEASE

24

HEALTH PROBLEMS IN INTENSIVE ANIMAL PRODUCTION

D.W.B. SAINSBURY
Department of Clinical Veterinary Medicine, University of Cambridge

Health criteria

It is difficult to exaggerate the importance of the influence of their environment and surroundings on the health of livestock. These influences are often
the most important factor affecting the economic viability of a unit. Such
statements require explanation and the ensuing section attempts a justification for them.

THE CHANGING DISEASE COMPLEX

The considerable strides made in the control of animal infections have led to
the effective elimination of many of the traditional causes of acute disease.
This has been achieved by a combination of appropriate vaccine usage, good
drug therapy and the development of disease-free strains of livestock. It has
made it possible to keep animals in much larger groups than hitherto and
also more densely housed, but the results have by no means been a dis-
appearance of infectious diseases altogether. On the contrary, there have
emerged complex diseases, difficult to diagnose and induced by a multiplicity of pathogenic agents. Whilst these may cause clinically overt disease
it is rather more likely that the effects will be to reduce the overall productivity of the livestock by, for example, slowing growth and reducing the food
conversion efficiency. Animals may not die or even show any obvious
symptoms so that the farmer may be unaware of what is happening unless he
keeps very careful records and uses them with more than the usual degree of
skill. It is also a very common phenomenon that intensive production on a
livestock unit starts efficiently but deteriorates with time so gradually that
the deterioration is not noticed until the consequences have become very
serious and control extremely difficult. The nature of these infections is of
especial interest and concern because the environmental and housing conditions have a profound effect on their incidence.

AN EXAMINATION OF THE DISEASE GROUPS

An important factor influencing the disease incidence in intensive livestock
production is the immaturity of livestock. Improved performance has
resulted in animals' reaching market weight much earlier whilst, for genetic

reasons, breeding animals are also younger on average than previously. There are thus more animals on livestock units than hitherto that are in a state of susceptibility to infectious agents whilst their progress towards immunological competence will normally take place over a prolonged period. This difficult state of affairs is further exacerbated by the large size of many livestock units. In many cases the young or growing stock have come in from widely separated areas and may have originated from parents of very different backgrounds. They therefore may have no resistance to local infections and will be susceptible to them, whilst at the same time contributing a new burden of pathogens to the unit they have entered. Altogether, the modern livestock unit may present at any one time a confusing immunological state, while the basic design and its management will influence the success or otherwise of disease control.

It is appropriate to record the major groups of infections that contribute to these problems. Possibly the most widespread are the respiratory diseases. Many respiratory diseases are subclinical, have a chronic debilitating effect, and are caused by a large number of different infective agents even in any one disease incident. They may not respond satisfactorily to vaccines, hyperimmune antisera, drugs or antibiotics, so that the only way to approach their control is by environmental means.

Another significant group of diseases are the enteric diseases. These, as in the case of respiratory diseases, have many different primary causative agents, ranging from parasites to viruses and bacteria. The reasons for their ascendency in recent years include those already listed but also the trend, with certain livestock, to eliminate the use of bedding, such as straw, which can be expensive to handle and store. Though the harmful effects of this may often be corrected by the use of good pen design, especially by the use of slatted or slotted floors, it is nevertheless rather more difficult to separate animals from their urine and faeces when the flooring is without litter, as the latter has a diluting and absorbent effect on the excreta. Several important bacterial infections have tended to increase in large intensive units. Examples of these are *Salmonella* and *Clostridia* bacterial spores, together with *Escherichia coli* and *Pasteurella*. Many forms of these organisms are normal inhabitants of the animals' intestines in small numbers but excessive 'challenges' causing disease may build up under unhygienic intensive conditions. Induction may be helped by poorly constructed surfaces of buildings which cannot be cleaned.

An excellent review of the factors influencing the dispersal, survival and deposition of airborne pathogens of farm animals has recently been published by Donaldson (1978). This chapter serves to emphasize the great dangers from disease inherent in large livestock units. One conclusion that may be drawn from the evidence of Pirie (1977) and Morzaria *et al.* (1979) is that to combat respiratory disease in cattle at the present time we must rely mostly on wisdom in our application of managemental, husbandry and housing techniques rather than vaccination or medication.

LIVESTOCK UNIT SIZE

In addition to the risks of the gradual build-up of disease-causing agents within a livestock unit, the dangers of livestock groups' becoming too large must be borne in mind. In all parts of the world where the development of large units has taken place, there have emerged viral diseases which tend to 'sweep' through areas with a high livestock population, leaving a trail of devastation that may be likened to a forest fire. There is likely to be considerable loss, often over a concentrated period, after which the animal population either develops a natural immunity or artificial immunities are promoted by the use of vaccines. It is now known that contagious particles can travel great distances from infected sites—certainly distances of 30 km have been proven (Smith, 1963), but they may well travel much further than this.

If livestock enterprises continue to grow in size then the dangers can only become greater. At present it is impossible to present soundly based objective advice as to the optimal unit size, as the factors involved are highly complex. However, there is some evidence that animals thrive less efficiently in large numbers, even in the absence of obvious clinical disease. For example, a survey of broiler growth (Sainsbury and Sainsbury, 1979) showed a variation in finishing weight ranging from 2.1 kg in groups of 20 to 1.4 kg in groups of 30 000, and an almost *pro rata* relationship with groups of 50, 100, 500 and 10 000. These differences occurred with birds with the same genetic background, eating similar food and without any obvious clinical disease. Emmans (1969) earlier reported a similar trend in broilers, though with less precise data on group size. He found that broilers grown in a location size of between 600 m² and 740 m² gave the best performance. For each doubling of location size beyond this, average bird weights decreased by 0.09 kg. The decline was almost certainly due partly to increased disease incidence.

With such accurate figures available, at least to the poultry industry, good use could be made of them in planning new units. Coles (1969) has given similar figures for the guidance of the commercial egg laying industry, to show how above a certain size of unit productivity tends to fall. However, it may be possible and economic to keep very much greater numbers of adult animals or birds together than young stock, since contagious-disease problems are less after the difficult growing stage and its immunological uncertainties are passed. As a good example of this, the results of a survey of large-scale egg production units by Richardson (1971) show that it was the relatively larger flocks that gave the highest productivity, which is the complete opposite to the position in young growing stock.

The maximum number of livestock that may advantageously be kept on a site will depend on a number of factors, apart from health considerations. Figures proposed by the author (Sainsbury and Sainsbury, 1979) which attempt to allow for all factors are as follows:

Dairy units	200
Beef cattle	1 000
Breeding pigs	500
Fattening pigs	3 000
Sheep	1 000
Commercial egg layers	70 000
Breeding poultry	3 000
Broiler chickens	200 000

Such figures are likely to require adjustment in the light of new developments in husbandry, housing and disease control, and especially with the anticipated trend to the use of specific-pathogen-free livestock.

DESIGN ESSENTIALS TO MINIMIZE THE DISEASE CHALLENGE

In addition to the total size of the livestock unit, there are a number of other factors which may help to provide the bases of good health. These will now be discussed.

Depopulation

One of the most important measures in disease control is to ensure periodic depopulation of a building or a site. The benefits of eliminating the animal hosts of disease-causing agents are well understood, and the virtue of being able to clean, disinfect and fumigate a building is also accepted. Nevertheless in practice the whole concept of the 'all-in, all-out' policy is more complex than the preceding sentences would indicate. It is for the young animal that it is extremely important to consider it, and far less so for the older animal, which has probably achieved an immunity to many contagious diseases that may challenge it. Much also depends on whether the herd or flock is a 'closed' one with few or no incoming animals or an 'open' one with a constant renewal of the animal population. If the latter is the case, then periodic depopulation is of much greater importance, as there is little or no opportunity for natural immunities to develop and the regular removal of the 'build-up' of infection is of great assistance in ensuring the good health of the livestock.

The health status

The policy on depopulation will also depend on the health status of the stock. At one extreme there are the so-called 'minimal-disease' or 'specific-pathogen-free' herds that have been developed to be free of most of the common disease-causing agents of that species. Here, the critical feature is keeping the animals isolated from outside infections. Because in most localities this danger is very serious, it is important to subdivide the animals in a unit into smaller groups, lessening the likelihood of a breakdown, or enabling isolation and elimination of a group that does become infected. At the other extreme are units that have a constant intake of new animals from outside and of unknown disease status. In this case there is a constant risk—and more usually a near certainty—that some of the newcomers will be either clinically infected with disease, or be carriers. Design specifications for such units should be quite different from those of the closed herd, so that defined areas of the unit, at least, should have groups of animals put through them in batches, after which the area can be cleared, cleaned and sterilized. It is obviously preferable if the whole unit can be so treated, since it ensures an absolute 'break' in the possible disease cycle.

Between these two extremes is the more usual arrangement of a herd or flock that is of reasonable health status, though certainly not free of the common diseases, and to which new livestock are added only occasionally. In such cases the precautions against disease 'build-up' and spread of infection can be more relaxed, but there should still be proper provision for the isolation of incoming and sick animals.

Group size

It is important to keep the animals in groups of minimal size. This may seem to be an outdated policy that would eliminate all those advantages from automation that large units can give us, but this certainly need not be the case. If groups are small, it is possible easily to match the animals in them for size, weight and age, and it is well established that it is under these circumstances that growth is likely to be most even and economical. Behavioural abnormalities, such as fighting and bullying, are also kept to a minimum in small groups and indeed they may be prevented altogether.

Fighting amongst animals seems a highly contagious condition, and under the most intensive management an almost casual accident that draws some blood can escalate into a 'blood-bath'. Pens that keep the animals in small groups will reduce the occurrence of such disasters.

There is yet another advantage in keeping animals in small groups. It is obviously good practice for a farmer to keep his livestock somewhere near the highest density that has been shown to be optimal for productivity. For example, it is known that a broiler chicken will grow to its maximum potential at a stocking rate of about 15 birds per square metre. If birds are kept at a density such as this they should spread across the house evenly, so that they occupy and use the whole area. In practice, however, this is very rarely the case, especially when large numbers are housed together without any subdivisions. The birds (or other livestock) may crowd in certain parts of the building, which become grossly overstocked. This is bad enough in itself, but it has further unfortunate side-effects which may be seen as well with beef animals, dairy cows, pigs and sheep. If livestock crowd in certain parts of a house, this area is likely to become polluted to an abnormal and harmful degree; the humidity becomes high, proper air movement is impeded and the animals may become sick. Sick animals feeling cold tend to huddle together so the vicious circle is perpetuated unless measures are taken to ensure a better distribution of the stock. When this problem arises under practical conditions it may often be impossible to subdivide the animals at once. An immediate response in the right direction—that is, encouraging the animals to spread themselves more uniformly over the house—can often be achieved by heating the building. There are excellent portable gas radiant heaters and oil-fired and electrical blower heaters available which can do all that is wanted if no permanent system is available. A survey of the effect of group size on the productivity of swine is given by Sainsbury (1978).

When animals are in large numbers, the effects of a fright caused by an unusual disturbance can be extremely serious. It is almost impossible to guard against all the extraneous sounds and sights that may affect the stock. The best safeguard, therefore, is once again to have all the animals in small groups, so that the effect of a panic movement will be limited and can never build up into dangerous proportions.

Floors

The profound effect of the floor surface on the wellbeing of intensively managed livestock cannot be exaggerated. In the times when bedding was almost invariably used and it was a cheap commodity to handle, there were apparently relatively few problems. Now that the farmer must turn to housing systems without bedding he finds himself facing a new set of problems. Often, slatted or other forms of perforated flooring are used in an attempt to produce economically a comfortable and clean environment for the animals. The reader is referred to the Proceedings of a Symposium arranged by the Cement and Concrete Association (1978) for an up-to-date and comprehensive study of all aspects of floor design which is essential reading for the designer and farmer. It is clear from this publication that we are not yet in a position to choose ideal litter-free flooring for many types of livestock.

Building design 'sophistication'

It is possible we are too naïve and uncritical in our attitude to so-called improvement in the design of livestock buildings. The author recently investigated the productivity of broiler chickens within an organization that grows a total of 15 million birds per annum. Their first buildings were erected in 1958 and the latest in 1978. The results were examined from four crops of broilers in the period January to October 1979, comparing 480 000 birds in the oldest site with an exactly similar number in houses with much 'improved' construction, thermal insulation and ventilation—incorporating the accumulated experience of 20 years both within the organization and outside. The results were as given in *Table 24.1*.

Table 24.1

	No. of birds placed	Finishing age (d)	Finishing weight (kg)	Food conversion efficiency	Mortality (%)
Site erected in 1958	480 000	49	1.89	2.1	3.7
Site erected in 1978	480 000	49	1.88	2.1	3.8

So far as is possible under commercial conditions, each site had similar chicks, management and food. The results were precisely the same in biological terms in the old and new site. The main advantages in the new buildings have been in economy in fuel and labour costs but not in productivity or health. These results illustrate the necessity for careful investigation of new systems under both experimental and practical conditions before they are advocated and adopted.

A housing classification in relation to health criteria

In a consideration of the relationship between health, environment and housing it is helpful to provide a classification that differentiates between the principal methods of housing, since they each require different schemes for environmental control. There are three essentially contrasting types of housing: 'climatic', giving only a cover and protection from the elements; 'controlled-environment', which regulates the microclimate as completely as is required for the particular stock being housed; and the 'kennel', which is in a sense a half-way house between the other two and gives two environments in the one building, allowing some choice for the animal.

The methods of use of each of these types and their suitability for different countries, climatic regions and forms of livestock vary enormously and must be carefully defined. In the author's experience some of the greatest errors are made in livestock housing by their incorrect application. It is therefore vital to specify the essential needs of the stock and the differences between housing. A summary of these is now given.

CLIMATIC HOUSING

The climatic house is most suited to the adult beast, which has developed a large measure of adaptability to climatic stress. The house can be cheap, as it is basically a cover only, but because of the lack of control of the climate the space given to the animals must be much more generous than in other forms of housing, especially since it is exceptional to have powered ventilation. In general, stocking densities tend to be half or less than half of those in the controlled-environment house. A major problem is created by agriculturists when they attempt to apply the highest stocking rates, suitable for the controlled-environment house, to the climatic house, as the building is unable to cope with the demands of the stock and poor productivity and serious disease problems can result. It is usually the correct choice for cattle over six months, for sheep of all ages, for adult pigs that are bedded and very occasionally for poultry. Climatic housing often requires deep bedding for its success in cooler climates. This may be especially important because it is usual in climatic housing always to have a surplus of air flow and little ventilation control. Even though this can be devised without draughts on the animals, it does involve widely fluctuating conditions.

THE CONTROLLED-ENVIRONMENT HOUSE

This type of building is the complete opposite of the climatic house. It may be used for all livestock but is especially appropriate for the young animal, the fattening pig, or chicken, the animal of almost any age housed without bedding, and livestock which require an environment with photoperiod or light intensity control. It is also most economically viable with animals that are fed largely concentrate food rather than substantial quantities of roughage, since the former is too expensive to be utilized as a form of energy. The housing is relatively expensive per unit area, and particularly so in areas of climatic stress and where cooling devices are required. Because of the cost it is usually necessary to stock the buildings as densely as is practically possible to make them viable economically, and this can put the animals under a great health risk. The rewards, at their best, can be great but the dangers are also enormous and both the planning of the building and management of the stock need to be of a higher standard than with the climatic house.

KENNEL ACCOMMODATION

Kennel accommodation is an increasingly popular system of housing and is a compromise between the climatic and controlled-environment systems. It attempts to combine the virtues of both at low cost and often succeeds. The essence of the system is that it keeps the animals in pens or groups which are sufficiently small to allow them to be closely confined, at least during their resting periods, without too great a danger from respiratory or other disease. The close confinement makes it possible to keep the groups warm and draught-free, generally by utilizing their own body heat and by good insulation of the kennel. This part of the accommodation approximates to the controlled-environment house. The rest is the climatic house, which is the area where the animals will be freely moving about and not normally lying. It is likely to contain the dung and so must be freely ventilated, usually not by artificial means. In cold and exacting climates this area will be covered but in

Table 24.2   AMBIENT TEMPERATURE RANGES AND HOUSING SYSTEMS SUITABLE FOR HOUSED LIVESTOCK

Type of animal	Ambient temperature range	Housing system
Adult milking cattle	Milk production optimum 10–15 °C but little effect on yield from −7 to +21 °C	Climatic housing usual and generally satisfactory
Beef cattle (from 3 months of age)	−7 to +15 °C the optimum range	Climatic housing appropriate but thermal insulation may be needed if bedding is absent
Calves	10–15 °C at birth, which may fall gradually thereafter. Higher temperatures, 15–21 °C, are used in veal houses	Controlled-environment housing or kennels required
Lambs	4–21 °C	Controlled-environment may be used if housing intensive; otherwise kennels satisfactory
Adult sheep	−7 to +30 °C	Climatic, or kennels
Adult pigs	4–30 °C	Climatic housing suitable where bedding is used, but not with individually housed adults without bedding when minimum temperature should be at least 10 °C higher
Fattening pigs	15–27 °C	Controlled environment or kennel housing required
Young piglets	21–27 °C	Artificial heating needed to supplement controlled-environment
Brooding poultry	30–35 °C	Artificial heating essential in controlled-environment housing
Broiler chickens	15–30 °C	Artificial heating essential in controlled-environment housing
Laying poultry	15–21 °C	Controlled-environment housing required with high-standard insulation or occasionally climatic housing with deep straw or other litter

milder regions there may be no need for this and the 'yard' can be left uncovered with benefit to health, probably to productivity, and without doubt to cost.

It is instructive to consider the arguments in favour of the kennel house. The capital cost can be as low as for any system, especially if the yarding is uncovered. Good health is promoted by separation of the animals into small groups. The separation of the kennels one from another should be as complete as possible, as this will limit the chance of a disease building up and spreading. It will also be of benefit to the health of the animals if the muck is not in the warmer or closely confined resting area, a virtue that is easiest to fulfil with the naturally clean pig but which is certainly achievable, at least in part, by other livestock. It is often possible as well, by good design, to keep the dunging areas separate, so that the muck from different pens does not come together until it has passed out of reach of the animals.

The cheapness of the housing is largely due to the fact that we are controlling the environment only where it is absolutely necessary. Controlled-environment housing is expensive and there is no possible justification for controlling the environment of the effluent from the animals, which not only incurs a considerable cost but is also likely to intensify the disease risk.

Table 24.2 is an attempt to summarize the foregoing paragraphs.

Disinfection

Under intensive systems of animal husbandry there is a greatly increased need for improved methods of disinfection. The disinfection of a building implies the elimination from the house of all micro-organisms that are capable of causing disease. This converts the place from a potentially infective state into one that is free from infection. A disinfectant is an agent that is capable of achieving this and in livestock farming it is usually a chemical agent. However, it should be emphasized that cleaning is an essential preliminary to disinfection. Organic matter has the power to reduce considerably the power of disinfectants, so that without cleaning, disinfectants may not be effective. It is thus an essential, from the hygienic viewpoint, that surfaces of stock buildings are made so they can be readily cleaned prior to disinfection. This requires smooth surfaces, with a minimum of cracks and crevices, especially in those parts in contact with the animals. With suitable construction as a basis, procedures can then be recommended for the disinfection of an animal house either after livestock has been through without a disease outbreak, or a more thorough programme if disease is present. The alternative procedures are as follows.

PROCEDURE FOR DISINFECTION WITH NO DISEASE PRESENT

All equipment and fittings that are removable should be demounted and taken out of the building. It is advisable for them to be soaked in a bath of disinfectant where the materials are able to stand up to this treatment.

Alternatively, they may be power-sprayed or steam-sterilized. Equipment such as poultry brooders will require fumigation after cleaning.

The roof and structural elements of the house should be dusted and cleaned, preferably with a vacuum cleaner. In a poultry house any litter may either be removed—which is generally the most practicable procedure—or stacked into heaps for at least 24 hours (and preferably three days). The temperature within this heap should reach 50 °C and thereafter the heaps should be rearranged so that the external layers are in the centre and it must again be verified that it reaches a temperature of 50 °C. The high temperature will greatly reduce the incidence of parasitic infections.

The lower part of the walls and the floors should be soaked and scrubbed with good detergent disinfectant. In the case of earth floors, these should be soaked in a solution of 0.5 litres of formalin in 50 litres of water, or a proprietary preparation of disinfectant of suitable quality. After the house and equipment have been cleaned, the building surfaces should be soaked with a disinfectant active against viruses, bacterial spores, fungi and insects. This is then followed, after any equipment has been remounted in the house, with formaldehyde gas (Scarlett and Mathewson, 1977).

Disease organisms may also be harboured within the structure of a house and particularly in the cavities associated with thermal insulation. This may be due to vectors such as beetles and mealworms (Eidson *et al.*, 1966; Harein *et al.*, 1970) and special treatment may be required to eliminate them.

DISINFECTION FOLLOWING A DISEASE OUTBREAK

A different procedure is advised after there has been an outbreak of virulent disease in the building. The building in this case should be closed and isolated from all visitors, then the litter and all other areas in contact with the stock should be sprayed with a strong disinfectant (a phenolic type is suitable and usual). The litter is subsequently removed from the building and may be burnt or buried so there is no possible contact with livestock. Portable equipment and fittings should be given the same treatment as previously suggested, preferably while in the house, and later taken out and aerated. The floor should be scrubbed and cleaned with a detergent disinfectant and then finally treated in the same way as in the previous procedure. Where the floor is of earth or other porous material it is a wise procedure to spread polyethylene or tarred paper over it before the new litter is put down, thus isolating the possibly still infected floor from the next batch. It may sometimes be advisable to skim off the top layer of soil around a heavily infected area. The approaches to the building should be treated with disinfectant and foot-dips should be provided.

An important question frequently and understandably posed by farmers is, for what period should a building or site be left empty before it is restocked? It is impossible to give a dogmatic answer to this question. If the disinfection process is less than thorough, then infective organisms, protected with organic matter, can stay viable for years. Leaving it empty may therefore achieve nothing, so that, provided the disinfection programme has been thorough, there is little object in leaving the house out of production for any period at all, other than to dry it out. In the detailed programme listed it is essential to depopulate the entire building since certain of the procedures advised, such as fumigation, cannot be carried out while animals are present. If it is impossible to carry out fumigation owing to the excessive size of the building, serious consideration should be given to subdividing the house into airtight sections. This will almost certainly have advantages in controlling disease spread in general.

Health and the disposal of manure

The method used for manure disposal has potentially important effects on health—both human and animal. Whilst the smell from composted solid manure is a little objectionable, it rarely creates a grave problem and there is little if any risk to the human or animal population from this form of manure under temperate climatic conditions. Slurry, however, is quite a different problem. Slurry placed straight on the land from the animal house or after being held in a tank anaerobically has an extremely offensive smell. While masking agents are possible, they are too expensive at present to be considered economic. The worst smell comes from the pipeline and gun spreader, because the droplet size is small and spray may carry considerable distances. The least smell arises from a tanker spreader because the slurry is much thicker and not spread by aerially dispersed small droplets. The most satisfactory way to prevent the slurry from causing offence is to treat it aerobically in some way before spreading. Human and animal health problems may arise from slurry and, in particular, dangers have been recognized from the *Salmonella* group of organisms and *E. coli*. In a survey it was found that potentially pathogenic bacteria were able to survive for up to three months in slurry kept under anaerobic conditions (Rankin and Taylor, 1969). There is no doubt that other, more resistant organisms, such as *Bacillus anthracis*, *Mycobacterium tuberculosis*, *Clostridial* species and *Leptospira* species could survive much longer. There are thus three possible problems from the distribution of anaerobically stored slurry: (a) smells objectionable to the human population; (b) hazards to human health, and (c) hazards to animal health. If the slurry enters a river or stream, the pollution may have still more serious effects. It is thus an essential of all enterprises with slurry as the disposal system that either the slurry is placed on land where it cannot be a nuisance or health risk, or it is treated to remove the risks.

The dangers from gases in farm buildings

Hazards associated with gases may occur in and around livestock farms, particularly owing to gas effusions from slurry channels under perforated floors. Fatalities due to gas intoxication have been recorded in livestock and even in man. In addition, there is mounting evidence that pollutant gases in livestock buildings may affect production adversely by reducing feed consumption, lowering growth rates and by increasing the animals' susceptibility to invasion by pathogenic micro-organisms.

The most serious incidence of gas intoxication arises from areas of manure storage in slurry pits or channels under the stock, usually but not always associated with forms of perforated floors (Molony, 1965; Brannigan, 1967). The greatest risk arises when the manure is agitated for any reason—usually when it is removed. Several cases of poisoning have been reported when sluice-gates are opened at the end of slurry channels and the movement of the liquid manure has forced gas up at one end of the building. There is also a further danger if a mechanical system of ventilation is used. If this fails and it is the only method of moving air in the house, gases may build up to dangerous concentrations.

High concentrations of gases, chiefly ammonia, may also arise from litter accumulated in animal housing. This is most likely in poultry housing, since the deep litter system is the most commonly used arrangement with broiler chicken and poultry breeders. The danger has undoubtedly been exacerbated within the last few years, owing to the necessity of maintaining relatively high ambient temperatures in order to reduce food costs, while at the same time there has been good evidence that higher temperatures than used hitherto are required for optimal productivity. The poultry farmer has often attempted to achieve such temperatures by restricting ventilation. In the absence of good thermal insulation of the house surfaces the result may often be harmful, if not dangerous.

The most popular form of heating has long been by gas radiant heaters suspended from the ceiling; well over half of all poultry housing in the UK is heated in this way. Because the cost of gas as a fuel has decreased in relation to that of others, the system is also being used more for pig housing and especially for piglets in the early weaning system. If combustion is incomplete, toxic quantities of carbon monoxide can be produced by such heaters. This is a risk with inexpert use, improper servicing and if the house is insufficiently ventilated.

THE GASES PRODUCED

The most important gases generated from stored manure are carbon dioxide, ammonia, hydrogen sulphide and methane; in addition, there are a

Table 24.3   ACCEPTABLE LIMITS
(THRESHOLD LIMIT VALUE, TLV)
FOR GASES IN THE AIR OF
LIVESTOCK BUILDINGS

Gas	TLV (ppm)
Carbon dioxide	5000
Ammonia	50
Hydrogen sulphide	10
Carbon monoxide	50

large number of trace organic compounds. Figures are given in *Table 24.3* for the limits of concentrations that are believed to be acceptable, known as *threshold limit values* (TLV) and given in parts per million (ppm). A threshold limit value is the maximum concentration to which industrial workers may be repeatedly exposed for an 8 hour day during a working lifetime without adverse effect. It may be argued that the TLV for animals should be lower than that for man, since they may be exposed to the gases continuously (Muehling, 1969).

Carbon dioxide

The recommended TLV is 5000 ppm, while concentrations of 2000 ppm can commonly be measured in normally ventilated controlled-environment houses. Normal air contains about 300 ppm of CO_2, and more is released by respiration of animals and by manure decomposition. Most of the gas in bubbles coming from stored liquid manure is also carbon dioxide. The gas in itself is not toxic, but large quantities can contribute to oxygen deficiency and asphyxiation. However, levels as high at 200 000 ppm may do no harm for short periods (Taiganides and White, 1968).

Ammonia

Ammonia is released from fresh manure and during anaerobic decomposition of organic matter. There is less of a problem with slatted flooring than with solid flooring because of the high solubility of ammonia in water. A TLV of 50 ppm has been set for man, to protect against irritation to the eyes and mucous membranes of the respiratory tract, though air containing 50–100 ppm can be inhaled for some hours without any apparent effect. At 100–200 ppm ammonia induces sneezing, salivation and loss of appetite. It is known that in the chicken such levels slow growth rate, induce kerato-conjunctivitis and reduce appetite, but even in the worst ventilated of poultry houses it is unlikely that levels higher than 50 ppm would be found (Valentine, 1964); though modern practice in reducing ventilation to conserve heat makes the possibility distinctly greater.

Hydrogen sulphide

Hydrogen sulphide, with its highly characteristic smell of rotten eggs, is produced from the decomposition of organic wastes under anaerobic conditions. It is a highly toxic gas and the TLV is 10 ppm. Dangerous concentrations can be released into a house if there is any form of agitation of stored slurry—concentrations of over 800 ppm have been found in livestock houses after slurry removal. At this level, unconsciousness and death in the human can result through respiratory paralysis (Day, Hansen and Anderson, 1963). Animals are made uncomfortable by prolonged exposure to low concentrations of hydrogen sulphide. If exposed continuously to about 20 ppm they display loss of appetite and hyperexcitability: concentrations of 50–200 ppm cause vomiting, nausea and diarrhoea. In acute poisoning, hydrogen sulphide acts so rapidly that there are few symptoms of impending danger. Pigs that recover from exposure may be susceptible to pneumonia (McAllister and McQuitty, 1965).

Methane

Methane is generated from the anaerobic decomposition of manure. Since it is lighter than air and insoluble in water, methane tends to accumulate near the ceiling in stagnant corners or in the top corners of tightly enclosed manure pits. It is not strictly a toxic gas and is not in itself harmful in any concentration likely to be found in an animal house, but when the air mixture reaches 50 000 ppm, any small spark can cause a dangerous explosion.

Carbon monoxide

It is unlikely that carbon monoxide will appear in a livestock building unless there is incomplete combustion in gas burners. The TLV is set at 50 ppm. There are a number of circumstances where this can occur, as with brooding chickens or turkeys, when the demands for heat are high but the need for ventilation is minimal. If management is careless, and ventilation provides insufficient oxygen for the burners, substantial quantities of carbon monoxide are produced rather than carbon dioxide. Similar effects may occur if the maintenance of the burners is neglected, so that deposits build up around the air inlets and starve the burners of oxygen.

There is adequate evidence that gases cause intoxication in and around animal buildings but there is little information as yet on the magnitude of the problem. Reports in recent years have tended to highlight the more extreme and spectacular dangers when, for example, ventilation fails, slurry is agitated or the combustion of burners is incomplete. Very much less is known about the effects on livestock of prolonged sublethal concentrations of gases, or even a lifetime's exposure to them. It has been the author's experience that in buildings with poorly designed perforated-floor systems, which allow the introduction of gases into the living area, there may be serious morbidity from respiratory disease and direct or indirect mortalities resulting on occasions.

In contrast to the known harmful effects of gases, there is little evidence that dust has any deleterious consequences on health at the normal concentrations found in piggeries (Curtis *et al.*, 1974).

Conclusions

In many respects, the great advances in nutrition, genetics and health control have not been paralleled by the same progress in animal housing. Whilst the most virulent and contagious animal diseases have been largely controlled, health problems are now centred on more subclinical and chronic infections which have a highly damaging effect on productivity. These infections are predominantly respiratory or enteric in nature and are best avoided by attention to the fundamental principles of animal housing and hygiene. These are concerned with the following factors: some limitation on the size of units; the ability to achieve periodic depopulation and the associated cleaning and disinfection; disposition of animals within buildings to ensure their uniform distribution; maintenance of 'comfortable' conditions that avoid any mechanical stress on the animals, and disposal of waste products, which prevents recycling of pathogenic or toxic products.

It is also essential that the fundamental divisions between basically different concepts of housing are appreciated and classified if the good health of the livestock is to be aided. The groups so described are the 'climatic', 'controlled-environment' and 'kennel' forms. The simplicity of the climatic house is generally ideal for the mature animal, which is reasonably adapted to environmental fluctuation, whereas the controlled-environment building is mostly appropriate for immature and fast-growing stock, with more exacting environmental and nutritional requirements, and where acclimatization may have an economic disadvantage. Between these two extremes are 'kennel' systems, which are the established compromise to satisfy economics and health criteria in appropriate cases. Many disasters occur in practice when these differences are not appreciated. For example, the stocking density in a climatic house must be about half of that in a controlled-environment building if the natural ventilation of the former is to cope at all seasons. In practice, this difference often goes unheeded and the overstocking of the climatic house causes dire disease problems—both respiratory and enteric. It is not that the information to guide the designer or user of buildings is not available, at least if the 'small print' of advisory and research publications is studied. This chapter can do no more than attempt to bring the essentials together and perhaps lead to a greater awareness of the health aspects of intensification.

References

BRANNIGAN, P.G. (1967). *Fm Bldg Prog.* **12**, 14–16

CEMENT AND CONCRETE ASSOCIATION (1978). *Animal Housing: Injuries due to Floor Surfaces.* Cement and Concrete Association, Wexham Springs, Slough, Buckinghamshire

COLES, R. (1969). *Proc. Cobb. Conference, University of Warwick*, pp. 71–75

CURTIS, S.E., JENSEN, A.H., SIMON, J. and DAY, D.L. (1974). *Proc. Int. Livestock Environment Symposium, 1974*, pp. 209–210. American Society of Agricultural Engineers, St Joseph, Michigan

DAY, D.L., HANSEN, E.L. and ANDERSON, S. (1963). *ASAE, Paper 63*, 920

DONALDSON, A.I. (1978). *Vet. Bull., Weybridge* **48**, 83–93

EIDSON, C.S., SCHMITTLE, S.C., GOODE, R.B. and LAL, J.B. (1966). *Am. J. vet. Res.* **27**, 1053–1057

EMMANS, C.G. (1969). *Poult. Rev.* **8**, 13–69

HAREIN, P.K., CASAS, E. DE LAS, POMEROY, B.S. and YORK, M.D. (1970). *J. econ. Ent.* **63**, 80–82

McALLISTER, J.S.V. and McQUITTY, J.B. (1965). *Rec. agric. Res.* **14**, 73

MOLONY, V. (1965). *Vet. Rec.* **77**, 944

MORZARIA, S.P., RICHARDS, M.S., HARKNESS, J.W. and MAUND, B.A. (1979). *Vet. Rec.* **105**, 410–414

MUEHLING, A.J. (1969). *Swine Housing and Waste Management.* Department of Agricultural Engineering, College of Agriculture, University of Illinois, Urbana-Champaign

PIRIE, H.M. (1977). *Vet. Rec.* **101**, 255–258

RANKIN, J.D. and TAYLOR, R.J. (1969). *Vet. Rec.* **85**, 578–581

RICHARDSON, D.I.S. (1971). *Economics of Scale of Egg Production. A Survey of 60 Large-scale Egg Production Units 1969–70.* Department of Agricultural Economics, University of Manchester

454   *Health problems in intensive animal production*

SAINSBURY, D. (1978). *Pig Housing*, 5th edn. Farming Press, Ipswich
SAINSBURY, D. and SAINSBURY, P. (1979). *Livestock Health and Housing*. Baillière Tindall, London
SCARLETT, C.M. and MATHEWSON, G.K. (1977). *Vet. Rec.* **101**, 7–10
SMITH, C.V. (1963). *Agricultural Memo. LX.* Meteorological Office, Bracknell, Berkshire
TAIGANIDES, E.P. and WHITE, R.K. (1968). *Origin, Identification, Concentration and Control of Noxious Gases in Animal Confinement Production Units.* Department of Agriculture, Ohio State University Research Foundation, Columbus, Ohio
VALENTINE, H. (1964). *Br. Poult. Sci.* **5**, 149–159

25

ANIMAL BEHAVIOUR AND WELFARE

I.J.H. DUNCAN
Agricultural Research Council's Poultry Research Centre, Roslin, Midlothian

Introduction

Environmental design in animal housing has, until recently, been concerned mainly with climatic control, labour-saving devices and hygiene; little attention has been paid to the effects of housing on behaviour. Recently, two issues have forced research workers to pay more heed to the behaviour of animals that are being housed in new systems. First, it has become obvious that animals do not always behave 'appropriately' in some of the new environments that are being designed for them. The husbandman would term behaviour 'inappropriate' if it interfered with productivity, although, of course, the behaviour may be perfectly appropriate in terms of the animal trying to adapt to its environment. Examples of behaviour patterns interfering with productivity are tail-biting in pigs, feather-pecking and cannibalism in poultry and licking and ingestion of hair in veal calves. Secondly, the welfare of animals housed intensively has generated an increasing amount of public debate.

Animal welfare

The attention of the British public was first drawn to the welfare of intensively kept animals by the publication of Ruth Harrison's book *Animal Machines* (Harrison, 1964). The public outcry was so intense that the British Government formed a committee under the chairmanship of Professor Rogers Brambell to investigate the welfare of animals kept under intensive husbandry systems. In its report (Command Paper 2836, 1965) the Brambell Committee stated, 'Welfare is a wide term that embraces both the physical and mental wellbeing of the animal. Any attempt to evaluate welfare, therefore, must take into account the scientific evidence concerning the feelings of animals that can be derived from their structure and functions and also from their behaviour.' Thus the Brambell Committee acknowledged the importance of understanding behaviour when trying to evaluate welfare. Animal welfare is therefore something that should be kept in mind when environments for livestock are being designed. However, it should be emphasized that judgements on welfare cannot be totally objective.

OBJECTIVE MEASUREMENTS AND SUBJECTIVE FEELINGS

Agriculture is the exploitation of plants and animals for man's benefit. The decisions as to whether or not we exploit animals and, if we do, to what extent we exploit them, are, in the final analysis, ethical decisions. They are therefore decisions that should be made by society at large and not by any one small sector of it. However, society should not be expected to make these decisions without knowing the facts, and the facts, or scientific evidence, can be provided by scientific research. Scientists should be expected to produce evidence on such things as the disease risk, the amount of fear, the degree of frustration and the severity of pain or discomfort that will be experienced by animals under particular systems or during specific procedures. These are facts. It is possible to be objective about them. Although a phenomenon such as 'fear' is really a hypothetical intervening variable it is still possible to define it operationally and measure it in the same way as hunger or thirst can be measured.

Of course, what we want to know ultimately is whether or not animals are suffering. The term 'suffering' implies a particular type of mental experience, a subjective feeling, and this is what the members of the Brambell Committee were trying to take account of when they referred to 'mental wellbeing' and 'feelings of animals'. Subjective feelings are not directly accessible to scientific investigation but that does not mean that they do not exist. Other human beings are generally accepted to have subjective feelings and mental experiences although, strictly speaking, we cannot prove it. However, if we observe their physiology and behaviour and if we listen to their description of what they say they feel, we probably accept that they have subjective feelings as we ourselves do. If we accept this to be true of other human beings without hard proof, then we should at least consider the possibility that it may be true of other species. Scientists who accept biological evolution in animals should not eschew the concept of continuity in animal mental experiences. Griffin (1976), in his book *The Question of Animal Awareness*, has argued cogently that evidence from animal orientation and navigation studies and from animal communication studies suggests that animals do have mental images, subjective feelings and intentions. Although objectivity is usually assumed to be the first principle of ethology, nevertheless even its founders have occasionally speculated on subjective feelings. In one of his classic early papers, Lorenz (1935a, b) acknowledged that he believed animals to have subjective feelings and he later explored this theme in more detail (Lorenz, 1971). Students of phenomenology such as Buytendijk (1958) have made a more comprehensive comparison of animals and human beings with regard to subjective feelings. In an elegant little paper entitled 'Toucher et être touché', Buytendijk (1953) showed that, when deprived of visual cues, some animals, such as octopus (*Octopus*), were able to distinguish between touching something and being touched. Other animals, such as starfish (*Asterias*), were unable to make the distinction owing to insufficient neurophysiological integration between receptor and effector control. Buytendijk interpreted this finding as demonstrating that animals above a certain phylogenetic level have a mental image of their immediate environment, the space occupied by their bodies in that environment and the stability (or otherwise) of stimuli in that environment.

It may be possible, then, by careful experimentation, to gain some knowledge of an animal's subjective feelings including whether or not it is suffering mentally. In the meantime, the behavioural scientist can still provide objective evidence on such things as fear, frustration, conflict, pain and discomfort, as stated previously. Also, freedom from mental suffering is only one facet of welfare, albeit an important one, and it would be a pity if difficulties involved in measuring this interfered with judgements based on obvious physical criteria.

The environment may adversely affect the welfare of animals in many direct physical ways. For example, it is self-evident that welfare will be reduced by anything that reduces health, by climatic conditions which depart far from the optimum or by badly designed flooring or equipment which causes physical injury, and many of these aspects are dealt with elsewhere in this volume. The present discussion will be restricted to the effects that environmental factors may have on behaviour or on the physiology or biochemistry of the animals, by being perceived through the higher senses.

STRESS

It is inevitable that in any discussion of welfare one must consider the concept of stress. In physiology the term 'stress' is used to describe the bodily state of an animal which results from exposure to noxious stimuli or 'stressors', and which involves release of adrenocorticotrophic hormone (ACTH) as described by Selye (1950). However, it is common for 'stress' to be used as a blanket term to describe any change—physiological, biochemical or behavioural—induced in an animal as a result of exposure to adverse conditions without, necessarily, the classical response with ACTH release.

There is a wide range of noxious stimuli or stressors; at the one end are unconditional stimuli which act directly, damaging and distorting form and function, and at the other end, conditional stimuli that have no direct noxious effect but act as signals usually perceived through the higher senses. The former types of stressor include such things as extremes of temperature, nutritional deficiencies, physical injury, disease infections, etc. As stated earlier, these factors are being dealt with elsewhere and this chapter will deal only with the effects of the latter type of stimuli which have no direct noxious effect but which act as signals. Some cases where environmental factors interfere (or appear to interfere) directly with behaviour patterns will also be considered.

Modern husbandry systems can influence behaviour and welfare either through their effects on the animal's social environment or by providing a very artificial physical environment. One of the problems associated with measuring changes in behaviour of a domestic species is in trying to decide what is 'normal' or 'natural'. Domestication has exerted its influence on the behaviour of species in two major ways. First, the species responds to the type of artificial domestic environment in which it is placed. Secondly, genetic selection of specific strains for certain desirable characteristics leads to even greater deviations from 'normal' behaviour (Kretchmer and Fox, 1975). However, we can get some idea of the relative importance of these effects by comparing the behaviour of a domestic species with that of its wild progenitor in an artificial domestic environment, or by looking at the behaviour of a domestic species in a wild or 'natural' habitat. As yet this has been carried out in detail with only a few species, such as the duck (Desforges and Wood-Gush, 1975) and the chicken (Duncan, Savory and Wood-Gush, 1978).

Social environment

All of our agricultural domestic animals are social species, which means that they normally live in groups with an organized social structure. It has generally been considered that most groups of domestic animals form a 'dominance hierarchy' with animals higher in the hierarchy having priority of access to any finite resource. This system has the advantage of reducing social friction, since animals will tend not to compete for resources but to give way according to their position in the hierarchy. The best-known example of this system is the peck-order of chickens, first described by Schjelderup-Ebbe (1922) and much investigated since (reviewed by Fischer, 1975). Turkeys also develop a peck-order (Hale and Schein, 1962) but little seems to be known of the social organization of domestic ducks in confinement (McKinney, 1975). In pigs, two types of dominance order have been described, namely the 'teat-order', which is established among the litter when the piglets compete for the preferred anterior teats (McBride, 1963), and the 'dominance hierarchy', which is established after weaning (Signoret *et al.*, 1975). Cattle may also form a dominance hierarchy (Hafez and Bouissou, 1975). There is little evidence of sheep establishing a dominance order; small groups will compete for limited food by pushing and shoving rather than threatening or butting. Nevertheless, there is social organization within a flock of sheep and this can probably be best described in terms of leader–follower relationships (Hulet, Alexander and Hafez, 1975). A dominance hierarchy normally develops in groups of horses maintained together (Waring, Wierzbowski and Hafez, 1975). However, in spite of the prevalence of dominance hierarchies among domestic species, their establishment may well be an artefact of confinement, as pointed out by Kiley-Worthington (1977). It should be emphasized that there is increasing evidence that the hierarchy of a group in one situation, for example, while competing for food, need not be the hierarchy in another situation, such as while competing for a mate, although usually some similarity exists. This has been demonstrated to be true in horses (Waring, Wierzbowski and Hafez, 1975) and domestic fowl (Hughes, 1977a).

Modern husbandry systems can influence the social environment of animals, and thus their behaviour and welfare, in many ways. For example, intensive husbandry systems often involve the following deviations from what might be considered as 'normal' or 'natural':

(1) The formation of the normal parent–offspring bond may be interfered with or prevented completely, e.g. artificially reared chicks have no contact with their dam or with a substitute dam.
(2) Young animals may be weaned early, e.g. fattening pigs.

(3) Animals may be kept in large groups, e.g. broiler chickens in a deep-litter house, or small groups, e.g. laying hens in cages.
(4) Animals may be kept at a high density, e.g. cattle in a yard.
(5) Animals may be kept in single-age groups, e.g. fattening pigs, or in single-sex groups, e.g. dairy cows.
(6) Group membership may be disrupted, e.g. dry cows removed from, and newly calved cows returned to a dairy herd.
(7) Animals may be isolated to some extent, e.g. stud boars and dairy bulls.

These factors will be considered in turn in the following pages.

FORMATION OF PARENT–OFFSPRING BOND

The bond between parent and offspring is particularly important between a female mammal and her young. Domestic birds appear able to develop normally without contact with the dam, and usually do so under commercial practice. However, it has been shown that the frequency of aggressive interactions is higher in non-brooded groups of domestic chicks than in groups brooded by a hen (Fält, 1978). The bond between a female and her young is formed very soon after parturition, when a dam will not usually attack any newly born young; later, unfamiliar young may be attacked. Occasionally, husbandry practices may require fostering of young and to be successful this should obviously be done as soon after parturition as possible (Hosman, 1971; Smith, Van Toller and Boyes, 1966). Neathery (1971) has shown that a single injection of a tranquillizing drug given to a ewe will allow her to suckle a foreign lamb and also accept the lamb after the effects of the drug have worn off. Kristjansson (1957) had similar success in using tranquillizers to prevent nervous and 'savage' sows from eating their young. These results suggest that attacks by dams on young may be motivated by fear. The formation of the mother–offspring bond is facilitated in ungulates by the parturient female separating herself from the rest of the herd (Fraser, 1968; Lent, 1974). Obviously, if there is no room for this to happen, there is a risk of interference by herd-mates. Also, some ewes show a premature onset of maternal behaviour and show great interest in the newborn lambs of other ewes and this can lead to 'lamb-stealing' (Alexander, 1960; Shillito and Hoyland, 1971). This situation will be exacerbated if large numbers of parturient ewes are confined in a small area. Bonding may not take place if maternal behaviour has been reduced by a long and difficult parturition (Alexander, 1960; Shelley, 1970) or by severe prenatal undernourishment (Thomson and Thomson, 1949). Absence of maternal bonding and reassurance was one reason put forward by Noyes (1976) for gnotobiotic piglets' uttering more squeaks and squeals than normal piglets and also having a higher mortality than expected.

WEANING

The breaking (or reduction) of the parent–offspring bond during weaning leads to some distress in the young. In many species during normal weaning, the mother switches from showing maternal care to behaving aggressively towards the young, which often results in their being hyperactive and vocalizing for a few days (Lent, 1974). Early-weaned lambs, even when given access to artificial teats, spend time sucking the scrotum, navel and ears of other lambs (Stephens and Baldwin, 1970). In another study, early-weaned piglets spent more time massaging and sucking other piglets than did control piglets with a sow. However, there were other differences between the treatments which could have accounted for the results (Van Putten and Dammers, 1976).

CROWDING

Of all the factors affecting welfare under modern intensive systems, population density and group size, often lumped together as 'crowding', are usually considered to be the most important. They have been shown to influence the social behaviour of our agricultural animals in a variety of ways but, unfortunately, most of the studies have failed to separate the two variables; this is discussed by Bryant (1972). In a comprehensive review of this topic as it affects poultry, Hughes (1975a) concluded that increased colony size depresses egg production, raises food consumption and increases mortality; while decreased area per bird depresses egg production, reduces food consumption, lowers body weight gain and increases mortality. These effects are independent and additive and there are marked strain effects. Hughes proposed a model in which feather-pecking, cannibalism, disturbance and competition play important roles in mediating these effects.

The theoretical principles underlying the spacing behaviour of domestic fowl have been discussed by McBride (1968, 1970); the same principles will apply to other agricultural species although the sensory modalities involved may be different. Each bird has an area or 'personal field' around it, which other birds avoid if they can. These personal fields do not have an equal radius in all directions, but are greater directly in front of the face. They also vary in size depending on the activity that the birds are performing. When two hens come into close contact so that their personal fields overlap, the dominant bird will threaten or peck the subordinate; the subordinate may retreat or remain in a submissive posture. This social control is further extended since the presence of a dominant bird inhibits a subordinate bird from threatening or pecking a third, even more subordinate bird. When population density is high, there will be continual overlapping of personal fields and unremitting social friction. Duncan (1974) has also pointed out that three birds at a high density have a much greater chance of arranging themselves so that their personal fields do not overlap than nine or ten birds at the same density. One of the predictions from this model of personal fields would be that agonistic interactions should increase with increasing density. This has been shown to be true of poultry, but only down to a space allowance of about 800 cm^2 per bird. When hens are given less space than this, there is a sharp reduction in aggressive behaviour (Al-Rawi and Craig, 1975) with threatening declining before aggressive pecking (Banks and Allee, 1957; Hughes and Wood-Gush, 1977). Hughes and Wood-Gush (1977) thought that this decrease in aggression might be explained by the fact that birds at very high densities would be *continuously* in each others' personal fields whereas perhaps only *moving into* a field stimulated an aggressive response. Also, the presence of a dominant bird would exert social control over *all* the birds within its personal field. However, there could be a third reason for the decrease in aggression. Prevention of escape has been shown to block learning (Seligman, 1975) and perhaps birds in very crowded conditions are exhibiting 'learned helplessness'. It is interesting to note that when given the choice, hens preferred to be on their own or with one other hen rather than with groups of four or five (Hughes, 1977b).

Much basic work still needs to be done in measuring the personal fields of each agricultural animal species during different activities. Only then will it be possible to predict the effects of group size and density.

It should be remembered that crowding may occur only at certain places or at certain times in the animals' environment, when there is competition for limited resources. It has been shown that decreasing the accessibility of food in stable flocks of domestic cocks (King, 1965) and goats (Scott, 1948) increased the frequency of aggressive interactions between animals. Since in all social species there is a tendency for members of a group to synchronize their activities, it follows that in an ideal husbandry system there should be enough room for all animals to take part in any one activity (and particularly the maintenance activities) simultaneously.

SINGLE-AGE, SINGLE-SEX GROUPS

Very little is known about the effects of keeping animals in single-age or single-sex groups. However, it could be postulated that both of these procedures are likely to increase social friction. In a mixed-age society such as a troop of wild baboons, juveniles will be recruited into the hierarchy gradually and the whole unit will have an inherent stability (Hall and De Vore, 1965). In a large group of animals of the same age such as domestic chicks being reared for egg production, all the animals will be attempting to establish dominance at the same time and this is bound to lead to much social friction at particular stages of development. It has already been pointed out that the presence of a mother hen reduces aggressive interactions among her brood (Fält, 1978) and McBride, Parer and Foenander (1969) have observed in a population of feral fowl that the presence of a dominant cock inhibits the expression of aggression among all birds within 6 m.

DISRUPTION

Another potential source of distress is when group membership is altered. This frequently results in intense fighting, often accompanied by a variety of physiological effects and a reduction in productivity. When strange chickens are mixed together they show adrenal hypertrophy (Siegel and Siegel, 1961), tachycardia (Candland *et al.*, 1969) and more agonistic interactions (Craig, Biswas and Guhl, 1969) than birds in stable flocks. Pigs also show more aggressive interactions when strange animals are mixed together

(Ewbank and Meese, 1971) and these interactions can be reduced in frequency by tranquillizers (Callear and Van Gestel, 1971; Symoens and Van
den Brande, 1969). The introduction of new cows to an established herd
increases aggressiveness but violent fights are infrequent (Bouissou, 1974a,
b). Also, when given a choice, hens prefer to be near hens they know rather
than strangers (Hughes, 1977b), which suggests that social mixing might be
distressful. Many studies have shown that there is a loss of economic performance among pigs, poultry and cattle when group membership is altered
(Bryant, 1972).

ISOLATION

Since all the common agricultural animals are social species, it follows that
separation from conspecifics for any prolonged period will be stressful.
Isolated cockerels show disturbed behaviour such as chasing their tails and
excessive aggressiveness as well as abnormal sexual behaviour when placed
with females (Wood-Gush, 1958). Also, when weanling pigs were placed in
individual pens in visual isolation they showed abnormal withdrawn behaviour and had a high incidence of arteriosclerosis (Ratcliffe *et al.*, 1969). It
is probable that the extreme aggressiveness and intractable nature of stud
boars and bulls is due to the isolated conditions in which they are all too
often kept.

Physical environment

As stated previously, it is obvious that a poor environment can adversely
affect the welfare of animals in many direct physical ways. For example,
Ekesbo (1966) has shown that the incidence of trampled teats and clinical
mastitis is higher when dairy cows are housed either without bedding or in
stalls less than 1.85 m in length. In a similar survey of fattening pigs,
Lindqvist (1974) found that health status was better in houses with 'production strictly in batches', less than 500 pigs per section, a house volume of at
least 3 m^3 per pig, a total floor area of at least 0.7 m^2 per pig, a solid manure
handling system in the house and free access to drinking water.

However, in addition to these types of direct effect, the environment can
seriously disrupt behaviour and reduce welfare in three ways (Duncan,
1974; Wood-Gush, Duncan and Fraser, 1975). First, the tendency to perform a particular activity may be blocked by some aspect of the environment: this is often called thwarting or frustration. Secondly, an artificial
environment may be lacking in key stimuli necessary for eliciting or 'releasing' certain behaviour patterns. Thirdly, the level of general stimulation may
be wrong. A complex and changing environment may lead to overstimula-
tion and high arousal (Johnson, 1975) whereas a barren environment may
lead to understimulation and 'boredom'.

FRUSTRATION

Severe confinement may lead to frustration. The Brambell Committee
thought that an animal should have at least sufficient freedom of movement

to turn round, groom itself, get up, lie down and stretch its limbs without difficulty. This amount of freedom would generally be accepted as a reasonable minimum. However, when husbandry systems are examined in detail, many do not allow even this. For example, tether-stalls for dry sows, farrowing crates for parturient sows and crates for veal calves do not allow all of these basic freedoms. Moreover, even some traditional systems, such as tying dairy cows by the neck in stalls throughout the winter, do not meet these requirements. In defence of traditional systems, it might be argued that there are differences between species and between age groups in this regard; that dairy cows have been selected over many generations for their suitability to be tethered whereas sows have not; also, since young animals are generally much more active and indulge in more play behaviour than adults (Welker, 1961) then calves are more likely to be frustrated by tethering than cows. Nevertheless, serious thought should be given to designing husbandry systems that allow animals these five basic freedoms.

In addition to obvious restraint, it has often been alleged that very artificial environments will frustrate certain behaviour patterns. For example, it was stated in the Brambell Report that, in the case of laying hens, 'Much of the ingrained behaviour pattern is frustrated by caging. The normal reproductive pattern of mating, hatching and rearing young is prevented and the only reproductive urge permitted is laying. They cannot fly, scratch, perch or walk freely. Preening is difficult and dust-bathing impossible . . . . The caged bird, which is permitted only to fulfil the instinctive urges to eat and drink, to sleep, to lay and to communicate vocally with its fellows, would appear to be exposed to considerable frustration.' This hypothesis can be tested experimentally. The feeding, drinking, nesting, incubating, brooding and sexual tendencies can be frustrated experimentally in many different ways and the responses observed. These responses can then be compared with those that occur in battery cages, and if they are similar we can conclude, with a fair degree of certainty, that hens in battery cages are frustrated. It was found that when hens were mildly frustrated experimentally they showed an increase in displacement preening; when severely frustrated they showed stereotyped back-and-forward pacing, and if two or more hens were frustrated simultaneously the dominant birds showed an increase in aggression towards the subordinates (Duncan, 1970; Duncan and Wood-Gush, 1971, 1972a, b). There was also evidence that severe frustration was very aversive to the birds (Duncan and Wood-Gush, 1974). However, with one exception, which will be discussed later, these responses are not seen in battery-caged hens. It can be concluded that, generally speaking, caging *per se* does not lead to frustration. Of course, that is not to say that some of the previously discussed factors, such as crowding, could not cause frustration in caged birds.

In addition to this systematic approach, there have been many studies involving a comparison of the behaviour, physiology and health of poultry kept on at least two systems, one more intensive or artificial than the other (Bareham, 1972, 1976; Brunner and Fölsch, 1977; Burckhardt and Fölsch, 1977; Eskeland, 1977, 1978; Fölsch *et al.*, 1977; Wennrich and Strauss, 1977; Hughes, 1978; Vestergaard, 1978). All of these studies have reported differences between systems, and frustration has often been suggested as the reason. However, it should be of no surprise that animals behave differently in different environments. This may simply demonstrate how adaptable they are. In particular, one cannot say that, because a particular behaviour pattern is missing, the system is *preventing* it.

This is best illustrated by reference to one example. It has been said that battery cages prevent wing-flapping (Hughes, 1973; Martin, 1979), and, of course, a commercial battery cage is not large enough to allow the full motor pattern to take place. However, there could be other explanations. For example, perhaps the battery cage does not stimulate or 'release' wing-flapping. Perhaps the bird in a cage is not motivated to flap its wings. Wing-flapping is often described as the bird stretching its wings, but this is a purely subjective description. It could be a stretching comfort movement but it could also be a sexual signal or a social signal or an intention movement to fly. Until we know what wing-flapping is, what causes it, what function it serves, how it develops and how it has evolved, we cannot say that caging *prevents* wing-flapping.

The ethologists who suggest that artificial husbandry systems will inevitably lead to frustration (e.g. Wennrich, 1975; Martin, 1979) are usually of the Lorenzian school of ethology, believing in a 'psychohydraulic' model of motivation (Lorenz, 1950). According to this model, action-specific energy accumulates with time and is liberated when the behaviour pattern in question is 'released' by a specific releasing stimulus. In the absence of specific releasers, the action-specific energy builds up until the behaviour pattern in question bursts out as a 'vacuum activity'. Although this model of motivation does seem able to account for a few behaviour patterns, it is not widely accepted by the majority of modern ethologists. It has been criticized on various grounds (Kennedy, 1954; Hinde, 1960; Manning, 1967), and other models of motivation such as that of Deutsch (1960) are now more widely accepted. Wood-Gush (1973) has discussed some of the consequences of using these two models in interpreting animal welfare problems. The psycho-hydraulic model predicts that 'hereditary behaviour patterns' must occur either through being released by suitable stimuli or as vacuum activities. If their occurrence is prevented physically or because of a conflict with other tendencies, then the action-specific energy will 'spark over' to another motivational system and be dissipated in the form of 'displacement activities'. For example, dust-bathing motor patterns and nest-building motor patterns are occasionally seen in battery cages. Because they occur as vacuum activities Martin (1979) states that they indicate reduced welfare, although the activities can, and do, occur. She also states that hens in cages spend much time pecking objects, such as the wire mesh, that are 'inadequate'. These objects are certainly inedible, but so too are many of the objects pecked at by hens in the 'natural environment'. Therefore, in the 'natural environment' hens spend much of their time pecking at objects that are both edible and inedible. In the battery cage they do exactly the same. Whether or not the objects are 'adequate' is pure speculation. Martin (1979) argues that in order for the welfare of hens to be protected, 'hereditary behaviour patterns' must be released by stimuli which occur 'in the natural environment'. If they are released by other stimuli or occur as vacuum activities or if the 'action-specific energy' is dissipated as a displacement activity, then, according to Martin, welfare is adversely affected. However, even if a psychohydraulic model is accepted, it could be argued that as long as the energy finds an outlet that is not damaging to the hen itself or its flockmates, then welfare will not be adversely affected.

LACK OF KEY STIMULI

It was mentioned in the previous section that with one exception hens do not show symptoms of frustration in battery cages. The exception is that certain strains of bird show stereotyped back-and-forward pacing during the pre-laying phase, when the birds appear to be frustrated because they cannot find a suitable nest site (Duncan, 1970; Wood-Gush, 1972). The fact that these strains in battery cages show an increase in aggression before laying (Hughes, 1979) is further proof that they are frustrated at this time. It appears that the key stimulus for the 'sitting' component of nesting behaviour is provided by the battery cage in some strains but not in others.

An example of the lack of a key stimulus in cattle has been reported by Selman, Fisher and McEwan (1967). The stimulus that leads a calf to find the teat seems to be nose contact with the highest part of the cow's underbelly. In beef cattle this reaction quickly locates the teat, but in dairy cattle, where the udder is low, the calf may spend hours nosing at the xiphoid, brisket and flanks. Genetic selection seems to have removed this particular key stimulus. Fraser (1975) reported that sows in tether stalls with straw spent much time chewing and manipulating the straw and more time lying down than sows in stalls without straw. The latter showed stereotyped activities such as bar-biting and licking chains and feeders and stood or sat motionless with drooping head to a greater extent. These results could be interpreted in terms of a key stimulus (straw) leading to frustration by its absence but the experiments were not as systematic as those for poultry, and other explanations are possible.

GENERAL STIMULATION

It has been suggested that animals will try to maintain an optimal level of sensory input (Hebb, 1955; Leuba, 1955). Sensory input can be increased by the animal's exploring its environment (Johnson, 1975) and can be reduced if the animal performs certain behaviour patterns or adopts certain postures that tend to 'cut off' the arousing stimuli (Chance, 1962). In a confined and barren environment the animal is restricted in the amount of exploration it can do and this may lead to what, in human terms, would be called 'boredom'. Bareham (1972) noted that hens in battery cages 'head-flicked' more than those in deep-litter pens and he interpreted this as attempts by the caged hens to increase their sensory input. Duncan and Hughes (1972) taught hens to work for food by pecking a disc in a Skinner box. When the hens were given access to free food, they still preferred to work for at least part of their diet. There was a trend over the whole experimental period for the mean proportion of food earned to increase, indicating that, at least in a Skinner box, hens may try to increase their general stimulation. If it is true that hens can be understimulated in barren environments, then it follows that the phylogenetically more advanced agricultural species, such as pigs, are at risk as well.

It is likely that most animals would habituate to a complex and changing environment if the changes were occurring continuously. Overarousal is more likely to be caused by sudden and intense changes in stimulation leading to a state that is normally termed 'fear'. Fear in domestic species has not been studied in any systematic way until recently. In domestic fowl, the conditions and stimuli that cause fear and the behavioural and physiological responses which the bird shows are now better understood. Extensive studies by Murphy and others at the Poultry Research Centre have shown that so-called 'flighty' strains of chicken show more fear of strange environments and of human beings than 'placid' strains (Murphy and Wood-Gush, 1978; Murphy and Duncan, 1977). 'Placid' strains, however, show more fear of novel food and unusual objects than 'flighty' strains (Murphy, 1977). Research on fear is being continued on young chicks, whose behavioural response seems to be a good indicator of their level of fear. Freezing indicates a high level, peeping an intermediate level and active movement a low level of fear (Jones, 1977a). Factors that influence fear have been examined and it has been found that males are more fearful than females (Jones, 1977b) and chicks which have kept in a barren environment are more fearful than those exposed to a number of different stimuli (Jones, 1977c).

A little work has been done on the effects of sudden noises, particularly sonic booms, on the behaviour of farm animals. Cattle and sheep in fields which were exposed to 28 sonic booms and 10 low-altitude subsonic flights during four days showed minimal behavioural reactions (Espmark, Fält and Fält, 1974). In another study, grazing cattle, sheep and ponies, dairy cattle in a parlour and intensively housed poultry all showed a startle response to a simulated sonic boom. However, they showed marked habituation after three or four exposures in one afternoon (Ewbank and Mansbridge, 1977).

The question arises as to whether animals that are showing behavioural symptoms of fear are necessarily stressed physiologically. This has been investigated recently at the Poultry Research Centre using radiotelemetry techniques to measure short-term physiological responses such as changes in skin temperature and heart rate. Shank temperature was found to be a good indicator of fright; frightened birds had cold feet (Duncan, Filshie and McGee, 1975). Also, 'flighty' strains of hen showed far more avoidance and panic to sudden visual stimuli than 'placid' strains. However, the heart rate of the so-called 'placid' birds rose almost as much and took longer to recover than that of the so-called 'flighty' birds (Duncan and Filshie, 1980). This suggests that some reappraisal is required of the classification of birds as 'flighty' or 'placid', since placid birds may be as frightened as flighty birds in physiological terms. Both strains habituated quickly to auditory stimuli, which agrees with the casual observation that bird scarers based on sudden loud noises are not very effective. It seems likely that telemetry studies will become more important in welfare investigations, since they enable physiological and behavioural changes to be monitored simultaneously with minimal interference to the animals.

Future research

One promising approach to evaluating both social and physical environments with regard to welfare is to give the animal a choice of environments and see which it prefers. This has already been done for poultry with different aspects of the environment (Hughes and Black, 1973; Hughes, 1975b, 1976, 1977b) and for complete environments (Dawkins, 1976, 1977). Although the methods have been severely criticized for various short-comings (Duncan, 1978), preference testing should, with refinement, become a useful tool.

A related method is to use operant conditioning techniques to see how hard animals will work to obtain, or to avoid, some aspect of their environment. For example, pigs placed in cold environments learned to press a switch in order to obtain a short burst of infrared heat as reinforcement. They were able to keep their metabolic rate within a range shown by the same animals in a thermoneutral ambient temperature. They also learned to press a switch in order to turn off a draught (Baldwin and Ingram, 1967). In general the pigs worked at a rate which was proportional to the heat reward (Baldwin and Ingram, 1968). However, in another study, young pigs offered a natural environment chose to spend little of their time in a warm shelter, although they did avoid draughts (Ingram and Legge, 1970). This suggests that in a rich environment other tendencies were competing with the one to choose warmth. This technique has also been used to investigate illumination preferences and sensory reinforcement in pigs (Baldwin and Meese, 1977), sheep and calves (Baldwin and Start, 1980) and also temperature preferences in poultry (Richards, 1976).

The best reason for the judicious use of preference tests is that the argument, 'the animal itself prefers . . .' carries a great deal of weight, particularly with lay people.

References

ALEXANDER, G. (1960). *Proc. Aust. Soc. Anim. Prod.* **3**, 105–114
AL-RAWI, B. and CRAIG, J.V. (1975). *Appl. Anim. Ethol.* **2**, 69–80
BALDWIN, B.A. and INGRAM, D.L. (1967). *Physiol. Behav.* **2**, 15–21
BALDWIN, B.A. and INGRAM, D.L. (1968). *Physiol. Behav.* **3**, 409–415
BALDWIN, B.A. and MEESE, G.B. (1977). *Anim. Behav.* **25**, 497–507
BALDWIN, B.A. and START, I.B. (1980). *Appl. Anim. Ethol.* **6**, 389–390
BANKS, E.M. and ALLEE, W.C. (1957). *Physiol. Zoöl.* **30**, 255–268
BAREHAM, J.R. (1972). *Br. vet. J.* **128**, 153–163
BAREHAM, J.R. (1976). *Appl. Anim. Ethol.* **2**, 291–303
BOUISSOU, M.-F. (1974a). *Annls Biol. anim. Biochim. Biophys.* **14**, 383–410
BOUISSOU, M.-F. (1974b). *Annls Biol. anim. Biochim. Biophys.* **14**, 757–768
BRUNNER, E. and FÖLSCH, D.W. (1977). *Tierhaltung* **2**, 1–64
BRYANT, M.J. (1972). *Vet. Rec.* **90**, 351–359
BURCKHARDT, C. and FÖLSCH, D.W. (1977). *Tierhaltung* **3**, 1–32
BUYTENDIJK, F.J.J. (1953). *Archs néerl. Zool.* **10**, 34–44
BUYTENDIJK, F.J.J. (1958). *Mensch und Tier.* Rowohlts Enzyklopädie, Hamburg
CALLEAR, J.F.F. and VAN GESTEL. J.F.E. (1971). *Vet. Rec.* **89**, 453–458
CANDLAND, D.K., TAYLOR, D.B., DRESDALE, L., LEIPHART, J.M. and SOLOW, S.P. (1969). *J. comp. physiol. Psychol.* **67**, 70–76

CHANCE, M.R.A. (1962). *Symp. zool. Soc. Lond.* **8**, 71–89
COMMAND PAPER 2836 (1965). HMSO, London
CRAIG, J.V., BISWAS, D.K. and GUHL, A.M. (1969). *Anim. Behav.* **17**, 498–506
DAWKINS, M. (1976). *Appl. Anim. Ethol.* **2**, 245–254
DAWKINS, M. (1977). *Anim. Behav.* **25**, 497–507
DESFORGES, M.F. and WOOD-GUSH, D.G.M. (1975). *Vet. Rec.* **96**, 509
DEUTSCH, J.A. (1960). *The Structural Basis of Behaviour*. Cambridge University Press, Cambridge
DUNCAN, I.J.H. (1970). In *Aspects of Poultry Behaviour*, pp. 15–31. Ed. by B.M. Freeman and R.F. Gordon. British Poultry Science, Edinburgh
DUNCAN, I.J.H. (1974). *Anim. Prod.* **3**, 9–19
DUNCAN, I.J.H. (1978). *Appl. Anim. Ethol.* **4**, 197–200
DUNCAN, I.J.H. and FILSHIE, J.H. (1980). In *A Handbook on Biotelemetry and Radio Tracking*, pp. 579–588. Ed. by C.J. Amlaner and D.W. Macdonald. Pergamon Press, Oxford
DUNCAN, I.J.H., FILSHIE, J.H. and McGEE, I.J. (1975). *Med. biol. Engng* **13**, 544–550
DUNCAN, I.J.H. and HUGHES, B.O. (1972). *Anim. Behav.* **20**, 775–777
DUNCAN, I.J.H., SAVORY, C.J. and WOOD-GUSH, D.G.M. (1978). *Appl. Anim. Ethol.* **4**, 29–42
DUNCAN, I.J.H. and WOOD-GUSH, D.G.M. (1971). *Anim. Behav.* **19**, 500–504
DUNCAN, I.J.H. and WOOD-GUSH, D.G.M. (1972a). *Anim. Behav.* **20**, 68–71
DUNCAN, I.J.H. and WOOD-GUSH, D.G.M. (1972b). *Anim. Behav.* **20**, 444–451
DUNCAN, I.J.H. and WOOD-GUSH, D.G.M. (1974). *Appl. Anim. Ethol.* **1**, 67–76
EKESBO, I. (1966). *Acta Agric. scand.* (Suppl. 15).
ESKELAND, B. (1977). *Meld. Norg. LandbrHøisk.* **53**, No. 7
ESKELAND, B. (1978). *Meld. Norg. LandbrHøisk.* **57**, No. 18
ESPMARK, Y., FÄLT, L. and FÄLT, B. (1974). *Vet. Rec.* **94**, 106–113
EWBANK, R. and MANSBRIDGE, R.J. (1977). *Appl. Anim. Ethol.* **3**, 292
EWBANK, R. and MEESE, G.B. (1971). *Anim. Prod.* **13**, 685–693
FÄLT, B. (1978). *Appl. Anim. Ethol.* **4**, 211–221
FISCHER, G.J. (1975). In *The Behaviour of Domestic Animals*, 3rd edn, pp. 454–489. Ed. by E.S.E. Hafez. Baillière Tindall, London
FÖLSCH, D.W., NIEDERER, C., BURCKHARDT, C. and ZIMMERMAN, R. (1977). *Tierhaltung* **1**, 1–72
FRASER, A.F. (1968). *Reproductive Behaviour in Ungulates*. Academic Press, London
FRASER, D. (1975). *Anim. Prod.* **21**, 59–68
GRIFFIN, D.R. (1976). *The Question of Animal Awareness*. Rockefeller University Press, New York
HAFEZ, E.S.E. and BOUISSOU, M.F. (1975). In *The Behaviour of Domestic Animals*, 3rd edn, pp. 203–245. Ed. by E.S.E. Hafez. Baillière Tindall, London
HALE, E.B. and SCHEIN, M.W. (1962). In *The Behaviour of Domestic Animals*, 1st edn, pp. 531–564. Ed. by E.S.E. Hafez. Baillière, Tindall & Cox, London
HALL, K.R.L. and DE VORE, I. (1965). In *Primate Behaviour*, pp. 53–110. Ed. by I. De Vore. Holt, Rinehart & Winston, New York
HARRISON, R. (1964). *Animal Machines*. Vincent Stuart, London
HEBB, D.O. (1955). *Psychol. Rev.* **62**, 243–354

HINDE, R.A. (1960). *Symp. Soc. exp. Biol.* **14**, 199–213
HOSMAN, L. (1971). *Sb. vys. Sk. zemed. Praze* **539**, 107–121
HUGHES, B.O. (1973). *Vet. Rec.* **93**, 658–662
HUGHES, B.O. (1975a). In *Economic Factors Affecting Egg Production*, pp. 271–298. Ed. by B.M. Freeman and K.N. Boorman. British Poultry Science, Edinburgh
HUGHES, B.O. (1975b). *Br. vet. J.* **131**, 560–564
HUGHES, B.O. (1976). *Appl. Anim. Ethol.* **2**, 155–165
HUGHES, B.O. (1977a). *Br. Poult. Sci.* **18**, 611–616
HUGHES, B.O. (1977b). *Br. Poult. Sci.* **18**, 9–18
HUGHES, B.O. (1978). In *1st Danish Seminar on Poultry Welfare in Egglaying Cages*, pp. 21–32. Ed. by L.Y. Sørensen. National Committee for Poultry and Eggs, Copenhagen
HUGHES, B.O. (1979). *Appl. Anim. Ethol.* **5**, 83–93
HUGHES, B.O. and BLACK, A.J. (1973). *Br. Poult. Sci.* **14**, 615–619
HUGHES, B.O. and WOOD-GUSH, D.G.M. (1977). *Anim. Behav.* **25**, 1056–1062
HULET, C.V., ALEXANDER, G. and HAFEZ, E.S.E. (1975). In *The Behaviour of Domestic Animals*, 3rd edn, pp. 246–294. Ed. by E.S.E. Hafez. Baillière Tindall, London
INGRAM, D.L. and LEGGE, K.F. (1970). *Physiol. Behav.* **5**, 981–987
JOHNSON, J.I. (1975). In *The Behaviour of Domestic Animals*, 3rd edn, pp. 63–72. Ed. by E.S.E. Hafez. Baillière Tindall, London
JONES, R.B. (1977a). *Behav. Proc.* **2**, 315–323
JONES, R.B. (1977b). *Appl. Anim. Ethol.* **3**, 255–271
JONES, R.B. (1977c). *Behav. Proc.* **2**, 163–173
KENNEDY, J.S. (1954). *Br. J. Anim. Behav.* **2**, 12–19
KILEY-WORTHINGTON, M. (1977). *Behavioural Problems of Farm Animals.* Oriel Press, Stocksfield, England
KING, M.G. (1965). *Anim. Behav.* **13**, 504–506
KRETCHMER, K.R. and FOX, M.I. (1975). *Vet. Rec.* **96**, 102–108
KRISTJANSSON, F.K. (1957). *Can. J. comp. Med.* **21**, 389–390
LENT, P.C. (1974). In *Behaviour of Ungulates and its Relations to Management*, pp. 1–24. Ed. by V. Geist and F. Walther. IUCN, Morges, Switzerland
LEUBA, C. (1955). *Psychol. Rep.* **1**, 27–33
LINDQVIST, J.O. (1974). *Acta Agric. scand.* (Suppl. 51)
LORENZ, K. (1935a). *J. Orn., Lpz.* **83**, 137–213
LORENZ, K. (1935b). *J. Orn., Lpz.* **83**, 289–413
LORENZ, K. (1950). *Symp. Soc. exp. Biol.* **4**, 221–268
LORENZ, K. (1971). *Studies in Animal and Human Behaviour*, vol. 2, pp. 323–337. Methuen, London
McBRIDE, G. (1963). *Anim. Behav.* **11**, 53–56
McBRIDE, G. (1968). In *Adaptation of Domestic Animals*, pp. 360–366. Ed. by E.S.E. Hafez. Lea & Febiger, Philadelphia
McBRIDE, G. (1970). In *Aspects of Poultry Behaviour*, pp. 3–13. Ed. by B.M. Freeman and R.F. Gordon. British Poultry Science, Edinburgh
McBRIDE, G., PARER, I.P. and FOENANDER, F. (1969). *Anim. Behav. Monogr.* **2**, 127–181
McKINNEY, F. (1975). In *The Behaviour of Domestic Animals*, 3rd edn, pp. 490–519. Ed. by E.S.E. Hafez. Baillière Tindall, London

MANNING, A. (1967). *An Introduction to Animal Behaviour*, pp. 64–68. Arnold, London

MARTIN, G. (1979). *Tierhaltung* **8**, 101–122

MURPHY, L.B. (1977). *Appl. Anim. Ethol.* **3**, 335–349

MURPHY, L.B. and DUNCAN, I.J.H. (1977). *Appl. Anim. Ethol.* **3**, 321–334

MURPHY, L.B. and WOOD-GUSH, D.G.M. (1978). *Biol. Behav.* **3**, 39–61

NEATHERY, M.W. (1971). *Anim. Behav.* **19**, 75–79

NOYES, L. (1976). *Appl. Anim. Ethol.* **2**, 113–121

RATCLIFFE, H.L., LUGINBUHL, H., SCHNARR, W.R. and CHACKO, K. (1969). *J. comp. physiol. Psychol.* **68**, 385–392

RICHARDS, S.A. (1976). *J. Physiol., Lond.* **258**, 122–123

SCHJELDERUP-EBBE, T. (1922). *Z. Psychol.* **88**, 225–252

SCOTT, J.P. (1948). *Physiol. Zoöl.* **21**, 31–39

SELIGMAN, M.E.P. (1975). *Helplessness: On Depression, Development and Death*. W.H. Freeman, San Francisco

SELMAN, I.E., FISHER, E.W. and McEWAN, A.D. (1967). *Proc. Soc. vet. Ethol.* **1**, 7–8

SELYE, H. (1950). *Stress*. Acta, Montreal

SHELLEY, L. (1970). *Proc. Aust. Soc. Anim. Prod.* **8**, 348–352

SHILLITO, E.E. and HOYLAND, V.J. (1971). *J. Zool., Lond.* **165**, 509–512

SIEGEL, H.S. and SIEGEL, P.B. (1961). *Anim. Behav.* **9**, 151–158

SIGNORET, J.P., BALDWIN, B.A., FRASER, D. and HAFEZ, E.S.E. (1975). In *The Behaviour of Domestic Animals*, 3rd edn, pp. 295–329. Ed. by E.S.E. Hafez. Baillière Tindall, London

SMITH, F.V., VAN TOLLER, C. and BOYES, T. (1966). *Anim. Behav.* **14**, 120–125

STEPHENS, D.B. and BALDWIN, B.A. (1970). *Br. vet. J.* **126**, 659–660

SYMOENS, J. and VAN DEN BRANDE, M. (1969). *Vet. Rec.* **85**, 64–67

THOMSON, A.M. and THOMSON, W. (1949). *Br. J. Nutr.* **2**, 290–305

VAN PUTTEN, G. and DAMMERS, J. (1976). *Appl. Anim. Ethol.* **2**, 339–356

VESTERGAARD, K. (1978). In *1st Danish Seminar on Poultry Welfare in Egglaying Cages*, pp. 11–17. Ed. by L.Y. Sørensen. National Committee for Poultry and Eggs, Copenhagen

WARING, G.H., WIERZBOWSKI, S. and HAFEZ, E.S.E. (1975). In *The Behaviour of Domestic Animals*, 3rd edn, pp. 330–369. Ed. by E.S.E. Hafez. Baillière Tindall, London

WELKER, W.I. (1961). In *Functions of Varied Experience*, pp. 175–226. Ed. by D.W. Fiske and S.R. Maddi. Dorsey, Homewood, Illinois

WENNRICH, G. (1975). *Arch. Geflügelk.* **39**, 113–121

WENNRICH, G. and STRAUSS, D.D. (1977). *Dt. tierärztl. Wschr.* **84**, 293–332

WOOD-GUSH, D.G.M. (1958). *Anim. Behav.* **6**, 68–71

WOOD-GUSH, D.G.M. (1972). *Anim. Behav.* **20**, 72–76

WOOD-GUSH, D.G.M. (1973). *Br. vet. J.* **129**, 167–174

WOOD-GUSH, D.G.M., DUNCAN, I.J.H. and FRASER, D. (1975). In *The Behaviour of Domestic Animals*, 3rd edn, pp. 182–200. Ed. by E.S.E. Hafez. Baillière Tindall, London

26

THE EFFECT OF THE WELFARE CODE PROVISIONS ON ANIMAL HOUSING

R. MOSS
Ministry of Agriculture, Fisheries and Food, Tolworth, Surbiton, Surrey

Introduction

This chapter reviews UK experience concerning the effects of livestock welfare codes in practice. In 1964 the Brambell Committee was appointed by the Minister of Agriculture and the Secretary of State for Scotland: 'to examine the conditions in which livestock are kept under systems of intensive husbandry, and to advise whether standards ought to be set in the interests of their welfare and if so what they should be' (Command Paper, 1965). One of the results of this committee's recommendations is that the Agriculture (Miscellaneous Provisions) Act 1968 allows Ministers, subject to parliamentary approval, to 'issue codes of recommendation for the welfare of livestock'. The Committee also made a number of detailed recommendations relating to individual species and stated that, in principle, animals should be provided with a husbandry system appropriate to their health and behavioural needs.

The Committee set out a number of basic requirements which it felt should be provided by each husbandry system. These requirements are embodied in the preamble to every one of the five Codes of Practice already published (HMSO, 1971, 1977, 1978); they are:

(1) readily accessible fresh water and nutritionally adequate food as required;
(2) adequate ventilation and suitable environmental temperatures;
(3) an adequate space to allow freedom of movement and stretching of limbs;
(4) sufficient light for satisfactory inspection;
(5) rapid diagnosis and treatment of injury and disease;
(6) emergency provision in the event of a breakdown of essential mechanical equipment;
(7) flooring that neither harms nor causes undue strain;
(8) the avoidance of unnecessary mutilation.

The Codes of Practice are based on these requirements and take into account available scientific knowledge and current farming practice. They are issued under the Agriculture (Miscellaneous Provisions) Act 1968, as already mentioned, and are sent to every known livestock farmer.

471

The effect of the Codes

It can be argued that the Codes could have a negative effect on building design, since they set out certain *minimum* parameters. However, I believe the effect to be beneficial as I hope to illustrate.

All five Codes are divided into four basic sections:

(1) housing; which includes the control of the ventilation and temperature of both climatic and totally controlled environment housing, lighting, mechanical equipment and services;
(2) space allowance for individual animals depending upon their species, age, sex and systems of husbandry;
(3) the provision of food and water, in both intensive and extensive systems;
(4) management; which includes the provision of isolation facilities, the loose housing of cattle and the housing of calves, farrowing quarters, sow quarters in piggeries, the tethering of sows, and the outside shelter of sows kept in extensive systems, the cleansing and disinfection of buildings, the facilities for handling cattle during routine tuberculin testing, vaccination, etc., and the provision of facilities to cover the risk of fire.

Space limitations made it necessary for this chapter to be confined to consideration of one or two representative items for each section. The examples chosen should serve to illustrate the contention that the effect of the welfare Codes of Practice on the design of housing and the provision of facilities for livestock is of a positive nature and not in any way restrictive, in the pejorative sense of that word.

Housing

VENTILATION AND TEMPERATURE

The provision of adequate ventilation and environmental temperature has received much attention over the years (HMSO, 1976a), and the Codes set out the basic criteria to be observed in the design of livestock housing. Ventilation, temperature and humidity are interrelated and their control figures prominently in intensive systems for keeping domestic fowl, pigs and calves.

To quote the Codes for the *domestic fowl*: 'Ventilation rates and house conditions should at all times be adequate to promote sufficient gas exchange. In particular, accumulations of ammonia, hydrogen sulphide, carbon dioxide and carbon monoxide should be avoided.' In the UK, poultry industry environmental control is widely practised, houses being windowless, insulated and with sophisticated ventilation systems. The recommended minimum temperature is 21 °C in laying houses. With typical stocking densities of about 20 birds per square metre this temperature is achievable under nearly all UK conditions without artificial heat. Insulation is, however, necessary; it both prevents excessive condensation and reduces the need for ventilation in winter. In the summer, ventilation rates have to cope with the removal of excess metabolic heat in order to prevent heat stress. Food intake is, of course, more uniform and predictable when temperatures can be held between narrow limits. The implications of these needs in a sophisticated system are, therefore, a well-insulated building with adequate reserve fan capacity, monitored to allow for outside changes.

The Code recognizes the need for regular maintenance, and where failure of any mechanical or electrically automated equipment could cause distress to birds an alarm system is essential. The Welfare of Livestock (Intensive Units) Regulations 1978 (HMSO, 1978) reinforce the Codes' recommendations by requiring that automated equipment, along with livestock, should be thoroughly inspected at least once a day.

These principles are equally applicable to *pigs* kept intensively, which have the same critical standards in relation to ventilation and temperature, requiring insulation of walls, roof and floor to retain body heat. The free-range or more extensively kept pig can face problems of a different kind. Older converted buildings rarely have adequate ventilation or insulation, which means that in the height of summer or depths of winter problems arise.

The Code for *cattle* states, 'For housed calves it is essential to avoid conditions which could produce chilling, particularly for animals up to four weeks of age . . . Effective ventilation and the avoidance of draughts are essential. Expert advice may be necessary to ensure correct temperature and humidity for the type of housing used.' To comply with this advice it is necessary when designing calf accommodation to keep the following in mind: that the air space per calf should be generous, as it can act as a buffer where air exchanges are inadequate and reduce the risk of draughts, and that the inlet area for ventilation should generally exceed the outlet and be located evenly throughout the house at a height that will prevent draughts (Mitchell, 1976). The exception is the Silsoe high-speed system, which uses very small inlet apertures. Air circulation must be designed in relation to the width of building and the building should be sited in suitable relation to factors affecting local climatic conditions such as prevailing winds, ground slopes and frost pockets. Adjacent buildings and trees can affect ventilation dramatically.

Dairy calves may be reared in conditions that range from the ad hoc through to intensive, each system having its associated problems. Respiratory disease associated with poor ventilation remains a major problem within the field of preventative medicine and welfare. There must be many non-specialist calf-rearing enterprises where in the autumn/winter period some 40–50 calves pass through the pens. Problems will arise when the number of calves builds up and air inlets are shut down to keep the house warm. The problem is obvious, but the solution often expensive, requiring better ventilation and insulation with the provision of a heat source for the younger calf.

LIGHT

Sufficient light for satisfactory inspection is a recommendation of all the Codes and in the case of domestic fowls and turkeys the need for a period of darkness in each 24 hour period is recognized.

For hens, the intensity of lighting has a direct bearing on the aggressiveness of birds within a particular environment. Day length has a considerable influence upon the physiology of the laying hen and plays an integral part in the egg laying process. Although cannibalism and feather pecking in poultry have many causes (HMSO, 1976b), light intensity is important and the recognized level of between 5 and 22 lx satisfies most situations. Provision should be made for lowering the level to the minimum of 5 lx, in the event of an outbreak of cannibalism or feather pecking, but at least 50 lx should be allowed for proper inspection by the supervisor.

The effect of light on pigs has not been fully investigated, but it has been suggested that the performance of pigs improves at reduced light intensities. Practical experience shows that pigs will often tend to dung in the lighter area of a pen—siting of lights is therefore important in relation to cleanliness.

FLOORS

'All floors, particularly slotted ones (or metal mesh in the case of the hen and turkey) should be designed, constructed and maintained so as to avoid injury or distress to the animal.' This is a requirement of all the Codes.

In the case of the domestic hen in cages, the slope, mesh size and gauge of wire are all important factors in the prevention of injury, particularly to the feet. Floors for pigs have received much attention over the years, as injuries have caused welfare problems and considerable economic loss. Smith and Mitchell (1977) suggested that slatted floors for sows and litters should satisfy certain basic requirements. They should (a) provide a non-slip, non-abrasive surface that is also easy for piglets to walk on; (b) provide an acceptable level of cleanliness for sow and piglets without manual removal of dung (at least in those areas where the sow is allowed to stand); (c) be durable, and (d) be of low total cost.

The thermal comfort provided by flooring is also a major consideration, and Bruce (1978) concluded that floor insulation should be as close to the animal as possible and that penning of single animals requires special consideration. Materials vary in their thermal resistance, and hence comfort varies with the material. For each type of flooring and animal species there is also a minimum group size for thermal comfort.

Turning to the housing of cattle on slats, Murphy (1978) concluded that the primary cause of septic feet lesion is abrasion on slatted floors. The extent of this abrasion is governed by housing factors such as the floor environment and slat design, and by animal factors including breed, behavioural characteristics and physiological composition of hoof horn. He confirmed that in Ireland the interaction of these factors causes a high incidence of foot disease in beef cattle fattened on slats.

A more direct illustration of how floors can affect wellbeing is described by Gjestand (1979). He compared the slipperiness of rubber mats and concrete floors for dairy cows tied in stalls. Teat injuries are caused in slipping, but the cows preferred rubber mats to either rough or smooth concrete.

The choice of flooring for comfort and prevention of injury is complicated by the wide range of floor types available. To ensure compliance with the Codes, critical examination of all factors should be made at the design stage. Too many floors require corrective treatment beyond the limits of normal maintenance.

SPACE ALLOWANCE

The space allowance is one of the most difficult areas, as it is an emotive topic. In consequence, the restriction of animals has been much discussed in recent years. Since man first started to domesticate animals there has been some degree of confinement. The current range is exemplified by the intensive battery hen system on the one hand and the extensive hill sheep on the other. Although one could argue that the Codes were introduced to formulate guidelines for intensive systems, animals kept semi-intensively or extensively demand the same consideration. The domestic fowl Code states: 'Irrespective of the type of enclosure or system of management used all domestic fowls should have sufficient freedom of movement to be able, without difficulty, to stand normally, turn around and stretch their wings. . . . They should have space to be able to perch or sit down without interference from other birds.' However, the space allowance cannot be considered in isolation. Consideration needs to be given to the relationship between 'usable space', heat output per bird and ambient temperature. We need to maintain the latter within an optimum range yet allow the birds the functional freedoms mentioned above.

In a review of the relationship between space allotment and profitability, Wilson (1977) has also considered regulatory codes which have set space requirements. He emphasizes that more space, obviously, is required as the bird grows. There may be a reduction in the production of meat and eggs as a result of overcrowding, which also results in increased mortality due to disease and cannibalism. Space allowances must vary with the type of house, i.e. whether the birds are on littered floors, slats, wire floors or in cages.

Stocking density should be governed by the size, weight, behavioural and functional characteristics of the animal. However, until the answers to some of the questions that relate to stress and behaviour are answered, it will be difficult to set parameters, other than for some extreme cases.

Three fundamental issues concerned with the intensive housing of poultry remain unresolved. First, are observed differences in the behaviour of poultry in intensive and extensive environments significant? Secondly, how do we measure stress? At present the only practical measurement (and that is imprecise and unacceptable to welfarists) is related to production and a satisfactory state of health. Thirdly, does confinement in a battery cage allow the hen to be exposed to a sufficient range of external stimuli for 'natural' living?

With the possibility of the phasing out over the next ten years of the keeping of laying hens in cages (as currently practised), alternatives have to be considered and developed. Two such systems are the 'aviary' and the 'getaway' cage. The former allows more freedom and counters some of the deficiencies of the battery cage, and at the same time stocking densities can be maintained. The consequential maintenance of a high temperature and low production costs can nearly match the costs of egg production in the battery system. Further work on this system is needed. The 'getaway' cage has also been investigated. This system puts groups of up to 10 birds in a cage that has a built-in nest box and perches, and consequently allows more 'normal' behaviour. Comparison of costs has shown there to be no economic disadvantage in terms of performance. Work on this system continues in West Germany.

The pig Code states: 'The total floor space should be adequate for sleeping and feeding and of such size that soiling of the lying area may be avoided' and 'Where sows or gilts are housed individually they should be able to feed and lie down normally . . . If tethers are used they should not cause injury or distress.' Aggressiveness in dry sows is a recognized problem and when these animals are kept in groups there is a need for a high standard of management. Tethering of sows allows individual attention to feeding and prevents injuries; but the Brambell principle covering movement is not satisfied. Stalls are currently designed such that the occupant cannot turn around, e.g. 0.61 m (2 ft 0 in) wide for gilts and 0.685 m (2 ft 3 in) for sows. It may be possible to preserve the benefits of individual penning or tethering, by combining individual feeding with freedom of movement throughout the remainder of the day.

Confinement of the early-weaned piglet in a cage has developed over the past 10 years and has resulted in greater production per sow. Due consideration has, however, to be given to space allowance and group size. Overcrowding and too large a group may result in vices reflecting the behavioural changes that have taken place in the natural activities of playing, fighting, sucking, nibbling and massaging.

In the case of the cattle Code the recommendation is that 'All cattle, whether tethered or in pens, should have sufficient freedom of sideways movement to be able to groom themselves without difficulty and sufficient room to lie down freely: thus the width of the pen for a singly penned animal should be not less than the height of the animal at the shoulder. Where tethers or ties are used, they should not cause injury or distress to the cattle. Consideration should be given to the adoption of a suitable loose-housing system.' The Code recommendations therefore recognize the existence of management systems that require that an animal cannot turn completely around, although they also point in the direction of the Brambell recommendations by suggesting that loose-housing systems should be considered.

Much attention has been focused on the singly penned veal calf. Pen widths of 0.46 m (18 in) have not been uncommon, but with the Code provisions in mind and allowing for a normal lying position for a calf up to 4 weeks of age, the minimum should be greater than 0.685 m (2 ft 3in). Some producers have recognized these deficiencies and are currently developing loose-housing for veal calves in preference to single penning. From a commercial point of view the system appears to be achieving satisfactory results in terms of growth rate, meat quality and texture. Group housing of 20 calves at a density of 1.67 m² (18 ft²) per calf, with automatic feeders, reduces labour requirement but at the same time requires a higher standard of stockmanship.

FOOD AND WATER

The requirements for food and water cannot be separated, because the type of feed will have a considerable influence on the water requirements. The Codes are consistent in requiring access to water, but differ in the application of the word 'availability' according to species and the system of management followed.

The Code covering domestic poultry states: 'Adequate fresh food and water should be available to all birds according to the system of management followed.' It adds that 'food and water should be easily accessible to the birds'. In dealing with availability the Code states, 'Systems which call for the complete withholding of food and water on any day should not be adopted.' The battery hen is mainly watered via nipples arranged at the front or rear of the cage. The majority of nipples supply the needs of two or more birds. Powell and Hill (1975) studied the way in which birds obtain water from nipple drinkers and concluded that birds may be able to adapt their way of drinking, so that nipple drinkers with more than two birds per nipple seem adequate for the hens' needs. The two remaining criteria are, however, that access should be unimpeded and that the supply should be fail-safe. Thus a group of birds in any one cage should have access to more than one nipple and regular checks on function should be made.

The sheep Code specifies that 'drinking water should be available to all sheep' but recognizes that for good husbandry reasons (the prevention of mastitis) water can be restricted when ewes are being dried off. They should not be deprived of food or water for more than 24 hours at a time. With the trend to in-wintering of the in-lamb ewe, provision of water is essential. The cattle Code recognizes the special needs of the calf in relation to food and water supply, but states that 'after cattle are weaned, fresh clean water should be accessible to them at least twice daily'.

Feeding behaviour and its relation to the access to food is also important. Laying hens exhibit a diurnal rhythm in feeding activity. This may need to be considered when designing feeding systems.

The aggressive behaviour of pigs when feeding is recognized in the relevant Code, which advises that 'When pigs are fed by a system which does not allow continuous and unrestricted access to food all pigs in the group should be able to feed at the same time.' One pig too many in a pen can result in excessive competition for access to food and underfeeding of the pig at the bottom of the social order. Although figuring prominently in welfare issues, nutritional quality of diets has little influence on housing design, except that cognizance of the diet may well influence floor design and the method of disposal of excreta.

'MUTILATION' OF ANIMALS

Mutilation is performed on animals kept both intensively and extensively. The Brambell Committee disliked the practice in principle and did not regard it as permissible simply because it may be judged necessary to counter a defect in the system of husbandry. It is on this basis that the avoidance of unnecessary mutilations may be related to housing.

Beak trimming (a much better and more accurate description than de-beaking) of the hen is allowed by the Codes in cases of outbreak of vice, but it is emphasized that the more fundamental cause should be tackled. Lighting level, as indicated earlier, has an influence on cannibalism. Broiler breeders laying their eggs on the floor in well-lit houses are more at risk than battery hens, and battery hens on the top tier, near the lights, are more at risk than those lower down. Intermittent feeding is more conducive to cannibalism than is feeding *ad libitum*. Group size and stocking density also influence the level of cannibalism as does the space available for feeding and watering. Corrective measures taking into account these factors will eliminate the need to beak trim. Similar provisions apply to the fitting of blinkers and dubbing. De-winging (and other operations that involve mutilation of wing tissue) and surgical castration are prohibited. Docking of pigs' tails will prevent tail biting but may merely divert attack to ears or snout. Where good management is practised, including adequate space and trough allowance and the provision of bedding, tail docking becomes unnecessary.

OTHER FACTORS

The provision of companionship is referred to in the Codes in relation to calves and pigs penned singly. The siting of pens or stalls in relation to one another is important in this respect. Should a sow or a calf look at a blank wall? The advantages of single penning of pigs have been described earlier. In the case of calves in solid-sided pens they are reduction of the spread of disease and the elimination of vice, such as navel or ear sucking.

The rapid diagnosis and treatment of injury and disease depends initially on the level of stockmanship, the management system and the frequency of inspection. All the Codes refer to these points; for example, in the case of cattle it is recommended that 'Housed calves should be closely inspected at least twice daily and it is desirable that other cattle should be inspected daily, for signs of injury, illness or distress. It is equally recommended that provision should be made for segregation of sick or injured animals.' Housing therefore should provide for and allow unimpeded access and the ability to see the animal from all aspects. For example, the singly penned calf which is unable to turn around should be visible from the rear, because enteric disorders are a main cause of illness. Segregation of the injured pig is essential to prevent cannibalism. Provisions at the design stage for accommodating sick animals should take account of the management system and the likely numbers of animals to be accommodated at any one time. Four-tier vertical battery cages present a problem in respect of the top and bottom tiers. The top is difficult to see because of height and often the bottom tier is inadequately lit. Stepped cages remove some of these visual difficulties.

The need to provide emergency arrangements if breakdown of essential equipment occurs is frequently forgotten. Systems described earlier, that involve a controlled environment, depend entirely on the proper functioning of automatic equipment. Mention has been made of drop-out panels in relation to hen batteries. However, these are not commonplace and the welfare of the animal depends on regular inspection of the animals and equipment. The need for the latter is reinforced by the Welfare of Livestock (Intensive Unit) Regulations. Automatic warning devices are available and should be more widely used and installed at the design stage.

Future trends

Whether animal housing becomes more or less intensive depends on many factors:

(1) On the economics of production. Over recent years, with increases in labour costs and fierce competition within the UK and from outside, production costs have become of paramount importance. More reliance has been placed on mechanical aids in controlling environment and nutrition. This is exemplified by the poultry and pig industries. The degree of intensification for each system *in economic terms only* will depend on the 'break point' when losses due to increasing stocking levels exceed the gains.

(2) On moral and ethical pressures. Intensive husbandry systems and certain husbandry practices are coming under attack more and more from organizations interested in animal welfare. We cannot stand back and quote productive performance alone as suitable criteria to judge the wellbeing of the animal. There is a continuing need for more information on animal behaviour, and on alternative systems that do not subject animals to husbandry methods which may horrify the public and the welfare organizations, yet provide the same productivity and profitability for the producer.

The basic right of man to keep animals to process vegetable protein to produce animal protein has been challenged by many people who hold strong and sincere views. On the other hand, there is a continuing demand for inexpensive animal protein. Striking the balance between these opposed views is the function of the politician.

(3) On the political pressures from welfare groups, new legislation and Codes. In the UK, the Farm Animal Welfare Advisory Committee (FAWAC) set up following the Brambell Report to advise government ministers on animal welfare has been displaced by the new Farm Animal Council (FAWC). This council has a wider brief than FAWAC, and has been charged by ministers with updating the Codes of Practice in accordance with the latest scientific knowledge and current practice.

It is impossible to compare present UK Codes of Practice with legislation in any other country. Other countries may have certain regulations (Denmark does on the battery cage) and all have legislation relating to cruelty to animals, but none has yet moved to our system of codes. There is a European Convention for the Protection of Animals kept for Farming Purposes within the Council of Europe. This applies to the keeping, care and housing of animals and in particular to animals in modern intensive stock farming systems. This Convention has been ratified by the UK. Articles 3

480   *The effect of the welfare code provisions on animal housing*

and 4 of the Convention state that 'environmental conditions shall conform to the animals' physiological and ethological needs in accordance with established experience and scientific knowledge.'

We are committed politically to these aims.

References

BRUCE, J. M. (1978). *Floors Symposium, Fulmer Grange, Slough; AOAS, Nov. 1978*

COMMAND PAPER 2836 (1965). HMSO, London

GJESTAND, K. E. (1979). *Appl. Anim. Ethol.* **5**, 294

HMSO (1971 and 1977). *Codes of Recommendations for the Welfare of Livestock*. MAFF. HMSO, London

HMSO (1976a). *The Climatic Environment of Poultry Houses*. ADAS, MAFF. HMSO, London

HMSO (1976b). *Cannibalism and Feather Pecking in Poultry*. AL480. ADAS, MAFF. HMSO, London

HMSO (1978). SI 1978 No. 1800: The Welfare of Livestock (Intensive Units) Regulations 1978

MITCHELL, C. D. (1976). *Calf Housing Handbook*. Scottish Farm Buildings Investigation Unit, Aberdeen

MURPHY, P. A. (1978). *Floors Symposium, Fulmer Grange, Slough, ADAS, Nov. 1978*

POWELL, A. J. and HILL, J. A. (1975). *Wld's Poult. Sci. J.* **31**, 247–248

SMITH, W. J. and MITCHELL, C. D. (1977). *Fm Bld Prog.* **49**, 13–14

WILSON, W. O. (1977). *Feedstuffs, Lond.* **49**, 22–23

POSTER SESSION

The following section contains short abstracts of the posters presented at the School. Posters related to their chapters in the proceedings were also presented by A. J. McArthur and C. M. Wathes.

AN ANALYSIS OF THE MECHANICAL PROBLEMS OF FEATHER WEAR IN BATTERY HENS

G. D. MACLEOD
University of Nottingham, School of Agriculture, Sutton Bonington, Loughborough

One of the technical problems of the poultry industry is that of feather wear in battery-caged hens. The thermoneutral zone of a poorly feathered bird is displaced to a higher environmental temperature and at poultry house temperatures usual in the UK the bird then requires a corresponding increase in food consumption. A poorly feathered bird may also produce fewer eggs as a hormonal consequence of the thermal stress.

The areas of the body most affected are the neck and breast; feathers in these regions are frequently abraded against the cage structure and inter-mittently wetted during drinking. Measurements have been made of the stiffness and tensile strength of contour feathers from the neck, breast and back of brown hens. Variations in stiffness with the relative humidity of the atmosphere were also measured, over a range from 23 to 97%.

When compared with flight feathers, contour feathers are mechanically inefficient structures and fail at proportionately lower loads. The differences in the mechanical properties of the feathers from the different tracts were small and therefore cannot account for the propensity of the neck and breat feathers to wear. At the highest humidity the stiffness of the feathers was reduced by up to an order of magnitude and the tensile strength by at least half.

Therefore, contour feathers are relatively weak and a cage design is needed which allows for this inefficiency and reduces plumage damage. A major aim should be to reduce feather wetting since this may make the feathers highly flexible and facilitate the eventual collapse of the plumage structure, thereby reducing its insulation.

This work was supported by a British Egg Marketing Board Research and Education Trust Scholarship.

AN INTERACTION BETWEEN FEATHERING AND EGG PRODUCTION IN LAYING HENS

G. D. MACLEOD
University of Nottingham, School of Agriculture, Sutton Bonington, Loughborough

The effect of loss of feather cover on heat loss and feed intake has been well documented. In laying birds, production is likely to be affected by the hormonal interactions between egg formation and feather growth. However, the direct effects of feather loss on egg production have largely been neglected.

The subjective feather scoring technique was first validated as a measure of insulation cover. Scores of 1 and 5 indicate complete and minimal plumages, respectively. A significant negative correlation ($P < 0.05$) existed between weighted thermal resistance values for plumage covering 70 per cent of the surface area of brown hens and simultaneous observations of feather score. Twelve individually caged brown layers were then feather-scored at frequent intervals throughout a 30 week laying cycle and during and after a forced moult. The number of flight feathers replaced during the moult was also recorded.

The data were fitted to a linear regression of egg numbers on weighted feather score for both pre- and post-moult periods. In each case the relationship was negative, but it was significantly so only for the pre-moult period ($P < 0.05$). The number of flight feathers replaced was positively correlated with both the highest weighted feather score during the moult, and post-moult egg production ($P < 0.05$). The results suggest that the maintenance of a feather coat of maximum insulation and the efficiency of a forced moult may be of major importance to egg production.

This study was supported by a British Egg Marketing Board Research and Education Trust Scholarship.

TEMPERATURE EXPERIMENTS WITH EARLY-WEANED PIGS

ANNE FEENSTRA
Danish Building Research Institute, Hoersholm, Denmark

Experiments have been carried out with pigs weaned at 4 weeks of age and kept for 8 weeks under different temperature conditions. In the first series of experiments the effects of different air temperature ranges were compared. In the next, the room temperature was kept constant and low, and the effects of covers for the pigs, with or without additional local heating, were studied. The weight gain, feed conversion rate, health status and the behaviour of the pigs were regarded as the most important variables in the experiments.

The differences in weight gains and feed conversion rates between control and experimental groups were very small. In the low-temperature experiments there were health problems and a high mortality. In the overall results no differences were found between 27 °C and 24 °C at the beginning of the experiments. A steep temperature decrease (linear temperature fall, end-point 12 °C) gave as good results as a moderate decrease (linear temperature fall, end-point 18 °C). A simple cover at a constant air temperature of 18 °C resulted in a lower mortality among covered than uncovered pigs. At a constant air temperature of 10 °C, a simple cover fitted with a 250 W heating foil gave results comparable to those for an air temperature gradually decreasing from 24 to 18 °C. The results were the same in both the pen types used in the experiments.

EGGSHELL POROSITY AND WATER LOSSES DURING INCUBATION AND STORAGE

K. F. LAUGHLIN
Agricultural Research Council's Poultry Research Centre, Roslin, Midlothian

Achieving the correct temperature and humidity is important for successful incubation. It is generally assumed that the optimum temperatures for incubation of different strains and species of domestic poultry are well defined. It is apparent, however, that the broad generalization of 60% relative humidity (RH) during incubation is not appropriate for all species and modern strains.

The purpose of controlling humidity in an incubator is to control the water loss from the egg, since this has an important effect on hatchability. The amount of water lost by an egg also depends on eggshell porosity. Porosity can be measured as the rate of loss of water by the egg in known, controlled conditions and is then called the *water vapour conductance*. Once this is known, the weight loss in any other environmental conditions can be predicted; so the conditions can be set that will give a specified weight loss. Workers at the PRC have measured the porosity of eggs from several different strains of poultry and have shown that although porosity varies greatly, even within batches of eggs from the same day's production, the average value is characteristic of each strain and age of flock. Within batches, porosity tends to increase with egg size in such a way that water losses in similar conditions will be similar in large and small eggs. Both egg size and porosity increase as the flock gets older. In broiler breeder eggs tested at the PRC, egg size and porosity increased proportionately. This showed that there would be little change in the percentage weight loss if the incubation conditions were kept the same as the flock got older. In layer breeder eggs, however, the increase was disproportionate, so that it is necessary to adjust humidity for eggs from older birds, perhaps by as much as 20% RH.

The research shows that the humidity required for incubation of eggs from different strains, species or flock ages can be predicted. The humidity required to achieve a specified weight loss over the whole of the incubation period can also be predicted by weighing the eggs during the early part of the period. Such measurements can be made easily in commercial hatcheries.

ENVIRONMENT AND CALF BEHAVIOUR

CLAIRE SAVILLE
Department of Animal Husbandry, University of Bristol, Langford, Bristol

Five systems of calf husbandry are being studied. These cover the range of environments experienced on farms and are, briefly:

(1) Single suckled calves with their dams at pasture.
(2) Replacement calves in small groups fed cold, acidified milk *ad libitum* via a teat, and weaned at about 8 weeks of age.
(3) Replacement calves in individual pens, bucket-fed warm milk twice daily to weaning and grouping at about 6 weeks old.
(4) Veal calves kept in groups on straw with free access via teats to warm milk from an automatic dispenser.
(5) Veal calves kept individually in small wooden pens with slatted floors and fed only warm milk twice daily from buckets.

The welfare of veal calves is of particular concern. The UK Ministry of Agriculture's Codes of Recommendation for Welfare define the basic need as 'the provision of a husbandry system appropriate to the health and behavioural needs of the animals'. With this in mind, several aspects of the calves' environment are being monitored at different stages of their development. Preliminary results of behavioural observations were demonstrated, made on approximately 25 calves (10 weeks old) for each of the five systems.

Veal calves in a straw yard spent much of the time lying. This appears to be associated with the short time required to ingest a concentrated, high-energy diet. Large differences in the time spent ruminating (*Table 1*) were correlated with the time spent eating and the proportion of the diet which was roughage. If calves kept outdoors with their dams are taken as the standard, then teat-fed veal calves were similar, but early-weaned calves

Table 1   A COMPARISON OF THE TIME SPENT RUMINATING BY 10-WEEK-OLD CALVES OFFERED DIFFERENT LEVELS OF DIETARY FIBRE

	Amount of fibre	Time over 4 h*
Single suckled at grass	Moderate	4.8
Early weaned: teat	High	14.1
bucket	High	13.3
Veal: teat	Moderate	8.2
bucket	Nil	1.0

*Expressed as a percentage

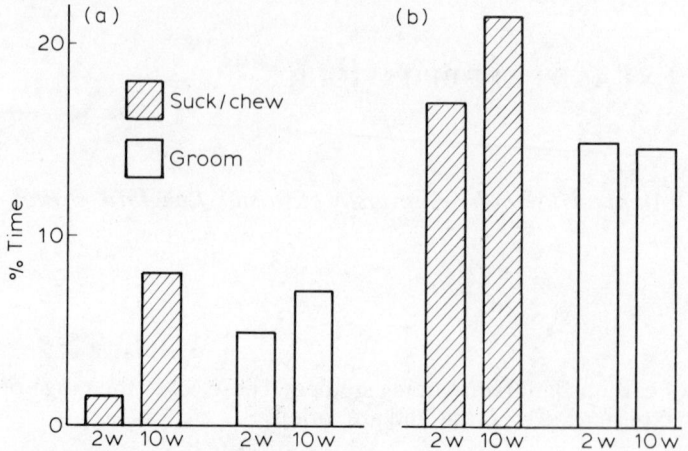

Figure 1 A comparison of the time spent on 'oral' activity for calves at 2 and 10 weeks of age and kept in two systems of veal husbandry

showed precocious rumination and crated, bucket-fed veal calves a small amount of rumination.

Figure 1 compares the time spent chewing or sucking non-food objects and grooming, for straw-yard veal calves (which were typical of the range shown by other groups) and bucket-fed, crated, veal calves. The latter spent about three times longer on these activities than calves in other systems. This abnormal behaviour is at least partly due to the lack of roughage in the diet, which also leads to abnormal gut development and the formation of hairballs, which may give rise to bloat.

Table 2   A COMPARISON OF THE REACTION TO HUMAN CONTACT BY 10-WEEK-OLD CALVES KEPT IN DIFFERENT SYSTEMS OF HUSBANDRY

		Score
Single suckled at grass		–48
Early weaned:	teat	–24
	bucket	– 4
Veal:	teat	+ 2
	bucket	–69

The more negative the score, the more fearful the response

The reaction of calves in different systems to human disturbance is shown in *Table 2*. Carefully defined actions were presented randomly to groups of calves. Their response was scored on a positive (friendliness/curiosity), zero (total indifference) to negative (fear) scale. The score shown in *Table 2* is the mean for several groups of calves of the average response score for all actions. Again, with the exception of crated veal calves, there seems to be a relation between the score and the amount of human (stockman) contact the calves had. The environment of the crated veal calves seems to be such that they can neither readily adjust to human contact nor use the flight response to fearful stimuli.

THE HEAT AND WATER BALANCE OF GOATS AT HIGH ENVIRONMENTAL TEMPERATURES

A. M. ABDELATIF
University of Nottingham, School of Agriculture, Sutton Bonington, Loughborough

The climatic elements act on homeotherms to evoke physiological responses that attempt to maintain energy and fluid balance and a relatively constant body temperature. In many tropical areas the productivity of farm ruminants is limited by climate because the animals fail to achieve thermal equilibrium with their environment. High air temperatures, high solar and terrestrial radiation and poor nutrition constitute the major factors that affect the energy budget of animals and on arid range land the heat load is maximal when water availability is minimal. Breed-specific potentials (which control metabolic conversion of food and water economy) and pelage characteristics (which modulate heat exchange with the environment) contribute to the tolerance and adaptation of various groups of animals to climatic stress. Investigations of the heat and water losses of the animals help us to determine the climatic conditions most suitable for their health and productivity.

Experiments were designed to determine the combined effect of air temperature, radiation load and feeding level on the thermoregulatory performance of different breeds of goat. The measurements are conducted in a climatic room in which air temperature can be controlled between 0 and 40 °C. Radiation flux densities comparable in magnitude with those experienced by animals in the tropics are produced using radiant lamps. Heat production of the goats is computed from oxygen consumption measured by an open-circuit respiration system using a Servomex paramagnetic oxygen analyser connected to a recording potentiometer. The evaporative heat loss is partitioned into losses from the external body surface and the respiratory passages. The levels and changes of the components are investigated using a ventilated hygrometric tent with separate compartments for the body and the head. The temperature and vapour density of the air entering and leaving the compartments are measured with wet and dry copper–constantan thermocouples connected to the recording potentiometer. Calibration of the system is checked regularly by evaporating water into the tent and the hood from heated beakers.

To provide information on body temperature regulation, the rectal temperature and skin temperatures at 22 sites (covering the head, neck, trunk and extremities) are measured potentiometrically using 38 s.w.g. copper–constantan thermocouples. Coat surface temperature is measured

493

using an infrared radiometer. Other measurements have included respiratory rate and heart rate.

The metabolic heat production and heat loss by evaporation can be measured at different air temperatures in a steady thermal environment in the climatic room. The non-steady response to a radiation load can also be measured.

This work is supported by a studentship from the British Council.

LIST OF PARTICIPANTS

Abdelatif, A. M.

Nottingham University, School of Agriculture, Sutton Bonington, Loughborough LE12 5RD

Alexandre, Dr C.

Nottingham University, School of Agriculture, Sutton Bonington, Loughborough LE12 5RD

Al-Fataftah, A. R. A.

Wye College, Nr Ashford, Kent

Ansell, Dr R. H.

c/o UNDP, Box 913, Khartoum, Sudan

Auty, N. S.

Golden Produce Ltd, Easton Road, Witham, Essex

Baker, G.

G. E. Baker (UK) Ltd, Heath Road, Woolpit, Bury St Edmunds

Berbigier, P.

INRA Domaine Duclos Bioclimatologie, 97170 Petit-Bourg, French West Indies

Berman, Prof. A.

Department of Animal Science, Faculty of Agriculture, Hebrew University, Rehovot, Israel

Bishop, C. F. H.

ADAS Mechanisation, Ministry of Agriculture, Block B, Brooklands Avenue, Cambridge

Bond, K. I.

J. Bibby & Sons Ltd, Stonebow House, York YO1 2NP

Brooks, Dr P. H.

Seale Hayne College, Newton Abbot TQ12 6NQ, Devon

Bruce, Dr J. M.

Scottish Farm Buildings Investigation Unit, Craibstone, Bucksburn, Aberdeen AB2 9TR

Buckingham, J. F.	Engineering Department, Edinburgh School of Agriculture, West Mains Road, Edinburgh EH9 2SG
Buckle, A. E.	Microbiology Department, Ministry of Agriculture, Fisheries and Food, Shardlow Hall, Shardlow, Derby
Buttery, Dr P. J.	University of Nottingham, School of Agriculture, Sutton Bonington, Loughborough LE12 5RD
Carpenter, G. A.	National Institute of Agricultural Engineering, Silsoe, Beds
Carpenter, J. L.	Seale Hayne College, Newton Abbot TQ12 6HN, Devon
Cena, Dr K.	Environmental Physics, Institute of Building Science, I/2, Technical University of Wrocław, Wyb. Wyspianskiego 27, 50–370 Wrocław, Poland
Cermak, J. P.	Ministry of Agriculture, Fisheries and Food, Land Service, Farm Buildings Group, Coley Park, Reading
Charles, Dr D. R.	ADAS, Shardlow Hall, Shardlow, Derby DE7 2GN
Charlesworth, R. P.	Marks & Spencer Ltd, 47 Baker Street, London W1A 1DN
Cherry, P.	Cherry Valley Farms Ltd, Divisional Offices, North Kelsey Road, Caistor, Lincoln
Clark, Dr J. A.	University of Nottingham, School of Agriculture, Sutton Bonington, Loughborough LE12 5RD
Clegg, F. G.	East Midlands Regional Veterinary Investigation Centre, Sutton Bonington, Loughborough, Leics
Close, Dr W. H.	ARC Institute of Animal Physiology, Babraham, Cambridge CB2 4AT
Clough, P. A.	National Institute for Research in Dairying, Shinfield, Reading, Berks

Cole, Dr D. J. A.	Nottingham University, School of Agriculture, Sutton Bonington, Loughborough LE12 5RD
David, A. J. G.	Woods of Colchester Ltd, Colchester, Essex
Dodd, Dr V.	Department of Agricultural Engineering, University College of Dublin, Upper Merrion Street, Dublin 2, Ireland
Duncan, Dr I. J. H.	ARC Poultry Research Centre, West Mains Road, Edinburgh EH9 3JS
Edwards, D. J.	Unilever Research Laboratory, Sharnbrook, Bedford
Feddes, J.	Department of Agricultural Engineering, University of Alberta, Edmonton, Alberta, Canada
Feenstra, Dr A.	Danish Building Research Institute, PO Box 119, DK-2970 Hoersholm, Denmark
Folk, Prof. G. E., Jr	Department of Physiology and Biophysics, University of Iowa, Iowa City, Iowa 52242, USA
Gatenby, Dr R. M.	University of Edinburgh, Centre for Tropical Veterinary Medicine, Easter Bush, Roslin, Midlothian EH26 9RG
Gibbons, D. F.	Ministry of Agriculture, Fisheries and Food, Veterinary Investigation Centre, Crown Street, Liverpool
Gloster, J.	Meteorological Office, London Road, Bracknell, Berks
Gustafsson, Dr B. S. H.	Department of Farm Buildings, University of Agricultural Sciences, Box 624, Lund, Sweden
Haartsen, Dr P. I.	Institute of Agricultural Engineering, Mansholtlaan 10–12, 6700 AA Wageningen, The Netherlands
Haynes, Dr N. B.	Nottingham University, School of Agriculture, Sutton Bonington, Loughborough LE12 5RD

Hearn, P. J.	Gleadthorpe Experimental Husbandry Farm, Meden Vale, Mansfield, Notts
Hiley, Dr P.	Nickerson Group, Rothwell, Lincoln
Hill, Dr J. A.	Gleadthorpe Experimental Husbandry Farm, Meden Vale, Mansfield NG20 9PF, Notts
Hodgetts, B.	Ministry of Agriculture, Fisheries and Food, ADAS, Woodthorne, Wolverhampton WV6 8TQ
Howles, C. M.	Nottingham University, School of Agriculture, Sutton Bonington, Loughborough LE12 5RD
Jones, Ms C. D.	Department of Animal Husbandry, School of Veterinary Science, Bristol University, Langford, Bristol BS18 7DU
Jongebreur, Dr A. A.	Institute of Agricultural Engineering, Mansholtlaan 10–12, 6700 AA Wageningen, The Netherlands
Kelly, T. G.	The Agricultural Institute, Economics and Rural Welfare Research Centre, 19 Sandymount Avenue, Dublin 4, Ireland
Kenyon, P. J.	BOCM Silcock Ltd, Basing View, Basingstoke
King, A. W. M.	ADAS, Ministry of Agriculture, Fisheries and Food, Kenton Bar, Newcastle upon Tyne NE1 2YA
Lamming, Prof. G. E.	Nottingham University, School of Agriculture, Sutton Bonington, Loughborough LE12 5RD
Laughlin, Dr K. F.	D. B. Marshall (Newbridge) Ltd, Newbridge, Midlothian
Le Dividich, J.	CNRZ Station de Recherches sur l'Elevage des Porcs, 78350 Jouy en Josas, France
Lundy, H.	ARC Poultry Research Centre, King's Building, West Mains Road, Edinburgh EH9 3JS

Lynch, P. B.	Pig Husbandry Department, Moore-park Research Centre, Fermoy, Co. Cork, Ireland
McArthur, Dr A. J.	Nottingham University, School of Agriculture, Sutton Bonington, Loughborough LE12 5RD
McCracken, Dr K. J.	Agricultural and Food Chemistry Research Division, Department of Agriculture, N. Ireland, Newforge Lane, Belfast BT9 5PX
Macfarlane, Prof. W. V.	University of Adelaide, Waite Agricultural Research Institute, Glen Osmond, South Australia 5064
McLean, Dr J. A.	Hannah Research Institute, Ayr
Macleod, G. D.	Nottingham University, School of Agriculture, Sutton Bonington, Loughborough LE12 5RD
Macleod, Dr M. G.	ARC Poultry Research Centre, King's Buildings, West Mains Road, Edinburgh EH9 3JS
McQuitty, Prof. J. B.	Department of Agricultural Engineering, University of Alberta, Edmonton, Alberta, Canada
Marriott, Miss P. K.	Ministry of Agriculture, Fisheries and Food, Animal Health Office, Quantock House, Paul Street, Taunton
Marsden, A.	Ministry of Agriculture, Fisheries and Food, St Mary's Manor, North Bar Within, Beverley HU17 8DN, North Humberside
Meltzer, Dr A.	Department of Animal Husbandry, Faculty of Agriculture, Hebrew University, Rehovot, Israel
Misson, B. H.	Houghton Poultry Research Station, Houghton, Huntingdon PE17 2DA, Cambs
Mitchell, Dr C. D.	Electricity Council, Farm-Electric Centre, National Agricultural Centre, Stoneleigh, Kenilworth, Warwicks

Monteith, Prof. J. L.	Nottingham University, School of Agriculture, Sutton Bonington, Loughborough LE12 5RD
Morris, Dr T. R.	Department of Agriculture and Horticulture, Reading University, Reading, Berks
Morrison, Prof. S. R.	Agricultural Engineering Department, University of California, Davis, California 95616, USA
Moss, R.	Ministry of Agriculture, Fisheries and Food, Animal Health Division, Tolworth, Surbiton, Surrey
Mount, Prof. L. E.	ARC Institute of Animal Physiology, Babraham, Cambridge CB2 4AT
Nimmermark, S.	Department of Farm Buildings, Swedish University of Agricultural Science, Box 624, S-22006 Lund, Sweden
Owen, J. E.	Department of Agriculture, Reading University, Early Gate, Reading RG6 2AT
Owen, Mrs J. M.	St Mary's Manor, Beverley, North Humberside
Packer, Miss J. A.	Ministry of Agriculture, Fisheries and Food, Gleadthorpe EHF, Meden Vale, Nr Mansfield, Notts
Palmer, G. J. H.	Smiths Industries Precision Fan Co., Witney OX8 5EE, Oxon
Parry, Mrs M. A.	Harper Adams Agricultural College, Newport, Shropshire
Pereira, Prof. N.	Facultad de Agronomia, Instituto de Produccion Animal, Maracay, Aragua, Venezuela
Perry, Dr G. C.	Department of Animal Husbandry, Bristol University, Langford House, Langford, Bristol BS18 7DU
Petchey, Dr A. M.	North of Scotland College of Agriculture, 581 King Street, Aberdeen AB9 1UD

Poczopko, Prof. P.	Institute of Animal Physiology and Nutrition, Polish Academy of Sciences, 05–110 Jabłonna, Nr Warsaw, Poland
Pringle, R. T.	Agricultural Engineering Division, North of Scotland College of Agriculture, Craibstone, Bucksburn, Aberdeen
Randall, Dr J. M.	National Institute of Agricultural Engineering, Wrest Park, Silsoe, Bedford
Robertshaw, Prof. D.	Colorado State University, Fort Collins, Colorado 80523, USA
Rowan, T. G.	Department of Animal Husbandry, Veterinary Field Station, University of Liverpool, Neston, Wirral L64 7TE
Sainsbury, Dr D. W. B.	Department of Clinical Veterinary Medicine, Cambridge University, Madingley Road, Cambridge CB3 0ES
Samson, Ms D. E.	Institute of Naval Medicine, Environmental Medicine Unit, Alverstoke, Gosport, Hants
Saville, Miss C. A.	Department of Animal Husbandry, University of Bristol, Langford House, Langford, Bristol BS18 7DU
Shipston, A. H.	RHM Animal Feed Services Ltd, PO Box 152, West Point, Slough SL1 1QF
Shipway, G. P.	ADAS, Ministry of Agriculture, Fisheries and Food, Woodthorne, Wolverhampton WV6 8TQ
Slee, Dr J.	Animal Breeding Research Organisation, Field Laboratory, Dryden, Roslin, Midlothian
Smith, A. T.	ADAS, Ministry of Agriculture, Fisheries and Food, Woodthorne, Wolverhampton WV6 8TQ
Smith, M. S.	Insulheat Ltd, 18 Guy Street, Leamington Spa, Warwicks
Smith, Dr W. K.	Poultry Department, West of Scotland College of Agriculture, Auchincruive, Ayr

Spencer, P. G.	Bernard Matthews Ltd, Great Witchingham Hall, Norwich
Starr, Dr J. R.	Meteorological Office, Ministry of Agriculture, Fisheries and Food, Coley Park, Reading RG1 6DT
Steinbock, M.	Bernard Matthews Ltd, Great Witchingham Hall, Norwich
Stenning, B.	National College of Agricultural Engineering, Silsoe, Bedford
Stombaugh, Dr D. P.	Agricultural Engineering Department, Ohio State University, 2073 Neil Avenue, Columbus, Ohio 43210, USA
Sykes, Dr A.	Wye College, Ashford, Kent TN25 5AH
Unsworth, Dr M. H.	Nottingham University, School of Agriculture, Sutton Bonington, Loughborough LE12 5RD
Valli, Miss B.	Butterworth & Co (Publishers) Ltd, Borough Green, Sevenoaks, Kent
Van Kampen, Dr M.	Laboratory for Veterinary Physiology, Alex Numankade 93, Utrecht, The Netherlands
Wathes, Dr C. M.	ADAS, Ministry of Agriculture, Fisheries and Food, Shardlow Hall, Shardlow, Derby
Webster, Prof. A. J. F.	Department of Animal Husbandry, School of Veterinary Science, Bristol University, Langford, Bristol BS18 7DU
Whelan, J. D. W.	J. Bibby Agriculture, Dunball Mill, Bridgewater, Somerset
Wheldon, Miss A.	Nottingham University, School of Agriculture, Sutton Bonington, Loughborough LE12 5RD
Wilton, B.	Nottingham University, School of Agriculture, Sutton Bonington, Loughborough LE12 5RD
Wiseman, Dr J.	Nottingham University, School of Agriculture, Sutton Bonington, Loughborough LE12 5RD
Young, Dr B. A.	Department of Animal Science, University of Alberta, Edmonton, Alberta, Canada

INDEX

503